Construction Planning, Programming and Control

Construction Planning, Programming and Control

Fourth Edition

Brian Cooke and Peter Williams

WILEY Blackwell

This edition first published 2025
© 2025 John Wiley & Sons Ltd

Edition History
1e 1998 by McMillan Education, UK; 2e 2004 by Blackwell Publishing, Oxford, UK; 3e 2009 by Wiley-Blackwell, Chichester, UK

All rights reserved, including rights for text and data mining and training of artificial technologies or similar technologies. [For any titles with third-party copyright holders, replace the previous sentence with: All rights reserved.] No part of this publication may be reproduced, stored in a retrieval system, or transmitted, in any form or by any means, electronic, mechanical, photocopying, recording or otherwise, except as permitted by law. Advice on how to obtain permission to reuse material from this title is available at http://www.wiley.com/go/permissions.

The right of Brian Cooke and Peter Williams to be identified as the authors of this work has been asserted in accordance with law.

Registered Office(s)
John Wiley & Sons, Inc., 111 River Street, Hoboken, NJ 07030, USA
John Wiley & Sons Ltd, The Atrium, Southern Gate, Chichester, West Sussex, PO19 8SQ, UK

For details of our global editorial offices, customer services, and more information about Wiley products visit us at www.wiley.com.

Wiley also publishes its books in a variety of electronic formats and by print-on-demand. Some content that appears in standard print versions of this book may not be available in other formats.

Trademarks: Wiley and the Wiley logo are trademarks or registered trademarks of John Wiley & Sons, Inc. and/or its affiliates in the United States and other countries and may not be used without written permission. All other trademarks are the property of their respective owners. John Wiley & Sons, Inc. is not associated with any product or vendor mentioned in this book

Limit of Liability/Disclaimer of Warranty
While the publisher and authors have used their best efforts in preparing this work, they make no representations or warranties with respect to the accuracy or completeness of the contents of this work and specifically disclaim all warranties, including without limitation any implied warranties of merchantability or fitness for a particular purpose. No warranty may be created or extended by sales representatives, written sales materials or promotional statements for this work. This work is sold with the understanding that the publisher is not engaged in rendering professional services. The advice and strategies contained herein may not be suitable for your situation. You should consult with a specialist where appropriate. The fact that an organization, website, or product is referred to in this work as a citation and/or potential source of further information does not mean that the publisher and authors endorse the information or services the organization, website, or product may provide or recommendations it may make. Further, readers should be aware that websites listed in this work may have changed or disappeared between when this work was written and when it is read. Neither the publisher nor authors shall be liable for any loss of profit or any other commercial damages, including but not limited to special, incidental, consequential, or other damages.

Library of Congress Cataloging-in-Publication Data

Names: Cooke, Brian, author. | Williams, Peter, 1947- author.
Title: Construction planning, programming and control / Brian Cooke, Peter Williams.
Description: Fourth edition. | Chester, UK : Wiley, 2025. | Includes index.
Identifiers: LCCN 2024011276 (print) | LCCN 2024011277 (ebook) | ISBN 9781119109457 (paperback) | ISBN 9781119109464 (adobe pdf) | ISBN 9781119109471 (epub)
Subjects: LCSH: Building–Planning.
Classification: LCC TH153 .C597 2025 (print) | LCC TH153 (ebook) | DDC 690.068 – dc23/eng/20240409
LC record available at https://lccn.loc.gov/2024011276
LC ebook record available at https://lccn.loc.gov/2024011277

Cover Design: Wiley
Cover Image: Peter Williams

Set in 9.5/12.5pt STIXTwoText by Straive, Chennai, India

Printed and bound by CPI Group (UK) Ltd, Croydon, CR0 4YY

C9781119109457_061124

Contents

About the Authors *xxv*
Preface *xxvii*
Acknowledgements *xxix*
Acronyms *xxxi*

1	**Construction Projects** *3*	
1.1	Introduction *3*	
1.1.1	Industry Reputation *4*	
1.1.2	Projects *4*	
1.1.3	Programmes *5*	
1.1.4	Portfolios *6*	
1.1.5	Mega Projects *6*	
1.2	Why Do Projects Go Wrong? *8*	
1.2.1	Looking for Reasons *9*	
1.2.2	Murphy's Law *10*	
1.2.3	Complexity *10*	
1.2.4	The Consequences of Late Completion *11*	
1.2.5	Construction Industry Reports *13*	
1.3	Managing the Risk of Delayed Completion in the 21st Century *14*	
1.3.1	Summary of Findings *14*	
1.3.2	Detailed Observations *15*	
1.3.3	Claims and Disputes *16*	
1.3.4	Project Management Software *16*	
1.3.5	Further Developments *16*	
1.4	The Latham Reports *16*	
1.4.1	Trust and Money *17*	
1.4.2	Constructing the Team *17*	
1.4.3	Post-Latham *17*	
1.5	The Egan Reports *22*	
1.5.1	Rethinking Construction *22*	
1.5.2	Accelerating Change *24*	
1.5.3	Post-Egan *24*	
1.6	The Wolstenholme Report *25*	
1.6.1	Egan Targets *26*

1.6.2	Wolstenholme Conclusions	26
1.6.3	Wolstenholme Blockers	27
1.6.4	Summary	27
1.6.5	Post-Wolstenholme	28
1.7	The Farmer Review	29
1.7.1	Farmer Conclusions	29
1.7.2	Recommendations	30
1.7.3	Farmer Initiatives	30
1.7.4	Post Farmer	31
	References	31
	Notes	32
2	**Project Environment**	**35**
2.1	Introduction	35
2.2	Industry Culture	36
2.2.1	Industry Customs and Practices	36
2.2.2	Separation of Design and Construction	37
2.2.3	Late Payment	37
2.2.4	Construction Supply Chain Payment Charter	38
2.2.5	Public Contracts Regulations 2015	39
2.3	Defining the Industry	40
2.3.1	Industry Classification	40
2.3.2	Industry Sector Output	41
2.3.3	Legislative Definition	43
2.3.4	Types of Work	44
2.4	Industry Clients	44
2.4.1	Definitions	45
2.4.2	Types of Clients	45
2.4.3	Professional Advisers	45
2.4.4	Funding	46
2.5	Construction Firms	46
2.5.1	Industry Output	46
2.5.2	Micro Firms	47
2.5.3	SMEs	47
2.5.4	Large Firms	47
2.5.5	Very Large Contractors	48
2.5.6	The Top 100	48
2.5.7	Margins	49
2.5.8	Suppliers	50
2.6	Industry Leadership	51
2.6.1	The Influence of Clients	51
2.6.2	The Construction Clients' Group	52
2.6.3	Main Contractors	52
	References	53
	Notes	53
3	**Project Risk**	**57**
3.1	Introduction	57

3.1.1	Risk	58
3.1.2	Uncertainty	58
3.2	Risk Management	59
3.2.1	Principal Risks and Uncertainties	59
3.2.2	Developers	60
3.2.3	Managing Risk	60
3.2.4	Crossrail	62
3.2.5	Heathrow Terminal 5	62
3.3	Risk Assessment	63
3.3.1	Risk Categories	63
3.3.2	Hazards	64
3.3.3	Assessing Risk	65
3.4	Design Stage	66
3.4.1	Feasibility	66
3.4.2	Funding	67
3.4.3	Design	68
3.4.4	Tender Documentation	69
3.4.5	Quantity Risk	69
3.5	Tender Stage	70
3.5.1	Tender Risk	70
3.5.2	Tender Risk Allowances	72
3.5.3	Programme–Time Risk	73
3.5.4	Method Risk	74
3.6	Construction Stage	74
3.6.1	Delay	75
3.6.2	Health and Safety	76
3.6.3	Fire Risk	76
3.6.4	Supply Chain Risk	77
3.6.5	Defects	78
3.6.6	Insolvency	80
3.7	Managing Project Risk	81
3.7.1	Risk Allocation	82
3.7.2	Reactive Risk Management	83
3.7.3	Proactive Risk Management	83
3.7.4	Dynamic Time Modelling	84
3.7.5	Risk Registers	86
3.7.6	Early Warning Registers	87
	References	87
	Notes	87
4	**Managing Construction Projects**	**91**
4.1	Introduction	91
4.1.1	Definition	92
4.1.2	Scope, Delivery and Deliverables	93
4.2	Projects and Programmes	94
4.2.1	Programme and Portfolio Management	94
4.2.2	Phased Projects	94
4.2.3	Project-Based Working	95

4.2.4	The Size of Projects	95
4.3	Management	96
4.3.1	The Principles of Management	96
4.3.2	Organisation	97
4.3.3	Organisation Structures	97
4.3.4	Operational Divisions	99
4.3.5	Business Units	100
4.4	Team Building	102
4.4.1	Complexity	103
4.4.2	Trust	104
4.5	Project Management	105
4.5.1	Decision Making	105
4.5.2	Project Managers	106
4.5.3	Client's Project Manager	106
4.5.4	Contractor's Project Manager	107
4.5.5	Project Manager Duties	108
4.6	Collaboration	109
4.6.1	Common Data Environment	110
4.6.2	Enterprise Resource Planning	112
4.6.3	Construction Management Software	112
4.6.4	Workflows	114
4.7	Design Management	114
4.7.1	Design Manager	115
4.7.2	Pre-construction Manager	116
4.8	Project Management Models	116
4.8.1	APM Competence Framework	117
4.8.2	PMBOK Guide	117
4.8.3	Prince2	117
4.9	Project Stages	117
4.9.1	Process Models	118
4.9.2	Types of Process Models	118
4.9.3	OGC Gateway Process	118
4.9.4	RIBA Plan of Work	119
4.9.5	The Guide to Rail Investment Process	120
4.9.6	The Project Control Framework	121
4.9.7	The Construction Playbook	121
4.9.8	TfL Pathway	122
4.9.9	Infrastructure and Projects Authority Cost Estimating Guidance	122
4.10	Digital Construction	123
4.10.1	Building Information Modelling	123
4.10.2	BIM Models	124
4.10.3	Benefits of BIM	124
4.10.4	BIM Standards	126
4.10.5	BIM Levels	126
4.10.6	BIM Level 3 Benefits	127
4.10.7	BIM Downsides	128
4.10.8	BIM Managers	128

4.10.9	Digital Twins	129
4.10.10	Augmented and Virtual Reality	129
4.11	Managing the Supply Chain	130
4.11.1	Supply Chain	131
4.11.2	Supply Chain Management	131
4.11.3	Supply Chain Integration	132
4.12	Managing Construction	133
4.12.1	Types of Subcontractors	134
4.12.2	Subcontract Scope	135
4.12.3	Attendances	135
4.12.4	Scoping of Work Packages	136
4.12.5	Invitations to Tender	136
4.12.6	Resources and Programme	137
4.12.7	Subcontract Coordinator	137
4.12.8	Lean Construction	138
4.12.9	Fast-track Construction	138
	References	139
	Notes	139
5	**Modern Methods of Construction**	**143**
5.1	Introduction	143
5.1.1	DfMA	143
5.1.2	Modern Methods of Construction	145
5.1.3	Definitions	145
5.2	MMC Categories	146
5.2.1	MMC Definition Framework	146
5.2.2	MMC Spectrum	146
5.2.3	Pre-manufactured Value	146
5.3	Modular Construction	148
5.3.1	Volumetric Modular – Structural	148
5.3.2	Volumetric Modular – Non-structural	151
5.4	Panelised Construction	152
5.4.1	Open and Closed Systems	153
5.4.2	SIPs	153
5.4.3	Cross-laminated Timber	154
5.5	Component-Based Systems	155
5.5.1	IKEA	155
5.5.2	Site Assembly	155
5.6	Hybrid Construction Systems	155
5.6.1	Low-Rise Construction	156
5.6.2	Medium-Rise Construction	156
5.6.3	High-Rise Construction	158
5.6.4	Civil Engineering	159
5.7	Digital Technologies	161
5.7.1	3D Printing	161
5.7.2	Drones	164
5.7.3	Wearable Technology	164
5.7.4	Concrete Monitoring	164

5.8	Planning for Modular Construction	165
5.8.1	The Early Stages	165
5.8.2	Construction Schedule	166
5.8.3	Erection Sequence	166
5.8.4	Installation	166
5.8.5	Site Work	166
5.8.6	Tolerance and Fits	167
	References	168
	Notes	168
6	**Procurement and Contracts**	**171**
6.1	Introduction	171
6.1.1	Definitions	171
6.1.2	Public Sector and Private Sector Procurement	172
6.1.3	Procurement Strategy	172
6.1.4	Design Procurement	173
6.1.5	Contractor Selection	174
6.2	Tendering	175
6.2.1	Tender Documents	175
6.2.2	Approved Lists	175
6.2.3	Frameworks	176
6.3	Special Relationships	179
6.3.1	Partnering	179
6.3.2	Alliances	180
6.3.3	Joint Ventures	181
6.4	Procurement Routes	182
6.4.1	General Contracting	182
6.4.2	Design and Build/Construct	183
6.4.3	Prime Contracting	188
6.4.4	Management Procurement	188
6.4.5	Public–Private Partnerships	189
6.4.6	Turnkey and EPC	191
6.5	Contractual Arrangements	192
6.5.1	Law of Contract	192
6.5.2	Letters of Intent	194
6.5.3	Forms of Contract	194
6.5.4	Subcontracts	194
6.5.5	Legal Jurisdiction	195
6.5.6	Housing Grants, Construction and Regeneration Act 1996	195
6.6	Standard Forms of Contract	196
6.6.1	JCT Contracts	196
6.6.2	NEC Contracts	196
6.6.3	FIDIC Contracts	197
6.6.4	ICC Contracts	198
6.6.5	CIOB Time and Cost Management Contract	198
6.6.6	IChemE Contracts	199
6.6.7	Other Contracts	199
6.6.8	Overseas Contracts	200

6.6.8.1	Australia	*200*
6.6.8.2	Hong Kong	*200*
6.6.8.3	South Africa	*201*
6.6.8.4	United States	*201*
6.7	Types of Contract	*202*
6.7.1	Lump Sum Contracts	*203*
6.7.2	Measure and Value Contracts	*203*
6.7.3	Cost Reimbursement Contracts	*204*
6.7.4	Target Cost Contracts	*205*
6.8	Time Management	*206*
6.8.1	Time in Contracts	*206*
6.8.2	Project Master Schedule	*206*
6.8.3	Site Possession	*208*
6.8.4	Progress, Completion and Extensions of Time	*208*
6.8.5	Liquidated and Ascertained Damages	*209*
6.8.6	Obligations Following Completion	*209*
	References	*210*
	Notes	*210*
7	**Estimating and Bidding**	*213*
7.1	Introduction	*213*
7.1.1	Definitions	*213*
7.1.2	Tendering	*214*
7.1.3	Estimating and Tendering	*214*
7.1.4	The Construction Market	*215*
7.2	Approaches to Estimating	*215*
7.2.1	Micro Companies	*215*
7.2.2	Small Companies	*216*
7.2.3	Medium-size Companies	*216*
7.2.4	Large-size Companies	*216*
7.2.5	Estimating Methods	*216*
7.2.6	Classification	*217*
7.2.7	Early Cost Advice	*217*
7.3	Measurement	*219*
7.3.1	Pricing Documents	*219*
7.3.2	Quantities	*219*
7.3.3	Link with Planning	*222*
7.3.4	Software	*223*
7.4	Estimating Software	*223*
7.4.1	Trade Apps	*224*
7.4.2	Price-a-Job	*224*
7.4.3	ConX	*224*
7.4.4	Cubit Estimating	*224*
7.4.5	Bidcon	*225*
7.4.6	CATO	*226*
7.5	Top-down Estimating	*227*
7.5.1	Definition	*227*
7.5.2	Traditional Procurement	*227*

7.5.3	Non-traditional Procurement	227
7.5.4	Principles	228
7.5.5	Floor Area Method	228
7.5.6	Functional Unit Method	229
7.5.7	Elemental Method	230
7.5.8	Hybrid Method	231
7.6	Cost Information	234
7.6.1	Historic Data	234
7.6.2	Sources	235
7.6.3	Indices	236
7.6.4	Inflation	237
7.7	New Rules of Measurement	239
7.7.1	Comparative Cost Planning	239
7.7.2	Cost Checking	240
7.7.3	Cost Reconciliation	241
7.8	Bottom-up Estimating	241
7.8.1	Modern Trends	242
7.8.2	Builders' Quantities	242
7.8.3	Bottom-up Estimating Methods	243
7.8.4	Unit Rate Estimating	243
7.8.5	Operational Estimating	244
7.9	Bidding	248
7.9.1	Estimating	248
7.9.2	Subcontract Enquiries	249
7.9.3	Bid Preparation	251
7.9.4	Integrated Software Platforms	252
7.9.5	Bid Submission and Formal Tender	253
7.9.6	Bid Documentation	253
7.9.7	Other Procurement Methods	255
7.9.8	Bid Management	255
7.9.9	Bid Manager	255
7.9.10	Bid Submission Documents	256
	References	257
	Notes	257
8	**The Planning Process**	**261**
8.1	Introduction	261
8.1.1	The Role of Planning	261
8.1.2	Client View	262
8.1.3	Project Team View	263
8.1.4	Contractor View	263
8.1.5	Types of Programme	265
8.2	Strategic Planning	266
8.2.1	Key Appointments	266
8.2.2	Key Dates	267
8.2.3	Statutory Approvals	267
8.2.4	Financing	267
8.2.5	Land Purchase	268

8.2.6	Design Procurement	268
8.2.7	Time and Cost	268
8.2.8	The Construction Period	269
8.3	Design Planning	269
8.3.1	The Design Team	269
8.3.2	Design Development	270
8.3.3	Design Management	270
8.4	Development Case Study	271
8.4.1	Project Master Schedule	271
8.4.2	Development Appraisal	272
8.4.3	Design Programme	272
8.4.4	Due Diligence	273
8.4.5	Hand-Down Document	273
8.4.6	Pre-construction Programme	274
8.5	Construction Planning	274
8.5.1	Tender Stage	274
8.5.2	Pre-construction	275
8.5.3	Construction Stage	276
	Reference	276
9	**Scheduling Techniques**	**279**
9.1	Introduction	279
9.1.1	Definitions	279
9.1.2	Work Versus Time	280
9.1.3	Gantt Charts	280
9.1.4	Scheduling Techniques	281
9.1.5	The Critical Path Method	281
9.1.6	CPM Software	283
9.2	Linked Bar Charts	285
9.2.1	Useful Features	285
9.2.2	Basic Principles	287
9.2.3	Worked Example	289
9.3	Arrow Diagrams	290
9.3.1	Basic Principles	290
9.3.2	Worked Example	291
9.4	Precedence Diagrams	291
9.4.1	Basic Principles	294
9.4.2	Worked Example	294
9.5	Arrow, Precedence and Linked Bar Chart Relationships	296
9.5.1	Common Features	296
9.5.2	Networks	296
9.5.3	Overlapping Activities	297
9.6	Line of Balance	302
9.6.1	Basic Principles	302
9.6.2	Worked Example	303
9.7	Time-Chainage Diagrams	306
9.7.1	Basic Principles	307
9.7.2	Worked Example	308

10	**Construction Sequences** *315*	
10.1	Introduction *315*	
10.1.1	Programme versus Work Sequence *315*	
10.1.2	Construction Methods *317*	
10.2	Resources *318*	
10.2.1	Labour *318*	
10.2.2	Construction Plant *319*	
10.2.3	Health and Safety *319*	
10.2.4	Continuity of Work *319*	
10.3	Temporary Works *321*	
10.3.1	Definitions *322*	
10.3.2	Legislation *322*	
10.3.3	Guidance *323*	
10.3.4	Management of Temporary Works *323*	
10.3.5	Temporary Works Designers *324*	
10.3.6	Temporary Works Coordinator *324*	
10.3.7	Temporary Works Supervisor *325*	
10.3.8	Types of Temporary Works *325*	
10.4	Work Sequences *326*	
10.4.1	Demolition *326*	
10.4.2	Façade Retention *327*	
10.4.3	Trenches *327*	
10.4.4	Soil Engineering *329*	
10.4.5	Pile Caps *331*	
10.4.6	Piled Walls *332*	
10.4.7	Basements *333*	
10.4.8	Core Walls *333*	
10.4.9	Top-Down Construction *334*	
10.4.10	Top-down, Bottom-up Construction *336*	
10.4.11	Offline Construction *338*	
10.5	Sequence Study *339*	
10.5.1	Sequence Diagram *339*	
10.5.2	Construction Cycles Times *344*	
	References *344*	
	Notes *344*	
11	**Method Statements** *347*	
11.1	Introduction *347*	
11.1.1	Definition *347*	
11.1.2	Uses of Method Statements *348*	
11.1.3	Clients *348*	
11.1.4	Consultants *349*	
11.1.5	Main Contractors *349*	
11.1.6	Subcontractors *350*	
11.1.7	Specialist Contractors *350*	
11.2	Format of Method Statements *351*	
11.2.1	Tabular *351*	

11.2.2	Written or Prose *352*
11.2.3	Generic *352*
11.3	Types of Method Statement *352*
11.3.1	Tender Method Statement *353*
11.3.2	Construction or Work Method Statement *354*
11.3.3	Safety Method Statement *355*
11.3.4	Lift Plans *355*
11.4	Preparing Method Statements *357*
11.4.1	Contents *357*
11.4.2	Layout *358*
11.4.3	Structure *359*
11.4.4	Safe Systems of Work *360*
11.4.5	Link with the Programme *360*
11.4.6	Changing Method Statements *360*
11.5	Worked Example *361*
11.5.1	Project Description *361*
11.5.2	Tender Method Statement *361*
11.5.3	Construction Method Statement *365*
11.5.4	Safety Method Statement *367*
	Notes *370*
12	**Planning for Safety** *373*
12.1	Introduction *373*
12.1.1	Human Factors *373*
12.1.2	Pressures on Health and Safety *374*
12.1.3	Definitions and Context *375*
12.1.4	Planning for Health and Safety *376*
12.2	Hazard and Risk *377*
12.2.1	Definitions *378*
12.2.2	Types of Hazard *378*
12.2.3	Common Hazards in Construction *379*
12.2.4	Identifying Hazards *380*
12.2.5	Guidance *380*
12.2.6	Persons at Risk *381*
12.2.7	Risk Evaluation *381*
12.2.8	Control Measures *382*
12.2.9	Risk Assessment *383*
12.3	Legal Framework *387*
12.3.1	Primary Legislation *387*
12.3.2	Subordinate Legislation *387*
12.3.3	Goal Setting versus Prescriptive Legislation *388*
12.3.4	Health and Safety Policy *390*
12.3.5	Approved Codes of Practice *390*
12.3.6	Guidance *392*
12.4	Managing Health and Safety *392*
12.4.1	HSG65 *392*
12.4.2	The Four Cs *392*

12.4.3	Preventative and Protective Measures	393
12.5	Planning the Work	393
12.5.1	Task-Based Approach	394
12.5.2	Hazard-Based Approach	394
12.5.3	Utilising the Programme	395
12.5.4	Statutory Duties	396
12.5.5	Safe Place of Work	398
12.5.6	Safe Systems of Work	398
12.5.7	Method Statements	399
12.6	Industry-Specific Legislation	401
12.6.1	European Union Directives	401
12.6.2	The CDM Regulations	401
12.7	The CDM Regulations 2015	402
12.7.1	Application	402
12.7.2	Notification	402
12.7.3	Duty Holders	403
12.7.4	Client Duties	403
12.7.5	Designers	404
12.7.6	Principal Contractors	404
12.7.7	Other Contractors	405
12.7.8	Principles of CDM	405
12.7.9	Welfare Facilities	405
12.7.10	Pre-construction Information	406
12.7.11	Construction Phase Plan	406
12.7.12	Site Rules	408
12.7.13	Health and Safety File	408
12.8	Health and Safety Training	408
12.8.1	Site Induction	409
12.8.2	Toolbox Talks	409
12.8.3	Task Talks	410
12.8.4	Walk-through/Talk-through	410
12.9	Measuring Performance	410
12.9.1	Monitoring	411
12.9.2	Audit	411
12.10	Enforcement of Legislation	412
12.10.1	Enforcement Action	412
12.10.2	Prosecution	413
12.10.3	Corporate Manslaughter	414
12.10.4	HSE Prosecutions Database	415
12.10.5	HSE Enforcement Notices Database	416
12.11	Accidents and Incidents	416
12.11.1	RIDDOR 2013	416
12.11.2	Accident Book	417
12.11.3	The Cost of Accidents	417
	References	418
	Web References	418
	Notes	418

13	**Planning the Project** *421*
13.1	Introduction *421*
13.2	Principles of Project Planning *421*
13.2.1	Getting a Feel for the Project *421*
13.2.2	Establishing Key Project Dates *422*
13.2.3	Completion *422*
13.2.4	4D BIM *423*
13.2.5	Establishing Key Activities or Events *424*
13.2.6	Work Breakdown Structure *425*
13.2.7	Activity Durations *425*
13.2.8	Programming Techniques *427*
13.2.9	Calendars *428*
13.2.10	Timescales *428*
13.3	Pre-construction Planning *428*
13.3.1	Pre-contract Planning *429*
13.3.2	Pre-contract Meetings *429*
13.3.3	Procurement *430*
13.3.4	Procurement Programme *430*
13.3.5	Procurement Programmes *431*
13.3.6	Lead Times *431*
13.3.7	Site Layout Planning *435*
13.3.8	The Master Programme *435*
13.4	The Baseline Programme *436*
13.4.1	Managing Change *438*
13.4.2	Progress *438*
13.5	Requirement Schedules *441*
13.5.1	Key Materials Schedules *441*
13.5.2	Plant Schedules *441*
13.5.3	Subcontract Schedules *442*
13.5.4	Information Requirement Schedules *442*
13.5.5	Requests for Information (RFIs) *444*
13.6	The Target Programme *445*
13.6.1	Legal Implications *445*
13.6.2	Practical Implications *447*
13.7	Contract Planning *447*
13.7.1	Subcontract Programmes *448*
13.7.2	Stage Programme *450*
13.7.3	Short-term Programme *450*
13.7.4	As-built Programme *453*
14	**Planning Cash Flow** *457*
14.1	Introduction *457*
14.1.1	Insolvency *457*
14.1.2	Cash Flow *458*
14.1.3	Working Capital *458*
14.1.4	Work in Progress *458*
14.1.5	Client Cash Flow *459*
14.1.6	Contractor Cash Flow *459*

14.1.7	Cash Flow Forecasts	459
14.2	Earned Value Analysis	461
14.2.1	Earned Value	462
14.2.2	Quarter–third Rule	463
14.2.3	Cumulative Percentage Value	463
14.2.4	The Bar Chart Programme	467
14.3	Cash Flow Forecasting	468
14.3.1	Preparing a Forecast	468
14.3.2	Credit Terms	471
14.3.3	Retentions	472
14.3.4	Payment	472
14.3.5	Payment Terms	473
14.3.6	JCT SBC	473
14.3.7	Other Contracts	473
14.3.8	Reflecting Payment Terms in Forecasts	474
14.4	Improving Cash Flow	476
14.4.1	At Tender Stage	476
14.4.2	During the Contract	476
14.4.3	Post-contract	477
14.5	Movement of Money	477
14.5.1	Money In and Money Out	477
14.5.2	Retention	477
14.5.3	Payment Delay	479
14.5.4	Credit	479
14.5.5	Late Payments	480
14.6	Working Capital	480
14.6.1	Forecasting Working Capital	480
14.6.2	Worked Example	482
	References	486
	Notes	486
15	**Project Control**	489
15.1	Introduction	489
15.2	Budgetary Control	489
15.2.1	Budgets	490
15.2.2	Reporting	490
15.2.3	The Reason for Budgets	490
15.2.4	Types of Budgets	491
15.2.5	Presenting Budgets	491
15.2.6	Construction Management Software	492
15.2.7	The Basis of Budgets	492
15.3	Establishing Budgets	493
15.3.1	Labour Budget	493
15.3.2	Plant Budget	493
15.3.3	Preliminaries Budget	493
15.4	Site Records	499
15.4.1	Site Diary	500
15.4.2	Labour Records	500

15.4.3	Instructions	501
15.4.4	Programme	501
15.4.5	Site Measurements	501
15.4.6	Variations	501
15.4.7	Delay and Disruption	502
15.4.8	Daywork	502
15.5	Meetings	502
15.5.1	Purpose	502
15.5.2	Types	503
15.5.3	Formal and Informal Meetings	504
15.5.4	Monthly Site Meetings	505
15.5.5	Weekly Progress Meetings	505
15.5.6	Subcontract Coordination Meetings	506
15.6	Key Performance Indicators	507
15.6.1	Types of KPIs	507
15.6.2	Uses of KPIs	508
15.6.3	Practical Applications	508
	Reference	508

16	**Controlling Time**	*511*
16.1	Introduction	511
16.1.1	Duty to Complete	511
16.1.2	Time for Completion	512
16.2	The Contractor's Programme	512
16.2.1	Milestones	513
16.2.2	Early Warning Systems	514
16.3	Progress and Delay	514
16.3.1	Delay	516
16.3.2	Delay Events	516
16.3.3	Change Control	516
16.4	Recording Progress	517
16.4.1	Progress Reporting	517
16.4.2	Cloud Based Reporting	519
16.4.3	Using the Bar Chart	520
16.5	Delay and Disruption	523
16.5.1	Reasons for Delay and Disruption	523
16.5.2	Definitions	524
16.5.3	Delay	525
16.5.4	Disruption	525
16.5.5	Types of Delay	525
16.6	Extensions of Time	526
16.6.1	Contract Provisions	526
16.6.2	Establishing Entitlement	527
16.6.3	Time and Money	527
16.6.4	Concurrent Delay	528
16.6.5	Mitigation	528
16.6.6	Float	529

16.7	The 'As-Planned' Programme	530
16.7.1	Shortcomings	532
16.8	The 'As-Built' Programme	532
16.9	Delay Analysis	532
16.9.1	Delay and Disruption Protocol	533
16.9.2	Delay Analysis Methodologies	534
16.10	Delay Analysis in Practice	534
16.10.1	Showing Delay on the Programme	535
16.10.2	Software	535
16.10.3	Worked Example	537
16.11	Project Acceleration	539
16.11.1	Time–Cost Optimisation	541
16.11.2	Terminology	541
16.11.3	Worked Example	544
	References	549
	Web References	550
17	**Controlling Money**	**553**
17.1	Introduction	553
17.1.1	Forecasting Revenues and Profits	553
17.1.2	Declaring Profits	554
17.1.3	Corporate Governance	555
17.2	Reporting Procedures	555
17.2.1	Small Firms	556
17.2.2	Medium-Sized Firms	557
17.2.3	Large Firms	557
17.2.4	Financial Control Methods	558
17.3	Earned Value Analysis	558
17.3.1	EVA Process	559
17.3.2	Limitations	560
17.3.3	Benefits	560
17.4	Cost Value Reconciliation	561
17.4.1	CVR Terminology	561
17.4.2	Reconciliation Date	561
17.4.3	CVR Process	561
17.5	Cost Value Reports	565
17.5.1	Cumulative Value	565
17.5.2	Reconciled Value	565
17.5.3	Reconciled Cost	566
17.5.4	Cumulative Value and Cumulative Cost	567
17.5.5	Provisions	567
17.5.6	Profit	567
17.5.7	Variance Analysis	569
17.5.8	Date of Report to Management	571
	References	571
	Note	572

18	**Controlling Resources** *575*
18.1	Introduction *575*
18.1.1	Targets *575*
18.1.2	Head Office *575*
18.1.3	Resources *576*
18.1.4	Preliminaries *576*
18.1.5	Resource Management *576*
18.2	Subcontractors *577*
18.2.1	Prequalification *577*
18.2.2	Subcontract Prices *578*
18.2.3	Subcontract Attendances *578*
18.2.4	Subcontract Programme *579*
18.2.5	Subcontract Liaison *579*
18.2.6	Benchmarking *579*
18.3	Labour Control *581*
18.3.1	Directly Employed Labour *581*
18.3.2	Self-employed Labour *583*
18.3.3	Resource Histogram *583*
18.3.4	Resource Levelling *584*
18.4	Materials Control *587*
18.4.1	Materials Management *589*
18.4.2	Waste *591*
18.4.3	Planning for Waste *591*
18.4.4	Improving Waste Management *592*
18.4.5	Duty of Care *592*
18.4.6	Site Waste Management Plans *594*
18.5	Plant Control *596*
18.5.1	Legislation *597*
18.5.2	Organisation *597*
18.5.3	Planning *598*
18.5.4	Inspection and Maintenance *598*
19	**Hotel and Commercial Centre** *599*
19.1	Project Description *599*
19.1.1	Project Details *599*
19.1.2	Site Constraints *599*
19.1.3	Procurement Strategy *600*
19.1.4	Project Organisation *602*
19.2	Design *603*
19.2.1	BIM *603*
19.2.2	Modular Design *603*
19.2.3	Developer/Client Role *604*
19.2.4	Advantages of Modular *605*
19.3	CIMS-MBS Off-Site Manufacture *605*
19.3.1	Modular Procurement *605*
19.3.2	Design Stage *606*

19.3.3	Prototype	607
19.3.4	Mass Manufacture	607
19.3.5	Payment	608
19.3.6	Installation	609
19.4	Construction	609
19.4.1	Construction Programme	610
19.4.2	Module Installation	613
19.4.3	Modular Interface	614
19.4.4	Defects	615
19.5	Legal Matters	615
19.5.1	Payments	616
19.5.2	Milestone Payments	616
19.5.3	Insolvency	617
	Reference	617
20	**Motorway Bridge Replacement**	*619*
20.1	Project Description	619
20.1.1	Project Background	619
20.1.2	Project Location	619
20.1.3	The Existing Bridge	619
20.1.4	Project Procurement	620
20.2	Project Scope	621
20.2.1	Enabling Works	621
20.2.2	Permanent Works	623
20.2.3	Off-site Works	623
20.2.4	Demolition	623
20.3	Project Organisation	624
20.4	Programme	626
20.4.1	Pre-construction Phase	626
20.4.2	Tender Period and Contract Award	627
20.5	Detailed Design Development	627
20.5.1	Design of the Permanent Works	627
20.5.2	Temporary Works Design	628
20.5.2.1	Bridge Abutments and Central Pier	628
20.5.2.2	Bridge Deck	629
20.5.3	4D BIM	630
20.6	Construction Planning	631
20.6.1	Major Milestones	631
20.6.2	Motorway Possessions and Road Closures	632
20.6.3	Bridge Abutments and Central Pier	633
20.6.4	Bridge Deck	634
20.6.5	Demolition	636
	Notes	640
21	**High Speed 2**	*641*
21.1	Project Description	641
21.1.1	Business Case	641
21.1.2	Project Delivery	642

21.2	Route	*642*
21.2.1	Oakervee Review	*642*
21.2.2	Integrated Rail Plan	*644*
21.3	Project Scope	*644*
21.3.1	Phase 1	*644*
21.3.2	HS2–HS1 Link	*645*
21.4	Legal Framework	*645*
21.4.1	Parliamentary Bills	*646*
21.4.2	Acts of Parliament	*646*
21.4.3	Planning Permission	*647*
21.4.4	The Secretary of State for Transport (SoST)	*647*
21.4.5	HS2 Ltd	*648*
21.4.6	Formal Agreements	*649*
21.4.7	Development Agreement	*649*
21.4.8	The Core Programme	*649*
21.4.9	The Wider Programme	*650*
21.5	Funding and Governance	*650*
21.5.1	Spending Oversight	*650*
21.5.2	Programme Governance	*650*
21.5.3	Programme Governance at HS2 Ltd	*652*
21.6	Roles and Cooperation	*652*
21.6.1	Secretary of State for Transport	*653*
21.6.2	HS2 Ltd	*654*
21.6.3	The Client Board	*654*
21.6.4	Programme Representative	*654*
21.6.5	Senior Responsible Owner	*655*
21.6.6	Programme Board	*655*
21.6.7	DfT Board Investment Commercial Committee	*655*
21.6.8	The Sponsor or Project Boards	*656*
21.6.9	Principal Accounting Officer	*656*
21.6.10	HS2 Ltd Accounting Officer	*656*
21.7	Order of Cost	*656*
21.7.1	National Audit Office Reports	*656*
21.7.2	Parliamentary Reports	*656*
21.7.3	Funding Envelopes	*657*
21.7.4	Oakervee Review	*657*
21.7.5	Forecasting Costs	*657*
21.7.6	Contingency	*658*
21.7.7	Schedule	*658*
21.7.8	Oakervee Review	*659*
21.7.9	Phase 1 Programme	*659*
21.8	Project Design	*660*
21.8.1	Design Vision	*660*
21.8.2	Design Panel	*660*
21.8.3	Design Review	*661*
21.8.4	Design Strategy	*661*
21.8.5	Design Stages	*662*

21.8.6	Design Critique	663
21.8.7	Station Design	663
21.9	Procurement	664
21.9.1	Design Procurement	665
21.9.2	Construction Procurement	665
21.9.3	Labour and Plant Procurement	666
21.9.4	Forms of Contract	666
21.9.5	Payments	667
21.9.6	Procurement Critique	667
21.10	HS2 Works Areas	667
21.10.1	Enabling Works	667
21.10.2	Main Civils Works	668
21.11	Construction Management	669
21.11.1	Area Organisation	669
21.11.2	Area North Joint Ventures	669
21.11.3	Administrative Support	670
21.11.4	Construction Planning	670
21.12	Construction Methods	671
21.12.1	Long Itchington Wood Tunnel	671
21.12.2	Colne Valley Viaduct	675
21.12.3	Old Oak Common Station	678
21.13	Innovation	682
21.13.1	Logistics Hub	682
21.13.2	Conveyors	683
21.13.3	Access Roads	683
21.13.4	Diesel-Free Site	684
21.13.5	Green Tunnels	684
21.14	Epilogue	686
	References	686
	Notes	687

Index *689*

About the Authors

Brian Cooke (MSc) is a former chartered civil engineer, chartered builder, quantity surveyor and principal lecturer in construction management. He has extensive industry experience and has lectured widely on management and financial topics in the United Kingdom and overseas. Brian is now retired.

Peter Williams (MSc) is a former chartered builder and chartered quantity surveyor with many years of experience in building and civil engineering contracting. He was also a principal lecturer in construction management, quantity surveying and construction health and safety management. Peter is currently a writer, researcher, lecturer and consultant.

Preface

The world has changed considerably since the previous edition of this book was published in 2009.

Not only have we all suffered a hugely damaging pandemic, the United Kingdom has left the European Union, and the world is seeing unimaginable levels of conflict and social change.

World supply chains remain unpredictable post-pandemic, and UK debt is more than £2.5 trillion – almost 100% of GDP – with industry output seriously impacted by reduced levels of public sector spending.

On a more positive note, a huge transformation in digital technology in the past 10 years has revolutionised communications generally and construction design and management in particular such that construction projects can now be seen in virtual reality before they are built. Building Information Modelling (BIM) has transformed the interface between design, planning and construction, thereby facilitating a 'what-if' approach to planning pre-construction that was hitherto not possible.

Mobile technology has also changed unrecognisably since the first iPad was released in 2010. Mobile devices have transformed the way in which construction projects are managed and progressed, meaning that site progress and requests for information can now be processed in real time.

This period of exciting development has taken place against a background of sobering change in the UK construction scene which has witnessed the collapse of Carillion, the United Kingdom's second largest contractor and the consequently seismic effect that this has had on the construction supply chain. Official reports suggest that significant numbers of insolvencies resulted from this catastrophe and call into question the robustness and reliability of auditing standards and practices. Since then, other large contractors have suffered grinding profits and financial instability – causing disquiet amongst shareholders – and other top 100 contractors have followed Carillion into insolvency.

Despite these winds of change, the aims of the first edition have not changed.

This is a book for students of construction-related professional and degree courses – construction management, building, civil engineering, quantity surveying and building surveying alike. Students and young professionals of all these disciplines need a basic understanding of the culture and methodologies of the industry. They need to know how construction projects are procured, planned and managed, and they need to appreciate the link between construction technology and the methods, order and sequence of work on site. This book aims to fulfil this need.

The fourth edition is a major rewrite necessitated by the tide of progress and changes in industry practice. The basic structure of the book is unchanged, but there are seven completely new chapters, three new major case studies and the remaining 11 chapters have been extensively rewritten and updated.

The ambit of the book has been broadened and deepened and the early chapters have been refocused on the management of projects and project risk. Major and mega-projects have been given suitable prominence in the book including the Shard, Crossrail, Heathrow T5, Hinkley Point C and HS2, and new material has been included covering joint ventures, frameworks and collaborative working.

The case studies and sequence studies in this edition cover a much wider variety of construction methods than previously and include modular construction, slip forming, tunnelling and top-down construction. In common with previous editions, health and safety and the provision of safe systems of work are integral to the book and retain their importance in a revised and fully updated dedicated chapter.

In this edition, particular emphasis has been placed on work sequences and method statements, and these chapters have been extensively revised and extended. The book is also illustrated with colour photographs for the first time to add realism to the case studies and the numerous worked examples.

Coverage of construction industry reports has been updated to include reflections on how the industry has responded to the challenges posed by Latham, Egan and so on. This includes the Farmer Review which has prompted a new chapter on modern methods of construction and a new case study dealing with modular construction. The topics of contracts and procurement have been amalgamated into a new, extensive and fully-revised stand-alone chapter. This includes coverage of a number of international contracts and the role of the programme in a wide variety of contract conditions.

Finally, this edition includes coverage of common data environments, enterprise resource planning, construction management software and the use of 4D BIM in construction scheduling.

Acknowledgements

I would like to acknowledge the invaluable generosity and support of several people who have helped me through the long and difficult task of writing this book. Some have provided access to the software used in the text and others have given permission to use copyrighted material, but all those mentioned here have given the most precious thing of all – their time. In this respect, I am particularly indebted to Chris Buckley, Project Manager for Sir Robert McAlpine Ltd, for his input to Chapter 20. Chris's great knowledge and experience of the industry informed other chapters too and my visits to site were always highly informative and good fun!

The process of writing a book is a solitary endeavour, but each personal contact – no matter how fleeting – has provided renewed inspiration to see it through to a conclusion and to make the end product the best that I can achieve. I can do no better.

Firstly, my sincere thanks go to everyone who has helped directly or indirectly with the technical aspects of the book – people who represent the generosity of a special industry with which I am proud to have been associated for 60 years.

Martin Belcher	4PS Construct
Tony Bolding	Cloud4 Ltd (QSPro)
Chris Buckley	Sir Robert McAlpine Ltd
Stephen Durkin	BSS Software (Cubit Estimating)
John Fozzard	Project Commander
Kieran Witsey	HS2 Ltd
Polly Catchpole	Elecosoft UK Ltd
Justin Kinsella	HTL.tech

Secondly, I would to like to record my particular thanks to my Wiley-Blackwell publisher Dr Paul Sayer, whose patience, support and unshakeable faith have made this publication possible.

Thirdly, throughout the ups and downs of several years since the fourth edition was commissioned, my dearest Jaqueline has been a rock of love and encouragement, a sounding board for ideas and the motivation to keep going when it seemed easier to give up.

Finally, but not least, the fourth edition of this book is dedicated to the memory of my dear friend and former colleague Paul Hodgkinson. Paul produced the many excellent illustrations in the three previous editions, several of which live on in this edition simply because they cannot be equalled. They represent a timeless tribute to a loyal and dedicated professional and an all-round great bloke who I am proud to have known.

Peter Williams
Chester

Acronyms

BAA	British Airports Authority
BCIS	Building Cost Information Service
BIM	Building Information Modelling
BQ	Bills of Quantities
BSS	Building Software Services
CAD	Computer-aided Design
CATO	Causeway CATO Suite
CDM	Construction Design and Management Regulations
CIOB	Chartered Institute of Building
CFA	Continuous Flight Auger
CIMC	China International Marine Containers
COSHH	Control of Substances Hazardous to Health
CPI	Consumer Price Index
CSCS	Construction Skills Certification Scheme
ERP	Enterprise Resource Planning
FIDIC	Fédération Internationale des Ingénieurs Conseils
GDP	Gross Domestic Product
GSK	GlaxoSmithKline plc
HGCR	Housing Grants, Construction and Regeneration Act
HMP	His Majesty's Prisons
HMRC	His Majesty's Revenue and Customs
HVAC	Heating Ventilating and Air Conditioning
ICC	Infrastructure Conditions of Contract
ICE	Institution of Civil Engineers
JCT	Joint Contracts Tribunal
KFC	Kentucky Fried Chicken
KPI	Key Performance Indicator
MEP	Mechanical and Electrical Services Book
MEWP	Mobile Elevating Work Platforms
M&E	Mechanical and Electrical

MOT	Ministry of Transport
NEC	New Engineering Contract
ONS	Office for National Statistics
PFI	Private Finance Initiative
QS	Quantity Surveyor
RC	Reinforced Concrete
RIBA	Royal Institute of British Architects
RICS	Royal Institution of Chartered Surveyors
RPI	Retail Price Index
SIP	Structural Insulated Panel
TBM	Tunnel Boring Machine

	Chapter 1 Dashboard	
Key Message		○ Major projects are complex. ○ The culture and methodologies of the industry condition the way projects are carried out. Appropriate procurement choices are vital. ○ Public sector clients tend to be excessively bureaucratic, and politicians and political decisions can severely disrupt the smooth running of projects.
Definitions		○ **Projects** may be standalone or part of a programme or portfolio. Projects are traditionally measured by the criteria of time, cost and quality. ○ **Programmes** are measured by the achievement of specific strategic objectives and benefits. ○ **Portfolios** are a means of structuring investment in properties or physical assets where a balance between investment and benefit is required. ○ **Building Information Modelling (BIM)** has improved project information exchanges and design clash detection as well as facilitating the 4D planning of projects.
Headings		**Chapter Summary**
1.1	Introduction	○ Construction is a hugely adaptable and inventive industry that undertakes an impressive array of projects from domestic scale to mega-infrastructure projects of cutting-edge complexity. ○ Many major projects have experienced poor out-turns regarding time and cost predictability. ○ Construction mega-projects are special due to their size, cost and complexity. They inform our approach to 'normal' projects and help to drive forward new ways of thinking and new technologies.
1.2	Why Do Projects Go Wrong?	○ Construction is a 'project-based industry' where the time, cost, quality, resources, problems and solutions are all geared to the project. ○ Construction work is complex because it involves the procurement and management of finite resources from a supply chain that often struggles to cope with demand.
1.3	Managing the Risk of Delayed Completion in the 21st Century	○ Construction projects – especially mega-projects – are frequently late and over budget. The many reasons for this include poor planning and management and the lack of dynamic scheduling, risk management and record keeping in order to control time effectively. ○ Many common forms of contract do not promote or encourage efficient time management. ○ Project management software can aid effective project planning and control and thereby minimise risk, delay and disputes in construction projects.
1.4	The Latham Reports	○ Commissioned to find ways to reduce conflict and litigation and encourage the industry's productivity and competitiveness.
1.5	The Egan Reports	○ Set up by Government to identify the scope for improving quality and efficiency in construction.
1.6	The Wolstenholme Report	○ Wolstenholme sought to establish what progress had been made since the second Egan Report in 2002.
1.7	The Farmer Review	○ Commissioned by the Construction Leadership Council to review the UK construction labour model and the poor performance of the industry. ○ Critical symptoms included low productivity, low predictability of time, cost and quality, structural and leadership fragmentation, low margins and adversarial pricing models.
Learning Outcomes		○ Understand the nature and complexity of construction projects. ○ Distinguish between projects, programmes and portfolios. ○ Appreciate the role of industry reports in understanding the culture and methodologies of construction.
Learn More		○ Read Chapter 1 Sections 1.3–1.7 to see whether the recommendations of official reports have had any significant impact on the way that the industry operates. ○ See also Chapters 2, 4 and 19–21.

1

Construction Projects

1.1 Introduction

Construction is like no other industry.

The built environment around us is testimony to the audacity and ingenuity of architects, builders and engineers over the centuries. The Egyptian pyramids, mediaeval castles and cathedrals, the canal and rail infrastructure of the nineteenth century and more recent projects such as the skyscrapers of New York and tunnels through the Swiss Alps and under the English Channel provide breathtaking exemplars that characterise the world of construction and engineering.

A particular feature of the history of construction around the world is that the buildings, structures, bridges, railways and tunnels are all essentially prototypes. This is true to this day, even where designs are identical. Repetitive housing, fast-food outlets, chain hotels and standardised factory-made components may well be 'jelly-mould' designs, but every construction site on which they are built is different, and each construction team will invariably be unique, assembled with different people from different socio-economic and cultural backgrounds. These are the people who turn design into reality in the tough and dangerous world of construction.

Construction is a hugely adaptable and inventive industry that undertakes an impressive array of projects from domestic-scale repair, maintenance and remodelling work to mega-infrastructure projects of cutting-edge complexity posing enormous technological challenges.

> The E39 highway in Norway is a good example – it is 1000 km (680 mi) long and crosses fjords up to 1.3 km deep with floating bridges and tunnels – a project of breath-taking scale and environmental sensitivity.
> https://youtu.be/HCT-FurFVLQ

Construction Planning, Programming and Control, Fourth Edition. Brian Cooke and Peter Williams.
© 2025 John Wiley & Sons Ltd. Published 2025 by John Wiley & Sons Ltd.

1.1.1 Industry Reputation

Paradoxically, the construction industry does not have the best reputation:

- It is widely recognised as being adversarial and slow to accept change.
- Easy entry into the industry encourages shoddy workmanship and so-called 'cowboy' builders.
- Many major projects have experienced poor out-turns regarding time and cost predictability including Wembley Stadium, the Scottish Parliament at Holyrood and Crossrail in London. England's current High-Speed Rail project (HS2) has suffered delays, overspending, extensive scope changes and widespread criticism.
- The industry generally suffers from a poor health and safety record.
- The Chartered Institute of Building (CIOB)[1] reports that construction underperforms in terms of inclusivity and diversity.

Conversely:

- Major contractors have encouraged top-down improvements in health and safety 'norms' which drive higher standards in the many subcontractors and smaller firms that operate in the industry.
- Building information modelling (BIM) has improved project information exchanges and design clash detection as well as facilitating the 4D planning of projects.
- Modern methods of construction, the concept of design for manufacture and assembly (DfMA) and the use of factory-built components and assemblies are being more widely integrated into mainstream projects with beneficial impact on time and cost certainty.

1.1.2 Projects

Construction is often referred to as a 'project-based industry' and, wherever there is a built environment, you will not be far from a tower crane, roadworks on a motorway or scaffolding around a building – indications of the presence of a construction project.

A construction project could be anything from a modest house extension to the £15 billion Crossrail project in London, one of the largest construction projects ever undertaken in Europe – but there is a common theme. The whole focus is on the project – the time, cost, quality, resources, problems and solutions are all geared to the project.

This brings enormous pressure on project teams, however, big or small they are, to ensure that the project is completed on time, on budget and to the correct quality standards.

The 'one-off' nature of construction creates additional pressures, however, because more or less every project has its own individuality and peculiarities depending on the site and location, the design and type of construction, the business arrangements between the parties and the hopes and expectations of all those involved. Projects may be defined as:

> Unique, transient endeavours, undertaken to bring about change and achieve planned objectives, which can be defined in terms of outputs, outcomes or benefits.
>
> (APM 2020)

This definition identifies the normal reasons why construction projects are carried out and the expected outcomes – a client satisfied with the finished result and completion within defined time, cost and quality expectations.

Whether a construction project is intended to provide an asset for personal use (such as a house) or for production or investment (a new factory or an office block) or to upgrade or maintain an existing asset (a house extension or repairs to a rail bridge), capital expenditure is normally required in the form of a loan, direct investment or public funding.

In some cases, public-sector projects are constructed with private-sector investment. The private finance initiative (PFI) used in the UK enabled the public sector to repay the capital cost of its projects over time according to the utility provided by the facility. This could be a toll bridge, a hospital or a prison, for instance. Concerns over value for money, however, led to PFI – and its successor PF2 (Private Finance 2) – being discontinued. Other forms of public-private partnership (PPP) have been developed in their place.

1.1.3 Programmes

In order to distinguish between a 'project' and a 'programme' take the example of a modern PPP between a local authority and the development arm of a large contractor.

> A joint venture was formed in order to build 109 new homes for sale and 69 for rent across two sites as part of a two-phase regeneration scheme that includes associated community facilities.[2]

This development is a 'programme' which are defined as:

> *Unique and transient strategic endeavours, undertaken to achieve a defined set of objectives, incorporating a group of related projects and change management activities. They can be defined as coordinated … combined to achieve beneficial change.*
>
> (APM)

A programme can therefore be described as a number of related projects brought together to achieve particular benefits in a more effective way than as a group of individual projects. Admittedly, each project in the programme may well be organised and managed individually but there might be shared facilities – such as an on-site concrete batching plant – and the same contracts manager may be in overall charge of all the projects comprised in the programme.

An important distinction between projects and programmes is that projects are traditionally measured by the criteria of time, cost and quality whereas programmes are measured by the achievement of specific strategic objectives and benefits which might otherwise have not been possible had the projects been managed independently.

A further notable difference between projects and programmes is that programmes are often punctuated by a number of milestones and are not always as strictly finite as a project. They also take far longer to complete than any of the projects within the programme and may, in some cases, have no specific end date at all.

HS2 is often referred to as a project, but it is, in fact, a programme with specific strategic objectives: to reduce journey times, increase connectivity and encourage investment. There are hundreds of individual elements to HS2 – stations, tunnels, bridges, viaducts and track, rolling stock and so on – that are geographically spread over some 400 km (260 mi), making it impossible to manage as one project.

However, HS2 also sits within a portfolio of public sector infrastructure investments in road, rail and major transport schemes.

1.1.4 Portfolios

Portfolios are used to select, prioritise and control an organisation's programmes and projects, in line with its strategic objectives and capacity to deliver (APM).

Local authorities, for instance, own and manage a wide variety of property, such as social housing, schools, care homes, waste and recycling centres, shopping and leisure centres, commercial property and municipal buildings, and so on, that require investment in order to maintain, adapt, replace or augment the estate. This investment usually takes place over an unspecified period, and the work involved has to be managed and prioritised.

Consequently, the related or unrelated programmes of work or stand-alone projects that arise need to be organised into a structured portfolio and managed according to urgency, need, budgetary and timing demands, usually in the form of an asset management plan.

Similar portfolios of work will be found in hospital or prison estates or in airport authorities that have large estates of properties to look after.

Portfolios are, therefore, a means of structuring investment in properties or physical assets, such as road and rail infrastructure, where a balance between investment and benefit is required and where projects and programmes are created and closed out accordingly.

The role of portfolios, and the interrelationship between projects and programmes, is illustrated in Figure 1.1 which distinguishes a stand-alone programme from that sitting within a portfolio and shows how a project can equally sit within a programme and a portfolio. It also depicts how an organisation can have stand-alone projects and programmes as well. The CIOB provides a useful summary:

- **Projects** are of relatively short duration measured in weeks/months – a new hospital for example.
- **Programmes** have longer durations measured in years with a finite end – such as upgrading a number of existing hospitals to meet modern standards.
- **Portfolios** are ongoing activities with no defined end – the repair and maintenance of a number of hospitals over an undefined or ongoing period, for example.

1.1.5 Mega Projects

The construction industry is renowned for its mega projects and there is no doubt that they have a beguiling fascination for their breathtaking scale, technical audacity, incredible timescales and enormous cost. Some recent examples of such projects are shown in Table 1.1.

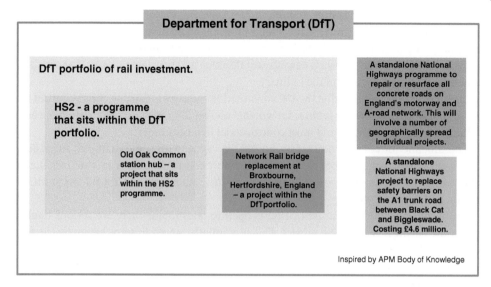

Figure 1.1 Projects, programmes and portfolios.

Table 1.1 Recent mega projects.

Project	Description	Indicative cost US$ billion	Indicative construction period Years
Al Maktoum International Airport, Dubai	Airport	82	5
Dubailand, Dubai	Theme Park and leisure project	64	12
South-North Water Transfer Project, China	Canal project for irrigation system	78	48
London Crossrail Project	Tube system	23	10
Linear Chuo Shinkansen, Japan	Ultra-high-speed railway using magnetic levitation technology	52	12 (Phase 1)

Realistically speaking, the time and cost figures associated with such projects do not really matter – they are broad estimates at best. Yes, these projects are planned and, yes, timescale and cost controls are in place, but everyone knows that time will overrun significantly, and cost will spiral way beyond original expectations – it is 'par for the course' for the vast majority of these huge projects.

This does not invalidate the need for planning and control – it is just that unknown, unforeseen and unexpected events override the process. Political delays, inflation, technological change and business failures mean that the risks are high, but so are the rewards.

Mega projects change the lives of millions of people, but they are the very tip of the construction 'iceberg' and might be considered 'abnormal' in an industry-wide context. One of the great benefits of mega projects, however, is that they inform our approach to 'normal' projects and help to drive forward new ways of thinking and new technologies that can filter down through the echelons of the industry.

Sir John Egan (1998) said that the UK construction industry is capable of carrying out *the most difficult and innovative projects* imaginable and another mega-project – High Speed 2 (HS2) – is one of the largest and most controversial in recent times.

The timeline for construction works for HS2 is vague, with some estimates suggesting that it might be 20 years before the project is entirely complete. Cost is another imponderable – originally estimated to cost £32 billion, predictions range from £56 billion to £100+ billion at the time of writing.

Politically speaking, the future of HS2 is open to question as there are serious doubts about the business case for the project. This is counterbalanced to some extent by the fact that HS2 is being modelled in a BIM environment, using the latest digital technologies. Consequently, the entire project will be realised virtually, from design through construction and occupancy, before physical work is carried out – thereby creating considerable economies.

This process should facilitate enormous savings in time and money compared to traditional methods – despite the eye-watering cost – and may well set the standard for future, more modest projects, in line with government aspirations.

1.2 Why Do Projects Go Wrong?

Late and over-budget major projects certainly grab the headlines:

- The Berlin Brandenburg Airport was completed nine years late.
- The Flamanville-3 nuclear power station in France is over 11 years late and more than five times over its initial €3.3 billion budget.
- Crossrail in London was similarly £billions over budget and several years late.
- In the United Kingdom, the first phase of HS2 is reportedly four years behind schedule and phase two was eight years late before it was cancelled.
- In its annual report 2022–2023, the UK Infrastructure and Projects Authority (IPA)[3] gave phases 1 and 2a of HS2 a 'red' rating. This means that successful delivery *appears to be unachievable* and that *there are major issues with project definition, schedule, budget, quality and/or benefits delivery, which at this stage do not appear to be manageable or resolvable.* The implication is that the project *may need re-scoping and/or its overall viability reassessed.*[4]

Reasons for such delays are easy to find and can include:

- Ineffective political governance.
- Lengthy statutory approvals processes.
- Suffocating bureaucracy and political meddling.
- Hugely complex chains of command – there are over 29 000 people engaged on HS2, for instance.

- Sheer scale – HS2 Phase 1 has over 350 major construction sites.
- Extensive scope changes and 'mission creep'.
- Poor accountability.
- Poor project management.
- Poor quality materials and inadequate workmanship.
- The COVID-19 pandemic and its aftereffects.
- Scarcity of resources – especially skilled labour.

A recent survey by the Project Management Institute (PMI) revealed that only 48% of major construction projects were completed on time in 2020 and that almost one-third of projects failed to meet their original aims and objectives.[5]

In an earlier CIOB survey[6] involving several thousand projects, simple, repetitive and low-rise projects (<6 storeys) were reported to have a high chance of success using traditional management processes. This contrasts sharply with more complex projects where the likelihood of completion on time, or within a short time after the intended completion date, was reported to be significantly lower.

1.2.1 Looking for Reasons

Whilst providing useful insights into why projects may be delayed, the results from the PMI and CIOB surveys are entirely predictable. Complex projects will face problems because they are just that – complex.

The CIOB survey tends to point the finger at contractors, project managers and at the standards of project management in the industry as the prime suspects when a project is delayed but there is much more to it than that:

- Construction projects invariably commence on site before the design is sufficiently well developed which can lead to late, incomplete or conflicting information being issued to contractors thereby compromising the proper planning of production on site. This issue has been raised in several official reports including the Latham Report (1994).
- Design changes are common in construction which leads to variations or changes to works information. This can lead to inefficiencies, reworking of completed work and delays awaiting confirmation of instructions.
- Many projects rely on design input from specialist subcontractors who are, invariably, appointed after the main contract has been awarded. This militates against commencement of work on site based on a complete design and can result in design clashes, conflicting or late information and potential delays should materials and components require long lead times before they can be delivered to site.
- Construction contracts often include provisional sums for work which is envisaged but not yet designed. Such work may well be defined in the contract but often this is inadequately detailed to enable the contractor to allow for the work in the schedule.
- Extensions of time allowed under construction contracts often provide inadequate recompense for the actual delay incurred on projects resulting from design changes, delayed instructions, unforeseen physical conditions or supply chain problems.
- In the event of delay, English law requires contractors to mitigate delay which can lead to inefficient working and poor use of resources.

- Contractors normally carry the risk of delay due to bad weather unless the weather is sufficiently inclement to be considered exceptional.
- Both clients and contractors are prone to being over-optimistic when planning their projects.
- Inflationary pressures and poor budgetary control.
- Disputes over contract payments can cause cash flow problems for contractors and sub-contractors. This can lead to delays on projects due to inability to secure and pay for the necessary materials and other resources needed to keep the project on schedule.
- Slightly tongue-in-cheek, there is also Murphy's law which says that:
  - Anything that can go wrong will go wrong.
  - Nothing is as easy as it looks.
  - Everything takes longer than you think it will.

1.2.2 Murphy's Law

The construction of the Scottish Parliament building in Holyrood, Edinburgh, Scotland, which was finally completed in 2004 – more than three years late – is a classic example of Murphy's law.

In a 271-page report by Lord Fraser,[7] a catalogue of reasons why the project cost so much money and took so long to build was given, including:

- The project cost was initially estimated to be £10–40 million.
- Outturn cost was £414 million.
- The Spanish architect Enric Miralles died before his vision could be completed.
- The architectural joint venture engaged for the project had different cultures and ways of working.
- Construction management procurement was chosen which placed full control of the project in the client's hands but also all the risk.
- The project was beset by accusations of poor and inexperienced management and criticisms of Members of the Scottish Parliament for meddling in the project and constantly making design changes.

It might be of some comfort that the building was generally acclaimed for its design which aimed to create a poetic union between landscape, people and culture and was awarded the 2005 RIBA Sterling Prize for architecture!!

1.2.3 Complexity

Construction projects – especially large projects – can run into difficulties simply due to the passage of time. It can take years to design a large project and the statutory approvals process – which might involve parliamentary consent – can take years or even decades. As time goes by legislation changes, project objectives and expectations can change and developments in construction materials, techniques and IT can also impact a project. The political and economic climate can change over time and changes of government, inflation or economic downturns can seriously affect the viability or scope of a project.

> Inflationary pressures on the predicted cost of HS2 mean that the project may be significantly scaled-back or abandoned entirely, mid-construction. The resultant impact on the industry is enormous with future order books decimated and an entire industry looking to find £billions worth of alternative work.

Construction work is complex because it involves the procurement and management of finite resources – materials, plant, labour, temporary works, specialist contractors and so on – from a supply chain that, especially at times of high industry output, struggles to cope with demand.

Additionally, projects are often undertaken in demanding conditions, subject to the vagaries of the weather, poor ground conditions, sometimes on confined sites with difficult access and where all manner of hazards are present beyond those directly related to the construction work itself (such as traffic, pedestrians, nearby buildings, underground services or tunnels).

A further layer of complexity is added to construction projects because they normally involve much more than the work on site:

- Justification for the intended project is normally needed in the form of a business case or some other criteria such as spatial requirements, modernisation of a facility, maintenance, etc.
- The necessary finance has to be arranged which, in some cases, can involve many billions of currency.
- Planning permission will invariably be required and, for very large projects of national importance, acts of parliament are needed.
- Construction projects take time – sometimes many years – before they can even start on site. In this time, economic circumstances can change, inflation can impact construction prices and unexpected supply chain issues can arise.
- Designs for architectural, structural, heating ventilating and air conditioning (HVAC) and so on have to be procured.
- In some cases, ground investigations, archaeological surveys and digs are required.
- There may be enabling work required such as demolitions, ground remediation, service diversions, access roads and highway diversions.

1.2.4 The Consequences of Late Completion

Research, empirical evidence and past history suggest that construction clients are more likely than not to be disappointed with the outcome of their projects from the perspective of time – if not cost and quality.

This is undoubtedly true, and it is also the case that many construction projects are poorly planned, badly managed and lack the necessary levels of dynamic scheduling, risk management and record keeping in order to control time effectively.

Conversely, it is also true that the industry can point to many examples of excellence in the time, cost and quality management of construction projects large and small. The PMI survey claims that 48% of complex projects were completed on time, and considering the complexity of such projects, this is not bad. Admittedly, it could be better.

Additionally, the CIOB survey suggests that low-medium rise projects with straightforward groundworks and simple HVAC services have a high chance of success. Considering that such projects represent a large proportion of the output of the industry, this is also encouraging.

As far as complex projects are concerned, there is no lack of expertise involved. They are carried out by the top contractors who use the latest digital technologies and have highly capable staff and risk management capabilities. Such projects are simply complex, and success or failure may depend upon influential factors that are beyond the control of clients or contractors. Very often – especially in the public sector – the levels of bureaucracy are suffocating, decision making is slow and inconsistent and politicians, frankly, meddle in things they do not understand!

The fact that construction projects can go wrong is hardly breaking news! In fact, since the advent of standard construction contracts in the mid-nineteenth century, there have always been provisions for dealing with delay, disruption and additional cost. Likewise, modern construction contracts all include clauses dealing with:

- Extensions to the time for completion with respect to matters such as:
  - Variations or changes to the works.
  - Late information, including design information.
  - Delays due to the presence of underground services or difficult subsoil conditions that could not have been anticipated.
  - Failure by the client to give possession of the site at the agreed time.
  - Suspension of the works by the client.
  - Archaeological or geological findings.
  - Exceptionally inclement weather.
- Damages for late completion:
  - Where this is the fault of the contractor.
  - Reimbursement of the client for the impact of delayed completion to the extent that money can compensate for its effects.
  - A fair assessment of the client's loss (hence common use of the term 'liquidated and ascertained damages' (LADs)).
- Loss and expense/compensation payable to the contractor in the event of:
  - Prolongation costs should delayed completion be the fault of the client.
  - The additional expenditure required to deal with risk issues that were not the responsibility of the contractor.
  - Delay or disruption to the works as a result of variations to the contract, late information or instructions or other reasons for which the contractor is not culpable.

The result of late completion is bad news for everyone:

- The client side may suffer:
  - Loss of revenue, rent or sales.
  - Loss of utility in the case of a house, hospital or prison.
  - Inconvenience to the public due to prolonged roadworks.
- The contractor side may suffer:
  - Loss of profits.

- Delays to other projects whilst awaiting resources tied up unnecessarily.
- Claims from subcontractors should their work be delayed or disrupted.
- Loss of reputation.

1.2.5 Construction Industry Reports

The search for answers to the failings of the construction industry stretch back into the mists of time.

In 1944, the Simon Report – *the Placing and Management of Building Contracts* – investigated how procurement methods could improve the efficiency of the construction industry. This report was followed by several other public reports which criticised the industry and its perceived poor record of client satisfaction and its failure to deliver projects on time, on budget and to the desired quality standards.

Official reports include the Emmerson Report (1962), the Banwell Reports (1964 and 1967), the Tavistock Report (1966) and the more recent reports by Sir Michael Latham (1993 and 1994), Sir John Egan (1998 and 2002), Andrew Wolstenholme (2009) and Mark Farmer (2016). These reports, and others, are shown in Table 1.2 and are discussed in detail by Murray and Langford (2003), who debate the extent to which government has tried to shape the performance and attitudes of the industry.

Table 1.2 Construction industry reports.

Report	Title	Year
Simon Report	The Placing and Management of Building Contracts	1944
Emmerson Report	Survey of Problems Before the Construction Industries	1962
Banwell Report	The Placing and Management of Contracts for Building and Civil Engineering Work	1964
National Economic Development Office (NEDO)	Action on Banwell	1967
Tavistock Report	Interdependence and Uncertainty	1966
Latham 1	Interim Report – Trust and Money	1993
Latham 2	Final Report – Constructing the Team	1994
Levene Efficiency Scrutiny	Construction Procurement by Government	1995
Egan Report 1	Rethinking Construction	1998
National Audit Office	Modernising Construction	2001
Egan Report 2	Accelerating Change	2002
National Audit Office	Improving Public Services through better construction	2005
Constructing Excellence	Never Waste a Good Crisis: Wolstenholme Report	2009
Construction Leadership Council	Modernise or Die: Farmer Review	2016

The Latham and Egan reports are, perhaps, the most well-known, but several of the reports published prior to these raised similar issues. Banwell even suggested that a common form of contract should be adopted for use on all construction projects which was 30 years before Latham made the same recommendation.

Official industry reports tend to be characterised by their strategic 'top-down' nature. Matters considered usually include leadership, client satisfaction, procurement methods, contracts, design and design briefing, dispute resolution and so on. Whilst these reports deal with issues that broadly impact project outcomes, they are pan-industry issues. They do not deal with the nitty-gritty of time management – the scheduling, resourcing and control of time (and cost) on individual projects.

To this end, a report entitled *Managing the Risk of Delayed Completion in the 21st Century* was commissioned by the Chartered Institute of Building in 2009. It was not sponsored by government but nonetheless adds considerably to the body of knowledge about the industry and its practices, especially as regards the subject matter of this book.

1.3 Managing the Risk of Delayed Completion in the 21st Century

This report investigated a significant number of construction projects in order to try to further the awareness of time management issues in the construction industry and to understand the importance of planning engineers and project schedulers in the management of time.

In doing so, the research also helped to identify the extent of unresolved delay in a variety of building projects and the level of understanding of project control techniques in the industry.

1.3.1 Summary of Findings

In a useful survey of some 2000 construction projects, the CIOB (2009), concluded that the common standard forms of contract used in construction *do not promote or encourage efficient time management*. A further observation was the trend towards developing contracts that are *increasingly punitive if not executed efficiently using good quality time management and project controls*.

Current industry trends were also reported to include:

- Demand for complex project solutions in shorter timescales and within tighter financial constraints.
- High demand for accurate completion dates.
- For contractors to undertake risks normally taken by the employer.
- A growth in Design and Build (D&B), Guaranteed Maximum Price (GMP) and Engineer Procure and Construct (EPC) contracts.

The report further concluded that simple, repetitive, low-rise projects have a high chance of success using traditional management processes but that the more complex the project, the less likely it is that it will be completed on or near to the completion date using traditional methods. Projects that are most likely to be delayed were reported to include:

- Low-rise hospitals, clinics and health-related buildings.
- Prisons and security buildings.
- Stadia and sports-related buildings.
- Railway stations.
- High-rise buildings.
- Complex engineering projects.

In a nutshell, the CIOB survey observed that:

- Most projects of a simple nature can be time-managed intuitively by competent practitioners.
- Complex projects cannot.

1.3.2 Detailed Observations

The CIOB report indicated that simple projects, such as petrol filling stations, school classrooms and repetitive housing, finished on or near the required completion date, but only 20% of complex projects finished on time, with over 60% being delayed by 6 months or more. In this context, the £150 million Holyrood Parliament building suffered £150 million in prolongation costs![8]

Further observations from the report were that:

- Only 10% of respondents were familiar with project management software suitable for satisfactory time and cost management.
- A further 10% stated that no project management software was used at all and that time schedules were prepared using spreadsheets.
- Over 50% reported that Microsoft Project was their software of choice.
- Only a third of respondents calculated the duration of activities on the schedule by applying the productivity of resources to the quantity of work to be carried out.
- More than half of respondents acknowledged that only the master schedule was used to manage time and that no short-term planning was employed.
- Nearly 20% of projects were procured on the basis of bespoke, non-standard contracts but this made no difference to the effectiveness of project time management.
- More than half of respondents used bar charts for the long-term planning of their projects but only one-sixth of these were familiar with fully linked critical path networks for managing the timing and sequencing of the work.
- 75% of respondents were familiar with short-term or 'look-ahead' programmes, but very few of these programmes were reported to be integrated into the master schedule in order to see the effects of the short-term programme on the overall timescale.
- Less than 10% of respondents kept site records in a relational database and more than 50% kept paper records only.

1.3.3 Claims and Disputes

A worrying feature of the CIOB Report is the extent to which claims were reportedly based on the master schedule, when the site was working on short-term programmes, and that progress reporting and delay notices were based on poor records measured against programmes that had not been updated. This observation may go some way to explaining why disputes develop and why many of them are ill-founded at best and spurious at worst. The Latham Report did not unearth this level of detail but was nonetheless critical of the industry's claims record.

1.3.4 Project Management Software

The CIOB survey makes a compelling argument for a greater uptake of project management software in order to aid effective project planning and control and thereby minimise risk, delay and disputes in construction projects.

In order to help in this process, the CIOB published a guide to good practice in 2010 (CIOB 2010).

1.3.5 Further Developments

Publication of the CIOB research provided the impetus for the development of an entirely new and novel form of contract drafted with the problems associated with the time and cost management of complex projects in mind. This contract was launched in April 2013 as the Complex Projects Contract 2013 (CPC2013).

CPC2013 was later amended following industry feedback and renamed the Time and Cost Management Contract 2015 as part of the Time and Cost Management Contract Suite of contracts. This suite also includes subcontract and consultancy appointment forms.

Turning to wider issues concerning why the construction industry generally underachieves its clients' expectations, it is necessary to look at government-sponsored reports – at least the more recent ones – and to try to establish whether they have had any significant impact on the way that the industry operates:

- The Latham reports (1993 and 1994).
- The Egan reports (1998 and 2002).
- The Wolstenholme report (2009).
- The Farmer report (2016).

1.4 The Latham Reports

Perhaps the most influential of all the reports concerning the industry and its problems was *Constructing the Team* written by the late Sir Michael Latham (1994) who was commissioned by both government and the industry to review the procurement and contractual arrangements in the UK construction industry.

Prior to final publication of his report in July 1994, Sir Michael produced an interim report in December 1993 called *Trust and Money*. This interim report encapsulated, perhaps more than the final report, the real problems in the industry, some of which remain unchanged 30 years later.

1.4.1 Trust and Money

This report raised concerns about the extent of mistrust between professionals and contractors and between contractors and subcontractors in construction. It also flagged up the endemic culture of late and conditional payments operating in the industry which the Housing Grants, Construction and Regeneration Act 1996 (commonly referred to as the Construction Act) has to some extent, but not entirely, resolved.

Finally, the prevailing atmosphere of mistrust and slow payments was reported to result in disharmony in project teams, poor standards of work and poor client satisfaction.

1.4.2 Constructing the Team

The purpose of *Constructing the Team* – better known as the 'Latham Report' – was to find ways to '*reduce conflict and litigation and encourage the industry's productivity and competitiveness*'. The specific terms of reference for the review were to consider:

- Current procurement and contractual arrangements.
- Current roles, responsibilities and performance of the participants, including the client.

The report took account of the structure of the industry and the need for fairness, accountability, quality and efficiency and paid particular regard to:

- Client briefing.
- Procurement methods.
- The design process.
- The construction process.
- Contractual issues.
- Dispute resolution.

1.4.3 Post-Latham

Whilst *Constructing the Team* made 30 main observations and recommendations, the principal emphasis was on 'teamwork' in order to achieve 'win-win' solutions. Additionally, Latham noted several issues which influence the ability of the construction industry to respond effectively to its customers' requirements. These are summarised briefly in Table 1.3 which also includes a commentary on developments since 1994.

Some of the main points made by Latham clearly have important consequences for the planning, production and control of construction and are therefore directly relevant to this book. These are included in Table 1.4 which also indicates the extent to which Sir Michael Latham's ideas have been adopted since his report was published.

Table 1.3 Latham issues.

	Factors governing ability of the industry to respond to customers' requirements	
	1994	**Now**
1	Sensitivity to changes in government spending patterns.	• Little change. • Industry deeply affected by public sector cuts during 2008–2013 recession. • More honesty required in public sector spending programmes.[a]
2	Intense competition for work.	• Margins under pressure.[b] • Impact of five years of downturn. • Reverse auctions driving down subcontract prices to unsustainable levels.[c]
3	Inability to respond to increased demand.	• Shortage of skilled workers in the industry.[d]
4	Lack of competency testing of firms/workers entering the industry.	• Great improvement due to Construction Skills Certification Scheme (CSCS) scheme. • CHAS (or equivalent) registration test for firms has raised health and safety standards.
5	Lack of training.	• Firms urged to make apprenticeship commitments.[e] • Lack of female students and shortage of funding for colleges could lead to future skills shortage.[f]
6	Mistrust between the participants in construction projects.	• Collaborative working yet to be widely adopted in the industry.[a] • Universities continue to perpetuate the divisive industry model of separate disciplines.[a]
7	Inadequate capital base (i.e. most contractors are under-capitalised).	• Most of the industry works on credit. • Over-valued work in progress and lack of working capital still endemic. • Perpetuated by low-margin/high-risk industry model.
8	Adversarial attitudes.	• Many supplier frameworks do not encourage collaboration.[a] • Risk is still passed down the supply chain.[a] • Unfair payment practices still common especially late and reduced payments to subcontractors.
9	Claims-conscious contractors.	• Partnering is skin deep with avoidance of risk and profit maximisation to the fore.[a] • Some clients abandoning frameworks in favour of competitive tendering.[a]
10	High levels of insolvency.	• Still high compared with other industries. • Large proportion of firms employing less than five people. • Peak period 1st quarter 2009.

References
a) Wolstenholme (2009).
b) Construction News 2/9/2014.
c) Ross and Williams (2013).
d) The Guardian 14/10/2015.
e) CITB 29/1/2016.
f) Building.co.uk 14/8/2015.

Table 1.4 Latham suggestions.

	Latham suggestions	
	1994	**Now**
1	The need for a set of basic principles for modern contracts.	• This has not happened. • JCT, NEC, ICC and Fédération Internationale des Ingénieurs Conseils (FIDIC) contracts have developed in different ways.
2	Greater use of the New Engineering Contract, which could become a common contract for the whole industry.	• An extensive 'family' of contracts has developed under the NEC banner. • Initial take-up was slow in the United Kingdom compared to internationally. • NEC contracts are now widely used.
3	Improved tendering arrangements and more advice on partnering arrangements.	• Frameworks have caught on in the industry. • Competitive tendering has made a come-back during the 2008-13 recession.[a]
4	Evaluation of tenders on quality as well as price.	• 2 and 3-envelope tenders now common. • Weighted evaluation based on price, quality and time now common.
5	Fairer treatment of subcontractors, with particular regard to tendering and teamwork on site.	• Reverse auctions are used in construction. • No evidence as to how widespread they are. • Can have the effect of driving down subcontractors' prices to unsustainable levels. • Main contractor-subcontractor relations on site largely governed by payment practices.
6	A real cost reduction target in construction of 30% by the year 2000.	• Only likely to have happened on Demonstration Projects.
7	Pay-when-paid contract terms to be outlawed.	• Conditional payment now prohibited.[b] • Only allowable if a third party (e.g. the employer) is insolvent. • Anecdotal evidence that 'pay-when-paid' is still common in the industry.
8	Adjudication to be the normal method of dispute resolution.	• An 'adjudication industry' has grown up post-Latham which is, unsurprisingly, populated by the legal profession. • Adjudication may be considered an expensive form of rough justice. • For small subcontractors, not to be entered into lightly, as losing can be costly.
9	Fair contract terms backed up by legislation.	• Unfair Contract Terms Act 1977 was in place prior to Latham. • Proposed changes under the Consumer Rights Act 2014 affect 'consumers' not 'contractors'. • Standard forms of contract are 'fair' because they are agreed multi-laterally. • Non-standard contracts are usually written in favour of the party offering the contract.

(Continued)

Table 1.4 (Continued)

	Latham suggestions	
	1994	Now
10	Insolvency protection by means of trust funds.	• Generally, no action except under the NEC. • A trust can be created under NEC Secondary Option Y, where there is a Project Bank Account (PBA). • The PBA is established and maintained by the main contractor. • Named suppliers/subcontractors sign a Trust Deed or Joining Deed and are paid by the project bank. • The Trust Deed provides insolvency protection.

References
a) Wolstenholme.
b) Construction Act 1996 Section 113.

One of the key issues considered by the Latham report was the productivity of the industry, and Sir Michael clearly considered that this is linked to the quality of design preparation and information. Inefficiency creeps in where designs are incomplete, or information given to the contractor is conflicting or too late to allow proper planning of production. The adoption of BIM will help to reduce these inefficiencies as the information exchanged between project participants will be quicker and more up-to-date. However, the full benefits of BIM will not be felt below Level 4 (iBIM) because:

- Only a federated model will allow seamless data exchange and clash detection.
  - Without a fully developed BIM model, other forms of design representation (such as digital or hard copy drawings) will be needed for those parts of the design that are not included in the model.
- Without a fully synchronised design, clashes will be inevitable resulting in the sort of design changes and variations that Latham complained of.

An issue of major importance is conflict in the industry both between clients and contractors and between contractors and their subcontractors. Latham suggested that considerable efficiencies can be gained by:

- Making changes in *procurement practice* and *contract conditions*.
- Introducing *tighter restrictions over set-off*.
- The *introduction of adjudicators as a normal procedure for settling disputes*.

Since 1994, procurement practice has changed considerably, with much more emphasis placed on developing frameworks, greater use of D&B and the adoption of far more stringent pre-qualification arrangements for both main contractors and subcontractors. Conditions of contract have also evolved since Latham with the JCT, New Engineering Contract (NEC) and Infrastructure Conditions of Contract (ICC) families having been considerably extended and modernised. Latham also concluded that the *most effective form of contract in modern conditions should include:*

---

1. *A specific duty for all parties to deal fairly with each other, and with their subcontractors, specialists and suppliers, in an atmosphere of mutual cooperation.*
   Update:
   (a) 'Mutual cooperation' clauses are common in some contracts – such as NEC3 – and in partnering contracts including PPC2000[1] and CPC2013.[2]

   References
   [1] ACA Standard Form of Contract for Project Partnering
   [2] Contract for use with Complex Projects, First Edition 2013 published by CIOB

2. *Taking all reasonable steps to avoid changes to pre-planned works information. But, where variations do occur, they should be priced in advance, with provision for independent adjudication if agreement cannot be reached.*
   Update:
   (a) Incomplete designs and variations remain common in the industry.
   (b) Some standard forms of contract provide for the pricing of variations on the basis of a contractor's quotation (JCT SBC2024, for instance).
   (c) NEC3 provides for 'compensation events' to value changes to the Works Information.

3. *That subcontractors should undertake that, in the spirit of teamwork, they will coordinate their activities effectively with each other, and thereby assist the achievement of the main contractor's overall programme. They may need to price for such interface work.*
   Update:
   (a) There is no formal contractual arrangement for subcontractors to coordinate their activities in such a way.
   (b) Good subcontractors do this informally.

The conclusions of the Latham Review were clearly extensive and led to the formation of the Construction Industry Board. This was subsequently replaced by the Strategic Forum in 2001 which also included the Construction Task Force, established in 1997 by the then Deputy Prime Minister John Prescott. The task force was responsible for the 1998 Egan Report whose chair, Sir John Egan, was appointed as the Strategic Forum's first chairman.

The Strategic Forum ceased to be a government-funded body in 2002 when it became an independent industry group. However, it has been the subject of much criticism for failing to speak on behalf of the entire industry.

In 2002, the Strategic Forum published *Accelerating Change* (the second of the eponymous Egan reports). One of its targets was that 50% of projects should be undertaken by integrated teams and supply chains by 2007. Despite some progress being made, this target was never achieved.

The present-day structure and working practices of the construction industry owe a considerable debt to the Latham Review:

- Some of the Latham recommendations were included in the Housing Grants, Construction and Regeneration Act (HGCR) 1996 which, *inter alia*, made conditional payment

(such as pay-when-paid) illegal and conferred the right on an injured party to suspend performance for non-payment.
- A statutory right to the adjudication of a dispute was also included in the HGCR Act 1996, which enables a dispute to be referred to an independent third party for resolution within a short timetable. Whilst the decision of the adjudicator is binding, the matter may be later resolved through litigation or, where the contract provides, arbitration.
- The Considerate Constructors Scheme[9] resulted from the work of the Construction Industry Council 'Latham Review Implementation Forum' in late 1994, the aim being to improve the image of the industry. This is a voluntary scheme founded with the objective of improving the relationship of construction companies with their neighbours, the public and the environment when running their sites.

In 1997, the (now Chartered) Institute of Building took responsibility for the implementation of the Scheme which was officially launched in June 1997. It is a non-profit-making, independent scheme whose members voluntarily register and agree to abide by the Code of Considerate Practice.

1.5 The Egan Reports

In common with Latham, there were two 'Egan' reports – *Rethinking Construction* was published in July 1998 and *Accelerating Change* in 2002. The first report informed the later Wolstenholme Report in 2009.

1.5.1 Rethinking Construction

Rethinking Construction[10] represented the work of a special task force which was set up by the government to identify the scope for improving quality and efficiency in construction. The task force was chaired by Sir John Egan, hence the popular title for the report – the Egan Report.

The Egan Report, at 40 pages, is certainly not as comprehensive as its predecessor, the Latham Report, but it was no less searching and probably considerably more controversial. It contained many 'home truths' but may also be said to have contained unfair criticisms, particularly with respect to comparisons with factory-based manufacturing industries, such as the motor industry.

Latham looked at designing an infrastructure for the industry aimed at removing the inefficiencies and inconsistencies, especially in terms of client briefing, better design management and more coherent project strategies. In *Rethinking Construction*, there was no industry 'blueprint' for change but the Construction Task Force, which produced the report, took the lead on a number of new initiatives including:

- Movement for Innovation (known as m4i) – a board of members whose task was to coordinate a number of demonstration projects, to disseminate best practice information and to oversee industry-wide benchmarking.
- The Construction Best Practice Programme which provided information for firms wanting to improve their performance.

- Inside UK Enterprise the idea behind which was for top-performing companies to have an 'open day' where other firms could visit and find out how things are done by the 'host' company.

The Egan Report undoubtedly recognised both the good and bad in construction and sought to build on those aspects of the industry which are excellent in a worldwide context. However, on balance, the conclusion of the report was that the industry as a whole is underachieving, and there should be radical change in key areas of its performance. These include quality, productivity, cost and time certainty and health and safety.

In the Executive Summary, the Egan Report made the following observations:

- The UK construction industry at its best is excellent. Its capability to deliver the most difficult and innovative projects matches that of any other construction industry in the world.
- There is deep concern that the industry as a whole is underachieving. It has low profitability and invests too little in capital, research and development and training. Too many of the industry's clients are dissatisfied with its overall performance.
- If the industry is to achieve its full potential, substantial changes in its culture and structure are also required to support improvement. The industry must:
  - provide *decent and safe working conditions and improve management and supervisory skills* at all levels and
  - design projects for ease of construction, making maximum use of standard components and processes.
- The industry must replace competitive tendering with *long-term relationships based on clear measurement of performance and sustained improvements in quality and efficiency.*

The Egan Report identified five key drivers of change needed to set the agenda for the industry:

1. Committed leadership.
2. A focus on the customer.
3. Integrated processes and teams.
4. A quality-driven agenda.
5. Commitment to people.

Among the year-on-year targets proposed by Egan were:

- 10% reduction in construction time from client approval to practical completion.
- 10% increase in productivity.
- 20% reduction in the number of reportable accidents.
- 10% increase in turnover and profits of construction firms.

One of the problems with the Egan Report was that the emphasis was placed on the 'top end' of the industry, whereas Latham looked at the fundamental problems of the entire industry. So, whilst Egan led to the development of several good ideas and worthwhile aims, the concepts were bound to take some time to filter down to the lower echelons of the industry. Over 20 years on, they clearly have not!

1.5.2 Accelerating Change

This report – the second Egan Report – presented the first year's work of the Strategic Forum for Construction (2002), which replaced the defunct Construction Industry Board that was set up following the Latham Report.

Accelerating Change identified ways of increasing the pace of change following the recommendations in *Rethinking Construction*, reported on the progress made to date and set out a strategic direction with targets. The report identified three main drivers to accelerate change in construction and introduce a culture of continuous improvement in the industry:

- The need for client leadership.
- The need for integrated teams and supply chains.
- The need to address 'people issues', especially health and safety.

The vision and aspirations set out in *Accelerating Change* emphasised the need for collaboration between the whole supply chain, including clients and manufacturers. The report represented a *manifesto for change* for all involved in construction, including government, schools and further/higher education and professional bodies.

1.5.3 Post-Egan

Despite identifying the problems, proposing solutions and stimulating debate about construction industry practices and procedures, most of the reports published over the years have had little influence on either government or the industry. The Egan reports are no different.

In the Foreword to the 2009 Wolstenholme Report, Sir John Egan identified that post-1998, there has been no revolution in the way the industry works, but what has been achieved is *a bit of improvement* and, at least, *people are now measuring performance*. In this context, Wolstenholme identified a number of blockages that need to change, as discussed later in this chapter.

The Egan Report probably had more publicity than the Latham Report and certainly there has been plenty of action as a result of the report, including the introduction of key performance indicators and demonstration projects exemplifying best practice. However, action does not necessarily mean results, and it is debatable whether the Egan reports have led to significant industry-wide change. There has been change nonetheless, particularly with regard to Egan's five key drivers:

1. **Committed leadership**: The UK government has certainly pushed forward its agenda to modernise and improve its procurement methods and supply chain integration, but with a certain lack of unity and unified direction.

   Since 2016, the government has also mandated the adoption of BIM Level 2 in all public-sector projects which has clearly influenced the way that the upper echelons of the industry operate.

   A similar agenda has been adopted by major industry clients and this has created a greater sense of unity of direction in both public and private sectors. Some clients have also taken a much more enlightened approach to the balance of risk in construction.

2. **Customer focus:** There is now much more emphasis on end-user satisfaction in the industry largely driven by the adoption of BIM and other digital technologies, including digital twins.

   Such advances have been enhanced by the evolution of Cloud collaboration platforms and standard messaging interfaces (SMS, WhatsApp, Snapchat, etc.) that address the communication gap between project teams and end-users identified by Egan.

3. **Integrated processes and teams**

   The Heathrow T5 project is a prime example of integrated project working – no doubt inspired by Egan – where a totally integrated project team was created between the client (BAA) and its entire supply chain.

   This was designed to reduce conflict, incentivise collaboration and foster positive problem-solving behaviour. Major risks traditionally undertaken by designers and contractors were retained by the client who took out project-wide insurance. Other risks were pooled into a programme-wide 'risk pot' which was managed by the entire project team.

   One of the largest infrastructure projects ever undertaken in Europe – London Crossrail – also developed a collaborative relationship with its principal contractors and the extended supply chain. This involved working with the supply chain in order to benchmark and achieve a 54% performance improvement for the completion of the stations and tunnelling work.

4. **A quality-driven agenda:** Improving quality is not simply about reducing defects and rework but also includes enhancing profit and driving improvement.

   Post-Egan, quality is seen as an outcome in construction that includes health and safety, reputation, the work schedule and time management, reduction of waste and the environmental impact of the project.

   Construction is a safer industry post-Egan – an agenda largely driven by major clients and large contractors – and it is demonstrably evident on most sites that there has been a significant improvement in the safety culture despite the industry retaining its reputation as statistically one of the most dangerous.

5. **Commitment to people:** A greater commitment to decent working conditions, fair wages and high standards of health and safety has been evident in the 20-or so years post-Egan but this is not reflected at all levels of the industry.

   Greater commitment to training has also been sporadic, but shining examples such as the Laing O'Rourke Apprenticeship + Programme are to be applauded.

The effectiveness of post-Egan initiatives is examined by Morton and Ross (2007) and, in particular, by the Wolstenholme Report (2009).

1.6 The Wolstenholme Report

Following a study of the impact of the Egan No 2 Report *Rethinking Construction,* Constructing Excellence published *Never Waste a Good Crisis* in October 2009. The Review Team was led by industry practitioner Andrew Wolstenholme of Balfour Beatty.

1.6.1 Egan Targets

The strategic targets identified in *Accelerating Change* are now well out of date, of course, but the 2009 Wolstenholme Report sought to establish what progress had been made since 2002 which also informs us today of how reactionary the construction industry is and how long it takes to change the status quo. First, the Egan targets set in 2002:

- By the end of 2004
    - 20% of construction projects by value to be undertaken by integrated teams and supply chains.
    - 20% of clients to adopt the principles of the Clients' Charter.
    - 10% annual improvement by adopting the Clients' Charter.
- By the end of 2007
    - These figures rising to 50%
- By the end of 2006
    - 300 000 qualified people to be recruited to the industry.
- By 2007
    - 50% increase in applications for built environment courses.
- No later than 2010
    - A certificated fully trained, qualified and competent workforce.

1.6.2 Wolstenholme Conclusions

Seven years on from *Accelerating Change*, Wolstenholme concluded that:

- Some progress had been made post-Egan but *not nearly enough*.
- Few of the Egan targets had been fully met whilst *most fell considerably short*.
- In many cases where improvement had been made, *commitment to Egan's principles was only skin-deep*.
- In the housing sector, there is *limited understanding of how value can be created through the construction process*.
- In the 10 years prior to the 2008 banking crisis and subsequent recession, *the industry has been sheltered by a healthy economy* with no incentive *to strive for innovation*.

The economic crisis facing the country post-2008 was seen by Wolstenholme as an opportunity for the construction industry to change its ways in the face of both private and public sector spending cuts and a long period of recovery from the recession.

In particular, the report suggested that *the era of client-led change is over*, and it is now time for the supply side (i.e. the industry itself) *to demonstrate how it can create additional economic, social and environmental value* through the principles of *innovation, collaboration and integrated working* espoused in *Rethinking Construction*.

The challenge for clients was to find ways to *reward suppliers* [such as contractors, specialists and consultants] *who deliver value-based solutions* through a more professional approach to their procurement of construction services.

1.6.3 Wolstenholme Blockers

In 2008, Wolstenholme gathered cross-industry opinion through an online survey about progress since Egan with the aim of contextualising data gathered about the construction industry Key Performance Indicators and Constructing Excellence demonstration projects. The findings were used to inform the Review Team and helped to identify and understand a number of 'blockers' that were thought to be preventing the industry from responding to change:

- Business and Economic Models.
- Capability.
- Delivery Model.
- Industry Structure.

The 'blockers' were then used to have a dialogue with the industry through a series of multi-disciplinary workshops and consultations with industry experts.

Wolstenholme's observations regarding the industry structure are of particular interest in the context of this book. Small and medium enterprises (SMEs) – firms employing less than 80 people – dominate the industry numerically, of course, but the large firms *still struggle to compete on a global level,* unlike some of the big UK multi-disciplinary design companies.

The implication here is that the UK market is still important for the large UK contractors who may be forced to tender for smaller, domestic, contracts than they would normally like, especially in a recession, as Ross and Williams (2013) observe.

The large number of small firms in the industry brings benefits and disadvantages. The industry is flexible and can cope with variations in workload to the extent that, when times are hard, lots of small firms 'go to the wall'. However, the lack of vertical integration observed by Wolstenholme means that subcontracting dominates the industry, and this creates horizontal interfaces that produce *yet another barrier to the free flow of information and innovation.*

A further observation by Wolstenholme is the lack of unity in the UK construction industry which has a relatively low profile at government level. This is despite the extent of capital spending in the public sector and the contribution that the industry makes to the overall economy.

The industry lacks real leadership, has few champions and has no coherent sense of direction compared with, say, aerospace, energy or the automobile industries. Wolstenholme felt that the appointment of a Chief Construction Adviser to the government in late 2009 was a step in the right direction.

1.6.4 Summary

In summary, the Wolstenholme Review Team concluded that the lack of progress in the industry is due to:

- Absence of a single, coherent voice for the industry.
- Lack of joined-up thinking by government and other key stakeholders.
- Too many industry bodies.

Despite identifying the many problems facing the industry, Wolstenholme was able to suggest a number of themes for future action:

1.	Understand the built environment	• Promote sustainability • Incentivise the creation of value
2.	Focus much more on the environment	• Embrace carbon efficiency • Encourage a 'green' recovery from recession
3.	Find a cohesive voice for the industry	• Greater collaboration between industry bodies and professional associations • Expand sector coverage of the UK Contractors Group or Construction Industry Council
4.	Adopt new business models that promote change	• Discourage short-term thinking • Incentivise long-term value creation
5.	Develop a new generation of leaders	• Especially at the top of the industry • Find leaders who can change culture and behaviours
6.	Integrate education and training	• Greater collaboration between professional bodies and the education sector • Better understand how built environment disciplines inter-relate
7.	Procure for value	• Procurement practices should focus on best value • Encourage bids based on innovative solutions
8.	Suppliers to take the lead	• Industry firms should demonstrate how they can create additional value • Encourage lean processes • Move away from lowest price tendering, negative margins and claims

1.6.5 Post-Wolstenholme

If Latham and Egan posed challenges for the construction industry, Wolstenholme's list of suggestions for future action was somewhat daunting!

In common with the Simon Committee report on building contracts 65 years earlier, Wolstenholme called for cultural change *to integrate and embrace the complex picture of how clients and contractors interact.*

Industry culture is clearly important to the way things operate but it is not easy to manipulate, change is slow, and, as Wolstenholme observed, there is *no cohesive voice for the industry.*

Some commentators also argue that the industry should, in fact, be viewed as three separate industries:

- Residential building
- Non-residential building and
- Engineering construction.

It could be said that each 'industry' has its own culture, with different types of clients, different approaches to design and different supply chain interactions. Each produces different products with different contributors using different processes, so it is important to recognise these differences and take them into account when determining policy and proposals for change.

Such differences make recommendations and policies directed to a single industry ineffective. It is also problematic that construction is very important to the gross domestic product (GDP) of the country (around 8% in the United Kingdom), but this means that policymaking is subject to political whim and public debate.

Post-Wolstenholme policies in the United Kingdom have, however, moved away from the culture debate towards improved productivity through better procurement and the use of BIM. This is evident through widespread use of procurement frameworks and the government mandate for Level 2 BIM as a minimum standard on public-sector projects. Pan-industry take-up is patchy, however, especially at the lower end of the industry, but great strides are evident, especially on major projects.

1.7 The Farmer Review

The Farmer Review was commissioned by the Construction Leadership Council in 2016 to review the United Kingdom construction labour model.

Mark Farmer, CEO of Cast Consultancy, an independent specialist construction consultancy, was also asked to look into alternative business models and new ways of working that could better support present and future skills availability in the industry. A further aspect of the review was to investigate the scope for greater use of off-site construction.

1.7.1 Farmer Conclusions

The review, entitled *Modernise or Die*, concluded *inter alia,* that the critical symptoms resulting in the poor performance of the industry included low productivity, low predictability of time, cost and quality, structural and leadership fragmentation and an industry trading on low margins with adversarial pricing models.

Modernise or Die was also critical of training, the size and demographics of the industry workforce and a lack of collaboration, investment and innovation in the industry.

One of the three main root causes of the industry's problems was found to be that the industry and its clients usually have non-aligned interests and that traditional procurement protocols and deep-seated cultural resistance to change reinforced the status quo.

The review highlighted that the industry and its labour model have reached a critical point, and that the industry could see a 20–25% decline in the available workforce by the mid-2020s. The review suggested that the industry is chronically under-invested and requires a wholesale and coordinated 'special measures' approach to drive transformational change that the industry is unlikely to initiate itself.

It was felt that clients would be the driving force for change and that public and private sector clients, government and the industry itself would need to work together to prevent the industry becoming seriously debilitated and unable to face the challenges of the future.

1.7.2 Recommendations

The Farmer Review made ten headline recommendations for action by government, industry and its clients, suggesting that:

1. The Construction Leadership Council should take a strategic role in implementing the recommendations of the review.
2. The Construction Industry Training Board (CITB) should be reviewed and reformed to provide a skills and training model more aligned to the future needs of the industry and its clients.
3. That industry, clients and government should work together to improve relationships and increase levels of investment in research and development (R&D) and innovation.
4. Government should act to stimulate innovation in the use of pre-manufactured solutions, especially in the housing sector.
5. Should a voluntary approach fail to achieve the step change the industry needs, government should consider introducing a charge on construction industry clients. The charge would be up to 0.5% of construction value unless clients could demonstrate how they were supporting skills development, pre-manufacturing or other forms of innovation and R&D in the industry.

The recommendations of the Farmer Review are not indistinct from the conclusions of other industry reports over the years. Where *Modernise or Die* differs is that the review was published at a time when the industry was rapidly reaching a crisis point in its current labour and skills provision.

The review highlights the fact that the current labour model is not sustainable and that different approaches to satisfying the needs of industry clients are required. In this respect, the Farmer Review highlighted several innovative approaches which, although at the top end of the industry, indicate how innovation, R&D and investment can contribute to making the industry more efficient and better able to meet the time, cost and quality predictability demanded by its clients.

1.7.3 Farmer Initiatives

The initiatives suggested by the Farmer Review include:

- **The use of factory-in-a-box technology**: This idea was developed by design and management consultant Bryden Wood for pharmaceuticals multinational GlaxoSmithKline plc (GSK). Shipping containers, packed with building components in reverse order for reassembly, can be shipped anywhere including to emerging markets in Africa and Asia.
- **The construction of a volumetric factory**: Major contractor Laing O'Rourke has constructed a purpose-built factory to support DfMA, to drive greater efficiencies in the use of resources and to help reconfigure traditional approaches to design, manufacturing, assembly, testing and commissioning of buildings.
- **The Singapore Building Control Authority**: This initiative works with industry to change the design and construction process, to increase construction productivity and to encourage the adoption of DfMA and prefabricated pre-finished volumetric construction.

1.7.4 Post Farmer

In July 2017, the Department for Business, Energy and Industrial Strategy[11] wrote to the then chair of the Construction Leadership Council (CLC), Andrew Wolstenholme OBE,[12] setting out the government's response to the Farmer Review.

The letter confirmed that the government had incorporated the review's findings and recommendations into policy development. It also confirmed the influence the review had had on the Housing White Paper to support increased housing supply and that it had helped inform the review of the CITB and proposed reforms to make it more responsive and focused.

In the annex to the letter, the view was expressed that the construction industry should step up to the challenges posed by the Farmer Review and that the CLC should have strategic oversight over the agenda set out in *Modernise or Die*.

Additionally, it was confirmed that the government was keen to see closer working relationships between the construction industry, its clients and government. In this regard, government is supporting more consistency in its procurement practices through the further development of BIM. Further commitments of support were also given concerning:

- Development of better models of client commissioning.
- Building longer-term collaborative relationships.
- Use of technology to improve performance and productivity.
- Development of the government's modern Industrial Strategy and working towards an understanding of the industry's priorities, removing barriers to innovation and development of a shared innovation programme.
- Improving industry image, recruitment and apprenticeships.

More tangible developments leading from the Farmer Review are set out in Chapter 5 – *Modern Methods of Construction* and in Chapter 21 – *High Speed 2* – where it can be seen that much progress has been made in bringing off-site production into mainstream construction.

> A further example is the new His Majesty's Prisons (HMP) Five Wells Prison, opened in 2022, for which 80% of the design was standardised by embracing a Design for Manufacturing and Assembly (DfMA) approach. Only 20% of the project required site-specific design and the benefit of off-site manufacturing led to 30% savings on site-based resources. The project, which was delivered 22% faster than traditional construction, comprised 15 183 precast concrete components and some 60 000 sub-components.

Such developments have facilitated great improvements in productivity, time savings and the environmental impact of construction activity as well as providing the means to produce high-quality factory-made components in the face of the current crisis in the availability of skilled site labour in the industry.

References

Andrew Wolstenholme (2009) *Never Waste a Good Crisis*, Constructing Excellence.
Association for Project Management (2020) *APM Body of Knowledge*, 7th edn.

Banwell Report (1994) Howard Banwell, *The Placing and Management of Contracts for Building and Civil Engineering Work*, HMSO.
Banwell Report (1967) National Economic Development Office, *Action on the Banwell Report*, HMSO.
Chartered Institute of Building (2010) *Guide to Good Practice in the Management of Time in Complex Projects*, CIOB.
CIOB (2009) *Managing the Risks of Delayed Completion in the 21st Century*.
CPC (2013) Chartered Institute of Building, *Complex Projects Contract*, 2013.
Egan, J. (1998) *Rethinking Construction*, HMSO.
John Egan (2002) Accelerating Change, *Rethinking Construction*, 2002.
Latham Report (1994) *Constructing the Team*, HMSO.
Mark Farmer (2016) *Modernise or Die*, Construction Leadership Council.
Michael Latham (1993) *Trust and Money*, Department of the Environment.
Morton, R. and Ross, A. (2007) *Construction UK: Introduction to the Industry*, Blackwell Publishing.
Murray, M. and Langford, D. (2003) *Construction Reports 1944-1998*, Blackwell Publishing, 2003.
Ross, A. and Williams, P. (2013) *Financial Management in Construction Contracting*, Wiley Blackwell.
Tavistock Report (1966) Tavistock Institute of Human Relations, *Interdependence and Uncertainty*, Tavistock Publications, 1966.

Notes

1 https://www.ciob.org/news/diversity-inclusion-charter-receives-100th-signature.
2 Luton Street development London, JV between Westminster Council and Linkcity (Bouygues).
3 https://assets.publishing.service.gov.uk/government/uploads/system/uploads/attachment_data/file/1171218/IPA-Annual-report-2022-2023.pdf.
4 https://www.constructionnews.co.uk/civils/hs2/major-projects-body-declares-hs2-unachievable-24-07-2023/.
5 Sunday Telegraph, 4 July 2021.
6 CIOB, *Managing the Risk of Delayed Completion in the 21st Century*.
7 The Rt Hon Lord Fraser of Carmyllie QC, The Holyrood Inquiry, Scottish Parliamentary Corporate Body, 2004.
8 Keith Pickavance, Webinar 23 April 2013, https://youtu.be/-2CBMYh-0t0.
9 https://www.ccscheme.org.uk/.
10 Egan, J. (1998) *Rethinking Construction*, HMSO.
11 Now split into the Department for Energy Security and Net Zero, the Department for Science, Innovation and Technology and the Department for Business and Trade).
12 Former Chief Executive Officer of Crossrail and now Group Technical Director of Laing O'Rourke.

Chapter 2 Dashboard

Key Message		Construction projects cannot be viewed in isolation from the environment in which they are carried out.
Definitions		**Culture:** The way things are done.
		Client: The entity, individual or organisation commissioning and funding the project, directly or indirectly (CIOB).
		SME: Small and medium enterprises.
Headings		**Chapter Summary**
2.1	Introduction	Construction has a global market of some US$15 000 billion.
		Construction operates within all sorts of languages, climates, physical conditions, economic circumstances and cultural and ideological regimes.
2.2	Industry Culture	Construction is a large and complex industry.
		The scale and diversity of the industry and its clients means that there is nothing 'typical' about its customs and practices.
		The planning and scheduling of construction projects takes place in the context of the culture of the country in which the work is to be undertaken.
		Information exchanges are notoriously difficult between the project participants, and this can impact the ability of a contractor to plan and control construction work effectively.
		Design is often separated from construction and late payment is endemic despite changes in law and good practice.
2.3	Defining the Industry	The construction industry has a 'long-tail' structure with a small number of large firms at the head and a tail of medium-size and small companies and sole traders.
		Construction may be defined by reference to the Office for National Statistics (ONS) Standard Industrial Classification (SIC).
		Industry output represents almost 8% of the UK gross domestic product.
2.4	Industry Clients	Construction projects are normally initiated by clients, promoters or sponsors who represent the focal point for consultants and contractors.
		Money flows from clients which is needed to pay for professional fees, construction costs, land purchase, legal fees, planning and building regulation fees, etc.
2.5	Construction Firms	Construction has a unique structure with a small number of very large firms and a very large number of relatively small firms.
		96% are micro firms employing less than 14 employees.
		In the UK, the list of top 100 construction companies is made up of firms with a turnover ranging from £80 million to over £3 billion,
		The average profit of these companies is less than 2% of turnover.
		Suppliers are influential in construction as they innovate and develop new products and are helping the industry to modernise and achieve greater sustainability and carbon reduction.
2.6	Industry Leadership	Construction is a highly fragmented project-based industry which has no single voice that speaks for the entire industry.
		Industry direction, if not leadership, comes largely from construction clients.
		UK major contractors are best placed to drive change in the industry simply because they are responsible for controlling such a large proportion of construction work.
Learning Outcomes		Understand construction industry culture.
		Appreciate the influence of cultural differences in countries around the world as they relate to construction work.
		Understand the role of clients and contractors in construction projects.
		Understand the influence of key participants in the construction process.
Learn More		See also Chapters 4 and 6.
		Chapter 12 explains the importance of a positive and active health and safety culture.

2

Project Environment

2.1 Introduction

Construction is a branch of commercial enterprise with a global market of some US$ 15 000 billion. In the United Kingdom, industry output is around £200 billion per year and the industry contributes some 8% to the Gross Domestic Product (GDP). It also provides employment for over 3 million people.

Being a worldwide industry, construction operates within all sorts of languages, climates, physical conditions, economic circumstances and cultural and ideological regimes. For construction projects, this means that the way of doing things – the culture – may be different depending on where the project is located. The culture in one country may not necessarily be the same as that in another. Even in countries that have many similarities – such as the United Kingdom, United States, Australia and New Zealand – construction operates differently according to the culture of each country.

Many large contractors, specialists, design practices and engineers from different countries work across the globe but must give due consideration to different legal jurisdictions, industry norms and working practices when working overseas.

Consequently, construction projects cannot be viewed in isolation from the environment in which they are carried out.

> A project to construct a sports stadium in the United Kingdom would be an entirely different proposition to the same stadium constructed in the Middle East. The project outcomes could well be equivalent, but the processes through which those outcomes are achieved may be entirely different because of differences in:
>
> - The political, religious and economic climate.
> - The industry culture and ways of working.
> - The procurement methods and contracts used.
> - The quality of management and the skill levels of the workforce.
> - The quality of training and the health and safety laws and standards.
> - The weather and its impact on working hours.

Construction Planning, Programming and Control, Fourth Edition. Brian Cooke and Peter Williams.
© 2025 John Wiley & Sons Ltd. Published 2025 by John Wiley & Sons Ltd.

2.2 Industry Culture

The planning and scheduling of construction projects takes place in the context of the culture of the country in which the work is to be undertaken. This requires consideration of the many factors that make up a particular culture such as:

- The number and timing of national holidays.
- Days of religious observance.
- The legal prohibition of work to safeguard the health of workers when temperatures exceed specified levels.
- Specific local conditions of contract that concern the contractor's master schedule.
- The degree to which industry reliance is placed on non-indigenous labour.
- Openness to new and more collaborative methods of procurement.
- The extent of state regulation of society generally and construction projects in particular.

2.2.1 Industry Customs and Practices

The customs, practices and working methods of the UK construction industry have been the subject of extensive research, investigation and analysis for over 70 years and numerous official and unofficial reports have revealed sufficient failings, shortcomings and bad practices to distinguish construction from any other industry.

> The way things are done in the construction industry defines its culture – *the set of shared attitudes, values, goals and practises that characterises an institution or organisation.*[1]

The scale and diversity of the clients, contractors, professionals and others who are involved in construction is enormous. A contractor may be a small 'sole trader' with an annual turnover of £100 000 or a large public company with a yearly workload of £1 billion. A client could be a domestic householder, a private sector corporation, a local authority or a government department. The project 'architect' could be a single practitioner, or a design-build contractor or a large, limited liability design partnership or a multidisciplinary public company.

This means that there is nothing 'typical' about the customs and practices in construction, which is a large and complex industry. It comprises a diverse range of clients, professional practices, contractors, specialist firms, suppliers and representative bodies, all of whom have their own 'agenda' and allegiances. This, together with the industry's unique system of contracts and procurement relationships, means that construction projects are conditioned by a set of complex interrelationships between participants whose behaviour, values and expectations are not necessarily aligned.

Information exchanges are notoriously difficult between the project participants, and this can impact the ability of a contractor to plan and control construction work effectively. This is especially the case for projects where design and production are separated by an intervening tender period. The Latham and Egan Reports recognised this feature of construction which is unlike other industries, such as manufacturing or aerospace.

Despite the popularity of design and build and other collaborative procurement arrangements, the separation of design from production is still common practice in the industry.

2.2.2 Separation of Design and Construction

The separation of design from production is a particular feature of general contracting or competitive tendering which discourages collaborative working and results in inefficiencies which can badly affect the contractor's programme and the efficiency of work on site:

- The opportunity to make the design more 'buildable' is lost without early contractor involvement.
- Designs are not fully developed at tender stage which results in:
  - Contractors' bids based on incomplete information.
  - The addition of time and cost risk allowances in tenders.
  - Uncertainty in terms of project planning.
  - Incomplete information pre-construction.
  - Design reiteration and requests for information during construction.
  - Inevitable variations which are disruptive and expensive.
  - Late information which can delay and/or disrupt the programme.
  - Disharmony in the project team leading to disagreement and disputes.
- The master schedule can suffer delay and disruption which impacts client outcomes and contractors' profitability.
- Arguments and claims often arise later in the project.
- Client representatives and contractors with different attitudes to profit and risk that are divisive and often work against the best interests of the client and the successful completion of the project.

Clearly, designers working in splendid isolation do not produce 'buildable' designs – designers are not builders – but it could also be argued that the polar opposite to this approach – design and build – is not perfect either. The design might be more buildable, but it might not entirely meet the employer's requirements.

Participative procurement methods – where the contractor is part of the design team – bring great benefits, but they are used mainly for larger projects. They include two-stage tendering, leading to a guaranteed maximum price, and construction management and management contracting. However, whilst two-stage tendering is now popular in the industry, construction management and management contracting account for only 2% and 1% of contracts, respectively, according to the RIBA Construction Contracts and Law Report (RIBA 2022).

2.2.3 Late Payment

Late payment is an endemic feature of the culture of the construction industry especially between main contractors and subcontractors and between contractors and suppliers such as builders' merchants and suppliers of aggregates, ready-mixed concrete and so on.

One of the reasons for this manifestation is that construction projects are generally based on credit.

> Credit terms vary, but typically:
> - The client pays the contractor for work in progress monthly in arrears plus the time taken to value the work and certify payment – typically six weeks in total.
> - The value of building materials delivered to site is invoiced at the end of the month and the contractor settles the invoice 30 days later, giving up to 60 days credit.

In practice, late payment to subcontractors can extend well beyond the contract terms and payments for materials can extend to 60 or 90 days or more. This provides contractors with a source of 'free' working capital.

Late payment has been highlighted in many industry reports, especially in Latham Report No 1, *Trust and Money*. This manifests itself in a number of ways:

- Industry clients normally enjoy a period of credit or deferred payment because most contracts provide for the contractor to be paid monthly in arrears. This means that the contractor 'cash flows' the project in its early stages, placing intense focus on the first and subsequent interim valuations or monthly payment applications.
- Disagreement, even disputes, can arise if the expected monthly payment is less than that anticipated. This can happen should the work in progress be undervalued, if variations and extra work are not valued soon enough to be included in the valuation or if the basis of valuation is disputed.
- The valuation of work in progress is a common area of disagreement. This is not a precise science and opposing views on the proportion of work completed can seriously affect cash flowing from the project. Whether or not the work carried out is in accordance with the specification may also be disputed, as might what qualifies as additional work.
- Retention is commonly deducted from payments to provide a fund for the future rectification of defects which:
  (a) reduces cash flow and increases demands on working capital and
  (b) provides opportunities for main contractors to abuse their position of power over subcontractors.
- Main contractors also enjoy a credit period from their subcontractors, but this privilege is commonly abused, resulting in subcontractors having to wait for their money well beyond the contractual payment period.
- Despite being illegal under the HGCR Act, conditional payments (pay-when-paid, pay-when-certified, etc.) are still common in the industry.

2.2.4 Construction Supply Chain Payment Charter

The problems of late payment, underpayment and capital lock-up due to withholding of retention monies have been recognised by the Construction Leadership Council (CLC) which published a *Construction Supply Chain Payment Charter* in April 2014. The Charter sets out 11 *fair payment commitments* including:

- A commitment to reduce payment terms to:
  - 60 days with immediate effect
  - 45 days from June 2015 and
  - 30 days from January 2018.

Other commitments made in the Charter include:

- Not deliberately delaying or withholding payment.
- Proportionate withholding of payment due to defects or non-delivery that is clearly, specifically and demonstrably justified.
- Not withholding cash retentions.
- Prompt and fair agreement and payment for contract variations.
- Making payments electronically.
- A transparent, honest and collaborative approach when resolving differences and disputes.

The Charter recognises, and is complementary to:

- The Late Payment of Commercial Debts Regulations 2013.
- A number of government procurement policy notes concerning invoices and payment.
- The Scheme for Construction Contracts (England and Wales) Regulations 1998 as amended by the Scheme for Construction Contracts (England and Wales) Regulations (Amendment) (England) Regulations 2011.
- Cabinet Office – A Guide to the Implementation of Project Bank Accounts in construction for Government Clients.

Signatories to the Charter, which includes several large contractors and developers, agree to apply fair payment commitments in their dealings with the supply chain and to be monitored for compliance against a set of agreed key performance indicators. Monitoring arrangements are currently being developed.

The Construction Leadership Council's eventual ambition for 2025 – that the construction industry's standard payment terms are 30 days and that retentions are no longer withheld – is symptomatic of the industry's problems which are especially hard felt further down the supply chain.

Most subcontractors would argue that they suffer from payment problems with their main contractors, but conversely, main contractors would argue that subcontractors' payment applications are often over-optimistic, inaccurate and lacking in substantiation. These complex issues are explored in more detail by Ross and Williams (2013).

Peter Hansford, the Government's Chief Construction Advisor, said:

> *This charter signifies the Construction Leadership Council's commitment to small and medium-sized business, and the important role they play in the construction industry.*
>
> *Through the Council, the government is working very closely with industry to give businesses of all sizes the confidence to invest – securing high skilled jobs and a stronger economy for everyone.*

2.2.5 Public Contracts Regulations 2015

As far as payments in the public sector are concerned, the UK government recognises that central and local government *should set a strong example by paying promptly*.

Under regulation 113 of the Public Contracts Regulations 2015,[2] contracting authorities must ensure that any contract awarded contains provisions to pay contractors no later than 30 days after an invoice is regarded as *valid and undisputed* unless the contract terms provide for a shorter payment period (e.g. JCT Standard Building Contract 14 days from certificate). Regulation 113 also requires that similar contractual provisions are 'stepped down' to subcontractors and sub-subcontractors.

The operative term is *valid and undisputed*. Payment applications in construction are rarely undisputed – especially subcontract applications – as there may be disagreement as to the value of work carried out, the valuation of variations to the contract and the applicability and/or authentication of dayworks (work valued on the basis of time and the cost of materials).

Where the work has been valued by the employer's representative, there is less likelihood of a dispute, but not every contract provides for the contract administrator/QS to value the work in progress. Even where there is no dispute regarding the main contract valuation, this is no guarantee that there will not be a dispute as regards subcontract applications.

2.3 Defining the Industry

Unusually, the construction industry has a 'long-tail' structure with a small number of large firms at the head and a tail of medium-sized and small companies and sole traders providing services to suit every size and complexity of project. It is made up of a wide variety of types and sizes or organisations:

- Micro-contractors and specialists.
- One-person design practices.
- Small- and medium-sized local companies.
- Large national and very large international contractors and professional practices, etc.

In an effort to more closely define what is meant by the 'construction industry', several approaches may be taken including:

- Industry classification.
- The value of industry output by sector.
- Legislative definition.
- Types of work undertaken.
- The types and sizes of firms that populate the industry.

2.3.1 Industry Classification

Construction may be defined by reference to the Office for National Statistics (ONS) classification of industries called the Standard Industrial Classification (SIC).[3] This provides a framework for collecting data about various industries and their activities.

Construction is officially defined in Section F of the SIC which distinguishes between new work, repair, additions and alterations and the construction of temporary buildings or structures.

There are three main divisions in Section F:

- The construction of buildings.
- Civil engineering.
- Specialised construction activities.

Each division is further divided into subcomponents, as shown in Table 2.1.

Table 2.1 Standard industrial classification.

		Section F: Construction	
		Example:	
41	Construction of buildings	41.1	Development of building projects
		41.2	Construction of residential and non-residential buildings
42	Civil engineering	42.1	Construction of roads and motorways
		42.2	Construction of utility projects
43	Specialised construction activities	43.2	Electrical installation
			Plumbing, heating and air conditioning
		43.3	Plastering
		43.9	Joinery
			Roofing
			Scaffold erection

Section F includes commercial and domestic building work, the construction of roads and utility projects and specialist activities including demolition, joinery, plastering, painting and so on, but not the professional, scientific and technical activities that support construction work.

Architecture, engineering, testing and analysis and consultancy activities are included in SIC Section M: Division 71.

The SIC definition of construction might be considered as somewhat misleading because it implies a structure – a degree of orderliness – that does not really exist. This is because construction is a fragmented industry comprising an eclectic mix of clients served by an even more diverse supply chain. There are no barriers to entry in construction, and anyone can set up in business as a builder.

2.3.2 Industry Sector Output

The construction industry sectors in Great Britain are defined by the ONS as residential and non-residential. This definition is further refined by distinguishing new work from repair and maintenance which is further categorised, as shown in Table 2.2.

The annual output of the UK construction industry varies according to the economic climate but is currently in excess of £200 billion. This represents almost 8% of the UK GDP. The value of industry output by sector is shown in Figure 2.1.

The proportion of total output represented by new work is 62%, as indicated in Figure 2.2 which also shows that repair and maintenance is a significant sector of the industry, contributing 38% to total output – around £76 billion in money terms. Figure 2.2 also highlights the importance of industry sectors that are particularly vulnerable to political influence and the vagaries of the economy:

- The proportion of all new work contributed by public and private sector housing is some 25% of total output, most of which (22%) comes from the private sector.

- Output from infrastructure and other public sector work, excluding housing and repair and maintenance, is 21% of the annual total.
- The health sector contributes around 12% to the total of all new work.

Table 2.2 Construction industry sectors.

Industry sector	Examples
Housing	
Public housing	Social housing for low-cost rental or ownership, married quarters for armed services/police, retirement homes, etc., including associated roads and services.
Private housing	Speculative and individual development of houses, apartments, etc.
Non-housing	
Infrastructure	Road, rail, airports, power supply, water supply and sewage disposal.
Public other new work	Hospitals, prisons, schools, etc.
Private industrial new work	Factories, warehouses, petrochemicals, manufacturing industries.
Private commercial new work	Hotels, shops, offices, garages, leisure buildings.

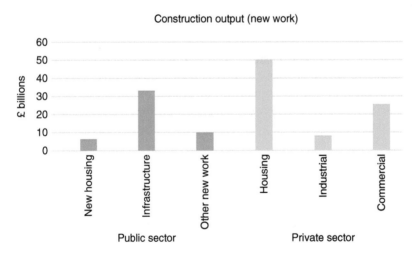

Figure 2.1 Construction output by sector.

From these statistics, it can be seen that private-sector housing is particularly important to the industry as it often acts as a driver for growth in total construction output and may be viewed as a general indicator of confidence in the outlook of the overall economy.

When times are good, the private housing sector booms. Reservations and completions for house sales go up, profits increase and, despite scarce resources and cost inflation, developers make money.

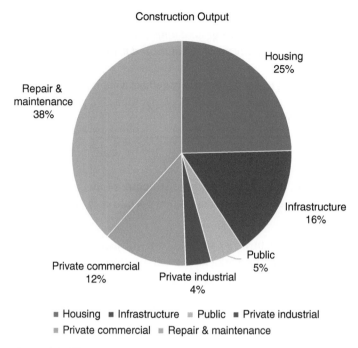

Figure 2.2 Construction output.

When times are bad, however, and reservations and completions decrease, developers' cash flows are squeezed, and profits are impacted by interest rates and by the costs of labour and materials which then forces down shareholders' dividends and share prices.

2.3.3 Legislative Definition

The Construction (Design and Management) (CDM) Regulations is important legislation in the health and safety management of construction projects. The regulations provide another definition of 'construction', the purpose of which is to determine the ambit of the regulations as far as they apply to construction work. Under the regulations, **construction work** means:

- *the carrying out of any building, civil engineering or engineering construction work and includes:*
- *the construction, alteration, conversion, fitting out, commissioning, renovation, repair, upkeep, redecoration or other maintenance, decommissioning, demolition or dismantling **of a structure**.*

In this context, **structure** means:

> *any building, timber, masonry, metal or reinforced concrete structure, railway line or siding, tramway line, dock, harbour, inland navigation, tunnel, shaft, bridge, viaduct, waterworks, reservoir, pipe or pipeline, cable, aqueduct, sewer, sewage works, gasholder,*

> *road, airfield, sea defence works, river works, drainage works, earthworks, lagoon, dam, wall, caisson, mast, tower, pylon, underground tank, earth retaining structure or structure designed to preserve or alter any natural feature and fixed plant.*

Whilst this important and far-reaching legislation helpfully defines what is meant by 'construction work', it does not define the construction industry other than giving a lengthy list of the sort of work that the industry carries out.

2.3.4 Types of Work

A narrower, but nevertheless valid, way of looking at the structure of the industry is to identify the type of work undertaken by the various firms that operate in the industry. For example:

- A large contracting company with a turnover of £1 billion will usually have a wide portfolio of clients and might structure its workload into such sectors as infrastructure, healthcare, education, heritage, accommodation, sports and leisure, industrial, retail and commercial.
- A medium-sized company, with a £500 million turnover might operate in sectors such as general building, civil engineering, rail, demolition and remediation.
- A company with a £30 million turnover might undertake smaller building and civil engineering work and specialise in formwork and reinforced concrete frame construction. Alternatively, a similar-sized company might involve itself with a more niche sector of the industry such as electrical installations and property maintenance services.
- A specialist subcontractor, with an annual turnover of £10 million, might undertake demolition, remediation, earthworks, small-scale civil engineering work and plant hire.
- Small businesses and sole traders might target specific sectors of the industry such as electrical engineering, plumbing and heating or plastering and screed laying.

These examples typify the multi-faceted nature of construction – it is an industry where different firms offer different services – but the sum of the parts makes up a coherent whole capable of satisfying the needs of any client on any project.

2.4 Industry Clients

Construction projects are normally initiated by clients, promoters or sponsors who represent the focal point for dealings with professional consultants and contractors and through whom money flows. This money is needed to pay for professional fees, construction costs and ancillary payments such as land purchase, legal fees, planning and building regulation fees.

Construction industry clients fall into two main groupings:

- **Private sector clients**: which include domestic clients, business and commercial enterprises and investors such as property developers and insurance companies.

- **Public sector clients**: often referred to as promotors or sponsors, they include local authorities, utility companies, government departments and government-owned companies or agencies.

2.4.1 Definitions

The CIOB defines 'client' as:

- *The entity, individual or organisation commissioning and funding the project, directly or indirectly.*

A different definition of 'client' is to be found under regulation 2(1) of the CDM Regulations 2015 which says that a 'client' is:

- *any person for whom a project is carried out.*

CDM takes this definition a little further by distinguishing the *domestic client* as:

- *one for whom a project is being carried out which is not in the course or furtherance of a business of that client.*

'Domestic' clients generally participate less in the affairs of the industry than, say, large industrial or commercial clients and government. They also tend to be more 'in the hands of' the builder when it comes to organising and managing a project, and arranging payment and programme, unless, of course, they engage professional consultants who have expertise and knowledge of how the industry works.

Large clients and government are very influential in how the industry operates and dictate their procurement arrangements with contractors.

2.4.2 Types of Clients

Private sector, 'non-domestic' clients, to the construction industry may be 'one-off' clients or may be regular or 'repeat-business' clients. The former are usually represented by professional consultants (such as architects, engineers, quantity surveyors and so on), whereas the latter may also be members of a representative body such as the British Property Federation or the Construction Clients' Group and may be regarded as 'expert' clients.

Public sector clients tend to have repeat business, and many are also members of the Construction Clients' Group. Members of the Construction Clients' Group include Acivico[4] Ltd., BAE Systems, Crossrail, EDF, Heathrow, Highways Agency, ProCure21+, etc.

2.4.3 Professional Advisers

Construction clients invariably engage professional consultants to advise and perform professional services such as project/programme management, design work, cost consultancy and site investigations/surveys.

The RIBA, ICE, RICS, etc., are the professional associations of such consultants which are represented at the 'top' of the industry by the Construction Industry Council. This is the representative forum for professional bodies, research organisations and specialist business associations which provides a single voice for professionals in all sectors of industry of the built environment.

2.4.4 Funding

A domestic client may commission a project – such as a house extension – but may not be directly funding the work as the money might be coming from a loan or mortgage. Similarly, a public sector infrastructure project, commissioned by a government agency, might be funded at the sanction of parliament via a sponsoring government department.

Private sector property development may also be funded indirectly on the basis of bank loans or venture capital funding or may be funded directly from shareholders' funds or cash reserves held in the business.

It is not uncommon to find a mixture of funding sources for construction projects, but external funding needs to be guaranteed via loan agreements or banking facilities in order that funds are available to pay bills as they become due.

2.5 Construction Firms

Construction has a unique structure with a small number of very large firms and a very large number of relatively small firms with few criteria limiting entry into the industry. There are over 350 000 construction firms operating in the United Kingdom:

- 96% are micro firms employing less than 14 employees.
- 3.5% are small and medium enterprises (SMEs) with between 15 and 299 employees.
- The remaining 0.5% are large or very large firms, of which approximately 50 employ more than 1200 people.

2.5.1 Industry Output

In terms of the £200+ billions of work carried out by the industry, statistics from the ONS indicate that:

- 44% of output is contributed by small/micro-sized firms.
- Large/very large firms carry out 18.5% of work.
- SMEs are responsible for the remaining 37.5% of output.

These statistics require a degree of interpretation because they give the impression that the very large contractors are only responsible for a modest proportion of the value of work done. In fact, the top 27 contractors alone are responsible for one-third of construction output in the United Kingdom.

The disparity in the figures is explained by the fact that major contractors sublet a considerable proportion of their workload to smaller firms and specialist contractors and this, of course, is represented in the value of output carried out at the lower echelons of the industry.

According to the UK Contractors Group (UKCG)[5] (now part of Build UK), 64p of every £1 sub-contracted by the top 27 UK contractors goes to SMEs. Consequently, a considerable degree of **control** is exercised over the affairs of the industry by medium- and large-sized firms, especially the very large firms, which is not reflected in the ONS statistics.

2.5.2 Micro Firms

Micro entities – which make up 96% of the industry – are defined in the Companies Act 2006 as having 10 or fewer employees and an annual turnover not exceeding £632 000. Very small firms employing less than four people make up 79% of the industry.

Of the total number of construction firms, some 16% have no employees. This could mean that they are either single tradespeople working on their own account or are sole proprietors who engage subcontractors to carry out all their work.

Some micro entities are sole traders, with unlimited liability for their debts, whilst others are limited companies and are, therefore, liable to file accounts at Companies House.

> A painting and decorating business has an annual turnover of £500 000 and has six employees and directors, of which one is an apprentice. The remainder of the labour force is self-employed and is engaged on a contract-by-contract basis according to the prevailing workload. The company carries out subcontract work for several major contractors and also has a portfolio of private domestic and small business clients.

2.5.3 SMEs

It could be argued that SMEs represent the backbone of the construction industry. Typically, they will employ between 14 and 299 people with an annual turnover of £2–50 million.

SMEs are a vital part of the construction supply chain because they represent all the 'trades' needed in building and civil engineering and are companies of greater substance than micro firms. They are, therefore, able to take on bigger subcontracts where the risks are greater and where their financial standing, availability of resources, health and safety record and quality assurance fits them out to be Tier 2 framework contractors.

Such companies may well undertake design responsibilities for work packages such as piling, structural frames, temporary works, cladding systems and so on.

> A company undertakes electrical installation and property maintenance services. It has an annual turnover of £40 million and a profit before tax and interest of £1.2 million. The company employs over 500 people, including directors and has a training hub where it trains 120 apprentices annually, both for the company and for other businesses.

2.5.4 Large Firms

Large firms typically employ between 300 and 1199 people. They represent 0.05% of the total number of construction firms in the United Kingdom – less than 200 out of the 350 000 total. The influence of large firms is, nevertheless, considerable, as they subcontract much of their work. Small firms (subcontractors) are thus required to demonstrate minimum standards of health and safety competence and must undergo tests to ensure their financial stability and technical competence to undertake the work required.

This makes the industry self-regulating to a certain degree because many of the smaller firms are engaged as subcontractors to large main contractors.

A residential property development company comprises four business units – build-to-rent, purpose-built student accommodation, affordable homes and accommodation management for residential rentals. The company has an annual turnover of £407 million and an operating profit before tax and interest of £24 million. Unlike many developers, the company carries out most of its own building work with the balance undertaken by third-party contractors. The company has 700 employees, including directors, half of which are involved in the construction, management and administration of the business, with the remainder dealing with accommodation management.

2.5.5 Very Large Contractors

Despite the high proportion of small firms in construction, the very large contractors (1200+ employees) enjoy a significant share of the construction market and wield considerable power over the affairs of the industry. They represent approximately 0.015% of the number of firms in the industry – around 50 companies in total. This includes well-known names such as Balfour Beatty, Kier Group, Laing O'Rourke, Amey and Sir Robert McAlpine.

The very large contractors take on the big risks – huge construction projects such as Hinkley Point C, Crossrail and HS2 – often working in joint venture (JV) with other similar companies.

The 1200+ employees category is somewhat misleading because some of the UK's largest contractors employ huge numbers of people worldwide. A top five contractor may well employ 30–35 000 people and carry a workforce of 15–20 000.

It is also a myth that such contractors sublet all their work as they undertake some operations with directly employed staff and may also undertake work originally sublet where the subcontractor fails financially or is unable to fulfil its contractual obligations.

A construction and infrastructure business operates in most sectors of the industry for both private and public sector clients and undertakes projects from £50k to £1 billion in value. It specialises in construction and infrastructure, fit-out, property services, partnership housing and urban regeneration. The type of work undertaken is wide-ranging and includes small works, repair and maintenance and large-scale building and civil engineering projects. The annual turnover is £3.6 billion with a profit before tax and interest of £139 million. The company employs almost 7000 people, including directors, 60% of whom are involved in construction and infrastructure.

2.5.6 The Top 100

In the United Kingdom, the list of top 100 construction companies is made up of firms with a turnover ranging from £80 million to over £3 billion, around 15% of whom have international origins. A couple of large firms specialise in prefabricated and modular buildings.

According to the latest full-year statistics,[6] the top 100 UK contractors undertook 1783 projects with a total value of £40.8 billion and an average project value of £22.9 million. In the same period, the top 10 contractors won a total of 771 contracts worth £15.5 billion, with an average project value of £20.1 million.

> The top UK contractor was awarded 173 projects with an average project value of £18.3 million, whereas the second largest contractor was awarded 300 projects with an average value of £7.4 million. Another top 10 contractor won only 30 contracts, but with an average value of £241.8 million. Some top 100 companies may have only one project underway in any particular year. They might be undertaking a very large project – such as nuclear power station – or could be engaged as a Tier 2 contractor with a huge tunnelling contract for projects such as Crossrail or HS2.

Some contractors are clearly more 'comfortable' with relatively modest contracts, whereas others have the capacity, resources and expertise to undertake mega projects. However, when it comes to big projects – over, say, £1 billion in value, even the very largest contractors will seek to spread the risk by engaging in a JV arrangement with other large contractors or designers. This might also be the case for more modest projects where a JV of firms with different expertise makes sense – where, perhaps, one contractor supplies the design expertise and the other undertakes responsibility for the construction work.

The portfolio of projects undertaken by a company is largely a question of risk. Taking on a small number of very large projects places a great deal of pressure on a company's management expertise and requires the support of a robust and reliable supply chain. If any one of these projects suffers from delays or becomes lossmaking, this could result in cash flow and resourcing problems, putting pressure on the rest of the business. More than one problematic project could be catastrophic.

2.5.7 Margins

In the upper reaches of the construction industry, the larger firms make substantial profits – counted in the tens of £millions. When compared to turnover, however, margins are modest. None of the top five, for instance, have profit margins of more than 4%, and the average profit of these companies is less than 2% of turnover.

On the other hand, when the profits of these firms are compared to the capital employed in the business – also called net assets – the returns are much healthier.

> One of the top five contractors returned a profit margin of only 0.5% when expressed as a percentage of turnover.
> With profit expressed as a percentage of capital employed, the return was 7.9%.

The financial metric of return on capital employed is commonly regarded as a measure of how well management has used the resources (capital) at its disposal. Another measure is return on equity – equity represents shareholders' funds. However, as some firms are highly geared – they have more debt/borrowings than shareholders' funds – and others have low gearing – this is a less reliable benchmark for inter-firm comparisons.

2.5.8 Suppliers

The construction industry has a complex and diverse supply chain of subcontractors, service companies and suppliers. Important though they all are, it is the suppliers who wield the greatest influence on the industry. They comprise:

- **Materials suppliers**: who are generally the manufacturers of specific materials and components such as precast concrete products, cladding and curtain walling, quarry products and ready-mixed concrete. They may deal with contractors directly or through a distributor or through another intermediary, such as a builders' merchant.
- **Builders' merchants**: who act as intermediaries between suppliers and contractors by supplying and delivering everyday construction materials such as cement, plaster products, bricks and blocks, timber doors and windows and so on.

Suppliers are influential in construction as they innovate and develop new products and are helping the industry to modernise and achieve greater sustainability and carbon reduction. Around 80% of all construction products used in the United Kingdom are manufactured in the United Kingdom. Exports amount to some £6.4 billion and so it is understandable that some of the industry's suppliers are bigger companies than the largest contractors in the United Kingdom. They consequently have considerable influence on the way the industry operates.

> The largest builders' merchant in the United Kingdom has an annual turnover of almost £5 billion. Its annual accounts state that its ambition is to be a leading partner to the industry and to help decarbonise and modernise construction. Another large builders' merchant – a subsidiary of a much larger private equity company – has an annual turnover of £4+ billion.

The Construction Products Association is the representative body for the construction manufacturing supply chain which champions manufacturers and suppliers of construction products and materials. It has a membership of some 24 000 companies with a collective turnover of more than £63 billion. This represents about 35% of total construction output.

Larger contractors tend to deal with suppliers directly because they have buying power and creditworthiness. This is an important relationship on projects as preferential prices and delivery times can greatly influence project success.

Builders' merchants tend to deliver relatively small quantities of goods. They provide a convenient service – especially for smaller builders and the trades – because a wide variety of construction materials can be obtained without the need to open a large number of trade accounts.

The relationship between suppliers and contractors is based on credit. This enables contractors, large and small, to function, as many do not have sufficient working capital to fund their activities without the benefit of deferred payment. Despite the risk of bad debts, suppliers are in a powerful position as they can deny credit should invoices not be settled on time.

The availability of credit and reliability of deliveries can have a serious impact on a contractor's ability to keep to the project schedule and can also have contractual implications with the client. In this regard, a major consideration for contractors is the lead time needed for supplies to be delivered to site.

Some supplies – such as elevators and HVAC systems – have always required considerable lead times for design and manufacture, but a combination of the pandemic, global economics and international conflict has made matters worse. Steel joists and decking, for instance, are reported to require a lead time of six to nine months whilst it might be four to eight months before certain roofing materials can be delivered.

Such lengthy delays can make life challenging and unpredictable for contractors when bidding for work and trying to deliver projects on time and within budget. Being heavily reliant on its suppliers, the construction industry needs to ensure that suppliers have standards of integrity, reliability, quality and environmental responsibility that align with those of industry buyers.

To this end, the British Standards Institute has developed *PAS 7000:2014 Supply Chain Risk Management – Supplier pre-qualification*. This is a standard that supports all businesses of whatever size, including businesses involved in construction.

PAS 7000 provides a framework for making informed procurement decisions about whether to engage with a potential supply chain partner based on their profile, capabilities and performance. It also helps to bring about consistency, and reduced risk, in the pre-qualification of supply chain partners.

2.6 Industry Leadership

Construction is a highly fragmented project-based industry which has no single voice that speaks for the entire industry. This contrasts with the professions, contractors and specialist contractors who have their own representative bodies,

The situation changed to a degree in 2013 with the creation of the CLC, the body set up to deliver the government's **industrial strategy for construction**. The CLC works between industry and government to identify and deliver the priorities and actions needed to build greater efficiency, skills, innovation, sustainability, growth and better communications in the construction supply chain.

The CLC, which is jointly chaired by a top executive from industry and a government Minister of State, has cross-industry representation and meets four times per year.

2.6.1 The Influence of Clients

On a day-to-day basis, industry direction, if not quite leadership, comes largely from construction clients – both public sector and private sector clients – and not from the contractors and specialists who carry out the work. It is the clients and their professional advisers who, to a large extent, dictate the procurement methods and forms of contract used in the industry.

The public sector – central and local government, devolved national and regional government, government departments and their respective agencies – has a major influence on the UK construction industry. The value of output in the public sector – new work and repair and maintenance – represents 37% of total industry output which amounts to around £75 billion worth of work annually. Public sector new work is worth approximately £49 billion.

The official statistics do not reveal whether public sector new work includes the capital value of privately funded public sector work – prisons, hospitals and so on – which is paid for over a number of years out of public revenue and not borrowings.

Public sector new work is made up of housing, infrastructure (such as road and rail) and other work such as investment in hospitals, prisons, schools, care homes and so on. The scale of this work makes industry clients such as the DfT, NHS and Network Rail very influential and such clients have a major 'say' in how the industry operates.

In the private sector, major clients also have considerable influence on the industry. These clients include 'household' names such as BAE Systems, Marks and Spencer, Nationwide, Tesco, McDonalds and Whitbread. The value of output in this sector of the industry is over £130 billion, with new work accounting for £84 billion of the total.

Some major clients, as well as contractors and consultants, recognise that construction projects can be procured more sustainably by effective management of the supply chain and the application of best practice guidelines. This thinking helped to develop the CIRIA *Guide to sustainable procurement in construction* (CIRIA 2010), which identifies that the British Standard for Sustainable Procurement, BS 8903:2010, can be used as a generic standard upon which to base procurement practice.

2.6.2 The Construction Clients' Group

The Construction Clients' Group has been a sector forum of Constructing Excellence since 2006 – formerly, it was part of the British Property Federation and changed its name from the Confederation of Construction Clients to the Construction Clients' Group in 2004.

Its history goes back to 1998 when, as the Construction Clients' Forum, it established the Clients' Charter which set out the minimum standards in procurement expected by clients. By registering for the charter, clients committed themselves to establishing a modern business culture with their suppliers, steady improvement of standards measured against nationally accepted criteria and the exchange of best practice experience.

The Clients' Charter has subsequently been rebranded as Client Commitments,[7] but nothing has changed in terms of the six areas where clients can make a positive difference to ensuring better value:

- Client leadership.
- Procurement and integration.
- Health and safety.
- Design quality.
- Sustainability.
- Commitment to people.

2.6.3 Main Contractors

Build UK is now the main representative body for major contractors and their supply chain partners operating in the United Kingdom. With their financial strength and leverage, the

UK's major contractors are best placed to drive change in the industry simply because they are responsible for controlling such a large proportion of construction work.

Being in a position of considerable influence, the industry's major contractors have lined up behind the Construction Skills Certification Scheme (CSCS) to help promote greater competence and attention to health and safety risks.

> CSCS is a voluntary registration scheme whereby trainees, workers, managers, supervisors, visiting professionals, etc., can obtain the relevant colour card if they pass a health and safety awareness test. The card enables them to gain access to UKCG sites.

UKCG was formerly known as the Major Contractors Group but is now part of Build UK following its merger with the National Specialist Contractors' Council in September 2015.

Build UK comprises 27 of the industry's largest main contractors and over 30 trade associations (including the Federation of Piling Specialists, the Glass and Glazing Federation and the National Federation of Roofing Contractors). It represents more than 11 500 specialist contractors and provides a strong collective voice for the contracting supply chain.

References

Ross, A. and Williams, P. (2013) *Financial Management in Construction Contracting*. Wiley-Blackwell.
RIBA (2022) *The RIBA Construction Contracts and Law Report*.
CIRIA (2010 Revised 2014) *Guide to sustainable procurement in construction*.

Notes

1 Merriam-Webster Dictionary.
2 http://www.legislation.gov.uk/uksi/2015/102/regulation/113/made.
3 http://www.ons.gov.uk/ons/guide-method/classifications/current-standard-classifications/standard-industrial-classification/index.html.
4 Acivico Ltd is a company created by Birmingham City Council to offer Building Control, Consultancy and Facilities Management services to the Council and to other public and private sector organisations.
5 UKCG figures.
6 Glenigan, www.glenigan.com.
7 https://constructingexcellence.org.uk/wp-content/uploads/2017/04/client-commitments-final&uscore;may-2014.pdf.

Chapter 3 Dashboard		
Key Message		Construction project risk can impact its participants and stakeholders in different ways.
		The extent of this risk depends on how risk is allocated and which party is best equipped to manage it.
Definitions		**Risk^1**: The threat or possibility that an action or event will adversely or beneficially affect an organisation's ability to achieve its objective.
		Risk^2: The consequence of the presence of a hazard.
		Uncertainty: The absence of information required for decision making.
		Hazard: Something with the potential to cause harm or loss.
		Risk assessment: A formal process carried out to control risk.
Headings		**Chapter Summary**
3.1	Introduction	The potential for project failure is commonly referred to as 'risk'.
		The threats (hazards) posed can impact clients, consultants, contractors, site operatives and the supply chain differently.
3.2	Risk Management	Management should identify and evaluate the risks faced and operate and monitor a suitable system of internal control.
		Risk management involves planning, organising, controlling, monitoring and reviewing the measures needed to prevent exposure to risk.
3.3	Risk Assessment	Risk is measured by evaluating the likelihood of something happening due to the presence of a hazard and the severity or consequence if it does.
		Risk assessments may be based on judgement (qualitative) or on a statistical analysis where a probability can be determined (quantitative).
3.4	Design Stage	Uncertainty is greatest at the early stages of a project.
		Time risk allowances should be included in the programme/schedule.
		Risk allowances included in the cost planning cover design uncertainty, potential changes to the concept design and inflation up to tender stage.
3.5	Tender Stage	It is essential that a realistic assessment of the construction period is backed up by a comprehensive pre-tender programme.
		A major consideration at the tender stage is the contractor's choice of construction method.
3.6	Construction Stage	Construction schedules are not robust and can easily be disrupted.
		Critical activities on the schedule – those with no float (or spare time) – are particularly vulnerable.
		Contract completion can be disrupted by design changes, late information or unexpected circumstances or events on site.
		Accidents on site, and especially fatalities, can be particularly disruptive and distressing.
3.7	Managing Project Risk	The CIOB suggests that the general standard of time and cost management of construction projects is rudimentary.
		Failing to act proactively to risk can cause unnecessary delay and extra cost.
		Early warning is a way of working dynamically in a spirit of mutual trust and co-operation to solve problems before they arise.
Learning Outcomes		Understand hazard, risk and uncertainty.
		Distinguish between quantitative and qualitative risk assessment.
		Identify common construction hazards and assess risk using a risk matrix.
		Understand the role of risk management during the design, tender and construction stages.
		Appreciate the significance of dynamic time modelling and risk/early warning registers.
Learn More		Chapter 6 explains a variety of procurement strategies and contractual arrangements from a risk management perspective.
		Chapter 6 also explores the importance of the programme in time risk management in the context of a wide variety of construction contracts.
		Chapter 12 covers health and safety risk management in detail.

3

Project Risk

3.1 Introduction

Every construction project carries an element of risk which can impact its participants and stakeholders in different ways:

- For clients the main risks are failure to complete on time, running over budget and failure to achieve the quality expectations as set out in the contract.
- Contractors may fail to achieve expected profits or may suffer a loss. Loss-making projects can lead to cash flow problems, lack of working capital and may even jeopardise the viability of the business.
- The construction supply chain risks late payment, under payment, non-payment or even the insolvency of a contractor or subcontractor.
- The workforce is at risk from the daily hazards that exist on construction sites including health, safety and welfare as well as possible loss of pay should a contractor or subcontractor become insolvent.
- Other stakeholders may be impacted by a project because of noise, pollution, traffic disruption or the blighting of an area during construction operations.

It is tempting to think about risk in silos – client risk, contractor risk, supply chain risk and so on – and it is true that each party carries an element of risk. The extent of this risk depends on how risk is allocated in the contractual arrangements between the parties and which party is best equipped to manage the risks involved.

Notwithstanding this, adverse risk outcomes can impact all participants irrespective of who is responsible for managing a particular risk:

- Clients are responsible for providing access to the works and for ensuring that the contractor is provided with the correct information from which to construct the works in good time. Failure to do so is compensated under the contract with extensions of time and reimbursement for loss and expense. Contractors frequently complain, however, that this is insufficient to cover the real costs of non-culpable delay and/or disruption.
- Contractors are responsible for completing the works on time and failure to do so can lead to the imposition of liquidated and ascertained damages by the client. This is monetary compensation for delay, but the inability to use the facility, or to meet an important opening date or to secure key tenants may be far more important to the client.

Construction Planning, Programming and Control, Fourth Edition. Brian Cooke and Peter Williams.
© 2025 John Wiley & Sons Ltd. Published 2025 by John Wiley & Sons Ltd.

- Serious accidents on site – and especially fatalities – cause distress, cost money, cause delay and can impact the lives of individuals and families immeasurably. Delay to the project means that client outcomes are adversely affected as can the contractor's profitability and reputation.
- When a main contractor becomes insolvent, site activity stops. As a consequence, the client suffers delays and additional costs, workers lose their jobs and subcontractors, suppliers and workers may not be paid.

The potential for project failure is commonly referred to as 'risk' but the threats (hazards) posed and their potential impact are different for clients, consultants, contractors, site operatives and the supply chain generally. Whether threats manifest themselves in undesirable outcomes depends on how well the risk is managed. This will be explored in this Chapter.

3.1.1 Risk

Risk may be defined as *the threat or possibility that an action or event will **adversely** or **beneficially** affect an organisation's ability to achieve its objective.*

This definition suggests that the outcome of a risk could be either good or bad. A contractor's tender bid, for instance, might result in a loss-making contract but could equally turn out to be extremely profitable and lead to additional work with the same client.

A further definition of risk is *the consequence of the presence of a **hazard***. Where roofing work is to be undertaken, for instance, the hazard is work at height. This could result in someone or something – such as tools or materials – falling.

The consequence or risk of working at height is defined by the **likelihood** of someone or something falling and the **severity** if this happens. A risk assessment provides a measure of the extent of the risk.

3.1.2 Uncertainty

BS ISO 31000 (ISO 2009) is the international standard for risk management which applies to most business activities including planning, management operations and communication processes. It is, therefore, relevant to construction. It defines risk as:

> *the effect of uncertainty on objectives* where uncertainties include events which may or may not happen as well as the lack of information or ambiguity.

It has been long understood, however, that there is distinction between 'risk' and 'uncertainty' and that, whilst they may be linked, they are not synonymous.

In a seminal work, Knight (1921) suggests that *'uncertainty must be taken in a sense radically distinct from the familiar notion of risk, from which it has never been properly separated'*. It is further suggested that risk is measurable, whereas uncertainty is not.

Winch (2010) also distinguishes between risk and uncertainty and suggests that:

- **uncertainty** [is] *the absence of information required* for decision making whereas:
- **risk** *is the condition where information is still missing, but a probability distribution can be assigned to the occurrence* of a particular event.

3.2 Risk Management

UK company law[1] requires companies to *prepare a strategic report for each financial year* and UK accounting standards require company directors to show that they have proper and ongoing procedures to manage the risks to which their organisations are exposed.

This means that *management should identify and evaluate the risks faced by the company...and operate and monitor a suitable system of internal control.*[2]

These requirements apply to companies – contractors, subcontractors, developers and businesses – who must make a statement in their published accounts to the effect that the directors have a strategy in place to identify the principal risks and uncertainties facing the business and how they are to be managed.

Specific legislation dealing with construction health and safety imposes duties on clients, designers and contractors to manage the risks to health and safety in so far as they apply to construction work.

3.2.1 Principal Risks and Uncertainties

Examination of the published accounts of a variety of companies reveals different risk perceptions according to the nature and size of the business and the type of work undertaken.

A top 100 contracting company, for instance, identifies the principal risks and uncertainties facing the business as:

- **Macro-economic impacts**: The construction market is closely linked to the general economy and has, historically, suffered from the 'boom and bust' of economic cycles.
- **Unforeseen contract risks and losses**: Contracting, especially, operates on tight profit margins which can be easily eroded by interest rate rises, estimating errors, unforeseen site conditions, bad weather, supply chain problems and so on. Such problems can aggregate into one or several loss-making contracts.
- **Cash flow and liquidity risk**: Cash flow is the difference between money coming into a business and money going out. It has nothing to do with profitability but has everything to do with maximising contract payments and minimising outgoings for supply chain payments. Liquidity depends on a company's ability to secure the necessary working capital to see the business through periods of negative cash flow.
- **Credit risk**: Deferred payment terms to clients is normal practice in construction contracts as it is throughout the supply chain. Should a client default on a contract payment, creditors and wages still need to be paid which can create cash flow difficulties.
- **Inflation risk**: Many contractors take on the risk of inflation on fixed-price contracts. Supply chain prices are variable, however, and may well rise above the contractor's inflation allowance. On large contracts, inflation risk is usually undertaken by the client, but contractors may suffer losses should the inflation reimbursement formula or indices used not reflect actual market price increases.

Other contractors may have a different perception of risk and uncertainty. A Top 5 contractor, for instance, might see that the principal risks and uncertainties facing the business include:

- Major project delivery failure.
- Major supplier failure.
- Health and safety issues.
- Failure to win the right type of work.
- Difficulties attracting and retaining suitable employees.
- Market fluctuations.
- Client failure.
- Loss of reputation leading to loss of future work.
- Environmental and regulatory changes.
- Cyber security and so on.

A major project delivery failure could include failure to meet contractual and legal obligations and failure to meet client expectations regarding time, cost, quality and safety. It could also mean adverse bidding conditions or contractual arrangements or failure to control the project or development effectively.

External factors, such as war, pandemic or significant political upheaval, could also lead to significant financial losses, delayed contracts, breach of contract claims from clients or the supply chain or liquidated damages claims from clients for delayed completion.

3.2.2 Developers

Developers will have a different attitude to risk compared to contractors, and this too will be explicit in the annual accounts.

Before commencing a development, finance must be in place to fund land acquisition, design and construction. This provides the working capital to fund site clearance, roads and sewers and construction of the works. The crunch time for property developers is when reservations and eventual sales should be happening – a point when the development may run into trouble. This is an entirely different scenario from that of contractors who rely on monthly valuations to cash flow the construction work.

On the other hand, subcontractors working for property developers do have to submit weekly or monthly payment applications and have the same cash flow issues as any other general contractor.

> A major housing developer, with an annual turnover of £4.5 billion and an operating profit of 20%, has a particular attitude to risk, as illustrated in Figure 3.1. Scrutiny of the risk categories in Figure 3.1 demonstrates the stark difference between contracting and housing development, especially regarding cash flow, liquidity and profitability risks.

3.2.3 Managing Risk

Clearly, the responsibility for managing risk lies where it should – with management – and this involves planning, organising, controlling, monitoring and reviewing the measures needed to prevent exposure to risk.

Risk Category	Annual change
A Government policies, regulations and planning.	▲
B Housing demand and mortgage availability.	▲
C Land bank and land availability.	▲
D Supply chain quality and availability.	▲
E Construction quality and reputation.	▲
F Health, safety and environment.	◀▶
G Climate change.	▲

Figure 3.1   Risks and uncertainties.

A structured approach is essential to the effective management of risk and normally follows some simple steps:

- Identification of the hazard
- An assessment of the extent of the risk
- The provision of measures to control the risk and
- The management of any remaining (residual) risk.

It should be emphasised, however, that risk assessment will not remove all risks on a project. Both clients and contractors must recognise that residual risks will always remain, no matter what provisions are made. The aim of risk management is to ensure that such risks are managed effectively by the party best able to do so.

Management clearly needs to be concerned about risk issues – both in client organisations and design offices as well as within contracting firms. The bottom line is the prevention of losses which are caused by exposure to risk.

The effects of exposure to risk include:

- Accidents to workers.
- Accidents to the public.
- Loss of skill and experience.
- Damage to property.
- Loss of time and production.
- Loss of money on contracts – both clients and contractors.
- Loss of reputation and future business.

The need to manage risk and uncertainty was seen as vital to the successful delivery of two major projects – the Elizabeth Line underground project in London (Crossrail) and Heathrow Airport Terminal 5 – but each adopted a different approach.

3.2.4 Crossrail

Crossrail was a huge programme of work – it cost £18 billion and the 100 km route comprised extensive upgrades to existing rail infrastructure, 42 km of new tunnels, eight new subsurface stations and two new above-ground stations – all individually large projects.

This required *a structured approach to the identification, assessment and treatment of risks to ensure project success by minimising threats and maximising opportunities* during design, construction, commissioning and handover (Crossrail Ltd 2011).

The Crossrail hierarchy of risks, shown in Figure 3.2, illustrates how risks could be escalated or delegated appropriately within or between the various levels of the hierarchy. This ensures that the risks were ultimately managed by the party best placed to manage them.

Levels of risk response within Crossrail were guided by a **risk appetite statement** which identified acceptable (low), undesirable (medium) and unacceptable (high) risks and how they should be prioritised:

- Is any single risk so significant that it should be escalated to the next level?
- Does the risk affect the wider programme?
- Can the necessary risk responses only be implemented at the next level of management?

3.2.5 Heathrow Terminal 5

The approach to risk management for the Heathrow Terminal 5 (T5) project was very different to that for Crossrail, as illustrated in Figure 3.3.

The £4.3 billion programme of work consisted of 16 separate major projects and 147 separate sub-projects. This included a terminal building with a capacity of 35 million passengers per year, a new transit system, a railway station, a major river diversion and earthworks that uncovered over 80 000 archaeological finds. The project took almost six years to construct.

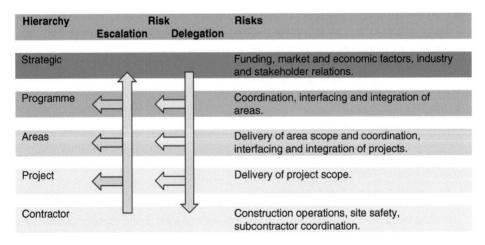

Figure 3.2 Crossrail risk hierarchy.

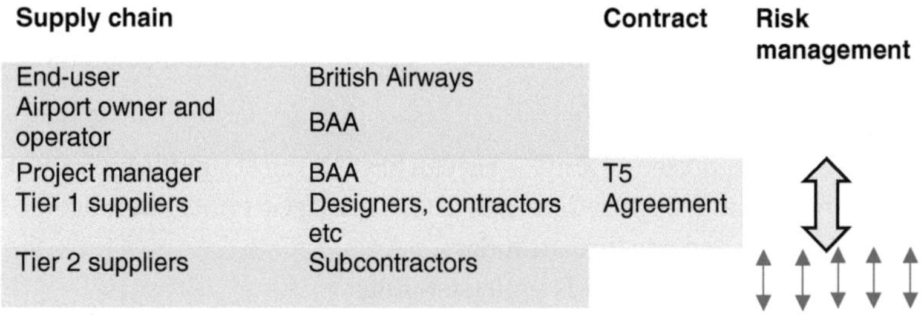

Supply chain		Contract	Risk management
End-user	British Airways		
Airport owner and operator	BAA		
Project manager Tier 1 suppliers	BAA Designers, contractors etc	T5 Agreement	
Tier 2 suppliers	Subcontractors		

Figure 3.3   T5 supply chain.

The project strategy adopted by British Airports Authority (BAA) was inspired by Sir John Egan's 1998 report *Rethinking Construction* and was based on integrated team working and a novel approach to risk management (Gil 2008).

The foundation of the project strategy was a partnering agreement (the T5 Agreement) where contracts with Tier 1 suppliers were based on incentivised cost-reimbursable target contracts with ring-fenced profits.

All financial and liability risk was undertaken by BAA – the airport owner and operator – who was also the project manager. This meant that insurance risks normally carried by contractors and designers were covered by BAA.

As a result of this strategy, BAA and the entire supply chain were dedicated towards managing risk based on jointly aligned objectives, openness and cooperation and contractors were incentivised to finish on time and on budget. Late opening of the terminal was not an option.

This necessitated anticipating, managing and reducing risks and eliminating unforeseen events collaboratively using a joint risk register. No individual supply chain partner was exposed to risk in the conventional sense, and the entire supply chain was focused on the work to be done without having to worry about dealing with risks individually.

The project was completed on time, within budget and is widely regarded as one of the United Kingdom's most successful modern complex projects.

3.3   Risk Assessment

In order to control risk, a risk assessment is carried out by firstly identifying hazards and who or what might be harmed. The extent of risk is then determined and control measures put in place. Residual risks can then be evaluated and managed.

Where there is no risk, there is no chance of a loss occurring, but in the real world, risk is all around.

3.3.1   Risk Categories

Risk falls into three categories:

1. **Known risks**: risks that are an everyday feature of construction activity.

2. **Known unknowns**: risks which can be predicted or foreseen.
3. **Unknown unknowns**: risks due to events whose cause and effect cannot be predicted (Smith 1999).

It is the third category – the unforeseen events – that can have the most significant impact, but risk, nevertheless, has to be recognised, assessed and managed rather than avoided which can lead to overcautious and negative attitudes.

Two additional factors may be added to this understanding:

1. Risk needs to be allocated – conventionally to the party best equipped to manage it and
2. Cost needs to be considered because clients and contractors, who pass risk along the supply chain, will inevitably suffer the consequence of higher prices.

3.3.2 Hazards

Risk is caused by hazards which BS 4778 defines as *a situation that could occur that has the potential to cause human injury, damage to property, damage to the environment or economic loss.*

> A roofer working at height has a chance of falling and suffering serious or fatal injury unless:
>
> - The work can be avoided (i.e. designed out), or
> - Suitable measures can be installed to prevent a fall, and
> - Suitable safeguards (such as safety nets) can be provided should the person happen to fall in any event.

Hazards are classified as:

- **Generic**: hazards that are present irrespective of the project type or nature of work involved.
- **Specific**: hazards that are specific to a particular project.

The noun 'hazard' is generally associated with health and safety risk but when a synonym is used – such as 'threat', 'exposure' or 'vulnerability' – the concept of 'hazard' in a commercial sense can be more readily appreciated.

The risks that contractors take when bidding for work should take into account 'threats' such as competition and market forces, supply chain factors, time for completion, cost and inflation, construction method, temporary works, contract conditions and liability.

Loss-making contracts also expose contractors to risk. The profits and losses made on contracts translate themselves into the annual accounts where losses impact on a company's ability to service its debt finance (interest payments), to provide the necessary working capital to fund the day-to-day running of the business and to pay a dividend to shareholders.

Just like health and safety risk, the contractor's exposure to commercial and technical risk must be assessed before committing to a tender price.

3.3.3 Assessing Risk

Risk is measured by evaluating the likelihood or chance of something happening because of the presence of a hazard and the severity or consequence if it does. This can be based on a hunch or by making an intuitive assessment of whether the risk is high, medium or low or by using a matrix of the likelihood of a hazard materialising and its severity.

> On a scale of 1–3, if the likelihood (chance) is 2 and the severity is 3, the risk is 6 out of 9. This would be in the high-risk category.
>
> However, it must be remembered that the prediction of likelihood and severity is difficult to do with any degree of certainty.

The matrix approach is a popular way to assess risk, but it is a **qualitative** method because it is based on judgement. Its use is not limited to assessing health and safety risk and can equally be used to assess commercial risk at tender stage or the time and cost risk of bad ground conditions when formulating a tender submission:

- Where there is a reasonable possibility that such ground might be present, the **likelihood** might be 3 on a scale of 5.
- The impact, or **severity**, of such ground conditions will be high as there is the possibility of collapse and the need to deal with groundwater. The rating could be 5 out of 5.
- Preventative measures will be costly, perhaps necessitating sheet piling, wellpoint dewatering, ground freezing and the like. The risk for the contractor will consequently be:

Likelihood		Severity		Risk rating	Risk assessment
$\frac{3}{5}$	×	$\frac{5}{5}$	=	$\frac{15}{25}$	Medium–High

Using other metrics, this risk could also be expressed as $6/10$ or $2/3$ or **orange** on a scale of green to red.

> Risk assessments of this nature are **qualitative** as they are not based on a statistical analysis where a probability can be determined. Numerical judgements of likelihood and severity are not calculations and are not quantitative risk assessments – they are qualitative.

It is well-documented that it is often unidentified hazards – for which no provision has been made – that have the most significant impact (Edwards 1995). Such hazards – or threats – can be missed during the design stages, during bid preparation, during pre-contract planning or during the construction stages.

Consequently – despite all the steps that designers and managers take – accidents and losses can and do happen because humans and human systems are imperfect.

It is also the case that risk is all too often ignored or dealt with in an arbitrary way on construction projects. For instance:

- Construction operatives are well known for taking 'shortcuts' to avoid safety management procedures because they think they know better or have always carried out a task in a certain way. This is a management failure.
- It is common practice in the industry to add a contingency allowance – typically a percentage – to a financial calculation such as an order of cost estimate or a tender bid. This is an arbitrary form of risk assessment based on hunch or previous experience but is not a considered calculation.
- A contractor may 'take a view' on the risks attached to a tender bid, taking into account prevailing market conditions and the extent of competition for the work on offer.

3.4 Design Stage

Uncertainty is greatest at the early stages of a project, and time and cost overruns can invalidate the client's business case by turning a potentially profitable venture into a loss-maker. Risks with the most serious effects for clients are:

- Failure to keep within the cost estimate.
- Failure to achieve the required completion date.
- Failure to achieve the desired quality and functional requirements.

Appropriate strategies are therefore necessary for the control and allocation of risk but, whilst risk cannot be eliminated through procurement choices, choosing the most appropriate contractual arrangements can greatly influence how risks are managed. This is illustrated in Figure 3.4 which provides a simplified view of the balance of risk according to the procurement strategy chosen.

3.4.1 Feasibility

The early stages of projects are defined by the Royal Institute of British Architects (RIBA) Plan of Work as:

- **Stage 0**: Strategic definition.
- **Stage 1**: Preparation and briefing.
- **Stage 2**: Concept design.

This is a difficult period for clients who must confirm the project business case, identify options and develop the preferred solution. The project must also be sanctioned – or the decision to build taken – whereupon major commitments are made in terms of finance, design, cost, procurement and construction arrangements. However, reliable time and cost planning early in a project is difficult to accomplish simply because there is relatively sparse information to go on.

Establishing lead times, procurement periods and the construction timescale is fairly easy. Most clients have previous similar projects to refer to or reliance can be placed on empirical data held by consultants or by referring to the Building Cost Information Service (BCIS) data base.

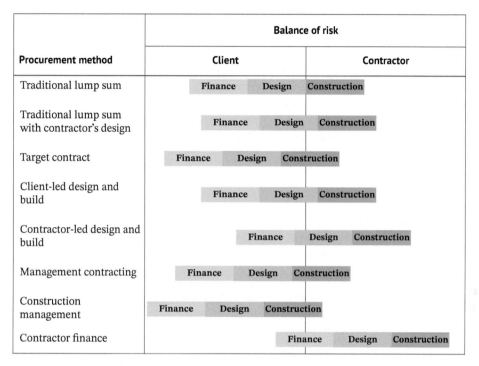

Figure 3.4 Procurement risk. Source: Inspired by Walker and Greenwood (2002) and Edwards (1995).

Design and statutory approvals are less easy to predict as many stakeholders could be impacted by a project which may result in several design iterations – and even public inquiries – before formal approval can be granted. The impact of such issues on the project programme is lessened, to some extent, by outline planning permission which, despite the possibility of reserved matters, can trigger serious progress on site albeit with some design issues to be settled later.

By the end of RIBA Stage 2, a project programme/schedule will have been prepared along with a design programme, order of cost estimate and, maybe, an outline or Formal Cost Plan 1 (as per the Royal Institution of Chartered Surveyors (RICS) New Rules of Measurement). Time risk allowances will have been included in the programme and risk allowances included in the cost planning to cover design uncertainty, potential changes to the concept design and inflation up to tender stage.

Construction stage inflation allowances will depend on the project timescale and the contractual arrangements chosen – the risk may be retained by the client or, for shorter duration projects, the contractor will make allowances in the tender bid.

3.4.2 Funding

The funds for a construction or development project must be arranged before contracts are entered into so that money is in place to meet contractual payments. Financial facilities must, however, allow for a certain degree of flexibility because there is no certainty that the tender price will be the same as the eventual outturn cost or final account.

The client's cost plan must, therefore, contain a contingency to cover for unexpected events, design changes and, where relevant, inflation.

The contractor's income is the client's negative cash flow and arrangements must be made for available funds which can be drawn down so that the client is able to make regular monthly or stage payments for work in progress. The client's quantity surveyor must be careful not to overvalue interim payments, however, which could expose the client to the risk of the contractor's insolvency.

3.4.3 Design

Business decisions are all about risk and reward, and the client must decide how much control is required over the design of the project and who will bear the 'design risk'.

Design risk is a question of liability – who is responsible should there be something wrong with the completed design:

- There may be structural issues such as cracks in reinforced concrete beams or columns.
- The building services may not function as designed – perhaps the number of air changes/hour is below expectations.

In such cases, the extent of design liability has been determined by case law which says that:

- Design liability for architects, structural engineers, HVAC engineers and the like is limited to the standards of *reasonable skill and care* expected of professional designers, whereas:
- Design liability for designer-contractors is one of *fitness for purpose* – effectively a guarantee of performance – which is a recognisably higher standard.

There are basically three choices for clients:

1. Engage professional designers – architects, engineers and the like – to prepare the concept design and to develop this into a fully detailed and specified technical design (pre-construction stage). Alternatively, if the client organisation has its own designers, the design could be carried out 'in-house'.
2. Engage a design-build contractor capable of undertaking both the design and construction of the project.
3. Engage professional designers to create the 'design intent' but leave the details as to how the design is to be constructed to the contractor or to specialist contractors. This is common practice in the industry where provisions are made in the construction contract for the contractor to take partial design responsibility for certain aspects of the project – such as piling, structural frame, roof, HVAC services, etc.

Choices 1 and 2 are extremes where either the client or the contractor undertakes the design risk. It is a difficult choice because, on the one hand, the client has complete control over the design and can make changes with least-cost outcomes. On the other hand, full contractor design might appear to meet the client's design requirements but as the contractor has complete control over how this is achieved, the client could be disappointed. To rectify the situation the client could ask the contractor to make design changes, but they will be expensive.

Choice 3 might appear to be a 'happy medium' but one where there is less clarity for the client should the design fail in some way. This is because liability for different elements of the design will be judged by different standards according to whether they were undertaken by professional designers or by designer-contractors. All designers carry professional indemnity insurance but the level of cover – and hence the client's payout – will depend on whether the expected standard is that of reasonable skill and care or fitness for purpose.

Full or partial contractor design has some benefits regarding the master schedule because detailed design is carried out concurrently with the construction phase based on the design intent. There may be benefits in terms of cost too as full or partial contractor design can be incorporated into the tender documents and therefore becomes a competitive element of the bidding process.

3.4.4 Tender Documentation

Traditionally, tender documentation consists of drawings and a specification accompanied by formal bills of quantities prepared by the client's quantity surveyor. The intention is to give tendering contractors a level playing field when competing for work. In this case, liability for the accuracy of the quantities rests with the client.

Williams (2016), however, reports that the use of formal bills of quantities is in decline which might be explained by the decline in traditional competitive tendering where:

- There are disadvantages for clients due to the separation of design and construction.
- Designs are rarely complete before going out to tender, leading to costly variations and delay.
- Bills of quantities may contain provisional quantities or provisional sums for certain elements of the design. This shifts the quantity risk even more towards the client.
- Defined provisional sums in the bills of quantities require the contractor to make allowances for this work in the schedule, even though there may be little information as to the precise nature of the work required.

It is now common to see tender documentation comprising drawings and specifications only where the risk of computing quantities shifts firmly to the contractor. This can create more variability in tender prices though – due to mistakes or misjudgement of the quantities – making it difficult for the client to compare bids.

On larger projects especially, some clients prefer to engage with contractors using more collaborative tendering procedures. Two-stage tendering, for instance, aims to select a preferred contractor based on soft issues – such as reputation, experience, financial standing and qualitative metrics – who can then contribute to design development and assist the client in developing a target or guaranteed maximum price.

3.4.5 Quantity Risk

Competitive tendering based on drawings and specification means that the contractor is responsible for determining the quantities of work. This risk is frequently passed down the supply chain to subcontractors. Alternatively, unquantified activity schedules may be used – prepared by the client or by the contractor.

The net result is that the quantity risk – the risk of mistakes in the quantities and therefore the tender price – rests with the contractor. Whether quantities are provided by the client or not, the contractor must assess the accuracy of the quantities at tender stage because margins can be eroded if the quantity of work is incorrect or subsequently reduced when the work is remeasured.

Where quantities have been prepared by the contractor, failure to include associated labour items may add risk to the overall bid. On design and build projects, the contractor is responsible for taking off the quantities but may not find errors until the contract has been awarded.

3.5 Tender Stage

Tendering/bidding for contracts is how most construction work is transacted. This means that contractors bid against each other in competition to win work, make profits and provide a return to owners or shareholders. There are some exceptions to this rule:

- Some construction work is negotiated where the client agrees a price and project time scale with a preferred contractor.
- Developer-contractors buy land or property and create work for themselves by building speculative housing, commercial or mixed development projects for sale or rent.
- Other developers similarly create their own market but may:
  ○ contract the building work to a builder or
  ○ project manage subcontractors who undertake the construction work or
  ○ engage their own in-house construction company to do the work.

The competitive tendering market prevails at all levels of the industry – large projects and small. This works slightly differently at the upper levels of the industry where procurement routes include not only traditional single-stage competition but also single and two-stage design and build, frameworks, negotiated and competitive dialogue arrangements and so on.

Most large contractors employ a bid manager to lead and actively manage the complexities of the bidding process. This role includes overseeing completion of pre-qualification questionnaires (PQQs) and the collection of pre-bid market intelligence. The bid manager will probably be involved in several bids at any one time and will implement and chair bid launch meetings to ensure that clear winning themes are developed for each tender submission.

3.5.1 Tender Risk

Part of the role of a bid manager is to work with the bid team to identify and implement technical solutions that could result in a winning bid. This will involve coordinating risk assessments, comparative costing exercises, analytical scoring matrices and organising answer planning workshops to ensure smooth development of the bid.

Bid managers will work with a variety of participants to decide how risks attached to the bid will be managed. Initially, this will include the marketing, estimating and commercial management departments but, as bid preparation advances, temporary and permanent works designers, planning engineers, buyers and procurement quantity surveyors will also contribute.

Risk management is, therefore, an essential part of the tendering process which requires consideration of many factors before a bid can be submitted satisfactorily. Bid submission is a big step because it effectively commits the contractor to a price and a programme based upon the contractor's view of how the works might be constructed, how long the project might take and what profit might be generated from the contract based on the tender documentation and an assessment of risk.

When a contractor receives a tender inquiry, it is crucial to carefully analyse the tender documentation and identify any risks that will influence the eventual bid:

- **Contractual risks**:
  - What is the form of contract and are there any non-standard or amended conditions?
  - Is a contract period or duration stated in the documents?
  - Is an amount stated for liquidated damages in the event of contractor delay?
  - Does the contract place the risk of inflation with the contractor?
  - Who is responsible for insurance of the works and are there any inflated insurance cover levels?
  - Is a performance bond or parent company guarantee required?
- **Document risks**:
  - How complete are the tender documents?
  - Are there any errors or omissions in the documents?
  - Who is responsible for preparing the quantities of work?
- **Design risks**:
  - Is the technical design complete (RIBA Stage 4)?
  - Is a full or partial contractor design required?
  - Where design and build is requested, how complete are the client's requirements and is there an intention to novate the client's design team to the contractor?
- **Construction risks**:
  - Is geological information supplied with the tender documents?
  - Is any warranty attached to this information?
  - Could alternative methods of construction be used to save time and money?
  - Is the project suitable for fast-track construction methods?
  - Is there scope for using novel construction methods?
  - Are any elements of the works suitable for adjacent-to-site, off-site or factory fabrication?
- **Commercial risks**:
  - Are there any aspects of the tender submission where losses could arise?
  - Are there any opportunities to capitalise on errors in the tender documentation?
  - Can the tender pricing document be structured to generate positive cash flows early in the project?
  - Could savings be made on resource costs once the contract has been won?

3.5.2 Tender Risk Allowances

The contractor's tender is effectively a budget figure that includes for:

- The necessary resource costs – including labour, materials, subcontractors and plant.
- The cost of supervising the works and running the site.
- Head office overheads.
- Profit.

The considerations – and allowances – made by a contractor when submitting a bid include both 'soft' and 'hard' issues:

- Soft issues cannot be measured, but a 'value' judgement can still be made:
    - Previous experience (good/bad) of working with the client team.
    - The financial stability of the client.
    - Market conditions and the level of competition for the contract.
- Hard issues can be evaluated, and money can be put against them. They include:
    - Inflation and whether a firm or fluctuating price is required.
    - Ground conditions and the balance of risk in the contract.
    - The completeness of the design and extent of provisional work in the contract.
    - Who is responsible for calculating the quantities, and which method of measurement (if any) is to be used?

As with any budget, tender figures are based on several assumptions, assessments and allowances, not least of which concerns the order, sequence and method of working decided upon. At tender stage, this is decided by the estimator, the planner and, ultimately, by senior management.

> The use of SIPS (structural insulated panels) for constructing the inner walls and roof of a school project added 5% to construction costs, but economies in foundation costs and time-related savings reduced the overall tender price by 10%.

It is quite likely that the site management team will have a different view of how to go about the job once the contract has been awarded. There may be a quicker, more efficient and less costly way of working than envisaged at tender stage, but any changes made must be accommodated within the tender figure and subsequent contract sum agreed.

> A contractor tendered for a major multi-storey office development based on using permanent contiguous piles as earthwork support for the deep basement construction. The site management team decided that the prevailing ground conditions would be more suited to a steeply battered excavation supported by a sprayed concrete lining and ground anchors. This solution was more economical and enabled the permanent basement works to be commenced earlier thereby reducing the overall project programme.

When competition for work is fierce, contractors have to find ways of winning work *and* making a profit at the end of the job. There are many ways that contractors can do this, including:

- Finding alternative methods of construction that are quicker and less expensive.
- Taking advantage of errors in the tender documentation, which could lead to variations or claims, by 'loading' certain rates and prices but not increasing the tender price.
- Taking into account buying 'muscle' on suppliers' and subcontractors' prices as a means of reducing the tender figure – this is often referred to as 'commercial opportunity' or 'scope' which relies on the 'carrot' of a contract award to sharpen up supply chain prices.
- Reallocating money in the bills of quantities, tender sum analysis or activity schedule without increasing the tender sum to generate profits and/or site oncosts earlier in the contract thereby improving project cash flow.

A contractor thinks that savings – commercial opportunity – can be made on materials and subcontract quotes should a winning bid be submitted. Some subcontract packages could be offered at a reverse auction to generate more competitive bids. These potential savings are taken into account at tender stage in order to reduce the tender price. This is clearly a gamble to win work as market forces may be different once the contract has been awarded.

Item	Saving %	Value £	Saving £	Overall tender price reduction £
Materials	Cash discount of $2^1/_2$% + additional 5% 'muscle' = 7.5%	240 000	18 000	
Subcontract	10% across the board	820 000	82 000	100 000

3.5.3 Programme–Time Risk

Should the time for completion be stated in the tender documents, the contractor may be at risk if the client/project manager has got it wrong. On the other hand, where the tender documents require the contractor to insert the time for completion, the contractor must make a judgement and gamble on being right.

It is essential, therefore, that a realistic assessment of the construction period is backed up by a comprehensive pretender programme. This programme is vitally important as it forms the basis for the contractor's pricing of the contract preliminaries – the site running costs. This represents a considerable element of the tender figure. The amount varies according to the size of project but normally lies between 7% and 20% of the tender sum.

When the contractor believes that the client's time assessment is incorrect, a risk allowance may be included in the bid as illustrated below based on the liquidated damages for non-completion that the contractor may have to pay if unable to complete on time.

Client duration (weeks)	Contractor duration (weeks)	Variance (weeks)	Liquidated damages/ week (£)	Time risk allowance (£)
76	84	+6	5 000	30 000

Adding £30 000 to the tender will, however, make the bid less competitive, and the contractor may well look at an alternative strategy. Maybe the programme could be revised by changing the construction method, or by introducing some concurrent working or by increasing resources on critical activities, for instance.

3.5.4 Method Risk

A major consideration at the tender stage is the contractor's choice of construction method. This will include choice of plant, cranes, access equipment and scaffolding as well as temporary works for excavations, concrete deck construction and so on. The method chosen may help the contractor to win the contract, but it must also be viable when work starts on site.

This can be risky because – for example – the ground conditions on site may be different to those expected and the type of earthwork support required may be more expensive than that allowed for in the tender. Alternatively, the estimator may have assumed that a certain number of bricklaying gangs will be available but resource issues on site may mean that the bricklaying activity will take longer than planned at tender stage.

Under certain conditions of contract, the contractor must seek the approval of the client's project manager or engineer as regards the order and timing of the works and the intended method of working. This may have to be revised, especially where the method of working impacts on the structural integrity of the permanent works design.

The chosen method of working determines the order and sequence of work on site. This translates into money via the associated method-related and time-related costs:

- Method-related costs are fixed costs such as construction of an access road, installation of a sheet-piled cofferdam or construction of the foundations for a tower crane.
- Time-related costs are variable such as the hire cost of a tower crane per week, the hire of support scaffolding as falsework to a bridge deck or the hire of a scissors lift or other mobile access equipment.

3.6 Construction Stage

When the client hands over possession of the site, the contractor's contractual obligation is to *proceed regularly and diligently* with the works (Joint Contracts Tribunal [JCT] contracts) or to *provide the works* (Engineering and Construction Contract [ECC] contracts) in accordance with the contract and within the contract period.

The expectation is also that the contractor will organise and manage the works and reasonably endeavour to prevent delay in progress or completion of the project. However:

- Construction schedules are not robust and can easily be disrupted.
- Critical activities on the schedule – those with no float (or spare time) – are particularly vulnerable.
- Time risk allowances in the schedule – especially for critical activities – can lead to inefficiencies as work expands to fit the time available.
- Misuse of the time risk factor can lead to financial risk such as additional time-related costs, reduced cash flow and loss of profit.
- Some 'known unknowns' can be anticipated but not predicted:
  - Delayed delivery of materials or components.
  - Design changes ordered by the client.
  - Inclement weather and so on.
- Others 'known unknowns' are not always obvious or predictable including:
  - The insolvency of subcontractors.
  - Serious accidents or incidents such as a crane collapse.
  - The discovery of antiquities on site – such as archaeological remains or ancient burial grounds.

3.6.1 Delay

The obligation to complete the project on time can be disrupted as a result of design changes, late information or unexpected circumstances or events on site for which the client is responsible under the contract. For the contractor, this represents non-culpable delay.

When the contractor is in non-culpable delay, an extension of time will be awarded under the contract. When delay is the fault of the constructor – culpable delay – this entitles the client to charge damages as prescribed in the contract.

These damages are called *liquidated and ascertained damages* and they represent a genuine estimate of the client's loss in the event of delayed completion – such as lost rental revenues from an office building or loss of sales from a shop, food outlet or supermarket.

Between these two extremes lies:

- The common law doctrine of mitigation of loss. This means that a contractor cannot seek an extension of time with costs without taking reasonable steps to reduce the delay.
- The contractual obligation for the contractor to *proceed regularly and diligently with the works* or *proceed with due expedition and without delay,* or words to that effect, depending upon which form of contract is used.
- The provision in some standard contracts that the contractor may be required to change the method of working, or change the accepted programme, where progress is too slow.
- The contractual right – in some contracts – enabling the client to ask the contractor for a quotation for acceleration if the works have been delayed through no fault of the contractor.

It can be seen that the risk of delayed completion could rest with either the client or the contractor and that standard construction contracts provide mechanisms to mitigate its impact depending upon who is responsible for the delay.

Where the balance of risk lies depends upon the circumstances, but collaborative contracts do help to mitigate delay. New Engineering Contract (NEC) forms of contract place a great deal of emphasis on the programme and upon the parties working *in a spirit of mutual trust and co-operation* to warn each other of potential problems and to manage their effect on progress and upon eventual contract completion.

3.6.2 Health and Safety

Health and safety legislation clearly makes the point that clients, designers and contractors all have statutory responsibilities for everyone directly involved in construction work as well as for those who might be affected by the construction process, such as visitors to site (including client representatives), the public and children.

However, the number of accidents and fatalities in construction continues to give rise for concern, despite the UK industry having a much superior safety 'record' than many other countries. Large, well-organised contractors are no less prone to suffer a fatal accident than a smaller company. Accidents on site, and especially fatalities, can have a huge impact on a construction project, causing delays, increased cost, loss of morale and, especially, distress to the individuals and families concerned.

> During construction of a £760 million football stadium, a subcontract worker suffered fatal injuries. This very sad and distressing incident brings immeasurable grief upon the family and sharply illustrates the dangers faced by construction operatives despite the modern safeguards and improved health and safety standards in the industry. Construction work on site is suspended until such time that the Police and Health and Safety Executive conclude their investigations which will include collecting evidence and conducting witness interviews to discover exactly what happened.

3.6.3 Fire Risk

Fire is an ever-present risk on construction sites, especially with respect to:

- Hot work such as welding, blowlamps, cutting and grinding.
- Heating appliances, especially gas bottles in welfare facilities.
- Litter, especially in rest rooms and drying areas.
- Arson.
- Smoking.
- Burning of waste on site.
- Stored materials, including adhesives and solvents.

The Construction (Design and Management) Regulations provide for such eventualities and require contractors to:

- Take measures to prevent the risk of injury from fire.
- Provide and maintain fire-fighting equipment, fire detectors and alarm systems.
- Give instructions to people in the use of fire-fighting equipment.
- Give instructions to people where their work activities involve a fire risk.
- Indicate fire-fighting equipment with suitable signs.

Standard forms of contract provide for insurance of the works during construction, including the risk of fire. Insurance of the works is normally the contractor's responsibility but where work is being carried out to an existing building, the client might take out the insurance cover.

The Joint Code of Practice, *Fire Prevention on Construction Sites* (Joint Industry Publications Ltd and Fire Protection Association 2015), provides, *inter alia*, that potential fire risks are considered at design stage and that a site-specific fire safety plan is developed by the principal contractor. This plan must include provisions for a comprehensive fire prevention regime including:

- Organisation and responsibilities for fire safety.
- Hot work provisions.
- Fire escape and communications.
- Fire drills and training.
- Emergency procedures.

The fire safety plan may be included in the Construction Design and Management Regulations (CDM) construction phase health and safety plan. It should be noted that adherence to the Joint Fire Code is a contract condition in some standard contract forms.

3.6.4 Supply Chain Risk

A contractor's supply chain is usually extensive but consists mainly of subcontractors, plant hire firms, suppliers of materials and components and builders' merchants. The supply chain operates on the basis of credit, where the client pays for construction work in arrears and the contractor settles its creditors' invoices following an agreed period of deferment – usually 30 days.

It is well known, however, that this is an imperfect system that frequently breaks down and – in *extremis* – fails altogether. Payment issues were highlighted by Latham (1993) in his report *Trust and Money* and are a major contributor to many of the insolvencies suffered by the industry.

Despite efforts by the government to overhaul its Prompt Payment Code (PPC) in 2021, £billions are regularly owed to businesses well beyond the target 60-day invoice settlement period which can cause great hardship, especially to small businesses.

When contractors and subcontractors enter into a contract, they are bound by its terms as regards payment and other matters, some of which can adversely affect cash flows:

- Payment for work carried out under a verbal instruction may be refused pending a confirmation of verbal instruction.
- Work instructed under a variation order may have been carried out, but payment or part payment may be refused because the work has not been valued by the quantity surveyor.
- Work to be paid for based on actual cost – usually called 'daywork' – may not be sanctioned for payment until the daywork sheets are signed and agreed by the contract administrator.
- Subcontract payment applications are habitually disputed by main contractors on the basis that they are over-valued, or lack confirmed instructions, or both.

The situation with suppliers is different because their standard terms invariably include a retention of title (*Romalpa*) clause which means that ownership of the materials rests with the supplier until they are paid for. Should a contractor or subcontractor become insolvent, materials on site cannot be used by an incoming contractor unless they have been paid for because they do not have title in the goods.

Materials are bought on credit and, as such, must be paid for within the agreed credit period – usually 30 days from monthly invoice. Contractors habitually abuse the system and may try to extend the credit period to 60 or even 90 days. If this persists, and payment is continually late, suppliers will suspend deliveries until payment is made and may even refuse to deal with the contractor in the future.

When materials cannot be obtained on credit, the contractor's cash flow is seriously affected and failure to acquire the necessary materials in time can cause inordinate delays to a project.

Plant hire and contract lift contracts also allow credit, but terms are commonly 30 days from weekly invoice. Contracts are usually based on Contractors Plant Association (CPA) standard terms. Failure to settle invoices on time might result in plant being removed from site with inevitable delays and disruption to work.

A particular complication when materials and plant are delivered to site is that some standard construction contracts contain a vesting clause. This means that ownership of any materials or plant on site is vested in or assigned to the client who has effective control of them and must give express permission for their removal, usually via the engineer/contract administrator. Vesting clauses are sometimes complemented by vesting certificates which formalise and clarify the contractual provision, especially where plant is hired and not directly owned by the contractor.

3.6.5 Defects

Construction projects conventionally consist of:

- A period for construction – the contract period.
- A completion or handover date.
- A period for the correction of patent defects – the defects correction period.

There are two different concepts of completion in the industry:

(a) **Practical completion**: where the works are complete and ready for use, save for the rectification of defects. Mainly found in building contracts.
(b) **Substantial completion**: where the works are complete save for remaining items of works that do not prevent use of the works and save for the correction of defects. Used in some civil engineering contracts.

This is illustrated in Figure 3.5 which shows two projects of equivalent length. It can be seen that:

- Contract completion takes place earlier in project (b) meaning that the first 50% of the contractor's retention monies can be released earlier than in project (a).

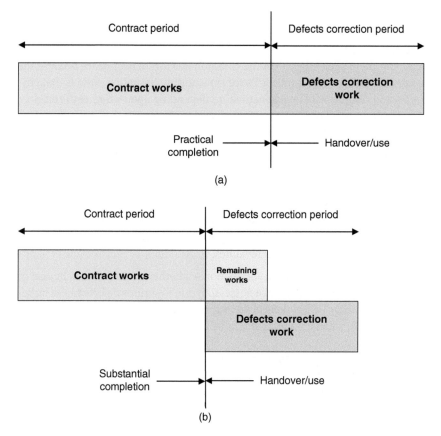

Figure 3.5 Completion. (a) Practical completion. (b) Substantial completion.

- In project (b) the contractor has remaining works to plan and schedule post-handover as well as defects correction works. This means that considerably more resources will be on site but also that productivity will be lower post-handover.
- Providing outstanding works and defects are completed as illustrated, the final 50% of retention monies will be released earlier on project (b).

> Substantial completion is typified by a highway project where the main carriageway and side roads can be opened for use but where minor drainage works, fencing, landscaping and testing of the works is outstanding.

During the defects correction period, the contractor's duty is to put right any defects certified by the architect or engineer – commonly referred to as the 'snagging list' in the United Kingdom or 'punch list' in the United States. Defective work also arises during the construction period when the client's remedy is non-certification of the work concerned, which must be put right before the contractor can be paid.

Some defects may not appear on the 'snagging list' and may not become evident during the defects correction period either. Indeed, it might be years later when such defects – called latent defects – appear.

In such cases, the client's remedies lie in the common law tort of negligence because the construction contract will have concluded. There are statutory limitation periods that govern the discovery and resolution of negligence claims depending upon whether the original contract was a simple contract or if it was executed as a deed.

The impact of defects is felt by clients in various ways:

- Defects discovered during the construction period may add considerably to the contractor's programme and, as a consequence, completion of the project will be delayed.
- Remedial work carried out in the defects correction period may well disrupt the efficient running and use of the completed project.
- Latent defects – especially structural problems – may require closure of the whole or part of the completed building or structure which might require suspension of normal use and consequent loss of utility or commercial revenues.

It is not uncommon for projects to reach completion – whether on time and on budget or not – only for the client to be unable to take possession due to serious defects in the building or other structure. Minor patent defects are put right in the defects correction period following completion but, when a building might be unsafe, this is another matter.

> A new £1 billion, 646-bed, PFI hospital suffered construction problems and delays and the eventual collapse of the main contractor. Serious cracking in concrete beams was discovered in the building and the cladding system failed fire safety tests. The replacement contractor had to strip out three floors of the building and carry out major structural steelwork and reinforced concrete remedial work in order to complete the project. The hospital opened five years later than planned.

3.6.6 Insolvency

Over the last 10 years, official Office for National Statistics (ONS) statistics reveal that construction consistently has the highest number of business insolvencies in any industry sector.

Anecdotal evidence, however, suggests that these statistics are somewhat skewed because of the disproportionately large number of small enterprises in construction which are usually under-capitalised and therefore vulnerable to the vagaries of the general economy.

Nonetheless, the industry is well known for its insolvencies which, over the years, has included a worrying number of top 100 contractors including, famously, Carillion in 2018.

> A top-30 contractor, with an annual turnover of £700 million and over 500 employees, is nearing completion of an £80 million football stand redevelopment project when it ceases trading and enters administration. The project has been beset with delays including the insolvency of a roofing and cladding subcontractor.

> An option to complete the project is to appoint a contractor to act as construction manager with all subcontractors and suppliers directly contracted to the football club. Alternatively, a new main contractor could be appointed to 'step into the shoes' of the insolvent contractor.
>
> Once a decision is made, work completed to date is assessed, any defective work is identified, and a new project completion schedule is devised. The site is then remobilised and the remaining work commenced. This includes cladding, brickwork, seating, concourse areas, fit out and external works. As many original subcontractors as possible will be re-engaged.
>
> Overall project completion is delayed by six months, meaning that important gate revenues are lost.

The reasons why large construction companies get into trouble are well documented – interest rate rises, overvaluation of work in progress, lack of cash due to late payment, the effects of inflation on fixed price contracts, labour shortages, spiralling costs and eroded margins – with the result that firms run out of working capital and are forced into administration or liquidation.

The net effect is that clients run the risk that the main contractor is unable to complete the contract which, in turn, means that:

- Construction work is halted and the site secured.
- Subcontractors and suppliers are left in debt.
- Further delays ensue whilst another contractor is engaged to finish the contract.
- Work to date has to be valued and a cost to complete calculated.
- Defective work has to be corrected at extra cost.
- The eventual final account is likely to be significantly higher than the original contract price.
- Project completion is delayed – possibly for several months or longer.

3.7 Managing Project Risk

Extensive research by the Chartered Institute of Building (CIOB) suggests that the general standard of time and cost management of construction projects is rudimentary. There is, seemingly, a widespread lack of familiarity and use of project planning software in the industry and few respondents said that they use fully linked critical path networks to manage the timing and sequencing of the works.

In its report – *Managing the Risk of Delayed Completion in the 21st Century* – the CIOB found that less than half of respondents use short-term planning, and most respondents only use the master programme for managing time on their projects. The standards of record keeping were also found to be poor, with more than 50% of respondents confirming that they keep paper records only.

The CIOB report gives no demographic as to the size of firms surveyed but, judging by the types of projects undertaken – housing, petrol filling stations, pharmaceutical and medical buildings, schools, sports stadia, prisons, railway projects, shops and shopping malls and high-rise buildings – the findings may be considered representative of large as well as small- and medium-sized enterprises.

Paradoxically, it is a matter of public record that large construction firms use sophisticated project management software packages – such as Asta Powerproject and Primavera P6 – and that they are also well-versed in Building Information Modelling (BIM) and other digital technologies. How effective they are in controlling their projects, however, is a valid question following the results of the CIOB report which prompted the Institute to publish its comprehensive guide to good practice in the time management of major projects.

3.7.1 Risk Allocation

The CIOB report is important because it questions the standards of time and cost risk management in the industry and also implies that there may be doubts about the standards of risk management generally in the industry. The suspicion is that construction project risk management is generally, at best, reactive.

This view is supported to some extent by the fixed nature of risk allocation in some of the standard forms of contract used in the industry. Such contracts are agreed and drafted by a pan-industry drafting committee which decides what is a 'fair' allocation of risk.

In such contracts, risk is allocated according to the agreed contractual arrangements which require the parties to formally notify each other in certain circumstances – such as when a design change is required or when physical conditions on site give rise to a claim for an extension of time and extra cost.

> Some problems are known about, or suspected, before they happen and before a formal notice must be given under the contract. Prior discussion before a formal notice is issued could avert unnecessary delay in a variety of circumstances:
>
> - A change in the design of rebar for ground beams.
> - The revised choice of facing bricks which might require a long lead-time for delivery.
> - A key subcontractor could be in financial trouble.

Failing to act proactively to risk can cause unnecessary delay and extra cost, whereas, in some contracts – the Infrastructure Conditions of Contract (ICC) and, notably, the NEC forms – a more proactive approach is taken.

The NEC approach is to enable clients/advisers to tailor the risk allocation whereby the client feels comfortable carrying certain risks or is happy to pay a premium for the contractor to do so. Risk is still allocated to the parties according to the agreed contractual arrangements, but risk is managed by working together to anticipate and resolve problems collaboratively.

At the heart of this is the concept of early warning, where a contractual undertaking is given to notify each other as soon as they become aware of a potential problem.

3.7.2 Reactive Risk Management

Conventionally, construction projects are monitored and controlled on the basis of progress meetings and by recording and reviewing progress using the contractor's master schedule. Progress meetings may be:

- **External**: monthly meetings involving the architect/engineer, consultants, quantity surveyors for both client and contractor, the site manager and, maybe, key subcontractors.
- **Internal formal**: monthly meetings usually attended by a senior manager, contracts manager or director along with the contractor's site manager and quantity surveyor.
- **Internal informal**: meetings held weekly, or even daily, that may be scheduled or ad-hoc. They are usually attended by the site manager, foreman and site engineer.

Progress meetings are held to discuss progress – of course – but normally only in relation to the contractor's master schedule and only to see whether the project is on-time or ahead of or behind schedule:

- External meetings are held monthly in arrears based on a pre-circulated agenda and are therefore reactive. Matters discussed will be progress, valuations, health and safety, variations to the design, outstanding information and upcoming issues for either the client or contractor representatives to resolve.
- Internal meetings are regular formal meetings to review site progress, health and safety performance, financial progress and any subcontractor issues such as resources, progress and any claims. They too are generally reactive.
- Internal informal meetings are usually short-term look-ahead meetings to plan upcoming work. This could be the next day, next week or a two-week look-ahead. Other issues discussed will include supply chain problems, health and safety audit results, internal cost-value reconciliations and so on, as well as upcoming toolbox/task talks with the workforce. Such meetings are reactive because the aim is to manage a given set of circumstances or problems but not to avert them.

3.7.3 Proactive Risk Management

The CIOB (2018) considers that proactive risk management requires:

- Active employer/client engagement in the strategic design of the project.
- A project strategy that optimises the balance of time and cost risks.
- A procurement method that encourages proactive management.
- A balanced allocation of risk between the parties.

Some major industry clients have taken this message on board and been rewarded with successful projects and desired outcomes. Central to this success has undoubtedly been use of the Engineering and Construction Contact in which early warning and the accepted programme (schedule) are central features of a proactive risk management culture:

- **Early warning**: a way of working in which the employer/client, contractor, project manager and supervisor act *in a spirit of mutual trust and co-operation* and contribute to a dynamic risk (early warning) register with the objective of solving problems before they arise. This enables cost, compensation, completion and performance of the works to be actively risk-managed as the project proceeds.

- **Accepted programme**: the contractor's schedule showing the order and timing of the works, float and time risk allowances, health and safety provisions and resources for each site operation. It also includes a statement of how the contractor intends to do the work and the key dates stated in the contract.
- **Revised programmes**: the contract also provides for the submission of revised programmes which must include actual progress on each operation and its effect on the timing of the remaining works. How the contractor intends to deal with delays and the correction of defective work must also be stated in revised programmes along with any proposed changes to the order or timing of the works or the contractor's method of working.

3.7.4 Dynamic Time Modelling

The programme and early warning provisions of the ECC are a good fit with the CIOB's dynamic time modelling guidance (CIOB 2018). A dynamic time model consists of:

- A high-quality critical path network.
- A master schedule based on:
  - A short-term look-ahead schedule for the first three months of the project with activity durations based on detailed, resource-based calculations (high density).
  - A medium density schedule for the following six months with activity durations based on limited calculation, subcontractors' suggestions and experience.
  - A low-density schedule for project activities taking place after nine months with activity durations based more on intuition and experience.
- A schedule based on a three-month 'rolling wave' principle, so that, at any point in the schedule, the next three months are scheduled in high density.
- The facility to update the high density part of the schedule from actual records of work done and resources used.
- A schedule that facilitates as-built rescheduling – calculating the impact of intervening events on the schedule according to reality.

Figure 3.6a–c illustrates the principle of dynamic scheduling:

(a) This shows how the master schedule is structured at the start of construction work. The first three months are scheduled in detail, but the level of detail diminishes after this point. Low-density scheduling is typical of most construction schedules where activities are represented by a single bar line with durations based on experience or previous similar projects.

(b) After one month of progress, the three-month high-density scheduling period moves in a 'rolling wave' fashion. Actual progress is added to the original schedule which can then be recalculated to see the impact of events to date.

(c) When six months have elapsed, the 'rolling wave' has moved on. The first six-month period will now furnish lots of as-built data which can be included in the schedule.

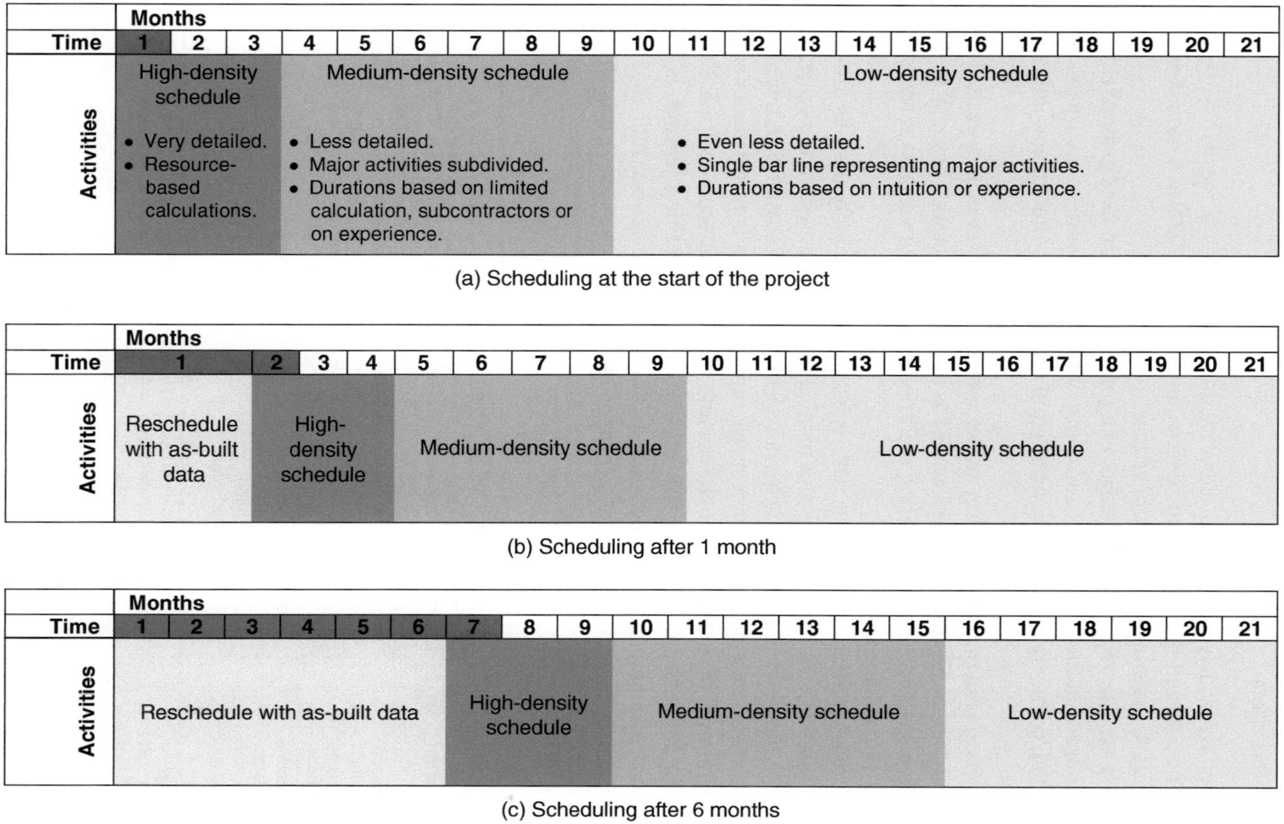
Months
Time
1 2 3 4 5 6 7 8 9 10 11 12 13 14 15 16 17 18 19 20 21
Activities
High-density schedule
• Very detailed.
• Resource-based calculations.
Medium-density schedule
• Less detailed.
• Major activities subdivided.
• Durations based on limited calculation, subcontractors or on experience.
Low-density schedule
• Even less detailed.
• Single bar line representing major activities.
• Durations based on intuition or experience.
(a) Scheduling at the start of the project
Months
Time
1 2 3 4 5 6 7 8 9 10 11 12 13 14 15 16 17 18 19 20 21
Activities
Reschedule with as-built data
High-density schedule
Medium-density schedule
Low-density schedule
(b) Scheduling after 1 month
Months
Time
1 2 3 4 5 6 7 8 9 10 11 12 13 14 15 16 17 18 19 20 21
Activities
Reschedule with as-built data
High-density schedule
Medium-density schedule
Low-density schedule
(c) Scheduling after 6 months

Figure 3.6 Dynamic time modelling.

If the project is ahead of or behind schedule, the three-month high-density scheduling period may appear earlier or later than is shown in the diagram.

3.7.5 Risk Registers

Risk registers are widely used in the construction industry as a means of predicting, evaluating and controlling risk. They are defined by the CIOB (2022) as *a formal record for risk identification, assessment and control actions.*

Contractors use risk registers at all stages of a project. They are used extensively during bid preparation and at the pre-contract and contract stages for a variety of purposes including contractual, commercial, technical, health and safety, subcontractor and inflation risk management.

Risk registers are not the exclusive province of contractors, however, and are used equally by clients and their professional advisers as well as by subcontractors. They may be devised at various points in time, typically during the briefing and design stages, at tender stage, pre-contract and at contract stage.

A type of risk register – the 'project risk register' – is commonly used as a means of identifying, assessing, communicating and managing risk issues within the project team. An example is shown in Figure 3.7 which evaluates risk as high, medium or low.

It is important to understand that project risk registers are not contract documents and, like the programme, should not find their way into the contract documentation. However, under certain contract conditions, a risk register, or early warning register, is defined under the contract and must be initiated and maintained by a designated individual – such as the client's project manager – who is responsible for ensuring that risk is effectively managed.

		Risk			Without controls				
Item	Description	H M L	Cost impact	Time impact	Other	Controls	Residual risk	Action	
1.	The architectural concept for Block C is not fully defined.		± £800/m²	Delay to completion of model		Design team workshop to develop alternative proposals	Planning may disapprove cladding design	Client Architect QS	
2.	Underground high-voltage cable crossing proposed access road.		£75 000	Could add 4 weeks to build schedule		Allow for diversion in cost plan	Potential delay by utility company	Architect QS	
3.	Awaiting client's decision on green roof to the multi-storey car park?		+£200 000 on cost plan.	3-week delay finalising the design	Planning application will have to be changed	Hold urgent meeting with client	Client may revert to roll-formed roof	Architect	
4.	Awaiting geological report for north–west corner of site.		Possible extra cost for piling of £750 000	Extra 5 weeks on project programme		Add contingency to cost plan	CFA piling may be unsuitable	Structural Engineer QS	

Figure 3.7 Project risk register.

3.7.6 Early Warning Registers

Risk registers have been a key feature of the ECC since it was first published in 1993 and are used in conjunction with the early warning protocols in the contract. They were renamed 'early warning registers' in the 2017 edition of the contract but are, to all intents and purposes, risk registers.

Under the NEC ECC contract, early warnings are recorded in an early warning register and are then discussed at a subsequent early warning meeting. This is a two-way dialogue which benefits both parties.

For instance, the contractor can raise potential problems that might arise regarding ground conditions or temporary works design and the client's project manager can equally raise early warnings of impending changes to the works information or question the contractor's resourcing of the project in terms of achieving the completion date. This enables appropriate action to be taken which can then be monitored until the problem is resolved.

References

CIOB (2018), *Guide to Good Practice in the Management of Time in Major Projects: Dynamic Time Modelling*, Wiley Blackwell.

CIOB (2022) *Code of Practice for Project Management for the Built Environment*, 6th edn. Wiley Blackwell

Crossrail Ltd (2011) *Crossrail Programme Controls, Risk Management Plan*.

Edwards, L. (1995) *Practical Risk Management in the Construction Industry*. Thomas Telford.

Gil, N. (2008) *BAA – The T5 Project Agreement*. The University of Manchester Business School

ISO (2009) *ISO31000:2009 Risk Management – Principles and Guidelines*.

Joint Industry Publications Ltd and Fire Protection Association (2015) *Fire Prevention on Construction Sites, Joint Code of Practice*, 9th edn.

Knight, F.H. (1921) *Risk, Uncertainty and Profit*. The Riverside Press.

Latham, M. (1993) *Trust and Money, Department of the Environment*.

Smith, N.J. (1999) *Managing Risk in Construction Projects*. Blackwell Science.

Walker, P. and Greenwood, D. (2002) *Construction Companion to Risk and Value Management*, RIBA Enterprises.

Williams, P. (2016) *Managing Measurement Risk in Building and Civil Engineering*, Wiley Blackwell.

Winch, G.M. (2010) *Managing Construction Projects*, 2nd edn. Wiley-Blackwell.

Notes

1 Companies Act 2006 Chapter 46 Part 15 Chapter 4A Section 414A.
2 UK Generally Accepted Accounting Practice (UK GAAP) and International Financial Reporting Standards (IFRS).

Chapter 4 Dashboard		
Key Message		Construction projects can vary enormously in size and scope, and this is a determining factor in how they are managed.
Definitions		**Project**: A unique set of coordinated activities, with definite start and finishing points, and defined schedule, cost and performance parameters.
		Scope: Project boundaries, goals, deliverables and deadlines.
		BIM: An intelligent 3D model based on team collaboration.
Headings		Chapter Summary
4.1	Introduction	Construction projects are transient endeavours that take place at discrete locations and are managed by temporary teams of people.
4.2	Projects and Programmes	A **project** is smaller and less complex than a programme and normally leads to an output (e.g. a completed building).
		A **programme** combines projects with change management to deliver benefits.
		A **portfolio** is a combination of projects and programmes and is designed to achieve strategic objectives.
4.3	Management	Management is concerned with seeing that the job gets done.
		This centres on planning and guiding the operations that are going on.
		This implies a unity of direction and a guiding hand at the helm.
4.4	Team Building	Team building in construction is difficult because individual members may not be involved in the project all the time.
		Teams are important because they serve as a means of building self-esteem and a sense of belonging among team members.
4.5	Project Management	Project management is a subset of management.
		Management is an ongoing process whereas project management has a final deliverable and a finite timespan.
4.6	Collaboration	The secret of successful project management is teamwork.
		The project manager's job is to bring organisations, people and data together, with a unity of purpose and direction.
		Common Data Environments are cloud-based spaces for project information that encourage collaboration.
4.7	Design Management	Design management is a subset of project management.
		Design managers coordinate design-related matters on construction projects.
4.8	Project Management Models	Project management models focus on the abilities and competencies needed by project managers at various stages of projects rather than tasks and workflows.
4.9	Project Stages	All construction projects – large and small – go through a process from initiation to completion and thence to use or occupancy.
		Process models – such as the Royal Institute of British Architects (RIBA) Plan of Work – guide project teams and help to define construction project work stages.
4.10	Digital Construction	Digital Construction is an evolution of Building Information Modelling (BIM).
		It facilitates the production and management of information relating to construction projects throughout their entire life cycle.
4.11	Managing the supply chain	The construction supply chain is complex and is made up of many different sorts of business enterprises.
4.12	Managing construction	This is concerned with the management of resources within the ambit of the scope, time and cost parameters of a project.
Learning Outcomes		Distinguish between projects, programmes and portfolios.
		Understand the principles of management.
		Differentiate between various subsets of management.
		Appreciate the significance of collaboration and team building.
		Understand the role of process models in the management of projects.
Learn More		Chapter 2 – it is vital to understand the environment in which projects are carried out.
		Chapter 3 – managing construction projects is about managing risk.
		Chapters 19 and 20 provide examples of project management structures and relationships.

4

Managing Construction Projects

4.1   Introduction

Construction projects are transient endeavours that take place at discrete locations commonly referred to as construction 'sites'. The site could be in any part of the country – a city centre, on the outskirts of a town, in the countryside, an existing building or at an existing location such as motorway or railway bridge, or it could be dispersed over a long distance where a highway or high-speed rail project is being constructed.

Projects are managed by temporary teams of people brought together by a variety of organisations – typically, the client organisation, firms of professional consultants, contractors and subcontractors. Each of these organisations will be represented on site by selected individuals, often randomly assembled on the basis of suitability, experience and availability. The glue that holds everything together is the project, which is represented by organisations, people, information, models, schedules, decisions and lines of communication, as illustrated in Figure 4.1.

It is rare to see the same team on successive projects – even for the same client – simply because people move on in their jobs, and different contractors are appointed based on competition and particular individuals may be unavailable because they are tied up on other projects.

In fact, it is quite usual to see project teams changing composition during the life cycle of individual projects. Contractors, for instance, often move their site teams on to other projects and bring in finishing teams to tie up all the loose ends.

Each of the participant organisations in a project – clients, designers, consultants, contractors, subcontractors, plant hire specialists and so on – will be organised differently, they will have their own culture and will have their own regular methods of working. There is no guarantee, therefore, that different teams following their own set processes and procedures will deliver a successful project. Project teams do not always coalesce successfully.

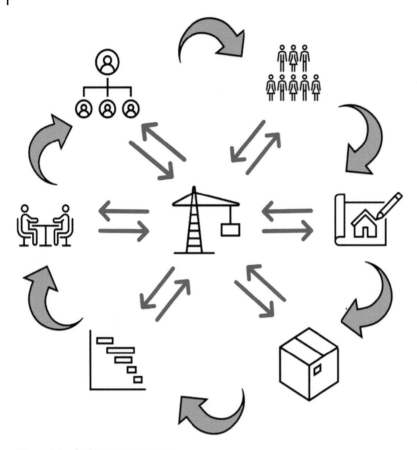

Figure 4.1   Project management.

4.1.1   Definition

A 'project' is succinctly defined in BS6079-1 as:

> a **unique** set of coordinated activities, with **definite** start and finishing points, undertaken by an individual or organisation to meet **specific** objectives within defined schedule, cost and performance parameters (BS1 2010).

Consequently, a construction project is made up of:

- Its scope as defined in the contract,
- Its commencement and completion dates and
- Its objectives in terms of time, cost and quality.

However, a common problem with construction projects is that the organisations involved all have their own objectives, and these objectives do not always align to those of the client. This issue is well documented in many construction industry reports, including the well-known Latham and Egan reports.

A project must, therefore, be distinguished from the 'processes and procedures' employed to bring it to a conclusion. All projects have their own individuality, as recognised by BS6079-1, and each has its own backdrop of risk and uncertainty within finite start and end dates.

4.1.2 Scope, Delivery and Deliverables

Achieving the objectives of a project *requires deliverables conforming to specific requirements, including multiple constraints such as time, cost and resources* (ISO21500).

Taking a 'high level' view, the **scope** of a project sets its boundaries and defines its goals, deliverables and deadlines. The scope of a project is, therefore, much more than the narrow perspective of work breakdown structure that normally defines the work activities on site.

In NEC4, 'scope' defines the services to be provided in the contract between the employer and other parties such as professional consultants and contractors. This could be for site investigation or design services or for work to be carried out by contractors such as demolition, piling or a complete project.

For construction work, scope will include the specification and description of works and any constraints to which the contractor is bound. The scope may be altered during the contract by a project manager's instruction.

In the normal ambit of a construction project – a school or factory for example – **delivery** refers to project completion and handover in accordance with the terms of the contract between the client and contractor. Apart from the correction of defects in the defects correction period, that is it.

In some cases, however, delivery refers to the handover of a fully functioning working facility – such as a hospital or prison – which is built and then run and maintained for a period of years. In a hospital, for instance, the service provider would provide all non-medical staffing whilst the hospital would provide the necessary medical services.

> For most construction projects, the main deliverable is the completed project – as defined by its scope – and this is normally achieved within a finite timespan – as determined by the project schedule. Projects are not always 'stand-alone', however, and may be part of a programme of associated projects or may sit within a portfolio or combination of projects and programmes.

Project **deliverables** are the project outputs set in accordance with client or stakeholder expectations. Deliverables – or goals – may be tangible or intangible outputs as defined in the project scope, agreed by all stakeholders, and they play a definitive role in accomplishing the project. An environmentally friendly building, a clash-free design, the absence of lost-time accidents, completion on time and the absence of defects are all examples of project deliverables.

Deliverables should not be confused with **milestones** which, whilst important, are simply a means of establishing key points in a project which help to track progress. They have no impact on project objectives.

4.2 Projects and Programmes

Construction projects come in all shapes and sizes – a new house, a shopping centre, an electric vehicle (EV) giga-factory or a high-speed rail line – and they are all referred to as 'projects'.

Strictly speaking, not all 'projects' are projects and both the Project Management Institute (PMI) in the United States and the Association for Project Management (APM) in the United Kingdom draw a distinction between projects, programmes and portfolios.

4.2.1 Programme and Portfolio Management

The APM takes a strategic view of the management of projects by recognising the development of both **programme management** and **portfolio management** in the ambit of its body of knowledge (BOK). The Association for Project Management body of knowledge (APMBOK) suggests that project management, programme management and portfolio management (P3) are not mutually exclusive but *are fluid and overlapping*. To help understand this concept, the APMBOK submits that:

- A **project** is smaller and less complex than a programme and normally leads to an output (e.g. a completed building such as a domestic-scale house extension, a new school or the London Shard).
- A **programme** *combines projects with change management* in order to deliver benefits such as Heathrow Terminal 5 (T5) – a step change needed to increase Heathrow's capacity by some 50% and to modernise the airport in response to competition from other continental hub airports. T5 comprised a large number of interrelated projects, which, at the time, was Europe's biggest programme.
- A **portfolio** is a combination of projects and programmes and is *designed to achieve strategic objectives* (e.g. Hinkley Point C – a new £30+ billion nuclear power station providing low-carbon electricity for 6 million homes and lasting benefits for the UK economy. The portfolio consists of a programme of construction projects, a nuclear commissioning programme and a programme of projects to connect the power station to the national grid).

It is not always the case that a programme is necessarily larger and more complex than a project or that a portfolio is larger and more complex than a programme. Sometimes, single large and complex projects may outweigh a programme of projects. Similarly, the strategic portfolios of relatively small organisations may be far less complex to manage compared with a large and complex project.

4.2.2 Phased Projects

A programme is not the same as a phased project.

Large housing projects are commonly phased enabling developers to start earning revenue from sales in order to cash flow the remainder of the construction work. There may be two or more phases, but each phase is managed by the same project team just like a single project.

Another type of phasing is where a large project is procured under a single contract but is completed and handed over in sections. In such cases, the contract will contain sectional completion clauses, with each section having its own completion date within an overall completion date for whole project. This arrangement enables the client to take over and use completed parts of the project before the entire project is finished.

Sometimes, access to parts of the site may be granted to the contractor at different times during the project. This might be to enable the client to decant existing buildings before refurbishment or demolition works commence or the client might be undertaking enabling works or decontamination works under a separate contract. This means that the contractor will have to work on other parts of the site until full possession is granted.

Partial possession or sectional completion arrangements will be written into the Contract Data (NEC contracts), Contract Particulars (JCT contracts) or the Appendix (FIDIC contracts).

4.2.3 Project-Based Working

For convenience, the APM BOK uses the terms 'project-based working' and 'the management of projects' to refer collectively to projects, programmes and portfolios whilst acknowledging that project-based working does not apply equally across all three concepts.

The way that work is managed depends upon factors such as scale, significance and complexity, but desired outputs, outcomes, benefits and strategic objectives are also distinguishing features.

This means that a project could be stand-alone or part of a programme of interrelated projects.

4.2.4 The Size of Projects

Construction projects can vary enormously in size and scope, and this is clearly a determining factor in how projects are managed. There is, however, no official classification for construction projects of different sizes.

Some standard forms of contract give clues as regards their suitability for various project sizes. The JCT family, for instance, distinguishes between:

- **Minor works**: smaller, basic construction projects where the work is of a simple nature.
- **Intermediate works**: projects involving all the recognised trades but without complex building service installations or other specialist work.
- **Large or complex works**: large or complex construction projects.
- **Major works**: large-scale construction projects involving major works.

The Chartered Institute of Building (CIOB) (Bahl et al. 2021), alternatively, makes the distinction between standard projects and major projects:

- **Standard project**: A project where construction works are estimated to cost less than £10 million pounds sterling.
- **Major project**: A project where construction works are estimated to cost at least £10 million pounds sterling.

Very large projects – often referred to as mega-projects – are not classified by the CIOB, but Flyvbjerg (2018) considers that they:

- Typically cost US$1 billion or more.
- Are large-scale and complex.
- Take many years to develop and build.
- Involve multiple public and private stakeholders.
- Are transformational and impact millions of people.

This definition may be contextualised by projects such as:

- **Tottenham Hotspur football stadium**: approximate cost US$1 billion (£850 million)
- **London Elizabeth Line (Crossrail)**: approximate cost US$24 billion (£19 billion)
- **Hinkley Point C**: projected cost US$40 billion (£33 billion)
- **HS2**: projected cost US$120 billion (£100 billion)

Flyvbjerg also considers that:

- Mega-projects are not simply magnified versions of smaller projects but are a completely different breed of project *in terms of their level of aspiration, stakeholder involvement, lead times, complexity and impact.*
- Conventional project managers should not lead mega-projects as they require reflective leaders *who have developed deep domaine experience in the specific field.*

4.3 Management

In a seminal text, Brech (1975) states that *management is concerned with seeing that the job gets done* and that this centres on *planning and guiding the operations that are going on.*

This statement is equally applicable to a government or local authority organisation, or a business or a project. The job to be done could be to deliver a policy or public service or guide a company to make profits in order to satisfy shareholders or it could be to run a construction project so that it is completed on time and on budget.

Walker (2015) suggests that:

- Management is the dynamic input that makes the organisation work.
- Organisation is the pattern of the interrelationships, authority and responsibility that is established between the contributors to it.

The underlying principles of management theory and practice are well-established and date back to the mid-nineteenth century.

4.3.1 The Principles of Management

The principles of management that we recognise today were established by Henri Fayol in the early twentieth century, who, along with Frederick Winslow Taylor, is widely recognised as the founder of modern management methods.

Fayol recognised that certain activities take place in an organisation including the activity of management. He also established that management comprises five functions:

1. Planning.
2. Organising.

3. Commanding.
4. Coordinating.
5. Controlling.

More recent thinking suggests that:

- Planning includes forecasting.
- Commanding and coordinating can be combined as meaning 'leading' or 'directing'.
- The ability of a manager to control requires feedback in order that variances can be identified and analysed and the necessary adjustments made.
- Effective managers need to be able to inspire and motivate people to do their best and to communicate clearly and effectively.

Underpinning the **functions** of management, Fayol suggested 14 **principles** of management, two of which stand out as particularly pertinent to construction projects:

- **Unity of direction**: Each group of organisational activities that have the same objective should be directed by one manager using one plan for the achievement of one common goal.
- **Esprit de corps**: Promoting team spirit will build harmony and unity within the organisation.

4.3.2 Organisation

Seeing that the job gets done tends to imply a unity of direction and a guiding hand at the helm who, as the need arises, delegates authority to an appropriate level. This is how organisations work and is why they are structured in a way that splits up the job to be done into manageable portions. This principle is illustrated in Figure 4.2 which shows how a construction company might delegate certain key roles to the various levels of management.

When it comes to construction projects, however, there may be many organisations involved, each with its individual sense of purpose. This may not align with the purpose of the construction client, and it is this conflict of interest that can impact projects adversely.

Each project participant will be organised differently and will have its own management structure, with someone at the top. It will have its own culture – way of working – and its own objectives, and the organisation will be set up in the best way possible to achieve those objectives.

4.3.3 Organisation Structures

All firms have a management structure and a means of communication suited to the way the organisation works. The size and complexity of this structure will vary according to the size of firm involved:

- Micro and small firms, with an annual turnover up to say £10 million, are characterised by a fairly flat organisation structure, as illustrated in Figure 4.3, where owners/directors are not separated from the workforce by several layers of management. This means that owners can micro-manage quality and maintain close relations with clients. It also means that there is little delegation which can be a limiting factor when trying to grow the business.

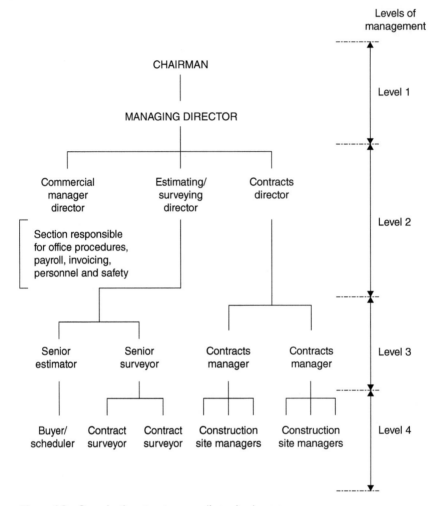

Figure 4.2 Organisation structure- medium-sized company.

- Medium-size firms, typically with a turnover of £10–50 million, need layers of management between the directors and the workforce, as illustrated in Figure 4.2. There may be several departments, such as estimating, commercial and construction, each led by a director or senior manager.
- Large firms are often public companies (PLCs) and typically require an extensive organisation structure sufficient to handle an annual turnover of up to £1 billion. Figure 4.4 illustrates that the company may comprise several operational divisions or business units, each with its own management structure.
- Very large firms – usually PLCs – will have an annual turnover well in excess of £1 billion. They are typified by very complex organisation structures. The PLC will invariably comprise a group of companies, and the PLC itself may be owned by an ultimate parent company. The PLC will probably consist of a number of wholly

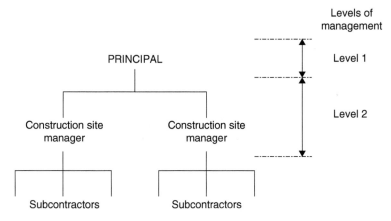

Figure 4.3   Project organisation – small company.

owned subsidiary companies, interests in companies registered in other countries and involvement in several joint venture companies on major projects.

4.3.4   Operational Divisions

Large firms tend to be involved in a wide variety of activities, and this often results in a corporate structure that is based on some sort of 'grouping' arrangement. There are several ways to do this and different methods may be combined:

- **Functional**: the firm is divided into departments such as marketing and human resources, and a production department is responsible for running the projects.
- **Product/Service**: where the company has special expertise in a number of markets such as design and build, major projects and civil engineering.
- **Geographical**: some large firms prefer a regional structure aimed at creating a local presence or they might acquire smaller local companies, often retaining the name to establish the local identity.
- **Customer**: where the company offers services aimed at particular customers such as design services, project management, highways and rail infrastructure.
- **Capital**: where the company has invested heavily in specialist equipment such as slip-forming, tunnelling, cranes and lifting services.
- **Project**: typified by the traditional 'building contractor' who undertakes a series of discrete projects with head office support for estimating, purchasing and administration, etc. Small firms are often organised this way too.

Figure 4.4 illustrates a large company divided into four divisions each being responsible for a particular type of construction activity. This would be a 'product' grouping, with each one managed by a main board director. If the product grouping was also organised regionally, there would be a regional director in charge of each product group reporting to the main board of directors.

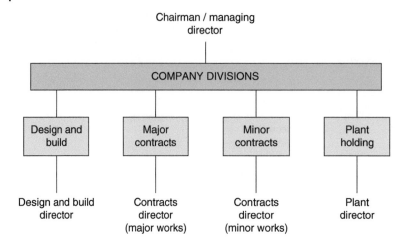

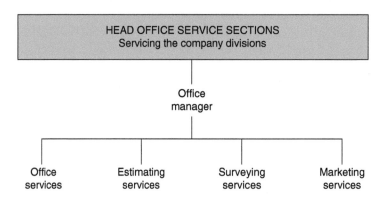

Figure 4.4   Organisation structure – large company.

Figure 4.5 shows how each of the four divisions organises the construction projects within its responsibility.

4.3.5   Business Units

Business units are an alternative to operational divisions. This is where divisions or parts of a company operate as separate businesses within the business as a whole.

The main difference between business units and operational divisions is that each business unit has its own managers and staff and is responsible for its own profit and loss. A business unit can be as small as a single product or service, or it can be an entire company within a group of companies. Head office services, such as marketing, design and pre-construction management, HR and quantity surveying, could be centralised and charged as an overhead to each business unit, or each business unit could have its own 'head office' facilities.

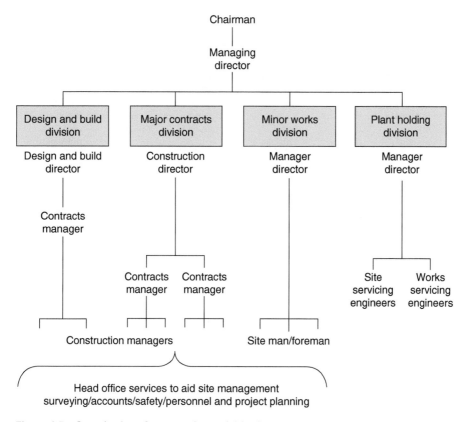

Figure 4.5 Organisation of construction activities in a large company.

Figure 4.6 shows a large company, with a £664 million turnover, organised into six business units. Each business unit operates nationally, but the major projects business unit is set up to service only HS2 Tier 1 contractors on a long-term framework arrangement. The turnover for each business unit is shown, with major projects contributing £56 million to the total for the whole business.

The aim of business units is to develop a team approach to managing projects, to promote corporate morale and identity and to maintain close relationships with clients. A business unit manager is responsible for each business unit, reporting directly to the commercial director. Each business unit has a small executive team, answering to the business unit manager, which deals directly with individual project or site managers.

Figure 4.7 illustrates how the business unit approach can be combined with operational divisions in a PLC with an annual turnover of £1.9 billion and a pre-tax profit of £51 million (2.75%). The main board of directors is responsible for the entire business which has four operational divisions. The construction division has six business units – such as major projects and transport – each of which conducts projects nationally.

Board of directors		Managing director			
	Deputy managing director	Finance director	Commercial director		
Business unit manager	Business unit manager	Business unit manager	Business unit manager	Business unit manager	Business unit manager
Building	**Civil engineering**	**Major projects**	**Rail**	**Sports and leisure**	**Demolition and land remediation**
Warehouses Logistics buildings Industrial buildings Fit out	Highways Lorry parks Bulk earthworks	Major earthworks	Stations and platforms Depots Infrastructure	Stadia Leisure centres Mixed use leisure, commercial and apartments	Rail bridges Site clearance Waste processing and recycling

Turnover	£260 million	£126 million	£56 million	£92 million	£120 million	£10 million
Total	**£664 million**					

Figure 4.6 Business units.

4.4 Team Building

Construction project teams are, at best temporary – they break up when the project is finished – and more often than not, they are transitory. There may be a small 'core team' leading the project but the other members come and go as the project develops – some may come back again later whilst others move on to the next job. Even members of the 'core team' may change as the site engineer is no longer required or maybe the general foreman is replaced by a finishing foreman.

Lavender (1996) makes the point that team building is difficult because individual members may not be involved in the project all the time. This is true in construction, where it is commonly the case that there will be a visiting planning engineer, quantity surveyor, setting out engineer and health and safety adviser. Whilst integral to the team, these people will almost certainly be head office based and will visit the site infrequently. The same could be said of subcontractors who come and go as the project evolves.

This scenario resonates with Winch (2009), who considers that unless team building takes place in the context of a partnering-type arrangement, there is a danger that the result will be a pseudo team which performs less well – more akin to a representative group than a team.

Teams are nevertheless important because they serve as a means of building self-esteem and a sense of belonging among team members but, conversely, some people are happier

	Main board of directors			
	Operating divisions			
	Development	Consultancy	Construction	Operation
	Mixed development	Project management	General contracting	Facilities management
	Town centre redevelopment	Cost consultancy	Construction management	
	Residential	Commercial management	Specialist services	
	Retail/leisure			

Major projects	Commercial	Fit out	Public sector	Technology	Transport
Power stations	Hotels	Commercial interiors	Health and education	Pharmaceuticals	Airports
Sports stadia	Offices	Developer fit out	Nuclear	Manufacturing	Rail
Office towers	Retail	Refurbishment	Justice	Telecoms	Ports
		Shell and fit	Defence		Highways

Figure 4.7 Operating divisions.

working on their own. This is quite understandable, and it is perfectly acceptable for a project to benefit from ideas generated outside the team as well as within it. Part of the role of the project manager is to work with all sorts of talented individuals and to get the best out of them for the benefit of the project as a whole.

Winch (2009) reminds us that many team-building programmes or events involve the participants getting cold, wet and muddy. Perhaps this is why so many project teams in construction are successful!

4.4.1 Complexity

In a thought-provoking lecture concerning the successes and failures of the Heathrow Airport Terminal 5 (T5) project, Tim Brady of the University of Brighton Business School reported that the University of Canfield found that 'complexity' was one of the main determinants of a successful project and that 'complexity' could be distilled into three main components:

1. **Structural complexity**: the physical scale, size and interconnectivity of the elements making up the project.
2. **Socio-political complexity**: the challenge of managing the various individual organisations and the complex mix of people, power and politics involved in the project.
3. **Emergent complexity**: the extent of uncertainty and change as the project develops which can disrupt the most thorough planning of the work involved.

Of these three factors, it was the management of people that was thought to be the most challenging.

4.4.2 Trust

It is clear from the many reports concerning the construction industry that a key frailty of the industry is the lack of trust between its participants. Clients, professional advisers, contractors and subcontractors tend to inhabit 'silos' within which self-interest is served, with the end result that project outcomes suffer as a consequence.

Despite trends towards partnering and greater collaboration in projects, the 'bottom line' is still that the profit motive dominates interpersonal relationships, and this perpetuates the 'blame culture' for which the industry is renowned. To an extent, this is understandable because all businesses must be profitable to survive, and construction is a highly competitive industry operating on wafer-thin margins.

> Some enlightened clients have understood this issue and have been brave enough to devise ways of creating a genuine atmosphere of trust, collaboration and 'win–win' approach to problem-solving on their projects whilst at the same time ensuring that project participants emerge with a profit consistent with their efforts and contribution to a successful project outcome.

One of the keys to the success of Heathrow T5 (construction phase) was the integrated team working culture initiated by the client, British Airports Authority (BAA), which was a feature of all sub-projects completed at Heathrow. However, Professor Brady suggested that there was no such culture at the interface between the occupier/user of the terminal (British Airways) and the owner/operator (BAA), and it was this lack of integration and cooperation that led to the problematic opening of the terminal.

One of the strongest messages to emerge from both the Latham and Egan reports was the need for the industry to work in a spirit of mutual trust and cooperation. This was the culture on T5 – driven by Sir John Egan, who was, at the time, Chairman of BAA.

The key documentation that made this possible was the T5 Agreement which was underpinned by the T5 Handbook. The T5 Agreement was a relational contract between BAA and all T5 first-tier suppliers.

In English law, this means that the explicit terms of the contract are only an outline, as there may be implied terms and understandings which are intended to determine the behaviour of the parties. Thus, whilst there is no doctrine of 'good faith' in English Law, a relational contract *may require a high degree of communication, cooperation and predictable performance based on mutual trust and confidence and involve expectations of loyalty* (Yam Seng Pte Ltd v International Trade Corp 2013).

The essence of the T5 Agreement was to formalise positive problem-solving behaviours in a culture that would not accept things going wrong as they invariably do on major contracts.

Added to this was the client's decision to retain all project risk and to pay for the works on a cost-reimbursement basis with ring-fenced profits and a profit-sharing arrangement to encourage innovation and efficiency. Tier 1 suppliers were expected to engage on similar terms with their subcontractors and suppliers so that the entire project supply chain adopted a problem-solving, cooperative and win–win attitude.

4.5 Project Management

Project management is a subset of management. The UK APM explains that management *is an ongoing process whereas project management has **a final deliverable** and a **finite timespan**.*[1]

Project management includes the integration of the various phases of a project's life cycle and the application of methods, tools, techniques and competencies to a project. This is accomplished through processes which Walker (2015) explains involves *the planning, coordination and control of a project from conception to completion.* This could be *on behalf of a client* or could apply equally to a contractor.

When things go wrong – as they inevitably do – then it is the quality and agility of management that differentiates successful projects from those that fail.

4.5.1 Decision Making

Throughout the project management process, innumerable decisions are made by many people at all levels of the project pyramid, including strategic decisions at client level and day-to-day operational decisions at site level. This is illustrated in Figure 4.8.

It can be seen from Figure 4.8 that decision making is a two-way process, as teams work much better by collaborating rather than responding to autocratic instructions. Each level of the hierarchy has its role to play but decisions and information that cascade down an organisation structure are invariably ill-informed or incomplete.

Inevitably, those at the construction workface will seek clarification and elaboration of practical issues – such as missing information, conflicts in drawings or models, unexpected events or inconsistent ground conditions – and will upload such questions through the project hierarchy as illustrated in Figure 4.8.

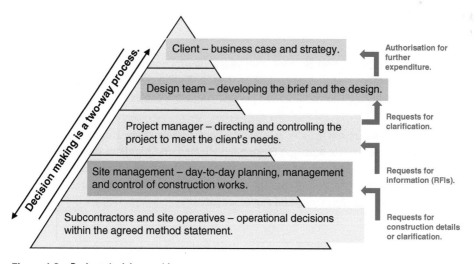

Figure 4.8 Project decision making.

> The person directly responsible for the project will be part of a team and will be responsible for managing the project team as well as being answerable to senior management. This applies to project managers for both client and contractor.

4.5.2 Project Managers

Successful projects need a driving force and a guiding hand – someone to champion the project. There may be several such champions, including those championing the design and 'soft landings' which ensure a smooth transition for the client from the construction period to occupancy and use.

> Vinci Building has been named as a building safety 'champion' by Building a Safer Future (BSF) – an initiative launched against the backdrop of the Grenfell tragedy aimed at raising standards and promoting culture change in the built environment.

For a project such as the London Shard, it was the developer – Irvine Sellar – who provided the drive, but the project needed an equally strong, determined and committed project manager – Bernard Ainsworth[2] – to deliver the completed scheme.

Not all projects are as successful and groundbreaking as the Shard, of course, and there are many examples of projects where weak leadership and ineffective project management have led to disappointing results for the client. In the public sector, especially, long chains of command, too many 'fingers in the pie' and excessive bureaucracy can suffocate a project and lead to lengthy delays and eye-watering overspending.

Even at a domestic level, clients can lack focus. They may not know exactly what they want to achieve, they make the wrong appointments, and they may take poorly informed decisions that directly impact on the project outcomes. This can be true of larger projects too.

> Construction industry reports over the years have consistently singled out the inefficiencies created by incomplete designs, lack of design coordination and too many design changes during the construction period. These are project management issues.

Project champions are not usually found at the construction stage. However, the contractor's project manager or site manager is often a dedicated and highly motivated individual whose interests extend not only to ensuring a successful and profitable outcome for the contractor but also to making sure that the client is happy with the completed project and with the contractor's performance and customer focus.

4.5.3 Client's Project Manager

A client project manager acts as an intermediary between the client and the project team. Their duty is to ensure that the client's needs are met whilst also making sure that the project is progressing in the right direction.

This includes looking after the client's interests and addressing any potential obstructions or misunderstandings before they become major issues. The client's project manager is focused on **managing expectations**:

- Liaising with stakeholders and third-party interests, such as neighbouring properties and local authorities.
- Design management and coordination.
- Value engineering and innovation.
- Estimating the impact of design changes and variations to the contract on the schedule and the client's budget.
- Advising the client on the projected cost to complete.
- Managing project risks.

4.5.4 Contractor's Project Manager

The role of the contractor's project manager/site manager is altogether different to that of the client's project manager. Whilst both are naturally focused on project delivery, the contractor's project manager is concerned with **managing resources**.

The contractor's project manager is the contractor's official representative on site through whom directions and information from the client's representatives are addressed. The job title may also be site manager, construction manager or site agent.

On large projects, there may be several site managers, site agents or sub-agents working under an ultimate project manager. In this case, the project manager's role is more strategic than that of the site manager.

> A £100 million development of 1300 purpose-built student accommodation (PBSA) units consists of six tower blocks and associated external works. A site manager is appointed for each block and another for the external works, all reporting to a project manager.

The project manager will be concerned with:

- Dialogue with the client's representative/contract administrator.
- The project schedule.
- Contractual issues.
- Financial performance of the project.
- Ensuring that the necessary resources are available to complete the project satisfactorily.
- Arranging for plant and temporary works such as scaffolding, formwork, access equipment, tower cranes and specialist lifting equipment.

The site manager – there may be several on a large project – will be concerned with the day-to-day management of the site:

- Short-term planning of activities.
- Making sure that progress is in accordance with the master schedule.
- Managing subcontractors and trades.
- Managing health, safety and welfare.

- Materials control and waste management.
- Site security, especially where there is a risk of plant theft.
- Keeping site records, especially concerning additional work, variations to the contract and delays due to bad weather.

Site managers often have little or no involvement in the financial side of projects, but invariably, the 'buck' stops with them in any case. Site managers normally report to a contracts manager or contracts director, who is ultimately responsible for the project outcome.

The site manager's job is a difficult one because, whilst contractors are naturally keen to deliver projects within their clients' expectations, there are competing interests. The overriding motivation for contractors is to achieve a profitable conclusion for the company. Construction businesses need to be profitable to survive, to satisfy shareholders and to develop and grow.

There is some common ground between the client's project manager and that of the contractor – the project schedule. Each party will be keen to avoid time slippage and consequent delay to completion which can equally impact the client's budget and the contractor's profit margin.

4.5.5 Project Manager Duties

Apart from the interpersonal qualities needed by a project manager and the ability to motivate, drive and control project outcomes, the project manager must ensure that all the work necessary to complete the project is done. This involves the interaction of many organisations, people and data, as illustrated in Figure 4.9.

For the client's project manager there will be considerable involvement in design briefing, establishing the project budget, dealing with a variety of stakeholders and ensuring that the design develops according to the procurement arrangements chosen. The project manager will also be concerned with contractor appointments, site mobilisation and construction right through to project handover/close out and occupation.

The project process generates a multiplicity of information of different types – specifications, quantities, 3D drawings, models and schedules, including the project schedule, the contractor's master schedule and numerous short-term schedules – which has to be managed.

Figure 4.9 illustrates these complexities and demonstrates that there is much more to a project than simply the physical work on site. For instance, some of the 'behind the scenes' work which the contractor's project manager might be involved with includes:

- Producing and updating the project schedule.
- Recording progress and dealing with delays.
- Determining and controlling the project budget.
- Dealing with information requirements that might not be clear from the drawings or Building Information Modelling (BIM) model.
- Ensuring that subcontractors are suitably resourced.
- Ensuring that materials and temporary works arrive on site at the right time.
- Attending meetings and responding to the outcomes.

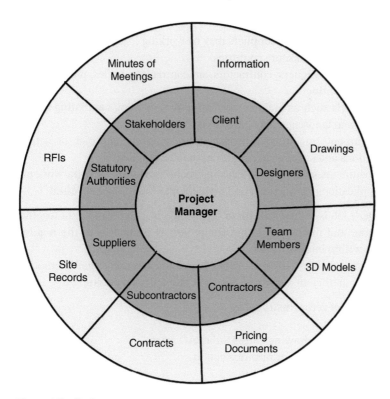

Figure 4.9 Project management.

- Requesting information (RFIs).
- Identifying potential problems before they happen and notifying the contract administrator.
- Making sure that extra work and variations are recorded and verified.

The last thing that a project manager needs is to be in Zoom meetings or be sat behind a computer all day, therefore, much of this work will be delegated. The project manager must nevertheless remain in touch with the many workflows involved because they are concerned with getting the job done.

4.6 Collaboration

The secret of successful project management is teamwork. With genuine 'buy-in' from team members – notwithstanding the natural self-interests of the participants (clients, contractors, etc.) – the aspiration of working in a true *spirit of mutual trust and co-operation* (NEC4 clause 10.2) – can be achieved.

Construction projects are complex, often involving multiple players as illustrated in Figure 4.9. For a modest industrial or commercial building project, for instance, there could be many contributors and many different sorts of data are generated in the project process which have to be effectively managed:

- 30+ businesses may be involved.
- Designers, contractors, specialists or suppliers may be working in separate geographical locations, or perhaps from two or more countries.
- 300–400 people, including designers, contractors, subcontractors, trades, plant hire and suppliers could be contributing.
- Several 3D design models such as architectural, structural, heating ventilating and air conditioning (HVAC) and landscape may have been created.
- 50 000+ shared documents including time schedules, bills of quantities or activity schedules, specifications and subcontract quotes have to be handled.
- 100 000+ formal documents such as architect's instructions, records of extra work and variations, RFIs, applications for payment and invoices may have to be processed.

The project manager's job is to bring all these organisations, people and data together, with a unity of purpose and direction, so that the project is completed to the required standard, on time and within budget.

This requires collaboration, but traditional paper-based and standard email methods of communication are very inefficient and waste time and effort. This includes:

- Ad-hoc communication of verbal instructions.
- Mountains of paper records.
- Dozens of lever-arch files and drawing chests full of A0 prints.
- Chasing subcontractors for quotes or resource details.
- Tracking down lost information.
- Submitting RFIs and confirmation of verbal instructions.
- Early warnings of potential problems on site.

Even where participants are using collaborative technology – such as OneDrive, SharePoint, Dropbox – this might not be on a project-wide basis. This does not provide a common collaborative platform and is, therefore, inefficient and prone to error.

4.6.1 Common Data Environment

Common Data Environments (CDEs) are Cloud-based spaces where all kinds of project information can be stored and made accessible to all project participants – including drawings, models, quantities, specifications, RFIs and reports. They can also be used to manage and collect information during construction and update BIM model data, ensuring that it remains aligned with reality – a Digital Twin.

CDEs ensure that all stakeholders – clients, designers, contractors, subcontractors etc. – have access to a single and up-to-date version of project data and provide a means for tracking and recording changes. Access depends on participants' requirements, level of authorisation and contractual obligations.

CDEs improve communication and collaboration between project team members and provide easy real time access to information at all stages of the project. This is a coherent way to connect people and data efficiently and comply with accepted standards (such as EN ISO 19650 and ISO 12006-2).

Each organisation – design consultant, Tier 1 contractor, subcontractor, etc. – has a secure individual workspace within the CDE, both for the duration of the project and as an archive following project completion. All individual workspace data can be controlled and shared with other stakeholders as required.

> CDEs provide a genuine single source of truth in a secure, auditable and easy to use platform. Strictly speaking, a platform that integrates several software packages in one place is not a CDE.

A simplified illustration of a CDE dashboard is shown in Figure 4.10 where it can be seen that a centralised project-wide interface can make many types of project data available in one place:

- BIM models.
- Documents, including drawings, specifications, bills of quantities.
- Cost data and site records.
- Bid documents including subcontract orders.
- Materials data sheets, installation manuals and Control of Substances Hazardous to Health (COSHH) information.

A particular feature of CDEs is the project-wide email interface and contacts directory where emails can be generated and answered within the data management system. Unlike ad-hoc emails, this facilitates complex email threads, various email types (advice, site instruction, RFI request, request for clarification, RFI response and transmittal), inclusion of a 'response-by' date and sophisticated mail search. A secure and standardised system provides much more consistency, increases productivity, reduces the chance of errors and reduces the chance that important messages are missed.

CDEs may be stand-alone or as a module in both Enterprise Resource Planning (ERP) Software and Construction Management Software.

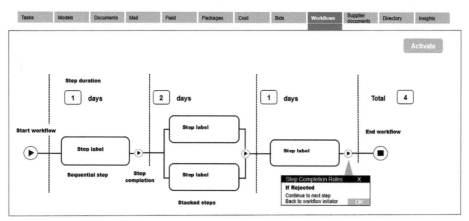

Based on Oracle Aconex

Figure 4.10 Common Data Environment dashboard – workflows.

4.6.2 Enterprise Resource Planning

ERP is a software/system that integrates a company's business processes – accounts, payroll, planning, estimating, procurement, risk management, supply chain management, project management, etc. – and allows them to be managed, often in real time.

ERP is a category of business management software that contains a suite of integrated applications that allow organisations to collect, store, manage, interpret and use data from a variety of business activities. Whilst Cloud access offers greater collaboration and efficiency, an Internet connection is not essential as some ERPs can be used locally.

ERPs facilitate integration of the numerous processes involved in running a business and provide a means for data to flow between them, thereby eliminating duplication and wasted effort and, importantly, data integrity and a single version of the truth.

> Whilst the construction industry is known to be reactionary and slow to accept change, some ERPs offer great benefits because they are construction industry specific and may be tailored to integrate existing software systems already being used. Consequently, third-party accounts packages, project planning software, estimating and measurement software and so on can be coded to the ERP system without loss of functionality.

Some ERP software is generic such as SAP S/4HANA, MS365, Syspro and Oracle Cloud ERP whilst others are specifically designed to serve the construction and engineering sectors, such as EQUE2 and 4PSConstruct, both of which harness the power of Microsoft Dynamics.

Modules within 4PSConstruct include:

- Estimating, bids and tenders.
- Project budgets.
- Valuations, variations and cost-value reconciliation (CVR).
- CIS and HMRC online interfaces for subcontract payments.
- Subcontract management.
- Plant hire and plant management.
- Project bank accounts and so on.

4PSConstruct also facilitates project management – including project planning, programming, budgeting and document management – and includes a collaborative CDE facility.

4.6.3 Construction Management Software

Whilst ERPs can be used to manage the business and its projects, construction management software is designed exclusively for project management but not business management. Such software will contain several modules, as illustrated simply in Figure 4.11:

- **CDE**: a common data environment where all project data is in one place.
- **Field**: real-time data capture offline on site which can be uploaded later.
- **BIM**: allowing visualisation of models throughout design, construction and occupancy of the project.

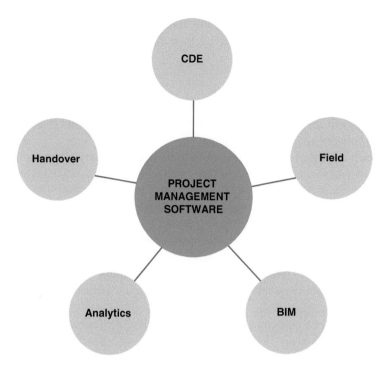

Figure 4.11 Construction management software.

- **Analytics**: permitting real-time reporting, coordinated decision making and project tracking.
- **Handover**: structured project data management, built asset information and health and safety files.

Construction management software provides a single repository for all project documentation and a toolset that facilitates project management tasks such as setting schedules and timelines, managing budgets, quality and safety management, completing reports and communicating and coordinating with all project stakeholders.

> Construction management software is to be contrasted with project management software. This is specific software – such as Powerproject and Project Commander – for project planning, scheduling, resource management and the management of change but not for document management or the broader aspects of managing project data.

Unlike most ERPs, construction management software is connected project technology designed specifically for the construction industry. Market leading software includes Oracle Aconex, Zutec, Autodesk Construction Cloud, Procore, Trimble e-Builder and ACCA usBIM.platform.

In common with ERPs, construction management software can have a CDE module embedded, and this gives the facility to allow client team access to parts of the software – estimating, accounts, subcontractor quotes and the like – on projects where open-book accounting or early contractor involvement procurement is used.

4.6.4 Workflows

Project data is increasingly managed by using document management software to make data handling more efficient and less prone to error. This creates a CDE thus enabling collaborative working between all project participants.

There are challenges in using these systems as some project participants may still be using old-fashioned methods of data management or they may well have their individual ways of working that do not align with the client or main contractor. Some organisations have no formal data management processes at all, or those that they have might be based around a project management tool, such as project scheduling software.

Data management processes bring considerable benefits as tasks, information and documents can be passed from one person to another for action according to a set of procedural rules. They remove the chance that important tasks might not be done at all and ensure that document transmittals are subject to a rigorous 'gateway' and audit trail regime.

Tasks within collaborative platforms are based on workflows which are formal rule-based processes for getting the work done.

To start a workflow a template is created which can then be used to create other workflows. Figure 4.10 illustrates the 'workflows' tab within a simplified CDE dashboard:

- A workflow consists of a series of steps which can be sequential or stacked as shown.
- Once initiated, the steps are punctuated by step-completion radio buttons which contain step-completion rules that determine how the workflow progresses.
- Once completed, the workflow is activated and sent to appropriate team members.

Workflows create a formality to the initiation, progressing and finalisation of tasks within the CDE. Such tasks could include the checking and approval of drawings and design changes, revisions to the project schedules or the verification of site records and so on.

Templates define the workflow steps and the rules applicable to workflow issue, review and final transmittal. It is vital that the project team buys-in to the workflow process which, if successfully executed, can result in seamless document transmittal and sign-off, leading to 'one version of the truth' on the project.

4.7 Design Management

Process models, such as the RIBA Plan of Work, identify the various phases of the project life cycle. Design is one of these phases, and it must be integrated with others if the project is to be successful.

Whilst clients require certainty over time and cost especially, managing construction projects is a complex process with many facets. This can make it difficult to achieve the desired outcomes. The design process is also complex with many intervening factors – planning permission, environmental issues, external stakeholder interventions, technical issues and even public inquiries. Consequently, it is usual practice for both client and contractor project management teams to include a design manager.

> Design management is a specialist form of project management which requires considerable direct or indirect experience in the design process. This is not simply a matter of coordinating design information but of driving the design process towards a conclusion. The design management role might be undertaken by qualified architects or engineers but could equally be carried out by other suitably experienced construction professionals.

The contractor's design manager will invariably be involved in the planning and scheduling of a project as well as with the development of design models (BIM). In this respect, BIM is much more about the management of information than aesthetics and is an essential element in the development of the contractor's bid.

On any given project, there may be several design managers working with the client's design team, with the contractor's design team or with specialist subcontractor design teams. Design management is a subset of project management and contributes significantly to the overall project outcome.

4.7.1 Design Manager

Design managers coordinate design-related matters on construction projects, which helps to keep design work running smoothly and ensure that design information is produced in a timely fashion. They may also be referred to as design coordinator, technical manager, BIM manager or coordinator, design leader, project manager, pre-construction manager or bid manager.

A design manager's precise role depends upon the size and scope of the project, the chosen procurement method and who the design manager is working for.

In design-bid-build procurement, the design is usually carried out by architects and engineers appointed by the client. The lead designer traditionally manages the design, but it is more conventional nowadays to appoint a design manager who works with the design team and the BIM manager.

A client-side design manager works closely with the client's design team in order to implement the client's brief. This includes representing the client's interests in design meetings and coordinating design details between the architectural, structural, HVAC and other design team members. The design manager will work closely with the BIM manager and report to the client's lead designer whose role is to steer the design in accordance with the client's expectations.

Designs are rarely complete at tender/contract stage and the information that the contractor initially receives is a design intent rather than a design suitably detailed for construction.

> Design intent provides the contractor with sufficient information to plan and organise the works but not enough to build from. Further information is needed in the form of product data sheets and the like and a limited degree of contractor design in the sense that any builder – or tradesman – is the designer of last resort.

It is, therefore, conventional for there to be an element of contractor design in traditional procurement – either carried out by the main contractor or by specialist subcontractors. This design work will either be detailing the design intent or can be the design of a specific element of the project – such as roofing/cladding, engineering services, suspended ceilings and so on.

This design work needs to be coordinated – the services and suspended ceilings designers need to talk to each other, for instance – and it is, therefore, usual for the contractor to appoint a design manager, design coordinator or pre-construction manager. The client's design manager needs to work closely with the contractor's equivalent.

There may, consequently, be several design mangers on any given project working for different parties. Specialist subcontractors – perhaps responsible for the design of piling, structural frame, lifts and conveyors and so on – may well have their own 'in-house' design managers as well.

4.7.2 Pre-construction Manager

'Pre-construction manager' is a commonly used term in contracting which is analogous to the contractor's design manager. There is no precise definition of the role, however, which can vary from firm to firm.

The job entails working with the multi-disciplinary teams involved in a project including both in-house and consultant designers and subcontract designers as well as those involved in the planning, scheduling and pricing of the work.

Pre-construction managers are usually involved both pre- and post-bid submission. They may also continue their involvement until construction is complete to ensure that the project runs smoothly and is delivered on time and within budget.

Their job not only entails driving the design and resolving issues but may include involvement in procurement so that the on-site construction team can make an immediate start to the construction works.

The role of pre-construction manager contrasts with that of the client-side design manager who coordinates the work of the client's design consultants and architects etc.

4.8 Project Management Models

Project management models focus on the abilities and competencies needed by project managers at various stages of projects rather than tasks and workflows. They should be distinguished from process models – such as the RIBA Plan of Work – which depict the tasks and workflows in a project.

4.8.1 APM Competence Framework

The UK APM Competence Framework establishes the 29 competences needed for effective project, programme and portfolio management.

It focusses on the abilities needed by project managers for successful project delivery including the ability to:

- Develop a business case.
- Procure the necessary resources.
- Deal with conflict.
- Undertake time-based planning.
- Monitor and manage supplier[3] performance.

4.8.2 PMBOK Guide

The PMI is a US organisation which publishes several practice standards related to the management of projects, programmes and portfolios including standards related to competencies.

Its project management body of knowledge (PMBOK) Guide is a well-known project management good practice guide which relates to project management principles rather than processes.

4.8.3 Prince2

Prince2 – which stands for **Pr**ojects **In** **C**ontrolled **E**nvironments – is a project management methodology that operates within a defined framework. It provides the mindset for managing projects and is used worldwide on projects large and small. It is not construction specific.

This structured methodology is based on seven themes, seven principles and seven processes that respectively provide the 'what', 'why' and 'how' of successful project management. The themes of Prince2 comprise business case, organisation, quality, plan, risks, change and progress.

The principles of Prince2 include learning from experience, managing by stages and management by exception and, if followed, helping to keep a project aligned with the Prince2 methodology. The seven processes that run alongside the Prince2 principles extend from project start-up through directing, controlling and managing project delivery to eventual project closeout.

Whilst Prince 2 is a scalable project management methodology, it is recognised that project management is a complex discipline requiring special personal qualities, education, qualifications and experience. Prince2 learning programmes can lead to graduated qualifications at foundation, practitioner and agile practitioner levels.

4.9 Project Stages

All construction projects – large and small – go through a process from initiation to completion and thence to use or occupancy. When they have outlived their usefulness, buildings or structures may then undergo adaption, extension, demolition or replacement.

For there to be the chance of a successful outcome, projects need to be planned in detail which means that planning the pre-construction stages – i.e. determining the clients' requirements, agreeing the brief and preparing designs – is equally as important as planning the construction stage.

4.9.1 Process Models

In order to guide project teams and to define construction project work stages, reference is often made to a process model – such as the RIBA Plan of Work or the Office of Government Commerce Gateway Process.[4]

Process models graphically represent the various steps in a project and, usefully, identify the tasks to be undertaken and the workflows needed to bring the process to a conclusion. They are commonly used in a number of countries around the world including several European countries, Scandinavia, Australia, New Zealand, South Africa and the United States.

The various models differ, sometimes significantly, but the main goal is the same – to provide guidance to clients and to promote a consistent approach for project teams. Some models include a tendering stage whereas others do not refer to procurement at all. In some cases, models do not address the post-construction and occupancy stages of projects, nor is there any overall consistency in the number of design stages included or the level of detail in the models.

4.9.2 Types of Process Models

A process model in construction is a visual depiction – usually diagrammatic – of the workflows that contribute to a project. There are several in common use:

- The Office of Government Commerce Gateway Process.
- The RIBA Plan of Work.
- The Project Control Framework used by National Highways (NH).
- The Guide to Rail Investment Process (GRIP).
- The Construction Playbook.
- The TfL Pathway – a project and programme methodology used by Transport for London (TfL).
- A cost estimating process model used by the UK Government's Infrastructure and Projects Authority to provide best practice guidance for cost estimating infrastructure projects and programmes.

4.9.3 OGC Gateway Process

Whilst the Office of Government Commerce (OGC) no longer exists, and its guidance has been archived, the OGC Gateway Process is still widely used to examine individual projects and programmes – such as capital investment programmes or schemes comprising several projects. The OGC Gateway Process is illustrated in Figure 4.12.

It can be seen that reviews take place at five key decision points in the project lifecycle. Their purpose is to look ahead so that one 'gateway' cannot be passed without assurance

4.9 Project Stages

Gateway review 0 →	Gateway review 1 →	Gateway review 2 →	Gateway review 3 →		Gateway review 4 →	Gateway review 5 →
Strategic assessment	Business Justification	Delivery Strategy	Investment Decision		Readiness for Service	Operational Review and Benefits Realisation
			Design Brief and Concept approval	Detailed Design Approval		

Figure 4.12 OGC Gateway Process.

that the project(s) can progress successfully to the next stage. This process represents best practice in UK central and local government, the health sector and in defence.

Some major contractors also use a 'gateway' approach to managing risk and opportunity throughout the project, from tender invitation to the end of the defect period.

This involves several mandatory approval and review gateways at chosen points in the process which can only be passed with senior management/board approval based on a proper assessment of risk and reward. In the early stages, Go/No-Go decisions are made at tender invitation and pre-tender submission stages prior to making a commitment to enter into a contract.

4.9.4 RIBA Plan of Work

The RIBA Plan of Work is, perhaps, the best-known and most widely used process model.

It is used mainly in building work and organises the project process into eight stages from inception to completion. This process model focuses on key tasks, sustainability checkpoints and information exchanges as the project develops. Uniquely, the RIBA Plan of Work identifies the stage outcomes, core tasks and information exchanges required at each stage.

This process model, illustrated in Figure 4.13, mimics the workflows or stages of a project rather than decision points, but the 2020 version is more flexible than previous editions. The model does not have a 'tender action' stage as it is recognised that several procurement routes could be employed for any given project and that there may be no specific tender period as such.

Both procurement and programme/schedule tasks are allowed to 'float' or overlap between the project stages to allow for alternative procurement strategies to be adopted. This is illustrated in Figure 4.14.

RIBA Plan of Work 2020							
Options:							
0	1	2	3	4	5	6	7
Strategic Definition	Preparation and Brief	Concept Design	Spatial Coordination	Technical Design	Manufacturing and Construction	Handover	In Use
←—— Procurement and Programme vary according to the Project Strategy ——→							

Figure 4.13 RIBA Plan of Work.

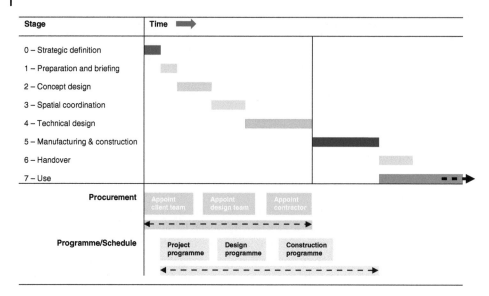

Figure 4.14 Overlapping project stages.

4.9.5 The Guide to Rail Investment Process

The GRIP is a project model[5] comprising eight stages. It is illustrated in Figure 4.15, where it may be compared to an equivalent process model used by National Highways.

Network Rail owns and operates some 20 000 mi (32 000 km) of track, 30 000 bridges, tunnels, viaducts and thousands of station-level crossings and signals in Great Britain (England, Scotland and Wales). This includes 20 of the country's largest stations.

Much of the UK rail infrastructure is at or near 'end of life' as many of its structures, buildings and earthworks were built between 1850 and 1920. Government funding for the operation, maintenance and renewal of the infrastructure amounts to over £40 billion covering a 5-year 'control period', and so it can be imagined that many projects, both large and small, building and civil engineering, are undertaken by Network Rail.

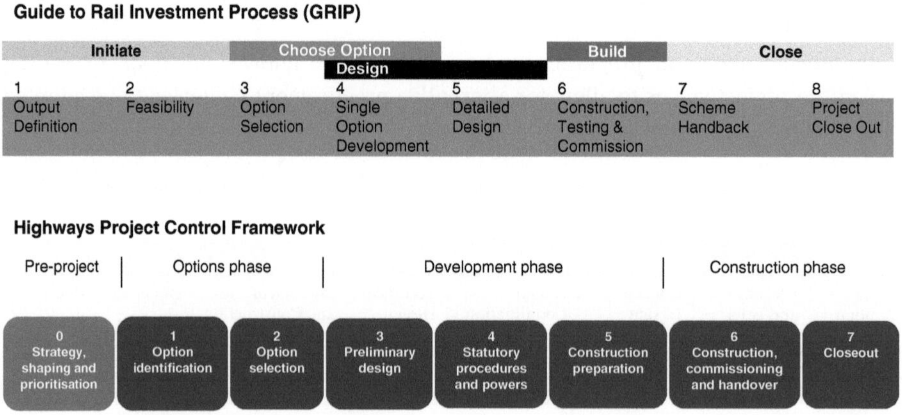

Figure 4.15 Infrastructure process models.

The GRIP model is used to manage and control such rail investment projects but, rather like the NH Project Control Framework, the underlying approach is product rather than process driven.

This means that emphasis is placed on matters such as objectives, scope, timing, option development and stakeholder involvement rather than on the workflows, key tasks and checkpoints used in the RIBA Plan of Work.

4.9.6 The Project Control Framework

National Highways is a government-owned organisation that is responsible for operating, maintaining and improving England's motorways and major A roads. It manages over £150 billion of assets with an annual expenditure on road schemes, operations, maintenance and renewals in the order of £4.5 billion.

The NH Project Control Framework – illustrated in Figure 4.15 – sets out how NH and the Department for Transport (DfT) manage and deliver major improvement projects costing in excess of £10 million.

There are four main phases to the project control framework:

1. **Pre-project phase**: this is when a strategic outline business case is produced for a transport problem (such as a bridge replacement) which takes place before a commitment to investigate options is made.
2. **Options phase**: this is when the preferred solution – such as a road widening scheme – is identified.
3. **Development phase**: this is where a commitment to develop a design for the preferred solution is made, the necessary statutory processes are followed and where a commitment to invest in construction is made.
4. **Construction phase**: this is the stage at which the preferred solution is built and handed over, at which point the project is closed down.

This framework provides a standard project lifecycle of eight stages, together with deliverables, processes and governance arrangements. It is used by the DfT, NH and other highways directorates and by NH project managers.

4.9.7 The Construction Playbook

The Construction Playbook is UK government guidance for sourcing and contracting public sector projects and programmes. It is not strictly a process model, although it aims to embed its principles and policies in the way projects are assessed, procured and managed.

The Playbook sets out five stages designed to achieve the successful delivery of public sector projects and programmes:

- Preparation and planning [of the project].
- Publication [tendering].
- Selection [due diligence].
- Evaluation and award [of the contract].
- Contract implementation [construction and close out].

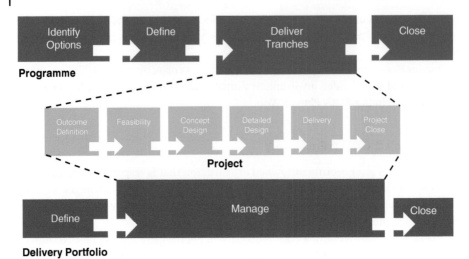

Figure 4.16 TfL delivery environment.

4.9.8 TfL Pathway

TfL is a large, public-sector organisation with an annual income in the order of £8 billion. It controls an investment programme of circa £2 billion which covers London Underground and Docklands Light Railway as well as tram and bus infrastructure, new river crossings, road improvements, station upgrades, cycle superhighways and so on. It also owns and runs the Elizabeth Line (Crossrail).

In order to control its many programmes and projects, TfL has developed the TfL Pathway[6] – a project and programme methodology – which is based on ISO 21500:2012: *Guidance on Project Management* and on an adaption of Maslow's hierarchy of needs.

The TfL Pathway comprises three key elements:

- **Organisational strategy**: business planning, strategy and service development, a 20-year look-ahead at London, its population growth and what needs to be done to meet future needs.
- **Operational business**: managing day-to-day TfL business operation and maintenance activities which keep services going.
- **Delivery environment**: project and programme delivery to meet TfL strategic requirements. The delivery environment is the TfL pathway project/programme delivery process.

The delivery environment identifies the programme, project and delivery portfolio life cycles and how they are managed from planning to realisation of benefits. This process model is illustrated in simplified form in Figure 4.16.

4.9.9 Infrastructure and Projects Authority Cost Estimating Guidance

A cost estimating process model used by the UK Government's Infrastructure and Projects Authority aims to provide best practice guidance for cost estimating infrastructure projects and programmes.

The model consists of eight steps – briefing, data collection, estimating methodology, base cost estimation, cost reporting, review and assurance, project sign off and use of the cost estimate in decision making. At the centre of the process is the equation:

$$\text{Base estimate} + \text{Uncertainty} + \text{Risk} = \text{Anticipated Final Cost} \pm \text{Variable range}$$

This guidance is intended for use on major road and rail infrastructure projects and recognises that deterministic, probabilistic and actual cost methods of estimating will be used as a project develops from brief, through business case development and on to estimate maturity.

In common with the Construction Playbook, this guidance is less process model and more good practice, but it does contain useful flow charts and diagrams to illustrate the cost estimation process.

4.10 Digital Construction

Digital Construction is an evolution of BIM. It facilitates the production and management of information relating to construction projects throughout their entire life cycle, helping to create digital and built assets.

There are enormous benefits to be gained from digital construction, including the enhanced connection of people, process and technology, improved consistency and decision making, smoother transition between the design, construction and operation stages of projects and reduced risk for all stakeholders.

4.10.1 Building Information Modelling

BIM, alongside other digital technologies, is gradually transforming the way in which construction projects are conceived, transacted, planned, managed and operated throughout their life cycle.

BIM models are built on 'information' and, therefore, BIM could equally mean Building Information **Management**. This information creates the basis of a common language which can be used to exchange data, and facilitate communication, between all participants in a project in a way that has never been seen before.

BIM can help to improve project coordination and increase buildability confidence using 3D and 4D imagery to demonstrate construction sequences and identify risks early in a project. Models can also be used from the early stages of a project to determine and communicate information concerning site access, storage and laydown areas, welfare facilities and site working areas.

A key advantage of designing in a BIM environment is the avoidance of conflicts in the design which traditionally occur when architects, engineers and other designers are working independently. Such conflicts or 'clashes' can theoretically be avoided if all designers are working on the same design or 'model'. There is, however, a caveat to this as much depends on which level of BIM is being adopted. It is far from the case in practice that designers will be working from a fully federated model or a single version of 'the truth'.

4.10.2 BIM Models

BIM models may be defined as:

> *A digital representation of physical and functional characteristics of a facility creating a shared knowledge resource for information about it and forming a reliable basis for decisions during its life cycle, from earliest conception to demolition.*[7]

A more succinct definition of BIM might be:

> *A means of developing an intelligent 3D model of a building or engineering structure on the basis of team collaboration.*

BIM models are a completely different proposition compared to 2D and 3D computer-aided design (CAD) designs. CAD drawings are based on lines and polylines whereas BIM models consist of parametric objects that carry technical, geometric and other data. The result of this is that BIM models are 'intelligent' whereas CAD drawings are 'dumb' because the lines that make up such designs do not carry any data.

There are various BIM acronyms – 4D BIM (3D parametric model + time schedule), 5D BIM (3D model + costs) and 6D BIM (3D model + facilities management).

> In the context of this book, 4D BIM enables construction planners to work with BIM models directly alongside the emerging project schedule and to run visual animations of the construction timeline.
>
> https://youtu.be/Sz6JK6_ZtGA

4.10.3 Benefits of BIM

BIM brings enormous benefits for project management, planning and scheduling. The ability to import BIM models into project management software – such as Asta Powerproject – means that a Gantt chart schedule can be developed alongside a 3D model of the building or other structure. Linking models to time schedules is known as 4D BIM.

In Asta Powerproject 4D, models can be imported from native software such as Revit, Archicad or Tekla directly into the scheduling software by using the Industry Foundation Class (IFC) file format. Figure 4.17 illustrates a BIM model linked to a Gantt chart in Asta Powerproject 4D and shows the schedule (left), Gantt chart with animation timeline (centre) and the BIM model (right). IFC model properties may also be displayed for dragging and dropping in order to create the schedule (not shown).

Asta Powerproject 4D facilitates the creation of a Gantt chart by dragging and dropping objects from the 3D view into the bar chart task view. Schedules can also be created by using structured templates with pre-defined searches that automatically link to the BIM model objects. Asta Powerproject 4D also enables a project timeline to be created which can then be played to simulate the build process linked to milestones and baselines. Animations can be exported to video.

4.10 Digital Construction | 125

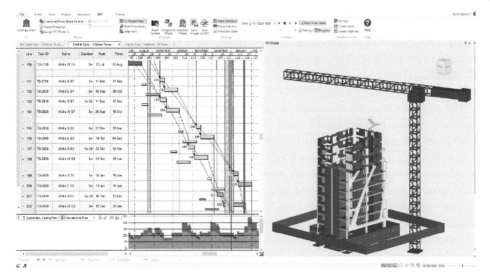

Figure 4.17 Asta Powerproject 4D. (Photograph courtesy of Eleco UK Ltd).

This facility can be used to simulate specific activities, elements or entire projects meaning that planned and actual schedules can be compared in 3D. Model views can also be saved showing how the project looks or should look at specific dates which can aid 3D pre-tender planning or progress control when on site. The IFC files used in Asta Powerproject 4D can be hosted in the Cloud or locally thereby facilitating greater collaboration. By importing costs from estimating software such as Bidcon, 5D BIM is facilitated.

With Asta Powerproject 4D, additional models can be imported into the software to reflect site objects such as hoardings, site accommodation, tower cranes and other construction plant.

Using 4D BIM, the project planner can 'play' with different build sequences to determine the desired construction order taking into account available resources, scaffolding and access requirements, crane use and so on. Site traffic movements can also be animated to demonstrate proposed site ingress/egress and plant and traffic movements around the site which is helpful for ensuring compliance with health and safety legislation.

As an alternative to using a single application such as Asta Powerproject 4D for planning in 4D BIM, the time tracking facility in BIM software such as Autodesk Navisworks or Bentley Synchro can import time schedules from project management software such as Primavera P6 or Microsoft Project. This allows BIM model objects to be linked to the time schedule which can then be exported back to the project management software package.

By linking 3D models with the project schedule, the build process can be animated to provide real-life walkthroughs, models can be filtered, coloured and decluttered by using the property information in the model to show only what is necessary and construction teams can simulate the build process to anticipate and resolve problems before work is commenced on site.

4.10.4 BIM Standards

Many countries around the world are developing BIM strategies and, in doing so have created, or are creating, their own standards to define project information requirements and the standards, methods, processes, deadlines and protocols needed to govern its production and review.

The purpose of such standards is to prevent wasted effort by the supply chain in generating **too much** information and to help clients, owners or project sponsors to avoid taking uninformed decisions about their projects/assets based on **too little** information.[8]

The British Standards Institute (BSI) *has been at the forefront of BIM since the start, developing a range of BIM standards covering all the key elements from design, information management, facilities management, and security.*[9] Until recently, the standards used in the UK were BS 1192 (2007) and PAS 1192-2.

Although some existing standards remain, these will eventually be replaced by BS EN ISO 19650 standards which, although based on UK standards, have been agreed internationally to reflect different approaches.

4.10.5 BIM Levels

The 2011 Government Construction Strategy introduced the requirement for fully collaborative BIM (BIM Level 2) by 2016 on all public sector projects. This mandate brought about increased awareness and an appreciation of the benefits associated with BIM adoption in construction.

At that time, BIM was represented as being in four levels as demonstrated by the Bew-Richards BIM Maturity Wedge (2008). The four levels (0–3) were:

- Level 0 represents 2D CAD.
- Level 1 is 2D CAD with a 3D conceptual model which facilitates visualisation of the completed design. Such designs are produced by single designers or contractors.
- Level 2 where 3D designs are prepared independently by members of the design team (including contractors). There is not necessarily a single model at Level 2, but individual models may be combined (federated) at some point in the process. The result is greater collaboration and Level 2 BIM was the minimum requirement stipulated by the UK government on all public sector projects.
- Level 3, being a fully digital design within a federated model, means that a fully collaborative design and construction process is feasible. Level 3 models enable rich data exchange between participants.

The current UK BIM strategy is defined by the UK BIM Framework[10] and is based on the emerging BS EN ISO 19650 standards and any remaining BS/PAS 1192 standards as they are phased out. Consequently, the UK government now requires all public construction projects to be performed by 'BIM Capable' organisations.

> Whilst no longer in official use, BIM Level 2 is still commonly referred to in the UK whereas ISO 19650 refers to the less easy to grasp stages of maturity of analogue and digital information management.

The UK BIM Framework recommends following ISO 19650 guidance. This means that information management and BIM model requirements for projects should be determined on a case-by-case basis, and according to specific consultant/contractor appointments. The objective is to ensure that there is no duplication or gaps in information delivery.

Even though ISO 19650 is based on UK standards, it contains a UK National Annex which reflects the changes in terminology and language needed to agree the documents internationally. ISO 19650 terminology refers to *appointed party*, *appointing party* and *appointment*, for instance, whereas UK convention is to use the terms *supplier*, *employer/client* and *contract/professional agreement*, respectively.

Another unusual ISO 19650 term is *information container* which, in the UK, means any form of unique file such as an ArchiCAD model, Excel spreadsheet, geometric model containing metadata, schedule, PDF document, photo or video, product data sheet and so on which, together, form an information model of a project.

The challenge now facing the industry is for BIM Level 2 to become 'business as usual' and for BIM Level 3 to develop from aspiration to reality:

- **Level 3A**: Enabling improvements in the Level 2 model.
- **Level 3B**: Enabling new technologies and systems.
- **Level 3C**: Enabling the development of new business models.
- **Level 3D**: Capitalising on world leadership.

Many designers, contractors and other organisations are already on the road to formal ISO 19650 (formally BIM Level 2) compliance which provides a means of validating their investment in BIM and their drive for business improvement via third-party assessment.

4.10.6 BIM Level 3 Benefits

Level 3 BIM means that not only is the same design model fully collaborative and accessible by all design contributors, but it is also accessible by the construction team. The benefits of this are clear:

- A design without clashes can be produced as a basis for tendering and project planning.
- Even though the design (model) may not be entirely complete, contractors and specialist subcontractors are able to develop the design as required by the terms of their respective contracts and they can contribute ideas for the overall buildability of the design.
- Planning of the construction phase can benefit from 3D visualisation, at any angle and in any plane, and the 3D model can be linked to scheduling software to:
  - Create 2D bar chart schedules directly from the elements comprising the model, either automatically or by using 'drag and drop'.
  - View a 3D model of the project alongside the Gantt (bar) chart.
  - Create and play a timeline of the project at the same time as seeing the model built from the ground up.
  - Run simulations of the project build to visualise the order and sequence of work in 3D and identify any inconsistencies.

- Add plant into the model – such as tower cranes, mobile cranes and vehicles in order to visualise plant movements and problems with reach or oversailing of cranes.
- Compare planned and actual progress.
- Save changes to the model, and the schedule, to the Cloud to facilitate data sharing and collaboration with all members of the project team.
- Derive data from the model, such as element properties and quantities.
- Share information between different software applications using the 'platform neutral' IFC file format.

4.10.7 BIM Downsides

Even though BIM has been around for some time, in the context of the entire construction industry it is still at an aspirational stage. Large design practices and contractors, however, are clearly fully on board, and many are at least at Level 2 and are combining the benefits of virtual reality as well.

In the case of smaller design practices and contractors, the story is a bit different as, in many cases, fees and tender prices do not justify the expense of Level 2 BIM. As the industry largely comprises small and medium-sized enterprises (SMEs), it is hardly surprising that historical issues – incomplete designs, design clashes, late information, poor project planning and time management – continue to plague the industry, resulting in disappointing outcomes for clients and inefficiency and poor profitability all around.

It might therefore be argued that large parts of the industry are in a BIM time warp, and it may take many years before the situation improves to any great extent. To make matters worse, empirical evidence suggests that smaller contractors and subcontractors tend to be unwilling to invest in the 'high-end' software scheduling packages capable of importing BIM models. The preference seems to be the use of more rudimentary methods of scheduling or, at best, less costly software packages that are not BIM-enabled.

4.10.8 BIM Managers

BIM managers – also known as BIM leaders – are increasingly employed by clients and contractors to act as collaborators between design teams, contractor teams and the supply chain. Their role involves overseeing the production of project information models containing 3D visualisations that enable data, drawings and schedules associated with the design and construction phase of a project to be brought together.

BIM Managers may also be responsible for delivering business technology strategies and for promoting the benefits of digital ways of working, ensuring that projects run smoothly and that project information models are delivered on time. They work closely with BIM coordinators who have the narrower responsibility for management of the digital process.

BIM managers may have many duties including:

- Acting as an intermediary between designers, clients and architects.
- Working closely with BIM coordinators who are responsible for producing project information models.

- Ensuring that BIM execution plans are produced and maintained for each project.
- Completing supply chain assessments.
- Carrying out audits and flagging up any non-compliances.
- Reporting to senior managers regarding BIM maturity and KPIs.

4.10.9 Digital Twins

A digital twin is an exact digital replica of a construction project, a building or group of buildings, a bridge, a highway or even an entire city. Digital twins may also be referred to as data twins, virtual models or next-generation as-built drawings.

Because it contains the necessary 4D (time), 5D (cost) and 6D (facilities management) data for a project, a digital twin can simulate the construction and operational phases of a built asset in virtual reality.

Such simulations are entirely relational and can be run in time-lapse so that many iterations are possible. This means that the viability of different methods and sequences of construction work can be tested to see where problems lie, thus enabling changes to be made.

> Two major engineering design practices are working together in a design joint venture on a mega infrastructure project.
>
> Due to the size and scope of the project, conventional means of collaboration and information management will not work.
>
> A digital twin is created to develop a connected data environment which enables all data to be managed and over 4000 design information models to be created, thereby saving time, enhancing design quality and improving project-wide collaboration and engagement with client and stakeholders.

Designs can also be tested to see what the impact might be of making changes. For instance, various solutions for the external envelope of a building can be tested to see the impact on the thermal performance of the building and its long-term energy consumption.

The digital twin can be employed during the occupancy phase of a project as well. Facilities managers can utilise real-time data from sensors built-in to the structure to monitor the performance of the building in order to optimise the operational phase of the facility.

4.10.10 Augmented and Virtual Reality

Augmented reality (AR) is a technology that facilitates the viewing of computer-generated 3D objects in a real-life environment. AR can be used on smartphones and tablets as well as on computers.

The main use of AR in construction is to showcase a realistic 3D model of a project interactively. This enables clients and project teams to both examine the design from any angle and to zoom in and out – just like being there!

By combining computer-generated information with physical surroundings, construction projects can be presented in real time with the obvious benefits for project planning, simulating different work sequences, hazard identification and planning for safe working.

https://youtu.be/3cFlNhTKtsg

AR can enhance BIM by providing the digital technology to place a BIM model into a real-world setting. This may seem confusing as BIM models are represented in 3D and can be accessed from any angle or at any plane or can even give a 'walk around' experience to the model viewer. However, AR transforms this experience into something akin to real life thereby transforming the user experience.

Virtual reality (VR), on the other hand, is a simulated virtual world typified by computer games. VR is completely immersive but only enhances a fictional reality and is completely virtual whereas AR is a combination of real world and virtual world.

Another difference between AR and VR is that VR requires a headset device, but AR can also be accessed through a smartphone, tablet or computer as well as by using headsets or special hard hats equipped with a computer and smart lens.

With the facility of AR, project teams can collaborate more effectively, solve problems and increase productivity in a hybrid environment with both real and virtual features:

- A 3D model is created using BIM.
- An AR device – such as headset, computer, tablet or smartphone – is used to show the user's actual surroundings, inside a building or on a construction site, for instance.
- The virtual data is overlaid on what appears on the AR device in real-time and in a real-world setting.

Practical applications on site can include:

- Visualisation of a piling sequence following reduced-level excavation.
- Simulation of a steel or concrete frame erection sequence from ground slab level.
- Visualisation of an HVAC system prior to installation.

AR can therefore help site managers to anticipate hazards, problems and work sequences before they happen. This is particularly useful for the short-term dynamic planning of projects promoted by the CIOB in its report *Managing the Risk of Delayed Completion in the 21st Century*.

The technology can also be used during planning and design, during manufacturing and during operation and maintenance as well as the construction phase of projects.

4.11 Managing the Supply Chain

The construction supply chain is complex and is made up of many different sorts of business enterprises. These include professional practices of designers and other consultants, contractors, subcontractors and specialist contractors, plant hire firms and suppliers, builders' merchants and other distributors of a wide variety of construction materials and components. The relationships between these businesses – and with the client – are defined by a complex system of procedures, protocols and contracts.

4.11.1 Supply Chain

Constructing Excellence[11] provides the following definition of 'supply chain' which is:

> *the term used to describe the linkage of companies that turns a series of basic materials, products or services into a finished product for the client.*

To achieve this requires design, planning, organisation and management which means that everyone involved in a project – designers, consultants, main contractors, subcontractors and suppliers – is part of the supply chain for that project. They are all increasingly referred to as 'suppliers' – because of legacy EU directives.

The project-based nature of construction means that there is no concept of a single industry-wide supply chain, and each project will have a different supply chain to that of the next project. This is not to say that every project supply chain is unique because some companies may appear in many different project supply chains.

Figure 4.18 illustrates a simplified example of a project supply chain. It shows that the client's supply chain is linked to that of the main contractor because of the contract between the two parties and the main contractor similarly connects to each subcontractor's own supply chain by virtue of its subcontract procurement procedures.

4.11.2 Supply Chain Management

Supply chain management may be defined as:

> *the formalised process that gives structure to the arrangements between companies that project participants 'supply to' and 'buy from'.*

Use of the term 'management' implies a degree of control, but no one participant in the project – including the client/employer and main contractor – has control over the entire project supply chain:

- Clients normally have control over designers, main contractors and any specialist contractors directly appointed by the client, but this is not the whole supply chain.
- Main contractors control their own subcontractors, specialist contractors and suppliers as they are in direct contract with them, but they do not control the client's supply chain.
- Subcontractors equally have their own supply chains of labour-only subcontractors, plant hire firms and materials suppliers, but there is no privity of contract between them and the main contractor, who, therefore, has no direct control.

Supply chains are commonly managed within framework arrangements in preference to the more traditional standing or ad-hoc approved lists. Frameworks help to reduce the bureaucracy of drawing up approved lists for each project. They are essentially a means of pre-qualifying a multi-supplier supply chain structure to enable clients to award contracts based on some sort of further competition (see also Chapter 6).

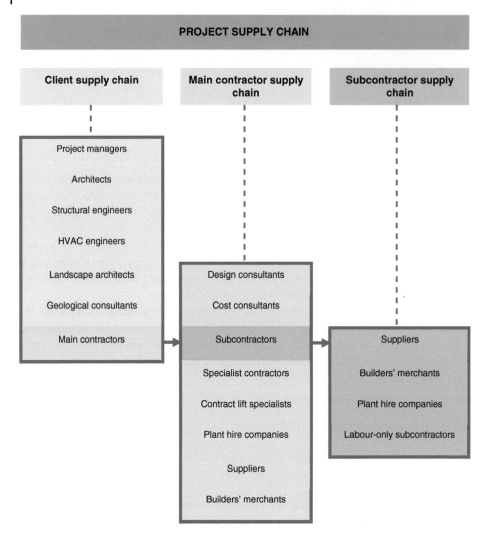

Figure 4.18 Project supply chain.

4.11.3 Supply Chain Integration

Construction projects have historically suffered from disconnected processes and data management. This has been reported in many construction industry reports which have particularly highlighted the separation of design from construction and the lack of coordinated project information as important reasons for inefficiency and poor project outcomes.

Electronic document management platforms can significantly improve project-wide communication by providing an end-to-end Cloud-based method of eliminating barriers and connecting teams on construction projects.

CDEs provide a complete project record captured on a single platform and an unalterable audit trail that helps to minimise disputes and keep projects on track by:

- Managing information and processes across projects.

- Providing tools and features to enable collaboration between clients, designers and contractors.
- Managing data and documentation.
- Providing complete transparency across all stages of a project lifecycle.

The architecture of CDEs is based on the principle of 'system neutrality' where all project participants have the same data access rights based on equality, security and trust. This applies equally to clients, designers, consultants, project managers, contractors and subcontractors.

With each organisation having its own space in the system, they can unilaterally decide who to share information with, either individually or collectively. Such platforms can also be configured to align with the project owner's specific requirements and workflows. Once invited, organisations can register, and their participation in the project will be subsequently activated. Participation ensures:

- Collaboration by exception.
- Every document is in one place.
- Every email conversation around those documents is retained.
- The drawing management system is always current.
- Defects (punch) lists can be accessed, items fixed and requests for re-inspection issued at a click.
- RFIs can be initiated and linked directly to an object in a BIM model.
- On site, a subcontractor can pull up the latest drawing on a tablet, check a design detail, take an on-site photograph and click off a variation request if an issue is spotted. The design can then be checked by the contractor or designer, and a variation can be approved if necessary, meaning that the project is not delayed.
- A digital operations and maintenance (O&M) as-built manual can be generated at the end of the project.
- The 'no delete' document framework provides a complete audit trail.

4.12 Managing Construction

Managing construction is concerned with the management of resources within the ambit of the scope, time and cost parameters of a project. As many contractors sublet most of the permanent works, the management of subcontractors is of considerable importance in ensuring successful project delivery.

Subcontracting is the vicarious performance of a contractual obligation. However, this does not relieve the main contractor from its responsibility to complete the works in accordance with the contract and within the time stipulated. Subcontracting protocols are:

- The main contractor enters into a direct contract with each subcontractor.
- Subcontractors may sublet their work to sub-subcontractors but may need the main contractor's agreement to do so.
- The main contractor retains contractual responsibility for the quality of the subcontractor's work and, ultimately, for rectifying defects in the work should the subcontractor default or become insolvent.

4.12.1 Types of Subcontractors

There is no official classification of subcontractors in construction and therefore custom and practice varies throughout the industry. Some standard forms of contract used in the industry refer to different types of subcontractors and the terms used have specific meaning.

Subcontractors may be referred to as work package contractors, trade contractors (in construction management), works contractors (in management contracting) or specialist contractors, depending upon the services offered. In some cases, subcontractors may be labour-only, where the main contractor supplies the necessary materials.

The different types of subcontractors are explained in Table 4.1, and the services they offer include:

- Single trades such as brickwork and blockwork, plumbing, electrical, plastering and screeds, painting and decorating – either labour-only or supply and fix.

Table 4.1 Subcontractors.

Domestic subcontractor	• A subcontractor chosen and engaged directly by the main contractor to carry out a particular trade. • May be labour-only or supply and fix. • Examples – groundwork, brickwork, roof tiling, plastering. • Some forms of contract require the main contractor to seek the contract administrator's approval before they can be appointed.
Nominated subcontractor	• A subcontractor chosen by the client to carry out specialist work. • Normally appointed on the expenditure of a provisional sum or prime cost sum included in the contract bills. • The nominated subcontractor enters into a contract with the main contractor. There may also be a collateral warranty between the nominated subcontractor and the client. • Used under FIDIC and Infrastructure Conditions of Contract (ICC) but not JCT or NEC contracts.
Named or listed subcontractor	• Subcontractors chosen, but not nominated, to carry out specific parts of a project. • Names are provided ('listed') for the main contractor to choose from. • The main contractor can add to the list. • The subcontractor becomes a domestic subcontractor of the main contractor.
Work package contractor	• Often used as a synonym for domestic subcontractor. • Might also imply responsibility for a specific section or package of the works, possibly including design work. • May provide a complete service with little or no reliance on main contractor attendances.
Labour-only subcontractor	• Self-employed individuals, partnerships or small incorporated firms. • May be engaged by a main contractor or subcontractor. • Materials supplied and paid for by the main contractor or subcontractor. • Typically include groundworks, brickwork/blockwork, fixing rebar. • May lead to higher materials wastage.
Trade contractor	• Contractors who carry out the work under a construction management procurement arrangement. • Engaged and paid directly by the client.
Works contractor	• Subcontractors appointed in management contracting. • Engaged and paid directly by the management contractor.

- Complete packages of work such as groundwork and drainage, formwork and structural frame, and hard and soft landscaping.
- Specialist work such as demolition, piling and geotechnical services, sliding formwork, HVAC installation and tunnelling.

4.12.2 Subcontract Scope

In order to place an order with a subcontractor, the work required has to be defined – 'scoped' – according to the trade or work package required. This is done by the main contractor.

However, whilst the aim is to ensure that subcontract packages are complete or 'self-contained', it is often the case that subcontractors do not – or prefer not to – carry out certain items of work necessary to enable the succeeding trade or work package to commence their work.

In such cases, an 'interface' is created whereby the main contractor will have a 'self-delivery' element to plan and resource. Interfaces can lead to the situation where none of the package contractors has included for a particular item of work, or duplication may occur where several firms are responsible for the same item. Either way, extra costs or disputes could result.

> A brickwork subcontractor is asked to price for the brickwork and blockwork package of a large housing development of 120 houses. The work package includes an electricity sub-station and some retaining walls. As this work requires a further visit to site towards the end of the project schedule, the subcontractor declines to price this element of the work.

Examples of 'self-delivery' items that the main contractor must deal with include:

- The provision of hard standings (piling mats) for piling rigs.
- The removal of excavation arisings from piling operations.
- Laying and blinding of hardcore post-earthworks package and pre-concrete/rebar package.
- Grouting of base plates post-steel erection.
- Builder's work (forming holes in walls, etc.) in connection with electrical and HVAC packages.

4.12.3 Attendances

It is quite common practice in the industry for main contractors to provide facilities to enable subcontractors to do their work. These facilities are called 'attendances':

- **General attendances:** use of mess rooms, site canteens and drying facilities, the provision of water and power, the use of standing scaffolding, the use of tower cranes and the removal of rubbish.
- **Special attendances:** such as offloading of materials, provision of specific access equipment or cranage, task lighting, etc.

General attendances are usually provided free of charge, although main contractors often make a contra-charge for waste removal.

4.12.4 Scoping of Work Packages

The scope of individual work packages can be determined in several ways according to the procurement arrangements of the main contract:

- Bills of quantities provide a list of item descriptions, and firm or approximate quantities, based on – or loosely based on – a recognised standard method of measurement (such as NRM2 or Civil Engineering Standard Method of Measurement [CESMM]).
- Activity schedules provide a list of unquantified construction activities, usually linked to the project programme.
- Schedules of rates are unquantified lists of item descriptions – similar to bills of quantities – and may be based on a standard method of measurement. They are used where the extent of work required is not known although an indication of the scope of work required might be given in a preamble to the schedule.
- Schedules of work are unquantified lists of composite work items. Being composite, the item descriptions may contain the work of more than one trade.

Figure 4.19 shows an extract from a bill of quantities for a structural steelwork package in which Item F relates to the fixing of holding down bolts. The holding-down bolts will be cast into the pile cap or concrete foundation well before the steelwork subcontractor starts on site. Therefore, the main contractor will carry out this work using its own labour or could include it within the groundworks or reinforced concrete work package.

4.12.5 Invitations to Tender

Bid invitations to subcontractors traditionally consisted of photocopied formal documents – pricing documents, specification extracts and drawings – delivered by post!!

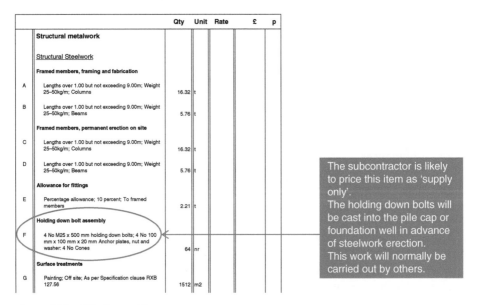

Figure 4.19 Bill of quantities.

Those days are long gone and now the most basic method of inviting bids is by email, attaching PDFs of the documents. More often that not, subcontractors receive no quantities – just drawings and a specification – meaning that it is they who take the quantity risk.

Alternatively, subcontractors may be invited to tender via an estimating portal or CDE – such as Causeway Estimating and Bidcon – where all documents can be accessed and bids submitted in one place.

These are cloud-based solutions, sometimes incorporating 4D BIM measure tools and 5D BIM (scheduling) capability, that facilitate collaboration across projects within a secure and auditable environment. They enable trade packages to be assembled within the software which can then be sent to multiple trade or work package subcontractors.

The software also enables the main contractor to see which subcontractors have responded to the bid invitation and carry out comparative analyses of the bids received.

4.12.6 Resources and Programme

Prior to placing any orders, main contractors will wish to discuss resources and programme with prospective subcontractors.

Price is not the only criteria for main contractors who will be keen to understand what resources subcontractors are able to devote to the project as this will directly impact the main contractor's master schedule and their ability to complete on or before the contract completion date.

It is important for subcontractors to appreciate that main contractors may expect additional resources to be made available during construction to keep up with general progress on site or to accelerate the schedule should the main contractor be in culpable delay.

4.12.7 Subcontract Coordinator

The control of subcontractors and specialist contractors – including contract lift specialists – is key to managing the project supply chain. Contractors often employ a subcontract manager or coordinator to do so, whose duties will include:

- Management of subcontractors across the project life cycle – design and development, bid process, pre- and post-contract and, where appropriate, post-occupancy.
- Management of the subcontractor procurement process from initial enquiry to contract appointment.
- Ensuring that subcontractors comply with the contractor's technical, schedule, cost and performance requirements for the project.
- Acting as a single point of reference for communications between the contractor and subcontractors, including requests for design and technical proposals (such as a lifting plan), subcontract pricing and tender evaluation, technical discussions, pre-contract negotiations, pre-start meetings and RFIs during construction.
- Subcontract negotiations, discussions regarding resources and scheduling of work, progress meetings and KPI monitoring.

4.12.8 Lean Construction

One of the great benefits of an integrated supply chain is the opportunity to apply lean thinking to the construction process.

The Egan Report – *Rethinking Construction* – advocated this principle which Cartlidge (2002) defines as *the elimination of waste from the production cycle* [which] *adds value to the process, leading to lower costs, shorter construction periods and greater profits.*

Part of the lean thinking approach is the use of just-in-time production where materials and components are manufactured and delivered to site as and when required without the need for long lead times and stockpiling on site. Lean thinking principles include:

- Elimination of non-value activities.
- Removal of waste from all activities involved in delivering the end product.
- Establishment of relationships with all members of the supply chain.
- Removal of delays in the design and production process using just-in-time management.

Lean thinking can be successfully applied to the construction process through innovative design and assembly, including the use of off-site manufacture, prefabrication, pre-assembly and supply chain integration.

4.12.9 Fast-track Construction

Many clients to the construction industry now demand early delivery of their project requirements. This is especially the case with commercial clients who are looking for shorter and shorter construction periods to ensure that the completed facility is online and earning revenue at the earliest possible moment.

As a consequence of this demand, procurement methods which overlap design, tendering and construction have emerged, but alongside these developments other approaches to the management of construction projects have evolved, including fast-track construction.

Fast-track construction is a management approach aimed at the early completion of the construction phase using a combination of innovative procurement methods, industrialisation of the construction process and the use of work package contractors in order to benefit from their expertise, especially as regards their design input.

The benefits of the system include:

- Overlapping of work packages both during design and construction.
- Less duplication of effort and reduced waste.
- Less uncertainty and inefficiency at work package interfaces.
- The use of innovative construction methods.
- Incorporation of cutting-edge technologies.
- Emphasis on the standardisation, pre-assembly and modularisation of the construction process.

The intensive construction programme which results from the fast-track approach requires high standards of planning, organisation and control. It is essential, therefore, to ensure that everything is right first time and that there are zero accidents to prevent unnecessary delays and disruption of the programme.

Fast tracking of projects does not happen by accident, and a great deal of planning and preparation is necessary to create the right client–supplier relationships. These are characterised by open and honest dealing, encouragement of innovation and the use of performance specifications to:

- Improve buildability by involving contractors in the design phase.
- Involve Tier 1 suppliers and works package contractors at an earlier stage.
- Reduce construction lead times through pre-ordering of key materials and components – such as structural steel, timber frame, concrete and cladding, partitions, ductwork, air handling units, modular components and so on.
- Arrange the direct supply of materials and components where repeat clients can benefit from volume discounts and higher quality by dealing directly with key suppliers.

References

Bahl, M., Taylor, P., McCall, K. (2021) *CIOB Planning Protocol*. Chartered Institute of Building.
BS1 (2010) *BS 6079–1 Guide to Project Management British Standard Part 1: Project Management*, BSI London.
Cartlidge D. (2023) *New Aspects of Quantity Surveying Practice*, Routledge.
Brech, EFL. (1975) *The Principles and Practice of Management*, Prentice Hall.
Flyvbjerg B. (2003) *Megaprojects and Risk – An Anatomy of Ambition*, Cambridge University Press.
Highways England (now National Highways) (2018) *The Project Control Framework Handbook*.
HM Government (2022) *The Construction Playbook*.
Infrastructure and Projects Authority (2021) *Cost Estimating Guidance*.
Lavender S. (2016) *Management for the Construction Industry*, Routledge.
Walker A. (2015) *Project Management in Construction*, 6th edn. Wiley-Blackwell.
Winch G.M. (2009) *Managing Construction Projects*, 2nd edn. Wiley-Blackwell.
Yam Seng Pte Ltd v International Trade Corp (2013) EWHC 111 (QB).

Notes

1. https://www.apm.org.uk/resources/what-is-project-management/.
2. Bernard Ainsworth and Turner and Townsend were joint project managers for the Shard.
3. In construction, 'supplier' may refer to the entire supply chain of designers, contractors, subcontractors and suppliers of materials and components.
4. Office of Government Commerce (OGC) Gateway Review for Programmes & Projects.
5. Governance for Railway Investment Projects (GRIP) Network Rail.
6. https://content.tfl.gov.uk/Item04-TfL-Pathway-PPP-May-2013.pdf.

7 Construction Project Information Committee (CPIC).
8 https://www.ukbimframework.org/about/.
9 https://www.bsigroup.com/en-GB/iso-19650-BIM/.
10 https://www.ukbimframework.org/wp-content/uploads/2021/02/Guidance-Part-D_Developing-information-requirements_Edition-2.pdf.
11 https://constructingexcellence.org.uk/wp-content/uploads/2015/03/supplychain.pdf.

Chapter 5 Dashboard	
Key Message	Challenges facing construction world-wide include an ageing and diminishing construction workforce, lack of skilled trades entering the construction sector and low productivity in the industry.
Definitions	Design for Manufacturing and Assembly (DfMA): A way of thinking about construction design by focussing on ease of manufacture and efficiency of assembly whilst retaining architectural freedom.
	Modern Methods of Construction (MMC): A range of offsite manufacturing and onsite techniques that provide alternatives to traditional methods.
Headings	**Chapter Summary**
5.1 Introduction	Several evolving technology-driven factory-based solutions are available that can be employed for specific construction markets.
	These methods are known in the industry as 'modern methods of construction'.
	Simplification of design makes it possible to manufacture and assemble the component parts of a building or structure in a more efficient way.
	This can reduce the overall time taken to complete a project and lower its cost.
5.2 MMC Categories	Modern Methods of Construction comprise seven genres or predominant materials from which MMC components or assemblies are made.
	This includes engineered timber, timber frame, concrete and cement derived materials and light-gauge and hot-rolled steel.
5.3 Modular Construction	Modular construction is much more than 'industrialised' building.
	It includes both structural and non-structural (pod) volumetric construction.
5.4 Panelised Construction	Panelised systems differ from volumetric construction in that they are two-dimensional elements.
	They provide a more efficient means of construction than on-site timber framing.
	They may be open or closed systems and include SIPS – structural insulated panels – and cross-laminated timber.
5.5 Component-Based Systems	Component based systems are based on a configurable kit of parts which are often referred to as 'platforms'.
	Their objective is 'design to value', supported by BIM methodologies.
5.6 Hybrid Construction Systems	Hybrid construction may be defined as the integration of systemised components and traditional construction methods.
	It is used in low, medium and high-rise building and in civil engineering.
5.7 Digital Technologies	Digital technologies are innovative site-based construction techniques that harness site process improvements falling outside the main pre-manufacturing categories.
	They include 3D printing, drones, wearable technology and concrete monitoring.
5.8 Planning for Modular Construction	Systemised construction requires contractors to adapt working practices to different methods of construction.
	This impacts the organisation, planning and control of projects.
	There are several systems on the market but no industry-standard design.
	Projects that contain a significant element of pre-manufacturing require different thinking and a degree of learning to plan and schedule the work successfully and achieve the desired project outcomes.
Learning Outcomes	Appreciate the need to find solutions to the construction industry's resource problems.
	Distinguish between modern methods of construction and traditional methods.
	Understand the interface between them and its impact on the planning and management of projects.
Learn More	Chapter 19 – a case study involving modular construction and traditional methods.

5

Modern Methods of Construction

5.1 Introduction

The construction industry worldwide faces many challenges emanating from a rapidly growing global population, consequent supply–demand imbalance in the housing sector and demand for more sustainable building methods in the face of climate change and environmental pressures. Added to this is an ageing and diminishing construction workforce, lack of skilled trades entering the construction sector and low productivity in the industry generally.

Whilst there is no single industry-wide silver bullet capable of fixing these problems, there are several evolving technology-driven factory-based solutions available that can be employed for specific construction markets such as housing, school buildings, the hospitality sector, hospitals and prisons.

These methods are known in the industry as 'modern methods of construction' (MMC), although this is an umbrella term that applies to other technologies as well as factory-based production.

In order to embrace and capitalise on the benefits of these technologies, new ways of thinking about the organisation and management of construction projects are required both in terms of design and construction. This encouraged a 2021 amendment to the Royal Institute of British Architects (RIBA) Plan of Work to include a Design for Manufacturing and Assembly (DfMA) 'overlay'. The overlay is illustrated in Figure 5.1, which identifies the DfMA core tasks, digital tasks and procurement impact.

5.1.1 DfMA

DfMA is a philosophy – a way of thinking – that resolves construction design in a way that focuses on ease of manufacture and efficiency of assembly whilst retaining architectural freedom.

DfMA is the combination of two methodologies:

- **Design for manufacture**: the design of the parts of a product for ease of manufacture.
- **Design for assembly**: the design of a product for ease of assembly.

This approach can be extended to buildings in use by factoring maintenance considerations into the design, thereby reducing life-cycle costs.

Construction Planning, Programming and Control, Fourth Edition. Brian Cooke and Peter Williams.
© 2025 John Wiley & Sons Ltd. Published 2025 by John Wiley & Sons Ltd.

5 Modern Methods of Construction

	RIBA Plan of Work 2020 – DfMA Overlay							
	Options:							
	0	1	2	3	4	5	6	7
	Strategic Definition	Preparation and Brief	Concept Design	Spatial Coordination	Technical Design	Manufacturing and Construction	Handover	In Use
Core DfMA Tasks	1. Consider opportunities for the seven MMC categories. 2. Consider impact on business case or client requirements.	1. Initiate DfMA thinking. 2. Incorporate the seven MMC categories. 3. Undertake R&D with manufacturers.	1. Embed appropriate MMC categories in concept design. 2. Identify DfMA solutions. 3. Include DfMA in project strategies. 4. Liaise with supply chain.	1. Update construction strategy. 2. Consider buildability. 3. Check warranties for MMC solutions.	1. Consider DfMA impact on building systems. 2. Develop DfMA components. 3. Consider manufacturing and assembly risks.	1. Update construction strategy. 2. Monitor quality of off-site manufacturing.	1. Provide DfMA feedback for future projects.	1. Consider DfMA feedback for future projects. 2. Monitor component performance and provide feedback.
Suggested Digital Tasks for DfMA	1. Analyse cost and programme data from previous DfMA projects.	1. Use BIM for feasibility studies. 2. Consider digital library including DfMA objects and components.	1. Develop digital information. 2. Validate model. 3. Consider DfMA tolerances.	1. Update digital information. 2. Validate model. 3. Use digital tools and technologies.	1. Update digital information. 2. Validate model. 3. Use 4D technologies.	1. Train site operatives. 2. Use digital tools to track progress.	1. Consider lessons learned.	1. Update digital asset information. 2. Use digital twin and smart building technologies.
Procurement Strategy	Appoint MMC adviser.							
MMC Categories:								
1, 2 and 4	Ensure client team has requisite MMC and DfMA knowledge.	Review possible subcontractors and manufacturers in relation to contractor appointment.						
3 and 5			Consider specialist subcontractors and constraints and embed into the design.					
6 and 7					Low impact on procurement.			

Figure 5.1 DfMA overlay.

Simplification of design makes it possible to manufacture and assemble the component parts of a building or structure in a more efficient way which can reduce the overall time taken to complete a project and lower its cost. This requires the adoption of digital technologies such as building information modelling (BIM), augmented and virtual reality, laser scanning and robotics, etc., which enable DfMA to be adopted in a productive way, thereby creating opportunities for digital design, manufacture and delivery to site.

The principles of DfMA are similar to those of lean construction:

- DfMA helps to identify, quantify and eliminate waste or inefficiency in the manufacturing and assembly process.
- It can also be used as a means of benchmarking construction products against those of competitors.

> Ideally, DfMA needs to be embedded in a project from the outset, and there are examples of construction projects around the world that have embraced this approach. Some major companies have invested heavily in DfMA – Laing O'Rourke and Berkeley Group have their own centres of excellence and production facilities.

A major benefit of DfMA is that off-site production of building components can be commenced well before work starts on site. This helps to avoid delays in the programme caused by labour shortages, problems with materials availability or long lead times for fabricated building elements.

DfMA, however, is in its infancy and there is more substantial evidence of projects adopting what might be called a subset of DfMA – MMC.

The RIBA recognises that DfMA will often – but not always – lead to MMC. It also suggests that the word 'modern' might be misleading because the methods and their underlying rationales and principles have been around for decades (RIBA 2021).

The reason that they are described as 'modern' is because they have yet to be adopted into mainstream construction.

5.1.2 Modern Methods of Construction

A simple Google search for MMC leads to the conclusion that this is a method of building using structural components built off-site largely geared towards solving the supply–demand imbalance for new homes. This is partly true, but MMC is a much wider term:

> *covering a range of offsite manufacturing and onsite techniques. MMC provides alternatives to traditional methods and has the potential to deliver significant improvements in productivity, efficiency and quality for both the construction industry and public sector.*
>
> (HM Government 2020).

MMC is also called 'smart construction', the aim of which is to use off-site manufacturing techniques in order to reduce defects, reduce costs and make the building process more efficient. This is not a new idea but, thanks to digital technologies, MMC are now emerging into the mainstream of the construction industry.

> At the present time, MMC is used mainly – though not exclusively – in conjunction with traditional construction methods. In hotel construction, for instance, it is common to see factory-built bathroom 'pods' delivered to site and craned into position. Off-site manufactured, fully fitted out, volumetric units are also used in house building and for schools, hospitals and prisons.

Factory-based production methods are applicable to other sectors of the industry too. There are many examples of such methods being used in road and bridge construction, water and sewage treatment works, power stations and so on. However, the significant difference is that these applications of off-site manufacture and prefabrication do not significantly reduce the demand for skilled labour on site.

5.1.3 Definitions

MMC remove the requirement for skilled trades and replace them with high-quality factory production methods, thereby enabling the components to be installed on site using unskilled labour. MMC may be considered under four headings:

- **Modular construction**
    - **Structural**: Factory-produced, pre-engineered structural building units that are delivered to site and assembled as large volumetric components or as a substantial part of a building.[1]
    - **Non-structural**: Commonly known as 'pods'.

- **Panelised construction**
  - Prefabricated sections of walls, floors or roofs that can be assembled at the building site as a panelised house.[2]
- **Component-based systems**
  - A configurable kit of parts based on common components that can be found in dissimilar buildings.
- **Hybrid construction systems**
  - Any construction system that combines two or more categories of MMC.

5.2 MMC Categories

In order to better understand the use of MMC in housing development and to support the interests of mortgage lenders, insurance companies and valuation surveyors, a UK government working group[3] identified seven categories of MMC.

These categories are now used in other construction sectors to improve communications and to encourage a better understanding of MMC.

5.2.1 MMC Definition Framework

In the published report (Cast 2019), *Modern Methods of Construction – Introducing the MMC definition framework* (2019), seven 'genres' or predominant materials were identified from which MMC components or assemblies are made. This includes engineered timber, timber frame, concrete and cement-derived materials and light-gauge and hot-rolled steel.

The list of materials excludes foundations or other parts of structures to which the materials may be fixed. This reflects the fact that MMC will always be reliant on traditional construction methods due to the nature and complexity of construction and to the individuality of individual construction sites.

5.2.2 MMC Spectrum

The seven MMC categories identified by the working group form an MMC spectrum as illustrated in Figure 5.2, which also provides examples of each category.

Categories 1–5 describe off-site and near-site pre-manufacturing methods. They are to be contrasted with categories 6 and 7 which refer to site-based process improvement rather than off-site manufacturing.

5.2.3 Pre-manufactured Value

The MMC spectrum also provides a means of analysing the proportion of a project made up of on-site labour, supervision, plant and temporary works in terms of pre-manufactured value (PMV).

The PMV of a project can be improved by increasing the proportion of pre-manufactured elements and/or by reducing the amount of site labour.

5.2 MMC Categories

MMC Category	MMC Spectrum	Description	Example
1	Off-site and near site pre-manufacturing	Pre-manufacturing (3D primary structural systems)	Systemised volumetric construction. May include fit-out. Whole building may be systemised or hybrid (traditional core and/or ground floor podium).
2		Pre-manufacturing (2D primary structural systems)	Systemised flat panel construction for floor, wall and roof structure. Commonly open panels with services, insulation, cladding and finishes installed on site.
3		Pre-manufacturing components (non-systemised primary structure)	Pre-manufactured structural members. Engineered timber, steel or concrete beams, columns, walls, core structures and slabs. May include substructure elements such as ground beams and pile caps.
4		Additive manufacturing (structural and non-structural)	Remote, site-based or final workface 3D printing. Structural and non-structural components.
5	Site-based process improvement	Pre-manufacturing (Non-structural assemblies and sub-assemblies)	Pre-manufactured non-structural walling systems, roof finish cassettes, mini-volumetric units (pods). Sub-assemblies exclude windows and door sets, etc. where included in conventional masonry construction. Includes volumetric podded assemblies (kitchens, bathrooms), panelised façade assemblies, roof cassettes and M&E assemblies.
6		Traditional building product-led site labour reduction/productivity improvements	Traditional single building products manufactured in large format, pre-cut configurations or with easy jointing features that reduce the extent of site labour. Large format external and internal walls. Large format roofing finishes. Easy site install brick slips, modular wiring and flexible pipework.
7		Site process-led site labour reduction/productivity/ assurance improvements	Innovative site-based construction techniques. Includes techniques falling outside categories 1–6. Also includes lean construction techniques, worker augmentation such as exoskeletons and assisted materials distribution, worker productivity planning tools such as GPS and wearables, site process robotics and drones, autonomous plants, etc.

Figure 5.2 MMC spectrum. Source: Adapted from Modern Methods of Construction – Introducing the MMC definition.

5.3 Modular Construction

Cynics might suggest that modular construction is another form of 'industrialised' building that had its origins in the prefabricated (PREFAB) single-storey houses developed after the end of the Second World War.

Long memories might also think about the infamous system-built tower blocks of the 1960s and 1970s which were designed on the basis of the then-popular Brutalist architecture movement. Many of these so-called 'slums in the sky' were badly designed and built, leading to cracks and structural decay, damp penetration and leaking roofs and even partial collapse.[4]

With MMC, nothing could be further from the truth.

5.3.1 Volumetric Modular – Structural

Volumetric modular units are large factory-built building elements constructed of steel or timber and may contain concrete elements, aluminium framing and a variety of composite building materials. The units can be linked together to form complete buildings.

Such buildings can be temporary (e.g. site accommodation and welfare), semi-permanent (e.g. school rooms) or permanent structures. They have a similar appearance to containers used for shipping, and some are just that – modified steel shipping containers. Other manufacturers prefer to use steel or timber structural elements whilst others may combine timber with concrete floors.

Volumetric modules can offer turnkey building solutions, but, in many cases, whilst internal finishes, mechanical and electrical (M&E) installations, windows and doors may be included, other work such as external insulation and external finishes/cladding may have to be carried out on site.

Figure 5.3 illustrates the factory to site process for composite modular units whilst Figure 5.4 shows the use of modified steel containers.

As well as housing, volumetric construction is quite common for building student or military accommodation, hotels and is used in the health sector. The modules can leave the factory with electrics, plumbing and heating installed and doors, windows and internal finishes can be included.

The units can be ordered to any desired degree of completeness and can arrive on site fully equipped with beds, bedding and toiletries if required.

In the health sector, units can be ordered to include complex mechanical services or medical installations which are tested off-site before delivery.

Volumetric units are transported to site and placed on prepared foundations and can be arranged horizontally, or stacked vertically, to the desired configuration. Once connected together, modular units need no additional superstructure, but a supporting framework may be required for taller buildings. Service connections, corridors, façades or cladding and a roof are usually carried out on site.

> By assembling volumetric modules in controlled factory conditions, defects are minimised, quality control is assured and cost is reduced as a result of the repeatability of the design and speed of construction.

5.3 Modular Construction

Factory conditions.

Volumetric units move around on bogeys.

Volumetric unit under construction.

Unit craned into position.

Roof cassette craned into position from low loader.

House structure complete.

Figure 5.3 Modular housing.

Steel modular unit lowered into position.

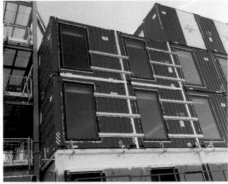
Windows and primary cladding rails.

Secondary cladding rails and scissor lift access.

Internal corridor.

Figure 5.4 Volumetric construction – steel.

A number of the main high street fast-food outlets in the United Kingdom, such as McDonald's, Burger King and Kentucky Fried Chicken (KFC), as well as Starbuck's and Costa Coffee, utilise modular buildings for constructing some of their premises. This is illustrated in Figure 5.5.

Once site works such as site clearance, foundations, utilities and drainage have been completed, fully serviced volumetric modules, factory-built roof sections, facade work and external works can be completed very quickly.

The units can be fully fitted out with food storage and preparation areas, toilets and seating facilities and, remarkably, can be cleaned and opened for business within 24 hours.

Burger King outlet on site.

Figure 5.5 Burger King.

5.3.2 Volumetric Modular – Non-structural

Non-structural volumetric units are better known as 'pods' and are pre-manufactured to create bathrooms, shower rooms, kitchens, utility cupboards, office washrooms, etc. They are commonly of steel frame or composite construction or may be made from glass-reinforced plastic (GRP).

The walls of steel-framed pods are normally lined with plywood or oriented strand board (OSB) and drywall/plasterboard construction. This means that future internal adaption of fittings and finishes can proceed as with traditional construction methods. Pods are assembled at the factory on wooden pallets, shrink-wrapped and loaded into delivery vehicles using forklift trucks.

As they are non-structural, 'pods' are usually craned into a position as the building superstructure is erected. They can adapt to any structural frame system be it timber, steel or concrete construction.

If pods cannot be craned directly onto the floors of the building, they can be moved into position from an external 'transfer deck' which forms a safe place of work. Pods can be moved with forklift trucks or on 'trolley' wheels attached to the external framework. At their final position, pods are carefully lowered into position with jacks and are levelled and fixed to the structure. Figure 5.6 shows some examples of bathroom pods and a transfer deck and also illustrates how pods are positioned, awaiting final placing and commission.

Pods can also be retrofitted via apertures in the external cladding – utilising transfer decks – or via lift shafts. Flat-pack assemblies are also available which are fully assembled at the factory and then disassembled into component panels for delivery. They can be disassembled in such a way that they pass through standard door openings and stairways and are then reassembled in their final position on site.

Bathroom pods at ground level and in position prior to installing the metal decking above.

Shrink wrapped bathroom pod.

Transfer deck.

Figure 5.6 Miscellaneous.

Pods are ideal for providing student accommodation and are also used in the construction of prisons and healthcare and educational premises. Morphologically, pods can be bespoke to the desired shape, size and specification and on-site connection is 'plug and play'.

5.4 Panelised Construction

Panelised systems differ from volumetric construction in that they are two-dimensional elements. The prefabrication of these panels facilitates a high degree of quality control which ensures that units are perfectly aligned, parallel and accurately sized.

They provide a more efficient means of construction than on-site timber framing whether to create the inside load-bearing element of a cavity wall or to construct internal partitions and load-bearing walls.

Panelised systems are to be contrasted with timber-framed buildings which have been in existence for many years and constitute some 70% of houses constructed worldwide. They offer a much higher standard of quality control, far less on-site assembly and speed of construction and reduced defects. They also provide much greater programme certainty and reduced costs.

5.4.1 Open and Closed Systems

Panelised construction may be either 'open' or 'closed' systems:

- Factory-made open panel timber frame systems are industry standard. External panels are usually sheathed on the outside with a vapour-permeable waterproof membrane attached. Structural elements, such as lintels, may be incorporated but the panels are not usually insulated. Once the building is watertight, first-fix electrical and plumbing can be installed, and the panels can be insulated and finishings added.
- As their name suggests, closed panels arrive on site ready-insulated as a minimum but might also be specified to include windows, services and plasterboard. They may also be finished on one or both sides.

Unlike volumetric construction, panelised systems are not 3D objects and are assembled on site or incorporated into existing structures.

This is illustrated in Figure 5.7, where it can be seen that the panels are sheathed externally but not insulated, which is site work. Panels can also be flat packed at the factory for delivery to site and then assembled.

5.4.2 SIPs

SIPs – structural insulated panels – are a subset of panelised systems and consist of two wooden boards bonded to a structural insulation core. The boards have integral 'splines' which facilitate the gluing and nailing of panels to form complete walls. SIP panels are distinguished by their integral solid-core insulation and their superior levels of thermal performance.

SIP panels can be made to bespoke requirements in a factory and are more adaptable than 3D volumetric units. Large format panels give flexibility on site as well as speed and quality. Openings/cut-outs for windows and doors are integrated as well as structural elements such as lintels and key element posts. Panels are tagged to enable tracking from factory to site.

Using SIPs obviates the need for cavity insulation, and this not only reduces cost but also takes a site activity off the contractor's programme. Compared to other timber frame solutions, SIPs provide the added benefit of a solid feel especially since plasterboard is fixed directly to the panels. No wet trades are necessary with SIP systems which also impacts the construction schedule positively.

SIPs panels are used for load-bearing external and internal walls, floors and roofs. Roofs do not require trusses, and additional useable space can therefore be created.

A great benefit of SIPs is that external and internal trades can start work earlier and, being relatively easy to erect, on-site construction time is reduced significantly. Buildings are also watertight sooner, and there is greater programme certainty because there are fewer site processes and fewer interfaces with other trades.

Digital technology plays a key role and whilst the production process is standardised, it is flexible enough to create bespoke panels of consistent quality. Stacking systems for orders make sure that the first panel offloaded at the work site is the first panel to be installed.

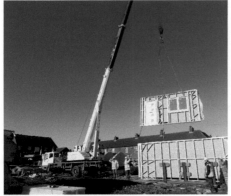
Uninsulated panel lifted from stack.

Pre-finished panel lowered into position.

Panels propped and restrained.

Ceiling and roof gable panels.

Figure 5.7 Panelised construction.

Installation on site is aided by the factory integration of lifting slings on each panel which saves time and makes site assembly safer.

5.4.3 Cross-laminated Timber

Cross-laminated timber (CLT) is used to construct walls, floors and roofs for a variety of structures. It is an engineered timber product where longitudinal and transverse layers of timber are glued together in a similar fashion to glue-laminated timber beams (Glulam).

CLT differs from glulam, however, as it is manufactured in panels rather than beams. With the layers of timber bonded perpendicularly to each other, the panels achieve structural strength in two directions, similar to concrete. The manufacturing process limits movement in the wood, and the resulting panels are dimensionally stable.

CLT panels can be used as a stand-alone building material or may be included as part of a hybrid structure. This material can be used in high-rise buildings for creating commercial office space and for housing development.

The versatility of CLT panels is achieved by the variability in available thicknesses and they are flexible and sustainable. CLT is sometimes referred to as 'super plywood'.

5.5 Component-Based Systems

Component-based systems are based on a configurable kit of parts which are often referred to as 'platforms'.

Global creative technology company Bryden Wood[5] is a world leader in the platform approach to DfMA in particular, and MMC in general. The company carries out research and development in the field of sustainable design solutions and has been instrumental in integrating design and automation in construction. Their objective is 'Design to Value'.

With a platform approach to DfMA, supported by BIM methodologies, it is possible to create multi-storey buildings characterised by:

- The automated assembly of structural frames.
- The elimination of all work at height.
- A multi-skilled workforce.
- Significant productivity gains.
- Net-zero carbon in construction and operation.[6]

5.5.1 IKEA

Platform construction takes its inspiration from the IKEA approach to creating various items of furniture using a limited selection of pre-manufactured parts and standardised methods of connection.

Designs are based on a library of digital 'core' components that have been found to be present in any structure. They are put together into a 'data set' which can be viewed in a BIM model or by using bespoke software.[7]

Advances in digital technology mean that BIM gives one view of a data set but not the only view.

5.5.2 Site Assembly

In the same way that IKEA furniture can be assembled using an easy-to-follow installation manual, and without any particular skills, component-based systems of construction are designed to be capable of site erection using unskilled labour, simple off-site manufactured components and colour-coded connections.

These methodologies can be augmented in various ways:

- Site assembly without the need for work at height.
- Simplification of site connexions, thereby eliminating delay and errors.
- Off-site fabrication of rebar cages and panels with robot tie-wiring.
- Concreting from below thereby eliminating the need for tower cranes for placing concrete.

5.6 Hybrid Construction Systems

Despite advances in DfMA, and the mainstream adoption of MMC, construction will, in most scenarios, always be a hybrid process. The reason is simply that each individual construction site has its own size, shape, topography and ground conditions with the result that traditional methods will invariably sit alongside new technology and new construction methods.

Hybrid construction may be defined as *the integration of systemised components and traditional construction methods*. Examples of this form of construction include:

- In situ concrete frame with precast concrete floor slabs.
- Steel frame and precast concrete elements.
- The combination of volumetric, modular and panel systems.
- Part systemised – such as volumetric – and part traditional construction such as a steel frame ground floor podium and a slip-formed concrete central core.

5.6.1 Low-Rise Construction

In the context of low-rise housing, hybrid construction can take several forms:

- Where the Category 1 primary structure is combined with traditional construction:
  o Volumetric modules placed on ground beams or trench foundations.
  o Mini-volumetric modules (pods) incorporated in steel or concrete framed buildings.
- Where two or more MMCs are used together, perhaps combining systemised Category 2 primary structures, non-systemised Category 3 primary structures and/or Category 5 pre-manufactured non-structural assemblies:
  o CLT wall and roof panels combined with glulam beams.
  o Pre-manufactured lightweight steel and glulam frame combined with pre-manufactured SIPs and pre-manufactured roof panels.
  o Precast concrete or glulam frames combined with CLT floors, roofs and staircases.
  o Unitised curtain-walling system or precast concrete cladding with integrated window assemblies.
  o M&E pre-packaged plant and prefabricated power and lighting distribution systems.

Figure 5.8 illustrates the installation of volumetric house units on traditionally constructed foundations. This shows:

1. Permanent polypropylene formwork and ground beam rebar.
2. Completed substructure and service ductwork.
3. Blockwork laid to receive volumetric units.
4. Installation of modules by mobile crane.

The extent of site preparation and groundworks in this example, prior to installation of the Category 1 modules, illustrates the challenge facing the industry. This is to find innovative site-based working methods that fall outside Category 1–5 off-site pre-manufacturing and Category 6 materials innovation and yet embrace Category 7 lean construction and digital technologies.

5.6.2 Medium-Rise Construction

The MMC Definition Framework explains that hybrid construction can mean part systemised and part traditional construction, where a 'structural chassis' is combined with a traditional core and/or ground floor podium.

Ground beams – rebar and permanent shuttering.

Completed ground beams.

Blockwork to receive volumetric unit.

Volumetric unit craned into place.

Figure 5.8 Hybrid systems.

In this context, a structural chassis is a volumetric unit that may:

- Require fitting out.
- Be complete with internal fit-out.
- Be fitted out complete with external cladding and/or roofing complete or
- Contain internal fit-out with 'podded' room assemblies such as kitchens and bathrooms.

Figure 5.9 shows part of a £100 million development that includes a five-storey hotel. The hotel is to be constructed using a structural chassis of 94 volumetric modules installed on a steel-framed commercial podium area with a steel-framed access staircase, as shown.

The modules are manufactured in China and arrive at UK docks in two batches for onward delivery to site. Each module arrives fully equipped with bed and bedroom furniture. Bedding and final room completion are carried out on site. All 94 modules arrive complete with fully finished luxury bathroom pods and are installed by a specialist team in 10 days.

External cladding consists of a combination of vertical aluminium-faced panels and renders with access provided by scissor-lift platforms.

Traditional steel framed podium and staircase.

First volumetric unit arriving on low loader.

Volumetric unit craned into position.

Installation of second level volumetric unit.

Figure 5.9 Hybrid – Holiday Inn.

5.6.3 High-Rise Construction

Figure 5.10 demonstrates the applicability of Category 1 systemised volumetric construction in a high-rise situation.

The project illustrated is a 29-storey, 580-bedroom student accommodation building comprising 679 modules of various sizes. The modules contain all plumbing and electrics as well as fully fitted-out kitchens, bathrooms and bed bases.

Module sizes vary in length from 4 to 5.6 m and are 2.8 m wide and 2.9 m high. External cladding consists of glass-fibre reinforced concrete (GRC) wall panels fitted from mast climbers, as illustrated. The use of mast climbers eliminates the need for external scaffolding and thereby reduces the risk of work at height.

Figure 5.10 also illustrates the lifting beam arrangement for raising the modules to the upper floors using a tower crane.

Each low loader delivery comprises two units which allows 11 modules to be installed per day and contributes to a 50% reduction in the two-year construction schedule. Amongst the benefits of using the modular approach is the smaller workforce and the consequential reduction in site accommodation and welfare requirements.

Modular units arrive on low loader.

Lifting beam, tower crane and mast climbers.

Module being lifted to the 25th floor.

Part front elevation showing GRC panels.

Figure 5.10 High-rise application.

High-rise volumetric structures are thought to be feasible up to 35 storeys but require different coupling arrangements compared to low-rise buildings as a consequence of differential structural movement caused by gradual settlement over time.

5.6.4 Civil Engineering

Factory-manufactured products have been an integral feature of civil engineering construction for many years. This includes the use of precast concrete bridge beams, tunnel segments, culvert and drainage components, earth retaining systems and so on.

Civil engineering also makes use of semi-automated site-based production methods – sometimes in combination with factory-made components:

- Concrete paving trains to lay continuously reinforced concrete pavement for highways and runways.
- Tunnel boring machines (TBMs) which excavate, support and remove spoil and permanently line the tunnel.
- Extruded concrete highway barriers.
- Incrementally launched bridge/viaduct decks.

An application of Category 1 hybrid methods of construction in civil engineering is illustrated in Figure 5.11. This shows the construction sequence for viaduct piers on the HS2 Phase 1 Wendover Dean Viaduct project.

The method combines the installation of bored piles and a reinforced concrete pile cap – constructed within a sheet-piled cofferdam – with factory produced precast concrete shell modules in a variety of configurations.

> The combination of 3D volumetric/modular systems, and more efficient site-based methods, reduces skilled labour requirements on site and improves site safety. Volumetric modules are cast under controlled factory conditions and are transported to site on a just-in-time delivery schedule. Site assembly is rapid and high-quality finishes can be achieved without the need for traditional formwork.

Bearing piles, pile cap and rebar constructed within sheet-piled cofferdam ready to receive precast concrete shell units.

Pier shell units installed ready to receive hammerhead section.

50Te precast concrete hammerhead shell unit lowered into place.

Hammerhead rebar lowered into shell.

Figure 5.11   Viaduct pier. Photographs courtesy of HS2 Ltd.

This process takes activities off the construction schedule, reduces project duration and offers cost savings whilst also making construction more sustainable.

5.7 Digital Technologies

As recognised in the MMC Definition Framework, Category 7 *is intended to encompass approaches utilising innovative site-based construction techniques that harness site process improvements falling outside the five main pre-manufacturing categories 1 to 5 or materials innovation in Category 6.*

The industry is already embracing technologies that align with these aspirations including:

- 3D printing.
- Drones.
- Wearable technology.
- Concrete monitoring.

5.7.1 3D Printing

Additive Manufacturing – or 3D construction printing (3DCP) – is an exciting modern method of construction with great potential to reduce the industry's reliance on skilled labour and to produce construction components and entire structures by incrementally laying down successive layers of a printing medium which creates the desired three-dimensional shape.

3DCP combines the power of BIM with automated site production methods resulting in a uniform, high quality and fast construction method. This method of construction is particularly suited to low- and medium-rise housing and is gaining traction throughout the world. It is also used for constructing commercial and industrial buildings, bridges and other structures.

3D construction printing is an infinitely flexible, sustainable method of building which facilitates the creation of imaginative architectural shapes that would otherwise require bespoke specialist formwork to achieve the same result. It would also take much longer to construct.

The magic of 3DCP is the process that links the initial design (CAD or BIM) with the printer head that deposits layers of material to gradually form a physical structure. This process is illustrated in Figure 5.12 which shows how the digital 3D model is translated into a language that the 3D mechanical printer can understand. This is done using slicing (translating) software to generate the geometric G-codes needed to operate the 3D printer's control system.

The printer head may be attached to a fixed or truck-mounted robotic arm or to a gantry as illustrated in Figure 5.13. Gantries are scalable according to the building footprint and comprise modular piers, motorised beams and a lightweight truss that carries the concrete delivery pipe and printer head. Printer heads can be fitted with a variety of nozzles according to the size, speed and finish of print required.

5 Modern Methods of Construction

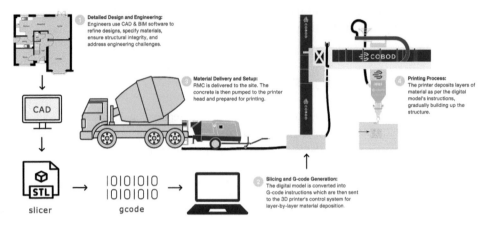

Figure 5.12 3DCP process overview. Image courtesy of HTL.tech.

Commencement of 3D printing showing modular pier, motorised beam, printer head, ready-mixed concrete truck, concrete delivery pipe and concrete pump.

3D printing of walls, showing gantry, motorised beam, printer head and concrete delivery pipe.

Concrete extrusion at printer head showing print layers and walls ties.

Traditional roof trusses craned into position following 3D printing of walls.

Figure 5.13 3D printing. Photographs courtesy of HTL.tech (https://www.htl.tech).

Concrete and mortars are commonly used print media. Harcourt Technologies Ltd (HTL.tech) uses ready-mixed concrete, with the addition of plasticisers and accelerators for flowability and buildability. Other 3D construction printing specialists, such as ICON and Peri 3D Construction, use specially developed proprietary mortars. These print materials require no temporary support (formwork) during the hardening or curing process.

Ready-mixed concrete produced at off-site batching plants offers several advantages including consistency of quality, scalability and reliability of supplies for continuous printing, supply-chain and cost efficiency and reduced environmental impact. Site mixed concrete is more labour intensive but can be attractive for smaller projects. Specialist mortars can be considerably more costly than concrete.

Once the printing medium has been delivered to site, this is pumped to the printer head which lays down the successive layers of material. Rebar, lintels and beams, box-outs, insulation, windows, doors and roofs are installed as required.

3DCP is an emerging technology that requires special skill sets and the expertise to plan and manage its integration with traditional construction methods in the build process. Based in Dublin and Manchester, HTL possess the expertise to design and manage such projects from inception to completion.

HTL can provide the architectural, 3D model rendering and project management expertise for 3DCP projects as well as training and the sale or lease of 3D printing hardware. HTL.tech have partnered with Danish 3D printing specialists COBOD International A/S and major building materials supplier Roadstone Ltd – as well as academic institutions – to research and develop 3D construction printing in the UK and Ireland, especially for low-rise housing.

Figure 5.13 illustrates the 3DCP process where a fixed gantry is erected over the building footprint.

Constructing walls for housing using 3D printing is essentially no different to traditional methods – a cavity wall is 'printed', wall ties are placed as the work proceeds and insulation is inserted manually. An accelerator solution is added at the print nozzle to stiffen the concrete as it is laid by the printer, thereby allowing layers to be printed in quick succession. The process is much quicker, less wasteful, more efficient and more flexible than traditional methods.

A small site team is required for 3DCP comprising pump operator, printer operator and a supervisor whose responsibilities include supervising the print path and making adjustments as needed, placing anchors and ties and interfacing with teams dealing with the conventional construction elements.

Despite being in its infancy, 3D construction printing has been adopted for several low-rise housing projects in Ireland, and North and Central America and also for constructing an impressively sinuous staircase for a footbridge over the M8 in Glasgow[8], a two-storey 6900 square foot public building in Dubai[9] and a 54 m × 11 m × 9 m high industrial building in Heidelberg, Germany.[10]

Major advantages of 3DCP are the speed of construction, the significant reduction of site-based labour, lower-carbon, reduced waste and the consistent quality of the work.

5.7.2 Drones

Drones or UAVS (unmanned aerial vehicles) offer many uses in construction, such as taking photographs and videos, the collection of geo-data and measurements, conducting thermographic surveys – such as heat loss from roofs –3D digital modelling, topographical surveys, roof inspections and so on. They also facilitate remote inspections, especially in confined spaces, site surveys and photogrammetry, site inspections and security and the delivery and movement of materials.

Data collected by drones can be exported into software such as Autocad, Revit or BIM which can then accessed through the Cloud by any stakeholder in any location.

Drones can also assist site layout planning, real-time progress recording and the detection and avoidance of site hazards, especially where humans would otherwise be required to work at height or be exposed to risks to health or safety.

5.7.3 Wearable Technology

Wearable devices – or 'wearables' – are basically 'smart' personal protective equipment (PPE), such as hard hats, safety glasses and high-viz safety vests. They include:

- GPS (global-positioning system) which combines wearable technology, mobile, wireless and RFID (radio-frequency identification) to track the whereabouts of people and assets in real time.
- Motion tracking sensors that measure slips, trips and falls and track a worker's location on site, in real time.
- Safety jackets and hard hats equipped with detectors that enable plant operators to detect anyone in the detection zone of the vehicle who is simultaneously warned by means of a loud audible alarm.
- Smart glasses that capture and share data with supervisors and managers to ensure that key decisions are based on the right information at the right time.
- Jackets equipped with a cooling mechanism to sustain a healthy body temperature by creating airflow so that perspiration is vaporised quickly.

5.7.4 Concrete Monitoring

A significant item on the contractor's programme is formwork, which is used to support concrete in numerous structural forms – foundations, suspended slabs, walls, columns, beams and the like.

Before removing (striking) formwork, the contractor must wait to achieve the desired concrete maturity which adds time to the programme, prevents the formwork being re-used elsewhere and costs money.

Traditionally, the decision to strike formwork is informed by testing concrete cube samples and taking an informed view on weather conditions, temperature etc. Using Cloud-based technology and sensors placed strategically in the concrete as it is poured, data regarding the temperature and strength maturity of concrete can help site managers to decide precisely when to strike the formwork thereby reducing both direct and indirect costs.

The sensors may be attached to rebar or simply placed in position during the pour. There are several types of monitoring systems available:

- Thermo-couple wires which emerge from the concrete enabling a transmitter to be attached which records the data.
- Wired sensors embedded in the concrete and connected externally to a data logger which can be connected to a hand-held device for simple analysis or transferred to a computer for full processing of the data.
- Wireless sensors embedded in the concrete which connect to mobile devices using wireless technology or, alternatively, to a hub.

Reports can be sent by PDF from site to office via a hub or site-based gateway or via the Cloud. Alternatively, data can be sent to a web portal.

5.8 Planning for Modular Construction

Systemised construction poses several challenges to contractors who must adapt their working practices to different methods of construction that impact the organisation, planning and control of their projects.

Whilst modular building systems, are, by definition, standardised, there are several systems on the market, and there is no industry-standard design for use – even in the housing market. Each system is manufacturer specific. Consequently, projects that contain a significant element of pre-manufacturing require different thinking and a degree of learning to plan and schedule the work successfully and achieve the desired project outcomes.

For developers, house builders and contractors who are new to this form of construction, there is a steep learning curve.

5.8.1 The Early Stages

Early contractor involvement at the design stage can be vital because adopting a system with which the contractor is familiar can greatly add to efficiency on site.

The ordering process must be conducted in such a way that factory-made components arrive on site on time and can be offloaded in construction order. This requires a sound appreciation of lead times for the fabrication of components. Planning and organising just-in-time crane lifts to offload and erect prefabricated components must also be considered at an early stage.

For the successful implementation of DfMA, the RIBA recommends the appointment of an MMC adviser to the project team. Their role could include advising on options for choosing the most suitable construction method or system to use depending on the project desired outcomes.

The MMC adviser's input could be beneficial throughout the design, manufacture and construction stages but is particularly important ahead of appointing a manufacturer or contractor.

5.8.2 Construction Schedule

The construction schedule is inevitably driven by the design, manufacture and delivery lead times for factory-made components.

This means that enabling works have to be planned, organised and scheduled to be ready for the first just-in-time delivery from factory to site.

This type of work is ideally suited to Pull planning which is a collaborative way of defining and sequencing tasks working backwards from pre-set milestones and eliminating inefficiencies.

5.8.3 Erection Sequence

Despite the dissimilarity of the systems in use, there is, nonetheless, some commonality. It may well be the case, therefore, that modules, panels and elements may be similar from one manufacturer to another, but the erection sequence and connection system may be quite different.

In some circumstances, it might be necessary to plan and prepare suitable lay-down areas to store components should site work fall behind schedule.

5.8.4 Installation

The pre-manufactured construction market offers a variety of services to its clients. Some manufacturers offer a complete delivery and installation service requiring no input from the main contractor. Other manufacturers only offer a partial installation, however. Therefore, whilst the use of lean processes and innovative methods promises superior standards of quality, sustainability, regulatory compliance and health and safety, the end results can be variable according to the expertise of the contractor on site.

The installation of volumetric units, panelised walls and roofs and the like is specialist work, and some manufacturers have their own fixing gangs who travel around the country from project to project. Failing this, the contractor must employ a specialist subcontractor or train its own team of skilled operatives to carry out the work.

In either scenario, the contractor must learn how to integrate other subcontract trades around the specialist work package brought in to undertake installation of the primary structural system.

5.8.5 Site Work

Modular construction demands greater attention to detail for preparatory work in readiness for the erection of prefabricated components. This means 'right first time' for foundations, steelwork and other traditionally constructed elements of the project.

With pre-manufactured components, the extent of work completed at factory production stage can differ widely depending upon the system used and the manufacturer/supplier.

For instance, panelised systems might normally ship with the external panels, windows and doors factory fitted, but, on occasion, the modules might be delivered to site without these elements in place. This may be due to the manufacturer's workload or simply the lack of manufacturing space.

In these situations, the contractor may be required to complete this work on site which means employing specialist fixing gangs or procuring the necessary resources and skill levels from the supply chain. This leads to more time and more cost.

5.8.6 Tolerance and Fits

Modular and panelised systems are precision made in a factory and require similar precision on site. This includes setting out and enabling work such as foundations, slabs and holding-down bolt assemblies, locating pins, seating plinths and shims. This is the secret to a smooth and speedy installation.

In addition, a great deal of site work may be required to finish certain pre-manufactured products, as illustrated in Figure 5.14. This demands high site workmanship standards to achieve the equivalent of a factory-quality finish.

Some manufacturers provide their own site-based personnel to monitor final connections and commissioning according to their own standards and procedures, but this is not always the case.

Other suppliers simply deliver, install and connect their product leaving the main contractor to complete outstanding work items. Not only is this a risk to finish quality, but it also adds cost because the work is 'bitty'. It also adds time to the construction schedule.

Insulation

Aquapanel and brickwork

Cladding

Roofing

Figure 5.14 Site finishing works.

Ideally, the main contractor will have access to key supply chain partners who understand the systems being used and can achieve the supplier's exacting quality requirements.

References

Cast (2019) *Modern Methods of Construction – Introducing the MMC Definition Framework*.
HM Government (December 2020) *The Construction Playbook*.
RIBA (2021) *DfMA Overlay to the RIBA Plan of Work*, 2nd edn.

Notes

1. Wikipedia.
2. www.dictionary.com/browse/panelised.
3. Ministry of Housing, Communities and Local Government's Joint Industry Working Group on MMC.
4. Ronan Point 1968.
5. https://www.brydenwood.com.
6. The Forge, 105 Sumner Street, London SE1, 2021, Architect: Brydon Wood, Contractors: Sir Robert McAlpine, Mace.
7. https://www.dfma.com.
8. https://constructionmanagement.co.uk/3d-printing/.
9. http://www.dailymail.co.uk/sciencetech/article-7975233/Worlds-biggest-3D-printed-building-opens-Dubai-6-900-square-foot-government-office.html.
10. https://www.peri3dconstruction.com/en.

Chapter 6 Dashboard			
Key Message		Selecting the right contractor is a crucial step in any project and requires great care, expert advice and careful evaluation.
Definitions		**Procurement**: The award, entry into and management of a contract. **Contractual arrangements**: The type and form of contract used.
Headings		**Chapter Summary**
6.1	Introduction	Procurement concerns the arrangements for organising and managing the design and construction of a project and, possibly, its operation. Procurement choices should be made based on issues that are important to clients.
6.2	Tendering	Most construction work is transacted by the submission of bids, usually in competition. Bids are normally based on tender documentation supplied by the client which will vary according to the procurement method employed. Tendering contractors are usually chosen from an approved list or a framework.
6.3	Special Relationships	Client to professional adviser and client to contractor are two common relationships found on construction projects. Other relationships include partnering, alliances and joint ventures.
6.4	Procurement Routes	Methods of procurement include general contracting, design and build/construct, management procurement, PFI/PPP and EPC/Turnkey. General contracting is where a design is prepared by an architect or engineer. Tenders are invited and a contractor is appointed to carry out the works. General contracting is also known as 'traditional', 'end-on' or 'design-bid-build' procurement and may be a single or two-stage process.
6.5	Contractual Arrangements	The procurement route does not define the entire strategy for a project. This also requires consideration of suitable contractual arrangements. It is important to recognise the basic principles of law as they apply to the formation of construction contracts which may vary according to the relevant legal jurisdiction.
6.6	Standard Forms of Contract	Standard forms of contract offer clients and their professional advisers various options designed to fit with different procurement routes and different types and sizes of project. Common contract families include JCT, NEC, ICC and FIDIC. Provisions regarding the contractor's programme differ according to the form of contract used.
6.7	Types of Contract	Contractual arrangements comprise the form of contract and type of contract. Lump sum, measure and value and cost reimbursement are the main types of contract. Each has a number of subsets.
6.8	Time Management	The preparation of programmes (schedules) is central to project procurement. The project programme (or master schedule) is an important part of the procurement strategy for a project. It is central to the procurement framework and sets the tone for the time management of the project.
Learning Outcomes		Understand the meaning and significance of procurement in construction. Know the different procurement routes available. Appreciate that contractual arrangements are part of a coherent procurement strategy. Be familiar with common forms of contract and their basis in the law of contract.
Learn More		Chapter 7 – Estimating and bidding. Chapter 4 and especially 4.11 – Managing the supply chain.

6

Procurement and Contracts

6.1 Introduction

Construction clients engage with the construction market for many different reasons. Some are looking to build – a house, a housing development, an extension to a business premises, a commercial property to lease to tenants, a new bridge or, perhaps, a road or tunnel to relieve traffic congestion. Other clients may not wish to invest capital in a construction project at all but may simply need a facility – a fully functioning hospital or prison for example – which they hope to pay for out of revenue.

In either case, the client's choices will hinge upon where the finance for the project is coming from, what degree of professional advice and support is needed, what control the client needs to have over the design process, project costs and supply chain integration and when the client wishes to appoint contractors.

Once the decision to proceed with a construction project has been taken, arrangements need to be made for organising and managing its design and construction and, possibly, its operation and maintenance. This is commonly known as *procurement*.

6.1.1 Definitions

The Procurement Act 2023 – which replaces the Public Procurement Regulations 2015 post Brexit – provides a statutory definition:

- **procurement** means the award, entry into and management of a contract;
- **covered procurement** means the award, entry into and management of a **public** contract.

This is a useful definition because it clearly identifies three elements to procurement:

- **The award of a contract**: the process of arriving at the point where an offer from one party has been accepted by another.
- **Entry into a contract**: the process of formalising the agreement into a legally enforceable contract.
- **Management of a contract**: the procedures and protocols to be followed when carrying out a contract and bringing it to a conclusion.

Construction Planning, Programming and Control, Fourth Edition. Brian Cooke and Peter Williams.
© 2025 John Wiley & Sons Ltd. Published 2025 by John Wiley & Sons Ltd.

Other definitions of procurement conclude that it is:

- *the framework through which the client engages with the market and the supply chain to deliver packages of work using one or more contractual instruments, reflecting the appropriate balance of cost risk and uncertainties the client contractor and suppliers are willing to bear* (Institution of Civil Engineers 2021).
- *essentially a series of considered risks* (Chartered Institute of Building 2010).

The procurement strategy chosen for a project – and the contractual arrangements between the parties to the contract – represent the framework within which construction projects are carried out.

These arrangements are complex because they not only involve client–contractor relationships but also encompass those with professional consultants, subcontractors, specialist contractors and suppliers. Frequently other stakeholders will have an interest in a project too, such as project funders, end-users, government departments, local authorities, local organisations, schools, shops and businesses.

6.1.2 Public Sector and Private Sector Procurement

In the private sector, clients are free to make their own procurement arrangements as they see fit, consistent with the law of contract. However, in the public sector, there are important requirements that contracting authorities and clients/sponsoring bodies must comply with:

- The Procurement Act 2023 introduces new and far-ranging legislation covering the procurement of public sector projects and frameworks. This includes greater flexibility for contracting authorities, more choice in selecting contract award criteria, greater transparency, a requirement that contracting authorities regularly publish key performance indicators (KPIs) and a move away from the familiar EU directive language.
- Public works projects and programmes are subject to government policies concerning how they are assessed, procured and delivered, including the *Construction Playbook* (2020). This contains 14 key policy reforms to enable 'faster, better, greener' construction. These policies include 'Effective Contracting', which aims to ensure that contracts are structured to support data exchange, collaborative working, improved value and to manage risk with the expectation of continuous improvement.

Following the UK's withdrawal from the EU (Brexit), contracting authorities and utilities are no longer required to advertise public sector projects in the Official Journal of the European Union (OJEU). Instead, from 1 January 2021, notices may be published on a new UK e-notification system – 'Find a Tender'.

6.1.3 Procurement Strategy

Conventionally, client procurement choices are made based on the supply-side procurement options available, such as general (or traditional) contracting, design and build/construct, management contracting, construction management and so on.

This approach can lead to loss of focus on client objectives which become blurred by a myriad of contractual and legal complexities and over-emphasis on the roles of project participants – designers, contractors, subcontractors and so on – rather than the simple issues that occupy the minds of most clients, such as:

- How will the project be funded?
- What will be the basis for calculating the price of the project?
- Who will manage the project and integrate the supply chain?
- How will the required expertise be chosen?
- Who will be responsible for design?
- What degree of collaboration is preferable?
- Who will coordinate the work on site?

By adopting a demand-side focus on procurement decision making – and by following the guidance and recommendations in BS8534 *Construction procurement policies, strategies and procedures* – procurement choices can be made based on issues that are important to clients.

The client's procurement strategy provides a means of:

- Finding suitable advice, design and other professional services as well as a competent team to construct and commission a project.
- Refining exactly what is required to complete the project and deciding when and how this should take place.
- Engaging the parties in a suitable legal relationship with appropriate duties, rights and obligations enforceable in law.

The legal relationships that result from the procurement process are defined in contracts – formal documents agreed and signed by the parties that set out their respective legal obligations.

6.1.4 Design Procurement

A crucial consideration when choosing the procurement route for a project is the project design. Decisions must be made at an early stage such as:

- Who will prepare the design?
- Who will bear the design risk?
- When, if at all, will the contractor be involved in the design?

Design is a complex and creative process requiring imagination, the aptitude to think out of the box, knowledge of the history of the built environment and the capacity to deal with a multitude of options, factors and uncertainties.

The ability to interact with clients and gain an understanding of their requirements is crucial in order to define a brief which best represents the clients functional, aesthetic, social and economic objectives. Design is also a collaborative process requiring the ability to interact with project team members.

Clients have several choices:

- Appoint consultants to prepare the design intent and to manage its development.
- Combine a consultant design with an element of partial contractor design.

- Appoint consultants to prepare the design brief and entrust the design development to a contractor.
- Involve contractors at an early stage in the project to help develop the design alongside the client's design team.
- Appoint consultants to prepare the outline design and novate the designers to a design-build contractor.
- Negotiate with design-build contractors to prepare the brief and develop the design.

The process models described in Chapter 4 identify 'design' within the project process and the RIBA Plan of Work, in particular, emphasises the timing, core tasks, statutory processes and information exchanges required. The RIBA Plan of Work also identifies the role of the design programme from Stage 2 (Concept design) to Stage 4 (Technical design).

The design programme should include key dates and timings for delivering the various design stages as well as provisions for cost planning, tendering and contract award and for the construction period. Important deliverables include various studies and reports, impact assessments, risk evaluation, design options, the development of drawings and models, structural and other calculations, preparation of specifications and cost studies.

6.1.5 Contractor Selection

Selecting the right contractor is a crucial step in any project and requires great care, expert advice and careful evaluation.

Expert clients may have 'go-to' contractors for projects of specific value and complexity but, in most cases, contractors should be pre-qualified following an expression of interest in bidding for the work.

Pre-qualification is the process of determining a contractor's experience, capacity, technical expertise and financial stability. Conventionally, this involves the completion of a pre-qualification questionnaire (PQQ) following initial short-listing. PQQs are notoriously bureaucratic, paper exercises that are invariably generic in nature. Frameworks – see later in this Chapter – avoid this repetitive and time-wasting process as, once established, framework members are pre-qualified for up to four years.

Other methods of establishing the credibility of a contractor include:

- Turnover ratio – where the capacity of the contractor to undertake a project is measured by the ratio of the contract value to the contractor's annual turnover. Prescribed limits may be set by legislation for public sector contracts or otherwise by 'rule of thumb'.
- The ability of a contractor to provide financial guarantees or bonds in the event of the contractor's default on a contract. They could take the form of parent company, bank or insurance company guarantees or performance, conditional or on-demand bonds provided by a bank or insurance company.
- The levels of insurance cover required for a specific contract relate to contract value and risk. The ability of a contractor to provide the required indemnity limits is a measure of financial capacity.

6.2 Tendering

Most construction work is transacted by the submission of bids, usually in competition. Tendering contractors base their bids on the tender documentation supplied by the client which will vary according to the procurement method employed.

Tendering in construction is either a single-stage or two-stage process:

- Single-stage tendering is a perpetuation of traditional procurement, where design is separated from construction and bids are invariably submitted based on an incomplete design. Once a suitable tender has been received, the client and successful contractor enter into a formal contract.
- Two-stage tenders facilitate early contractor involvement (ECI) – a collaborative tendering process involving both contractors and designers. They are particularly beneficial compared to single-stage tendering when the nature of the works is complex, or when time deadlines are critical:
  - Initial tenders are invited from a small number of contractors, usually based on outline information.
  - This establishes an order of cost.
  - One (or two) contractors may then be chosen for Stage 2 which concludes with a formal detailed tender offer and subsequent contract award.

6.2.1 Tender Documents

Construction contracts are invariably awarded on the basis of competition using a variety of pricing documents as the basis of the contractor's bid.

Common pricing documents include:

- **Bills of quantities**: either formal or informal.
  - Formal bills of quantities (BQs) are traditionally prepared by the client's quantity surveyor based on a formalised method of measurement such as NRM2 or the CESMM.
  - Informal BQs are prepared by contractors (or subcontractors) and may be loosely based on a method of measurement.
- **An activity schedule**: *a list of unquantified construction activities ... often ... linked to the contractor's programme* (Williams 2016).
- **Schedule of rates**: used when the scope of work is unknown or uncertain. They resemble bills of quantities in format, but there are no quantities. Tenderers price the items based on an indicative scope of work.
- **Schedules of work**: a list of composite items of work, usually containing several building trades in one item. A single item may include brickwork, plastering, joinery and painting for example. No quantities are given. They bear a resemblance to activity schedules.

6.2.2 Approved Lists

In traditional procurement, clients appoint an architect or engineer and other designers, perhaps a project manager and a cost consultant. Clients, or their advisers, then draw up lists of contractors suitable for the work envisaged. The selection criteria includes the size of firm, in relation to the value of project, their reputation and financial standing, technical competence, health and safety record and track record of successful project delivery.

Contractors are then invited to complete PQQs and, maybe, attend for interview. Tender lists may be:

- Standing approved lists lasting for defined periods. It is usual to have approved lists for a variety of project types and sizes determined by their value. This means that an appropriate size contractor can be appointed to undertake the work. For a specific project, a limited number of contractors will be invited to submit bids – usually between four and six firms. As a standing approved list may contain a significant number of contractors, tender invitations are rotated to ensure fairness to all firms on the tender list.
- Ad hoc approved lists assembled for one-off projects. In the private sector, clients are free to choose their preferred contractors and may repeatedly invite the same firms to tender. In the public sector, public money is at stake, and contracting authorities need to show fairness to all contractors wishing to tender for work. Consequently, there may well be some rotation of tender lists in the public sector.

6.2.3 Frameworks

An alternative to an approved list is a framework agreement. They are widely used in the construction industry – especially in the public sector – and are defined under Section 45(2) of the Procurement Act 2023:

> A 'framework' **is a contract** between a contracting authority and one or more suppliers that provides for **the future award of contracts** by a contracting authority to the supplier or suppliers.

It is clear from this definition that there is a distinction between framework 'agreement' and a framework 'contract' which suggests that frameworks are assembled by pre-qualifying a multi-supplier supply chain structure in a two-stage process:

- **Stage 1**: Award of a contract, without guarantee of work, with selection based on soft criteria, such as reputation, quality management, health and safety record and past performance. This contract is a 'framework agreement' which is an understanding between two parties that establishes the necessary agreed terms and conditions for possible future contracts. The agreement is not an enforceable contract as there is no consideration.
- **Stage 2**: Award of a contract, at some future date, for professional services or works or other service based on hard selection criteria, such as price, time and value added. The 'framework contract' could be awarded by negotiation or by mini competition between 'suppliers' on the framework. The consideration required to establish the contract could be a fee percentage, a fixed fee or a lump sum price or it could be a pricing formula such as a schedule of rates or a target price.

This process enables clients to award contracts for design services, site investigation, demolition, enabling works, supplies of materials or components and construction services – more or less anything that the client wants.

Framework agreements are relevant where a client has the prospect of a significant amount of future work such as a national highway maintenance or renewal programme, a programme of maintenance works for a portfolio of premises or a large infrastructure programme comprising many individual projects:

- Frameworks are not suitable for 'one-off' projects because there is a considerable and costly administrative burden in setting up the framework.
- Once a framework is established, additional supply chains for further projects may be put together using the same framework but not necessarily using the same suppliers.

The parties to a framework agreement agree to be bound by its terms for a fixed period (typically four years) but are free to negotiate or compete for individual contracts under the framework. There are two facets to frameworks:

1. The overarching general agreement that sets out how the framework will operate.
2. Individual contracts that are entered into based on those pre-agreed terms and any other terms that might be jointly deemed appropriate by the parties.

Framework agreements establish the terms governing a long-term relationship where one party (such as the client/employer) intends to enter into future contracts with one or more 'supplier'. The agreement establishes the 'framework' for a strategic partnering relationship which can provide continuous improvement and efficiency gains for the employer and the prospect of long-term turnover for the various suppliers.

> The framework supply chain model is conventionally expressed in levels or tiers with all the members of the supply chain frequently referred to as 'suppliers'.
>
> This follows the EU procurement directive which classes everyone engaged by a client as a 'supplier' – architects, engineers, project managers, quantity surveyors, contractors, subcontractors – they are all 'suppliers'.

There can be any number of 'tiers' in the model, but three is quite normal, as illustrated in Table 6.1. In a multi-tier framework, it is normally only in the first tier where the client makes the appointments:

- Tier 1 suppliers are those with a direct commercial relationship with a client. They include project managers, designers and other consultants as well as main contractors.
- Tier 2 is where subcontractors and suppliers have a direct contract with the Tier 1 main contractor. This can include designers working directly for contractors and novated designers – designers initially engaged by a client who, at some point, are transferred (novated) to the main contractor.
- Tier 3 includes those subcontractors and suppliers engaged directly by Tier 2 subcontractors.
- There may be several other tiers beyond Tier 3 where there are sub-sub-contracts, etc.

Table 6.1 Multi-tier framework.

	Suppliers	
Tier 1	Professional consultants Architects; Structural engineers; heating ventilating and air conditioning (HVAC) engineers	Main contractors
Tier 2	Specialist contractors e.g. Piling; Structural steelwork; M&E	
Tier 3	Trade contractors e.g. Groundworks; Plastering; Painting	

Tier 1 suppliers	Are engaged by the client/employer. This includes all professional appointments and those better known as 'main contractors'. In this scenario, the main contractor might be a single company or, for larger, riskier projects, there might be a joint venture of two or more 'main' contractors.
Tier 2 suppliers	Are synonymous with traditional 'subcontractors' but might also include companies that provide geotechnical advice or carry out site investigations or undertake demolition or asbestos removal services. Tier 2 suppliers are normally engaged by main contractors.
Tier 3 suppliers	Are engaged by Tier 2 contractors and represent sub-sub-contractors in the traditional sense.

In certain cases, Tier 2 suppliers may be engaged directly by a client and then novated to the main contractor once the main contract has been awarded. This can happen when a client is able to benefit from substantial economies of scale due to their buying power for certain materials. National highway authorities, for instance, procure huge quantities of macadam products annually and utility companies can similarly place substantial orders for cast iron drainage pipes and fittings, specialist equipment and so on.

Framework agreements have their own vocabulary:

- **Call-off**: The award of a contract under the framework agreement. The contract does not have to be advertised or tendered but could be awarded following a mini-competition.
- **Pricing document**: This is the pricing document returned by the 'supplier' to the client/contracting authority. It could be a schedule of rates, a priced bill of quantities, an activity schedule or simply a lump sum or percentage fee offer.
- **Contracting authority, employer**: In the public sector, a contracting authority is the public authority – such as local authority, government department or government agency – responsible for awarding contracts under the framework.
  In the private sector, the term 'employer' would be used in the framework contract as in conventional contracts.
- **Supplier, provider or economic operator**: Terms used in framework agreements, or relevant legislation, to identify organisations that provide services, work or goods (materials or components) under the framework.
- **Task**: A specific service or work package to be carried out under the framework.
- **Enquiry**: An invitation to tender from the employer to a prospective provider to carry out a task.

6.3 Special Relationships

By its very nature, the procurement process creates a great many relationships. Most of these are well-known and understood:

- **Client to professional adviser relationship**: design services, cost advice, project management under a professional services agreement.
- **Client to contractor relationship**: the provision of works services under a main construction contract.
- **Contractor to subcontractor relationship**: the provision of works services under a construction subcontract.
- **Contractor/subcontractor to supplier relationship**: the supply of goods, materials and components under a supply contract governed by the Sale of Goods Act 1979.

Other relationships can be created in construction where the parties involved wish to commit to a special arrangement based on *mutual trust and cooperation* (NEC4) and an alignment of values, collaborative effort, the sharing of skills and expertise and the common goal of successful project delivery. These relationships include:

- Partnering.
- Alliances.
- Joint ventures.

6.3.1 Partnering

Partnering may be defined as:

> *a structured methodology for organisations to set up mutually advantageous commercial arrangements, either for single projects or in long-term strategic relationships, which help their people work together more effectively.*
> (Construction Industry Board 1997).

It is an entirely voluntary arrangement which operates within the chosen procurement method, but it may become formalised in the contract should the parties so desire. The essential components of partnering are mutual objectives, problem resolution and continuous improvement, leading to:

- 'Win–win' solutions.
- Long-term profitability.
- Openness, where no one benefits from the exploitation of others and innovation is encouraged.
- Awareness of the needs, concerns and objectives of others and in helping their partner to succeed.

Partnering arrangements are quite normal in construction and may take the form of:

- **Strategic partnering**: where a relationship is developed over an indefinite period and where there is a long-term commitment to the partnering approach.

- **Project partnering**: where partnering is felt to be appropriate for a particular project or length of time, such as a three-year maintenance contract.

> Partnering arrangements may be based on a formal contractual arrangement between the partners, but some project teams do not want their partnering arrangements to become legally enforceable as this might be contrary to the very principles of partnering.

As an alternative to a contract, there can be a non-legally binding relationship based on a charter with a mission statement and a set of common objectives. The charter could be drawn up specifically for the project, or an 'off-the-shelf' document could be used. Whether or not there is a formal agreement, there needs to be a basis for agreeing prices and conditions of contract that are fair to both parties. There may also be a need to incentivise the contract such as having a target cost with a guaranteed maximum price (GMP) arrangement which can lead to:

- Cost savings through value engineering and improved buildability.
- Improved communications and reduced site correspondence.
- Good relationships and repeat business.
- Effective cost management of change.
- Zero accidents.

6.3.2 Alliances

An alliance is an agreement where the parties to it undertake to act in a certain way to achieve a common objective. The parties might agree to a basic partnering arrangement, or the agreement could extend to a fully integrated 'pure' alliancing model or even to a long-term 'strategic' relationship. An alliance may include varying levels of collaboration and risk sharing.

A pure alliance is a multi-party agreement between the key project stakeholders – client, designers and consultants, contractor and key subcontractors.

> Alliances are to be distinguished from partnering because the parties agree to form a unified entity that shares the risks and rewards of a project based on an agreed formula and within the ambit of a multi-party contract.

In partnering, the bi-lateral nature of the contracts used means that the parties retain their independence and may individually profit or suffer losses from the arrangement. Members of an alliance, conversely, 'sink or swim' together.

Alliances enable projects to start quickly and create a structure within which alliance members can react to difficulties and change with the least delay. They are suited to large and technically complex projects and encourage a collaborative, non-adversarial approach based on a contract that promotes openness, trust and alignment to project delivery instead of self-interest.

Common collaborative contracts include:

- **NEC Engineering and Construction Contract (ECC) Secondary Option X12**: facilitates partnering and collaborative working. It is an overlay rather than a true alliance contract because contracts between clients, consultants, contractors and subcontractors are bi-lateral.
- **The NEC4 Alliance Contract**: a multi-party contract aimed at creating an alliance to deliver a major project or programme of work. All parties sign up to a single contract.
- **The JCT Constructing Excellence Contract**, which is geared towards partnering, collaboration and integrated working.
- **FAC-1**: a framework alliance contract – developed by King's College London – that can be used alongside JCT, NEC, FIDIC and other contract forms.

6.3.3 Joint Ventures

A joint venture (JV) is a means of creating a relationship between two or more firms with the objective of undertaking economic activity together.

JVs are beneficial when two or more partners have complementary skills, expertise or commercial strengths, and they also provide a means of spreading commercial risk between the parties to the JV, especially on large or complex projects. JVs are common between contracting companies and between contractors and support service providers. They may also include design firms, geotechnical experts and so on.

A JV can be created for a specific project, or the parties might have intentions of working together long-term. The relationship between the parties can be formal or informal. JVs may be:

- **Incorporated:** where a separate company is formed for the purpose, and each JV member will own a percentage of the share capital.
- **Unincorporated:** where joint parties execute a project without a separate legal company being formed for the purpose. This is known as a joint operation.
- **By formal agreement:** where the parties enter into a contract to undertake a project, or a series of projects, without the need to create an incorporated company. The contract establishes the duties and obligations of each member of the JV and is subject to the normal law of contract of the relevant jurisdiction. The contract determines the balance of risk between the parties and how profits or losses will be shared.

A name for a JV is normally agreed by its members, and they will also agree the terms of the agreement. Other matters to be agreed include the JV's place of business, its management structure, the capital contributions (cash or property) of members, responsibility for the preparation of accounts, the timing and conduct of meetings and liability and insurance.

JV partners are jointly and severally liable for their contractual obligations. Should one partner become insolvent, then the remaining member(s) must complete the contract unless a replacement contractor can be found.

Figure 6.1 illustrates how HS2 Phase 1 has been organised into areas: North, Central and South. Each area has been awarded to separate JVs due to the size of the project and its geography. Apart from the Tier 1 JV appointments, there are several other JVs at Tiers 2 and 3 with contracts to carry out the various lots and sub-lots.

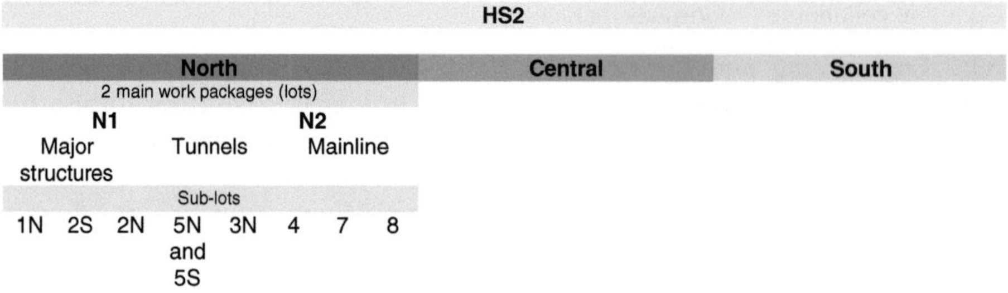

HS2								
North					**Central**		**South**	
2 main work packages (lots)								
N1			N2					
Major structures		Tunnels	Mainline					
Sub-lots								
1N	2S	2N	5N and 5S	3N	4	7	8	

Figure 6.1 HS2 joint ventures.

HS2 Area North has been awarded to the Balfour Beatty/VINCI (BBV) joint venture. It is divided into two main lots, each of which comprises major structures, tunnels and mainline projects and is further subdivided into eight sub-lots. Area North comprises over 300 individual projects.

There are several other joint ventures working on Area North, including:

- Mott MacDonald and SYSTRA – a design joint venture for lots N1 and N2.
- LM – a joint venture of Laing O'Rourke and J. Murphy & Sons Ltd undertaking enabling works to prepare the route ready for full construction to commence.

6.4 Procurement Routes

Procurement differs from purchasing because, once in place, procurement arrangements facilitate the placing of orders with one or several suppliers for one or several products or services, as required. Purchasing is a subset of procurement and operates at a much more granular level than procurement which is, essentially, strategic in nature.

There are several methods of procurement in the construction world including:

- General contracting.
- Design and build/construct.
- Management procurement.
- Private Finance Initiative (PFI)/Public Private Partnership (PPP).
- Engineer Procurement and Construction (EPC)/Turnkey.

6.4.1 General Contracting

General contracting is where a design is prepared by an architect or engineer, tenders are invited from contractors and a contractor is appointed to carry out and complete the works.

It is also known as 'traditional', 'end-on' or 'design-bid-build' procurement and may be a single or two-stage process.

For single-stage tendering, the intervening tendering period separates design from construction with no contractor input which can impact the cost efficiency and practicality of the design. Alternatively, two-stage tendering could be employed which has the dual benefit of accelerating the process and of contractor involvement with the design. In this case, one or two preferred bidders may be short-listed to discuss the project and enter into competitive dialogue prior to the submission of final bids.

For price certainty, the project should be fully designed before tender, but this is frequently not the case with traditional procurement. This can lead to design changes during construction and problems with the coordination of project information, leading to extra costs and, possibly, disputes.

General contracting usually leads to:

- A lump sum contract based on:
  - full documentation with bills of quantities supplied by the client.
  - drawings and specification where the builder prepares the quantities.
- An admeasurement contract – often referred to as remeasurement contract – based on approximate or notional quantities, where the contract price is determined at the end of the contract by measuring the actual work carried out.

The architect is normally regarded as the leader of the design team for building projects and the engineer in civil engineering. They coordinate the input of specialist designers and cost consultants etc., as well as undertaking duties regarding the administration of the client's contract with the contractor. For larger projects, clients may engage a design and/or Building Information Management (BIM) coordinator who works with the lead designer.

Figure 6.2 illustrates the relationships between the client, the client's design team and the main contractor in traditional procurement and Table 6.2 contrasts the advantages and disadvantages.

6.4.2 Design and Build/Construct

Design and build procurement – or design and construct in civil engineering – is a means by which clients can benefit from the construction expertise of contractors to produce a buildable design and also transfer some, or all, of the design risk to the contractor. Its advantages and disadvantages are contrasted in Table 6.2.

Contractors usually bid for the work in a limited competition, as tendering costs are significant, and will respond to the client's requirements by submitting contractor's proposals and a price. Once appointed, the contractor will be responsible for delivering the completed project in accordance with the agreed design.

Should there be a difference between the client's requirements and the contractor's proposals, they would need to be resolved through discussion and price negotiation or, where more than one contractor's proposals are on the table, by competitive negotiation.

The basic idea of design-build is that there is a single point of responsibility for both the design and construction of the project. The client benefits from the contractor's expertise during the early stages of the project, and there is also a common law fitness for purpose liability for the contractor's design. This benefit is often lost under some standard contract conditions which limit the contractor's design liability to that of an architect (i.e. the lesser standard of reasonable skill and care).

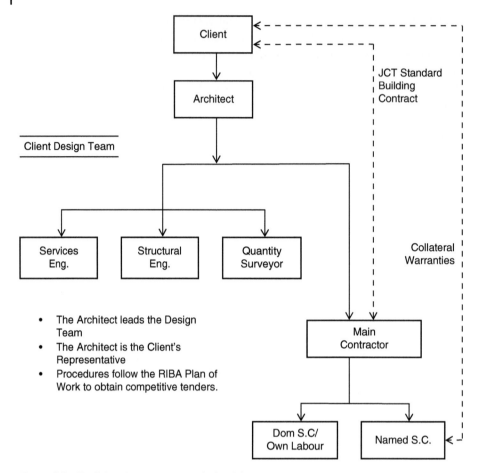

Figure 6.2 Traditional procurement relationships.

There are several variants of design and build:

- **Client-led design and build**: This variant is often used in the public sector where a client has specific design requirements. The design will be developed to an advanced stage – possibly RIBA Stage 4 – by the client's design team before contractors are involved. There may also be prescriptive or performance specifications for specialist installations such as heating, ventilating and air conditioning available at this stage.

  Bids are invited based on the client's design and the preferred contractor inherits the design and develops the detail required for the production stage. The contractor would need to be appointed at an earlier stage – perhaps RIBA Stage 3 – to be able to contribute anything imaginative to the design.

  Figure 6.3 illustrates the relationships in this variant, which is also referred to as *develop and construct* because the contractor develops an already substantially complete design.

Table 6.2 Procurement methods compared.

Procurement method	Advantages	Disadvantages
Traditional	- Price certainty before committing to a contract. - A firm date for completion. - Client control over design and quality. - The best market price for the full scope of the works. - Variations and contract changes accommodated within the contract.	- Client takes the design risk. - Potential for delayed completion due to variations and extra work. - The final account may exceed the contract price. - Potential for disputes.
Design and build	- Main contractor provides single-point responsibility for design and construction. - In most variants, major risks lie with the contractor. - Price certainty and less risk of price changes during design development. - Contractual completion dates are fixed early in the design process.	- The contractor assumes greater financial risk, and this is often reflected in the price. - Difficulties arise for the employer in matching like-for-like prices at tender stage. - The tender period and negotiation tend to be much longer. - It is more problematic to control design and quality. - Changes can be disruptive to the contractor and expensive for the client. - The project can become price driven at the expense of quality.
Management	- The expertise of the management contractor benefits buildability and value engineering. - Maximum overlap between design and construction. Once initial work packages are designed, work may commence on site. - A high level of supervision and quality control can be achieved at site level. - Variations can be kept under control as the price of changes can be agreed before work packages are let. - Relationships are less adversarial as the contractor is part of the project team. - Advantageous for complex projects where the design can be developed in stages.	- The client has no commitment from the contractor on price certainty. - Strict control of the cost plan/work package budget is essential. - The total cost of the project is not usually known until the project is well into the construction phase. - Blue-chip work package subcontractors are often chosen with no incentive to reduce costs. - Early letting of packages may lead to extensive design changes. - Damages for delay are difficult to pin on one subcontractor and they are expensive to negotiate. - The client takes all the risk, particularly in the construction management variant.

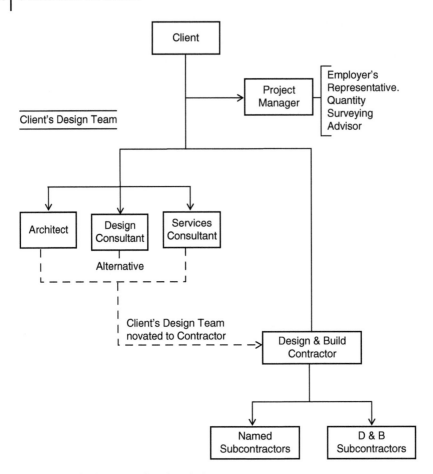

Figure 6.3   Design and build client-led.

- **Novated design and build**: This variant is a twist on develop and construct, where the design team is novated to the contractor once the contract has been awarded. Novation ends the client-design team contract and creates a new legal relationship between the design team and the contractor. The reason for the novation is to transfer design risk from client to contractor and to enable the original design team to work with the contractor to produce the detailed aspects of the design – RIBA Stage 4. The novation arrangement should be agreed with the design team when first appointed.
- **Contractor-led design and build**: Figure 6.4 illustrates how this variant is structured which is often referred to as *traditional design and build*. Tendering contractors are provided with minimal client's requirements – an outline brief or list of spatial requirements and, possibly, conceptual drawings depending upon whether the client has engaged a designer. Contractors will formulate a design – perhaps to RIBA Stages 2 or 3 – based on their interpretation of the client's requirements and will submit a bid on this basis. Whether the design satisfies the client, however, is a matter for conjecture and further

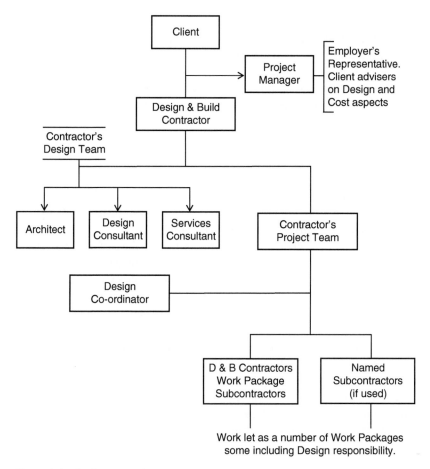

Figure 6.4 Design and build contractor-led.

negotiations with contractors may be needed. In any event, full design liability will be taken by the contractor.
- **Design and manage**: This is very similar to management contracting, except that both the construction works contractors and the design team are engaged directly by the contractor. The design–manage contractor is often selected in competition, usually on the basis of a design and management fee and, possibly, the contract preliminaries. Since there is scant firm information at tender stage, the client has very little control over eventual design quality. As the contractor's design emerges, packages of work can be prepared and prices can be obtained under the scrutiny of the client's cost adviser. The contractor undertakes the design risk using this method.
- **Design, manage and construct**: This is a similar method to design and manage, but the contractor undertakes some of the work packages as well as coordinating the other package contractors on site.

- **Package deal**: Package deals were one of the original methods of design and build where contractors offered standard building designs which were then adapted to suit individual clients and site conditions. Some of the designs were system buildings. This method of design and build still suits some clients.

    Package deals are sometimes called turnkey contracts, but true turnkey projects provide the client with a fully functioning building, including the provision of staff working inside, and the method is mainly used for overseas projects.

6.4.3 Prime Contracting

Prime contracting is similar in concept to design-build (D&B), design-build-operate (DBO) or design-build-operate-maintain (DBOM) arrangements where a contractor or JV is solely responsible for the design, and delivery of a construction project.

This method of procurement is often used where clients require the performance related operation of a facility within a pre-agreed performance specification and cost model. The level of design liability under the contract is one of fitness for purpose rather than reasonable skill and care which might be the case with some D&B contracts.

Selection procedures can be similar to standing approved lists or frameworks where 'prime contractors' are carefully chosen, usually via several stages of selection based on both 'hard' and 'soft' criteria. Prime contractors must be able to provide all the project deliverables for capital projects, including design, planning and cost control.

Prime contracting can be used for property maintenance as well as capital works and, once chosen, the prime contractor may become part of the client's project team for several years. This enables the client to take a 'hands-off–eyes-on' approach to their projects but with the partnering principles of mutual trust and cooperation very much as the focus.

6.4.4 Management Procurement

Management procurement is a means of engaging with a contractor at an early stage in a project based on a negotiated or competitive fee rather than a tender for the whole of the works. Once the client has appointed consultants, a project manager and cost manager, the contractor can join the project team. Under this method of procurement, the construction work is carried out by separate works or trade contracts who are appointed once the design is sufficiently advanced.

This arrangement effectively connects the design and construction phases and means that the client can benefit from the contractor's expertise regarding the practicability of the design, alternative construction methods, costs and value engineered solutions and the programming of the works packages. A cost plan is developed as the design advances which can be checked against competitive work package tenders once the design is sufficiently well developed.

Unlike traditional procurement and design-build/construct, the main contractor is not responsible for the unilateral procurement of subcontractors or for carrying out any of the physical work on site. The various works/trade packages are scoped and procured within the client team, of which the contractor is part, and the contractor is paid a fee based on a professional service.

The key advantages and disadvantages of management procurement are shown in Table 6.2 and may be distinguished from general contracting or design-build/construct because:

- The main contractor is part of the client team.
- A fee is paid for the services provided.
- The main contractor does not profit from the procurement of subcontractors as this is managed by the client team and within the client's cost planning of the project.

There are two common variants to this method of procurement which both benefit from the early involvement of the contractor:

- **Construction management**: This is where the construction manager acts essentially as a consultant to advise the client team and to plan and supervise the work on site. The construction manager does not directly undertake any of the construction work which is carried out by work package or trade contractors. The client enters into contracts directly with the various work packages/trade contractors and is directly responsible for payments and contract administration. The client plays a major role in the direction of the project.
- **Management contracting**: This variant is similar to construction management, but the management contractor is in direct contract with work packages/trade contractors, and they are contractually accountable to the management contractor. The management contractor provides the site infrastructure – site security, offices and welfare facilities, materials compound and so on – as part of the fee and also coordinates the work on site. During the pre-construction phase a programme of works packages is developed which are then let by competitive tender.

Figures 6.5 and 6.6 contrast the role of the contractor in management procurement and illustrate the relationships between the parties in both variants.

6.4.5 Public–Private Partnerships

Public–private partnerships are a means of securing private investment in major public building and infrastructure projects. They operate all around the world and enable governments to reduce public sector borrowing and transfer the risk of funding, designing, constructing and operating public facilities to the private sector.

The UK PFI and PF2 schemes are examples of a particular type of public-private partnership schemes, where long-term contractual arrangements were made between a public sector entity and a private sector provider (concessionaire). Projects were operated over a concession period of between 20 and 30 years, thereby providing an income stream for the concessionaire and use of the facility by the public sector client. Examples of income streams include:

- Selling electricity from a new nuclear power station.
- Collecting tolls from drivers using toll roads or bridges.
- Facilities fees for the use of a school, prison or hospital, for example.

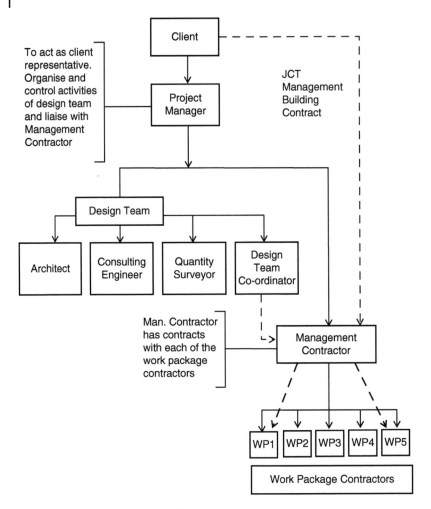

Figure 6.5 Management contracting arrangement.

Whilst there are several hundred PFI and PF2 schemes still running, they have been found to be costly, complex, inefficient and poor value and have no longer been used in the United Kingdom since 2018.

A variety of arrangements may be used for such projects depending upon the extent of design involvement, operation, maintenance or transfer of ownership required. They include:

- **BOOT**: Build Own Operate Transfer.
- **DBFO**: Design, Build, Finance Operate.
- **DCMF**: Design, Construct, Manage, Finance.

Other forms of public-private partnerships are also used both at government and local government levels. These might be contractual, corporate or collaborative arrangements designed to meet the requirements of specific public sector projects, especially in housing and regeneration.

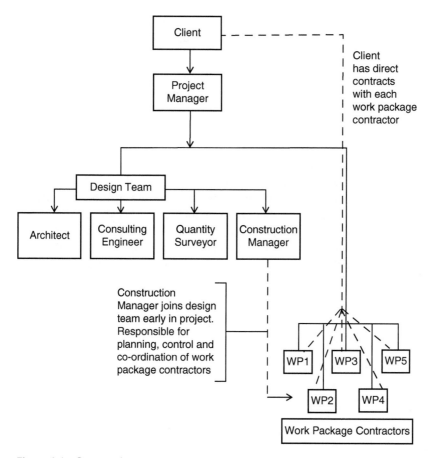

Figure 6.6 Construction management arrangement.

Several local authorities across the country partner with housing associations to build new and affordable homes. Such partnerships are based on a JV arrangement with capital funding derived from a variety of sources, including central government, local authorities, social lenders and major banks and financial institutions. Construction work is undertaken by local contractors from a framework which also comprises architects, engineers, marketing consultants and so on.

The focus for public sector procurement in the UK centres on the extensive use of framework agreements and:

- The principles of the government's *Construction Playbook*.
- Modern construction delivery.
- Sustainability and social value.

6.4.6 Turnkey and EPC

Turnkey projects are structured so that a client pays on completion for a fully functioning, staffed and operating facility. Such projects are initially financed, designed, managed and built by a contractor/developer who undertakes all the project risk.

An EPC contract differs from turnkey in that it might not include the complete project scope or might not include the commissioning and start-up of the facility.

> A client wishes to procure a 100 000 m² fully staffed and functioning distribution warehouse on a turnkey basis. The connecting road system is to be undertaken as a separate contract. This is an EPC, not a turnkey project.

Both turnkey and EPC methods give clients a choice of various management structures, different contractual arrangements and varying degrees of risk.

Turnkey procurement offers clients greater price certainty than other methods, reduced change management risks and a single-point responsibility for the right project outturn. The added benefits to clients include a shorter project timescale, a collaborative approach and a transfer of cost, time, quality and performance risk to the turnkey contractor.

Clients must be aware that the project must be fully scoped at the outset to avoid costly design changes and must be sure that the turnkey contractor is trustworthy and competent to deliver the desired outcomes.

6.5 Contractual Arrangements

Choosing the procurement route does not define the entire strategy for a project, which also requires consideration of appropriate contractual arrangements. The contractual arrangements chosen must be those most suited to:

- The needs of the client.
- The size and complexity of the project.
- The degree of risk that the client wishes to bear.

Contractual arrangements comprise:

- **The form of contract**: a bespoke or standard contract such as JCT, NEC, ICC or FIDIC.
- **The type of contract**: lump sum, admeasurement, cost reimbursement, GMP and so on.

6.5.1 Law of Contract

It is important to recognise the basic principles of law as they apply to the formation of construction contracts which vary according to the relevant legal jurisdiction – England and Wales, Scotland, United States, South Africa and so on.

Great attention must be paid to the formation of construction contracts as it is surprising, in practise, how frequently the parties get this wrong. Agreements often lack clarity – and have to be tested in the courts to see which terms apply – and, sometimes, there may be no contract between the parties at all despite the contractor having started work on site. Contracts provide three basic things:

- **Certainty**: A properly drafted contract sets out the rights and obligations of the parties and what exactly is to be provided under the terms of the contract. This enables clients and the businesses they engage for their project to plan the work, knowing the scope of work to be delivered and the standards expected. Certainty in a contract enables the courts to determine the meaning of the contract should a dispute arise.

- **Risk management**: Contracts establish the balance of risk between the parties and who carries the burden of managing those risks. This impacts how projects are planned, the working methods to be used and who pays the bills.
- **Change management**: Contracts not only set out what is to be supplied but also the means of dealing with any issues that might arise such as change orders, unexpected events, weather delays and how differences of opinion between the parties are to be dealt with.

It is not always the case, however, that the contract is clear or – in some cases – that there is a contract at all. Construction contracts can be poorly drafted, they can be overly complex, they may lack clarity as to their terms and conditions, and they might not be clear as to which terms apply to the contract.

The basics for a legally binding contract are:

- **Offer**: This is where one party offers to perform something on certain terms at the request of the other party. Offers must be clear and certain and be capable of acceptance. An invitation to tender is not an offer but an invitation to treat. An offer might be accepted or there might be a request for further clarification of the offer. The offer could also be rejected, or a counteroffer might be made.
- **Acceptance**: Once an offer or counteroffer is accepted, this must be unequivocally communicated to the other party either verbally, by conduct (e.g. by starting work) or, preferably, in writing or by email.

Sometimes, it is not clear which offer has been accepted. In the *battle of the forms,* it is the terms of the party who 'fires the last shot' that prevails. If a subcontractor offers to carry out work on its terms and conditions but the main contractor accepts the offer on its own terms and conditions, and the subcontractor accepts, the 'last shot' has been fired by the main contractor. The leading case is *Tekdata v Amphenol* [2009] EWCA Civ 1209.

- **Consideration**: A basic principle of English law is that ***each party*** must provide consideration (Uff 2021) if a contract is to be binding. Common forms of 'valuable' consideration are the payment of money in exchange for the provision of goods or the performance of work.
- **NB**: The principles of offer, acceptance and consideration apply to simple contracts which can be verbal or in writing. Contracts can also be executed as a deed which are normally enforceable even in the absence of consideration. However, they must be in writing and follow certain formalities. The statutory limitation period for simple contracts is six years from the cause of action.[1] An advantage of deeds is that the statutory limitation period is extended to twelve years from the cause of action.[2] Another advantage of deeds is that offer and acceptance is not necessary to make the contract valid. This is useful where the parties wish to be legally bound but there is no acceptable offer on the table at the time or where work must be started but its scope is vague.
- **Intent**: To be legally binding, a contract must have the intention to create legal relations irrespective of offer, acceptance and consideration.

6.5.2 Letters of Intent

Legal intent should not be confused with Letter of Intent (LOI) which are often used because of a 'grey area' in the procurement process. This 'grey area' sits between procurement Stages 1 and 2:

- **Procurement Stage 1**: brings the parties to the point where the contractor can make an offer to the client/employer which is capable of acceptance.
- **'Grey area'**: when contract documents are being prepared.
- **Procurement Stage 2**: where a legal contract can be signed on terms and conditions that are acceptable to both parties.

The 'grey area' arises because construction clients frequently want to start work on site "yesterday" which often means that site operations are commenced without a full contract in place. It can take a considerable time – several weeks, maybe – for the contract documents to be drafted, exchanged and signed by all parties.

This is why LOIs are common because they can be issued quickly for a quick start on site. LOIs are contracts but are sometimes called 'if-contracts' because, although they might express an intention to enter into a contract in the future, no liability as to any future contract is created.

It is surprising how many construction projects start based on an LOI and even more surprising that construction work often starts with a verbal order and nothing in writing at all.

6.5.3 Forms of Contract

Contractual arrangements include, *inter alia*, the choice of a suitable contract for a project. This might be one of the many standard forms available or could equally be an amended standard form of contract, an in-house form of contract or a bespoke contract specifically drafted for the particular project.

Before making a choice, it is important to consider how the project will be managed and:

- How, and to what extent, the programme will be used as a management tool.
- How the contract manages change:
  - Early warnings and risk review meetings.
  - Variations to the contract and the expenditure of provisional sums.
  - Loss and expense (compensation events) and the like.
- The impact of delay and disruption on the project outcomes.
- Payment provisions and their impact on cash flow.
- The timing and responsibility for issuing contractual notices regarding delay, information exchanges, applications for payment and extensions of time.

6.5.4 Subcontracts

Sub-contracting is *the vicarious performance of a contractual obligation*. However, when a main contractor sublets any part of the contract works, responsibility for that work remains with the main contractor under its contract with the client.

Subcontracts define the legal relationship between main contractors and subcontractors and are separate from the main contract agreement. The main contractor is free to agree the terms of the subcontract with the subcontractor which, in practice, are frequently quite different to the terms of the main contract.

When using standard contracts, such as JCT or NEC contracts, standard subcontracts are available which contain 'back-to-back' conditions equivalent to the main contract terms. However, main contractors frequently amend these contracts or use their own in-house subcontracts.

> A client awards a framework contract to a Tier 1 joint venture company based on a design and build arrangement and a target cost contract with a gain-pain mechanism.
> An enabling works package is let to a Tier 2 subcontractor by competitive tender on a lump sum fixed price basis.

6.5.5 Legal Jurisdiction

The standard forms of contract used in the United Kingdom are well known – such as JCT, NEC and ICC.

Whilst these contracts are drafted to be generally applicable throughout the United Kingdom, it is important to recognise that both Scotland and Northern Ireland are different legal jurisdictions. This means that there are some important differences in law and procedure in these countries compared to those in England and Wales.

Scotland has its own versions of JCT contracts, and there are adaptions for use in Northern Ireland. NEC contracts, which include a Northern Ireland secondary option, are drafted for use in any legal jurisdiction. ICC contracts can be used internationally, provided that the governing law is stated in the contract.

Another important difference is that the Housing Grants, Construction and Regeneration Act 1996 applies to construction contracts in England, Scotland and Wales, but Northern Ireland has its own provisions under the Construction Contracts (Northern Ireland) Order 1997.

6.5.6 Housing Grants, Construction and Regeneration Act 1996

The Housing Grants, Construction and Regeneration (HGCR) Act – commonly referred to as the 'Construction Act' – defines 'construction contracts' and 'construction operations' in the context of payment protocols, the right to suspend performance for non-payment and dispute resolution. This is important regarding cash flow, the solvency of subcontractors, potential delays to progress resulting from payment disputes and the resolution of such disputes through an adjudication process.

These definitions are wide-ranging and cover most building and engineering activity. The ambit of the Act covers contracts for the provision of professional services and project management as well as those for construction work.

This means that contracts for any such activity must either contain provisions that comply with the Act or, where not, the provisions of the Act will be implied into contracts included in the statutory definition of 'construction operations' through a statutory 'Scheme for Construction Contracts' under Section 114 of the Act.

There can be instances where construction operations are partially covered by the provisions of the Act and, sometimes, the entire works may not be covered at all. In both cases the parties should consider whether they want to expressly include the statutory provisions in the contract. The case of *Severfield (UK) Limited v Duro Felguera UK Limited* [2015] EWHC 3352 (TCC) refers.

6.6 Standard Forms of Contract

Standard forms of contract – including families such as JCT, NEC, ICC and FIDIC – offer suites of contracts that present clients and their professional advisers with a variety of options designed to fit with different procurement routes and different types and sizes of project. This includes design and build contracts, target contracts and contracts to facilitate Early Contractor Involvement (ECI) in the project.

6.6.1 JCT Contracts

Produced by the Joint Contracts Tribunal, JCT contracts are the most popular contract by number in the United Kingdom, especially in the private sector. There is a contract for every type and size of project including domestic projects. They are geared towards building work and building repair and maintenance.

The JCT is a pan-industry body and so the contracts they produce represent a balance of interests from client bodies, professional institutions and contractor and subcontractor representative bodies.

There are 12 JCT families of contracts and over 100 separate publications including the major, standard, intermediate and minor works forms as well as subcontract and sub-subcontract forms and domestic contracts. There is a standard framework agreement and a variety of other agreements, warranties and guides.

Whilst JCT contracts can be amended – and frequently are – they cannot be specifically tailored in quite the same way as NEC contracts which have a number of options to choose from to create a bespoke contract.

> **Programme**
>
> JCT contracts do not include specific programme requirements apart from requiring the client to set dates for contract start and completion, access and any sectional completions.
>
> In the JCT Standard Form, the contractor is required to submit its master programme, but only if one exists. There is no requirement on the contractor to submit a programme.

6.6.2 NEC Contracts

NEC contracts offer a 'hands-on' approach to the time and risk management of projects. They are used for building and civil engineering projects both in the United Kingdom and

Table 6.3 Main options of ECC.

Option	Brief description
A	A priced contract using an activity schedule (a list of activities linked to the programme).
B	A priced contract using bills of quantities (e.g. NRM2 or CESMM4).
C	A target contract using an activity schedule.
D	A target contract using a bill of quantities.
E	A cost-reimbursable contract using schedules of actual cost (similar in principle to daywork).
F	A management contract (with works package contractors engaged directly by the management contractor).

internationally largely, although not exclusively, in the public and infrastructure sectors. They are not as widely adopted numerically as JCT contracts but are the most used by value being used on very large projects such as the London 2012 Olympic Games, Crossrail, Hinckley Point C and HS2.

There is an extensive, and growing, portfolio of NEC contracts for professional services, frameworks, main works and subcontracts, DBO and so on including short versions for smaller, less complex projects. NEC contracts are infinitely adaptable to specific client requirements thanks to their unique system of core clauses and main and secondary option clauses. As a result, the ECC – which is the main NEC contract – is six contracts in one as illustrated in Table 6.3 which shows the main options available.

> **Programme**
> The ECC has extensive provisions for the 'accepted' programme and for keeping it up to date. This interacts with the early warning register that actively anticipates problems before they arise.
> In the short version of this contract – used for smaller, less complex contracts – the requirements for submitting a programme, and keeping it up to date, are significantly watered down.

The ECC programme requirements stipulate, *inter alia*, that the programme must include details of the order and timing of the works and the resources planned for each operation. Additionally, the contractor must show provisions for float, time risk allowances and health and safety in the programme.

In the ECC short version of this contract, the works information must state whether a programme is required and what form it should take along with requirements regarding its contents, submission and updating. Programme or not, the contractor is required to provide a weekly forecast of the date of completion.

6.6.3 FIDIC Contracts

FIDIC contracts are published by the Fédération Internationale des Ingénieurs-Conseils – the International Federation of Consulting Engineers. Responsibility for the documents lies with the FIDIC Contracts Committee and all drafts are reviewed extensively by many persons and organisations around the world.

These contracts are the most used in the world and are widely adopted for international construction contracts. FIDIC contracts are intended to be used in any legal jurisdiction but are not used a great deal in the United Kingdom.

FIDIC contracts are distinguished by the colour of their covers such as the famous Red Book which is meant for building and engineering works designed by the employer. Other options include the Green Book (short form of contract) for straightforward contracts, the Yellow Book for contractor design, the Silver Book for EPC/Turnkey contracts and, for DBO projects, there is the Gold Book.

> **Programme**
>
> Under the Red Book conditions, the contractor is required to submit a detailed time programme and to submit revised programmes consistent with actual progress.
>
> The programme shall include the intended order of the works including the timing of each stage of the design (if any).

6.6.4 ICC Contracts

The Infrastructure Conditions of Contract is a suite of contracts designed for infrastructure works. They replace the old Institution of Civil Engineers (ICE) Conditions of Contract and are sponsored by the Association for Consultancy and Engineering (ACE) and the Civil Engineering Contractors Association (CECA).

For traditional contracts, there is a measurement version and other options are available for design and construct contracts and for target contracts. ICC contracts are drafted with international contracts in mind too.

Although the ICC suite has a subcontract with terms back-to-back with the main contract, in practice, there is a tendency to use the CECA subcontract.

> **Programme**
>
> The ICC Measurement Version requires the contractor to submit a programme within 14 days of contract award. The contract prescribes that the programme has a critical path network and is accompanied by a general description of the contractor's proposed arrangements and methods of construction.
>
> The programme must be accepted by the engineer who is entitled to require amendments in order that it meets the requirements of the contract. Revised programmes may be required by the engineer should progress not match the accepted programme.

6.6.5 CIOB Time and Cost Management Contract

The Chartered Institute of Building (CIOB) Time and Cost Management Contract, together with a subcontract and consultancy agreement, is suitable for use by the public and private sectors in any country irrespective of whether the design is client-led or partly or wholly contractor-led.

This contract can also be used for turnkey contracts as well as construction management and management contracting.

> **Programme**
>
> The CIOB Time and Cost Management Contract is unusual as it has sophisticated time management provisions and requires competence in critical path network modelling, resource allocation and productivity analysis.
>
> A further feature is that design consultants, contractors and subcontractors work collaboratively with a Time Manager to avoid the unnecessary consequences of time risks.

6.6.6 IChemE Contracts

The Institution of Chemical Engineers publishes two suites of contracts for process plants – such as an oil refinery or gas processing plant – one for use in the United Kingdom and the other for international projects.

The red, green and burgundy books are intended for lump sum, reimbursable and target cost contracts, respectively, and there are documents for subcontracts, professional services and minor works.

> **Programme**
>
> The Institution of Chemical Engineers contracts are unusual:
>
> - They are performance-based contracts with two programmes – one for the construction works and one for takeover testing.
> - Completion of construction is an important point for assessing delay and applying liquidated damages but also marks commencement of a further programme of takeover testing but with no damages for delay.
> - The works are not handed over on completion but when they meet specifically scheduled criteria at which time 'takeover' happens.
> - The client/purchaser is entitled to terminate the contract, without cause, for convenience whereupon the contractor will be reimbursed its costs, but not lost profits.

6.6.7 Other Contracts

Many other contracts are used in the UK construction industry – especially by large public bodies – including:

- Network Rail which has its own portfolio of contracts, including amended versions of JCT and ICC conditions.
- Transport for London (TfL) which has bespoke contracts for services, purchase orders and professional services. For major projects, NEC contracts are used.

6.6.8 Overseas Contracts

Countries around the world have their own legal systems – some based on the English common law system – and consequently have their own approach to construction contracts. Contracts in some parts of the world – the Far East and Southern Africa, for example – have historically been based on UK contracts and nowadays NEC contracts, particularly, are used internationally.

From a project planning and management perspective, it is crucial to understand the legal jurisdiction in which projects are carried out.

In terms of the forms of contract used, some may be familiar, such as FIDIC or the NEC, but others may be entirely indigenous forms which have different approaches to procurement, project management and use of the programme as a management tool.

6.6.8.1 Australia

In Australia, several indigenous contracts are used including:

- Australian Standards (AS) contracts which provide for construction only, design and construction and minor works.
- Australian Building Industry Contracts (ABIC) – prepared by the Australian Institute of Architects and Master Builders Australia – which are used for projects superintended by architects.
- Contracts are also published by the Housing Industry of Australia and the Masters Building Association, but both tend to heavily favour the builder.

Of particular interest are the AS *General Conditions of Contract* – prepared by the Joint Standards Australia and Standards New Zealand Committee.

> **Programme**
>
> This contract has some unusual provisions not found in UK contracts under the clause dealing with *Programming*. This states that:
>
> - The superintendent (the contract administrator/project manager) may direct the order and timing of the various stages or portions of the work under the contract.
> - The construction programme is a written statement (more like a method statement than a programme) showing the dates or times within which the various stages or portions of work under the contract are to be carried out or completed.
> - The construction programme shall be deemed a contract document (contrary to legal precedent in the United Kingdom).

6.6.8.2 Hong Kong

For public sector and infrastructure works in Hong Kong, NEC contracts are widely used.

Indigenous Hong Kong contracts include the *General Conditions of Contract for Building Works* which contains specific provisions regarding the time management of projects.

Programme

A programme must be furnished, usually within 14 days of the contract award. This shall show the sequence, method and timing of the work including that of specialists and utilities.

Additionally, if required by the architect, particulars of the contractor's arrangements for carrying out the works shall be provided along with the constructional plant and temporary works that the contractor intends to use.

6.6.8.3 South Africa

The most commonly used standard forms in Sub-Saharan Africa are FIDIC, NEC and the Joint Building Contracts Committee (JBCC) series.

In the Republic of South Africa, the General Conditions of Contract for Construction Works (GCC) are also used.

Programme

Under the GCC form, the contractor must deliver a programme to the employer's agent before commencing the work. As well as the timing and sequence of the works, the programme must show the resources for the works, float and time risk allowances and a detailed cash flow forecast.

An adjusted programme is required when the accepted programme does not reflect actual progress.

FIDIC, NEC and GCC forms of contract can be used on all types of engineering and construction contracts. The JBCC contracts are, however, confined to building works.

6.6.8.4 United States

In the United States, standard construction contracts include those published by the American Institute of Architects such as the *General Conditions of the Contract for Construction* (A201-2017).

Programme

The A201-2017 contract requires the contractor to submit a contractor's construction schedule (programme) for the work promptly after contract award.

The schedule must not exceed the time limits stated in the contract, shall be revised at appropriate intervals and shall provide for expeditious and practicable execution of the work.

Another series of contracts and other documents used in the United States is prepared by the Engineers Joint Contract Documents Committee (EJCDC) which is co-sponsored by the American Society of Civil Engineers (ASCE), the American Council of Engineering Companies (ACEC) and the National Society of Professional Engineers.

These contracts are used for infrastructure and public works construction with an emphasis on collaboration, ECI and fast-tracking of construction alongside design development.

The EJCDC approach is referred to as the Construction Manager at Risk Project delivery method (CMAR) and is a development of design-build. It is a project delivery method in which:

- The owner (client) engages a construction manager (CM) to oversee the project from design to construction completion (close-out).
- The CMAR works initially as a consultant alongside the owner, architect and lead contractor during the design and planning stages.
- The CMAR, who could also be the lead contractor, acts as a *single point of failure* and provides a GMP after the project scope is fixed and before bids are received.
- The CM is responsible for paying the difference should the budget be exceeded and must closely manage the project budget and schedule to keep within the GMP.

> **Programme**
>
> The CMAR is required to submit a construction schedule within 30 days of the notice to proceed showing each trade and operation.
>
> The schedule should show the order of the work, the available float with each activity and the critical path shall be clearly identified.

The CMAR project delivery method should be distinguished from Construction Manager Agency (CMA), where the CM legally represents the client and makes project-specific decisions along the lines of construction management procurement in the United Kingdom.

6.7 Types of Contract

As with the procurement route, there are several options available when choosing contractual **arrangements** – the **form** of contract used to legally bind the parties (the contractual **agreement**) and the **type** of contract preferred which determines price and how the contractor will be paid.

The **type** of contract is the second element when deciding on appropriate contractual arrangements for a project. There are three main types of contract available, each of which has a number of subsets:

1. Lump sum contracts.
2. Measure and value contracts.
3. Cost reimbursement contracts.

The main distinguishing features between each type of contract are the different options they offer for:

- Establishing the contractor's price for carrying out the works.
- Paying for work carried out.
- Tendering arrangements, tender documentation and, especially, the choice of pricing document to accompany the contractor's bid.

In deciding which type of contract to select, the client must choose between several options depending upon the degree of price certainty required, the tolerable degree of price risk and who carries the risk for quantities and inflation.

Price certainty does not come without price risk resulting from factors such as uncertain ground conditions, design changes, the vagaries of the weather, insolvency and so on. The only time when price certainty with zero risk can be achieved is where there is an entire contract, a fixed price and no latitude for design changes. This can be achieved for small projects, but it is rare.

6.7.1 Lump Sum Contracts

Lump sum contracts are common in the construction world because, *prima facie*, they give the greatest price certainty and the least price risk for the client. They are most associated with general contracting or 'traditional' procurement, where a price is established by competitive tender or negotiation based on a 'complete' design.

Tenders can be invited using single-stage or two-stage open or selective tendering or by using a framework. It is, however, a fallacy to think that lump sum contracts are the least risk solution for clients simply because it is a rarity for work to start on site with a fully completed design. This can lead to delays, variations and contractual claims which impact the client's budget and project programme.

> When the contractor's programme of works is disrupted, this means that non-critical activities can lose some or all of their float with the consequence that it becomes more difficult and expensive to balance resources.
>
> In the case of critical activities, extensions of time may be granted but contractors always complain, with some justification, that loss and expense claims/compensation events, never fully compensate for the true costs incurred.

Lump sum contracts are traditionally based on bills of quantities – usually prepared by the client – which are priced by tenderers to arrive at a lump sum bid. However, it is now common practice for lump sum contracts to be based on drawings and specifications only, where the tenderers take on the quantity risk. They may be fixed price, firm price or fluctuating price, as illustrated in Table 6.4.

6.7.2 Measure and Value Contracts

Measure and value contracts – often referred to as remeasurement or, more correctly, admeasurement, contracts – rely on approximate quantities which are priced at tender stage and later admeasured to determine the contract sum.

Sometimes, there are no quantities at all – just a schedule of the contractor's rates – with the quantities being determined as the work progresses. This is common practice for repair and maintenance works and for term contracts.

Admeasurement contracts are more commonly found in civil engineering than in the building sector. However, it is frequently the case that approximate quantities – especially for groundworks – are found in building contracts despite there being a lump sum contract.

Table 6.4 Lump sum contracts.

Lump sum contracts	Description	Example
Fixed price	There is no provision in the contract for the client to change the scope of the works. This would be subject to a separate contract. The contractor is not reimbursed for the effects of inflation on the price.	• Common for small projects. • Used where the work is well defined. • Price reflects the extent of contractor risk. • Variations may be expensive.
Firm price	The contract allows the client to make changes (variations) as defined. The contractor must include for the risk of inflation except where this stems from changes in taxation and the like.	• For medium-sized contracts <12 months duration. • Inflation risk allowance could be significant. • Provisions for valuing variations included in the contract.

Fluctuating price	Variations to the contract are allowed and the contractor is reimbursed for inflation, both for tax increases and for increases in the costs of materials, labour and plant.	• Used for larger contracts >12 months duration.
		• Client bears inflation risk.
		• Inflation calculated using published indices:
		  • Retail or Consumer Price Indices.[a]
		  • Price Adjustment Formulae Indices administered by BCIS.[b]

a) Commonly used but not recommended for construction work.
b) PAF Indices available for building, civil engineering, highways and specialist engineering works.

This is a further example of why lump sum contracts can add price risk for the client, why it is difficult for contractors to programme the works and why the so-called price certainty of lump sum contracts can be an illusion.

6.7.3 Cost Reimbursement Contracts

Cost reimbursement contracts are based on the principle that the contractor is paid the actual cost of the work done. With this type of contract, there is usually a fee – which might be tendered in competition – that covers the contractor's general costs, overheads and profit. The fee can be a lump sum or a percentage of the value of the work done.

Such contracts expose the client to a low level of price certainty as neither the scope nor the price of the work can be predetermined. Paradoxically, however, cost reimbursement contracts can carry less price risk as there are, by definition, no variations as with conventional contracts, unless the client strays beyond the agreed scope of the works. There are no claims and no surprises because the contractor is paid its costs.

As regards the contractor's programme, there is a school of thought that suggests cost reimbursement contracts are inefficient. However, it is not in the interests of contractors to be inefficient – which contractor would not want repeat business on a cost reimbursement + fee basis from a happy client?

Cost reimbursement contracts repay the contractor for the actual costs of constructing the works:

- The contractor may be paid a fee – a fixed sum or a percentage – to cover for overheads and profit.
- In some instances, a GMP or target price may be negotiated. There may be a 'gain-pain' arrangement should actual costs be more or less than the GMP or target.

6.7.4 Target Cost Contracts

Target cost contracts are a subset of cost reimbursements contracts. They require greater collaboration – and more trust and openness – than conventional cost reimbursement arrangements and can be used between the client and the consultant team as well as between client and contractor. The purpose of target cost contracts is twofold:

- To enable the client team and contractor to engage at an early stage in a project – typically RIBA Stages 2 and 3 – to produce a cost plan in line with the developing design. The cost plan is then refined to the point when the contractor can agree a firm target cost or GMP.
- To provide a financial incentive that encourages value-engineered and buildable designs, efficiency, collaboration, early warning and effective management of the project schedule.

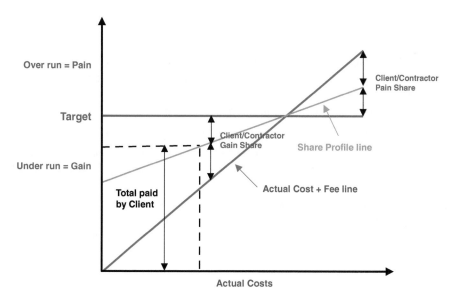

Figure 6.7 Target cost contracts. Source: Based on the work of Perry et al. (1982) and Broome (2013).

Target cost contracts have been popularised by NEC ECC Options C and D which are generally used on large public sector projects. The basis of Options C and D are activity schedules and bills of quantities, respectively.

Once work on site commences, actual costs are monitored by both parties, and any cost savings or overruns shared on the basis of an agreed formula. The share formula is commonly 50/50 but can vary in practice. Bonus or penalty payments may also be capped.

The fact that contractors can benefit from or be penalised by the target cost share formula has coined the phrase *gain-pain share* in the industry.

The principles of target cost contracts are illustrated in Figure 6.7 which depicts an out turn cost below target and the ensuing 'gain' shared between the client and contractor. Should a cost overrun occur, the 'pain' would equally be shared which should be sufficient to motivate the parties to seek below-cost solutions.

6.8 Time Management

The preparation of programmes (schedules) is recognised as a core task in the RIBA Plan of Work and is central to project procurement.

This includes preparation of the project programme – often referred to as the master schedule – and the design programme and construction programme.

As one of the key elements in most construction contracts, time is an important project deliverable – a primary criterion of success, along with cost and quality, for clients and crucial for contractors in terms of profitability and potential liabilities under the contract.

6.8.1 Time in Contracts

Clients invariably require their project to be carried out within a predetermined timescale but, when emergency work is required or when the scope of works cannot be defined, timescales are inappropriate.

Time is recognised in construction contracts in several ways:

- There will invariably be a clause similar to FIDIC Red Book clause 8.1 which states that *the contractor shall commence the execution of the works as soon as is reasonably practicable after the commencement date and shall then proceed with the works with due expedition and without delay.*
- Start and finish dates, or a period for construction, will be stated in the contract particulars (JCT contracts) or in the contract data (NEC contracts).
- Contracts may also incorporate partial completion times or dates and phased or sectional handovers.

6.8.2 Project Master Schedule

The project programme (or master schedule) is an important part of the procurement strategy for a project as it is central to the procurement framework and sets the tone for the time management of the project.

RIBA Plan of Work 2020								
Options:	0	1	2	3	4	5	6	7
	Strategic Definition	Preparation and Brief	Concept Design	Spatial Coordination	Technical Design	Manufacturing and Construction	Handover	In Use
Core Tasks		**Prepare Project Programme**	**Prepare stage Design Programme**	**Prepare stage Design Programme**	**Prepare stage Design Programme**	**Monitor progress against Construction Programme**		
Procurement Route	① Appoint client team. ② Appoint design team. ③ Appoint contractor.							
Traditional					③			
D&B single stage		①	②		③			
D&B two stage		①		③	③			
Management and Construction Management			③					
Contractor-led					③			
Information Exchanges		**Project Programme** Procurement Strategy						

Figure 6.8 Procurement strategy and programme.

The project programme incorporates the pre-design and design phases, the tendering period (if any) and the construction period.

Figure 6.8 illustrates that the project programme sits within Stage 1 of the RIBA Plan of work – right at the outset of the project. Preparation of the project programme is a core task – along with preparation of stage design programmes – and represents an important information exchange alongside the project procurement strategy.

Development of a procurement strategy naturally raises some key time-related issues to consider which can impact the contractual aspects of project procurement. These include:

- Possession and commencement.
- Progress, completion and extensions of time.
- Liquidated and ascertained damages.
- Obligations following completion.

The key dates in the project programme are somewhat aspirational in the early stages of a project as there are still many issues to resolve at this point in time. Eventually, these dates will be firmed up and find their way into the contract documentation prior to contractor appointment.

6.8.3 Site Possession

For clients, the date for possession of the site is a vital point in the procurement process as this is the date when the contractor legally takes over the site and commences work. There are, however, many issues for clients to resolve before this can happen, including:

- Resolution of public enquiries.
- Agreeing project financing.
- Obtaining planning and other approvals.
- Dealing with issues relating to historic or listed buildings, sites of special scientific interest or of archaeological or environmental importance.
- Design review in terms of building regulations.

6.8.4 Progress, Completion and Extensions of Time

Many standard contracts do not require a programme because they are geared towards small or modest-size projects. For larger projects, however, a programme will invariably be required, and this needs to be factored into the procurement arrangements for the project.

The significance of the contractor's programme comes into sharp focus when the schedule is delayed or disrupted, but this is too late. Time management of construction contracts should be carefully considered at the procurement stage.

Standard construction contracts provide for the eventuality of delays:

- through the inclusion of provisions referred to as *relevant events* (JCT contracts) or *compensation events* (NEC contracts) entitling the contractor to extensions of time and/or extra cost and
- the inclusion of damages provisions to reimburse the client for the delay suffered.

It is common practice for standard contracts to be amended – sometimes heavily – and issues such as provisions for adjustment of time should be considered when formulating the project procurement strategy and when deciding upon the most appropriate form of contract – and contract amendments or options – to use.

It is important to note that whilst a contractor's programme might be required under the contract, it should not be incorporated into the contract as this would make the programme a contract document. This could lead to unwanted legal repercussions as in the leading case of *Yorkshire Water Authority v Sir Alfred McAlpine & Son (Northern) Ltd* [32 BLR 114 QBD].

6.8.5 Liquidated and Ascertained Damages

When start/finish dates or contract periods are included in a contract, failure on the part of the contractor to complete on time will usually trigger an entitlement to damages. This provides the client recompense for the delay.

These damages are commonly referred to as 'penalties' but, to be enforceable in law, they cannot be punitive and must be a genuine pre-estimate of the client's loss. This is the role of liquidated and ascertained damages (LADs) which are intended to represent – in money terms – the loss of utility of the unfinished project due to delayed completion or loss of rents, sales or other revenues.

Whether or not to include LADs in the contract is a procurement issue because contractors need to be aware of their presence in the contract (or not) as they can represent a significant commercial risk at tender or bid submission stage.

LADs do not apply where the delay is the fault of the client.

6.8.6 Obligations Following Completion

Practical or substantial completion of a construction project signals the commencement of the defects correction period when the contractor is obliged to rectify the patent defects listed in the 'snagging' or 'punch' list prepared by the contract administrator.

The likelihood that there will be defects in the completed work is a procurement issue because significant defects might delay the client's ability to fully occupy or use the completed facility.

If the client considers that this would be unacceptable, then a procurement method geared to collaboration and defect-free work should be considered at the outset of the project:

- Early warning provisions in the contract.
- A blame-free culture.
- Ring-fenced profits for contractors and subcontractors.
- Focus of all parties on client outcomes.
- Use of a Cloud-based common data environment to improve communications and ensure that everyone is working to 'one version of the truth'.

After the fact is too late.

References

Broome J. (2013) *NEC3: A User's Guide*. ICE Publishing.
Chartered Institute of Building (2010) *Procurement in the Construction Industry*. CIOB.
Institution of Civil Engineers (2021) *Civil Engineering Procedure*, 8th edn. ICE.
Perry J., Thompson P. and Wright M. (1982) CIRIA Report No 85, CIRIA.
Uff, J.(2021) *Construction Law*. Sweet and Maxwell.
Williams, P. (2016) *Managing Measurement Risk in Building and Civil Engineering*. Wiley Blackwell.

Notes

1 Limitation Act 1980, Section 5.
2 Limitation Act 1980, Section 8.

		Chapter 7 Dashboard
Key Message		o Submitting successful bids provides the work that any construction business relies upon to make money, satisfy shareholders, to grow and to survive.
Definitions		o **Estimating**: The technical process of predicting the costs of construction. o **Tendering**: The process of preparing and submitting an offer thus converting the estimate into a bid. o **Bidding**: A preparatory process during which a tender figure is formulated prior to formal submission.
Headings		**Chapter Summary**
7.1	Introduction	o Planning and estimating are closely related and cannot be viewed in isolation, irrespective of the size of project. o Construction costs have a significant time-related element which is closely linked to the methods and sequence of work envisaged when pricing the work.
7.2	Approaches to Estimating	o Methods of estimating vary according to the size and sophistication of the contractor or subcontractor involved. o Methods of estimating used depend upon the type of pricing document. o The principal estimating methods are top-down and bottom-up.
7.3	Measurement	o Quantities are commonly prepared for most projects except where the project scope cannot be defined in sufficient detail. o The common pricing documents used in the industry are bills of quantities, activity schedules and schedules of rates.
7.4	Estimating Software	o Computer-based estimating is commonly used in the industry. o A variety of both proprietary and bespoke systems are used. o Spreadsheets are popular for cost planning and analytical estimating.
7.5	Top-Down Estimating	o Top-down estimating is more formally referred to as cost planning or design cost estimating. o This involves preparation of an initial budget which is then expanded into a detailed plan of expenditure commensurate with the developing design. o The floor area, functional unit and elemental methods are commonly used.
7.6	Cost Information	o Top-down estimating relies upon historic data from previous projects. o This data is extracted from priced bills of quantities. o BCIS and builders' price books are widely used sources. o Indices are used to update historic cost data and to allow for inflation.
7.7	New Rules of Measurement	o The RICS New Rules of Measurement is suite of measurement rules. o This includes NRM1: *Order of cost estimating and cost planning for capital building works*.
7.8	Bottom-Up Estimating	o Bottom-up estimating is a resource-based method. o Rates and prices are calculated by predicting the cost of labour, materials, plant and equipment. o This method of estimating can only be used where detailed quantities are available or where there is a schedule of rates with full item descriptions. o The main methods are the unit rate method and the operational method.
7.9	Bidding	o Bidding is a preparatory process and tendering is a formal offer to carry out work. o The bidding process is in two stages – estimating and bid preparation followed by bid submission and formal tender. o A bid manager is often engaged to manage the bid from initial enquiry to bid submission.
Learning Outcomes		o Distinguish between cost planning and analytical estimating. o Understand how order of cost estimates and cost plans are prepared. o Be familiar with top-down and bottom-up methods. o Understand the use of location factors and construction indices. o Understand the bidding process.
Learn More		o Chapter 18 – Controlling resources. o The CIOB Code of Estimating Practice.

7

Estimating and Bidding

7.1 Introduction

Estimating and bidding is the foundation of any contracting organisation. Submitting successful bids provides the contracts that any construction business relies upon to make money, satisfy shareholders, grow and survive in what is, after all, a very competitive market.

Planning and estimating are closely related aspects of construction projects and cannot be viewed in isolation, irrespective of the size of project in question. Construction costs have a significant time-related element, and the project timescale is closely linked to the methods and sequence of work envisaged when pricing the work.

7.1.1 Definitions

The CIOB (2018) explains that a bid is an *offer to carry out work for a price* based on an estimate of cost and provides the following useful definitions:

- **Estimate**: the net estimated cost of carrying out the works for submission to management at the settlement meeting.
- **Estimating**: the technical process of predicting the costs of construction.
- **Settlement meeting**: a timetable for the preparation of an estimate, all necessary supporting actions and for the subsequent conversion of the estimate into a tender.
- **Tender**: a sum of money, time and other conditions required to complete the specified construction work. For design and build, the term 'tender' includes the contractor's design proposals, price and contract sum analysis.
- **Tendering**: the process of preparing and submitting for acceptance a conforming offer to carry out work for a price, thus converting the estimate to a bid.
- **Bidding**: a preparatory process during which a tender figure is formulated prior to formal submission.

7.1.2 Tendering

Tendering is a formal offer to carry out work or to supply goods and/or services. In terms of the usual client–contractor relationship, this means that:

- The contractor prepares an estimate and submits a *bid*.
- The client receives a *tender* from the contractor which is the formal offer to treat.

An offer to treat is an offer capable of being converted into a legally binding contract which normally comprises three constituent parts:

- **An offer**: to carry out certain prescribed works,
- **Acceptance of the offer**: formal acceptance by the offeree (usually a client or client body) and
- **Consideration**: an amount of money.

Other types of consideration capable of creating a legal contract include:

- A schedule of rates or prices.
- A guaranteed maximum price (GMP).
- A percentage or lump sum management fee.
- A service charge to provide a defined facility over an agreed period such as a school, hospital or prison.

7.1.3 Estimating and Tendering

Estimating and tendering procedures in construction have changed considerably in recent years, particularly since the previous edition of this book was written. This period has observed a marked shift away from traditional competitive tendering – certainly in the higher echelons of the industry – with far greater emphasis now on design and construct contracts and early contractor involvement (ECI) in the design process.

Even when traditional procurement is the chosen option – either by a client or in the main contractor supply chain – the paradigm has shifted away from providing detailed bills of quantities with the tender documents towards the use of drawings and specifications, or activity schedules, which places the measurement risk with the tendering contractors. This element of risk is consequently passed down the supply chain to Tier 2 contractors and below.

The construction industry has also seen the consolidation of subcontracting to an extent not seen before, and many main/principal contractors rely almost exclusively on specialist contractors to price and execute most of the work involved in projects.

The impact of these changes has meant that methods of estimating used by contractors have evolved to the extent that pricing techniques now used reflect the absence of detailed bills of quantities. Where early cost advice or a design and build tender is required, methods of estimating that were historically the province of the professional quantity surveyor (PQS) are now commonly used by contractors.

These changes in contractors' estimating methods are explored in this chapter, along with the different ways that tenders for construction work are sought by the industry's clients nowadays.

7.1.4 The Construction Market

The construction market is extremely varied, and the value of projects carried out by the industry can range from a few hundred pounds to multi-billions worth of expenditure.

Projects for private clients tend to be driven by the test of affordability which will require a budget to be set and contracts awarded within the budget to prevent the client from overspending. Speculative developments, however, are driven by the need to make a profit – whether this be from the sale of houses, the sale or leasing of office space and the like or revenue streams from commercial developments such as restaurants, shops or places of entertainment.

Public sector projects, on the other hand, such as schools, hospitals, defence installations, power stations and road and rail infrastructure, are driven by the need to ensure that public money is spent in such a way that it is not wasted and provides value.

Consequently, approaches to estimating construction costs vary across the industry according to the size, complexity and value of the project in question and to the size of the organisations carrying out the work.

7.2 Approaches to Estimating

The nature and detail of the information provided by a construction client is influential on the methods of estimating used by a contractor or subcontractor. For instance, the pricing of small projects, especially at a domestic scale, can be based upon little more than a discussion with the client whereas for large projects contractors may be able to base their pricing upon detailed drawings, specifications and even bills of quantities.

Methods of estimating construction costs can also vary considerably according to the size and sophistication of the contractor or subcontractor involved.

7.2.1 Micro Companies

Sole traders, partnerships and other small traders and contractors tend to price on the basis of the wages that they require, plus materials, plus an on-cost for the provision of a van, tools, equipment, etc. This gives a baseline price.

However, there is a tendency, at this level in the industry, to 'price condition' clients according to what price the trader thinks the client is willing to pay. This means that the actual price quoted is often significantly higher than the baseline.

This is a sort of 'what can the market afford' type of elevated pricing which increases the profit return for the trader/contractor.

7.2.2 Small Companies

Slightly larger contractors and subcontractors may add a level of sophistication to their pricing by formalising their calculations or by having standard rates for different types of work.

They may even have a simple standard method of pricing related to or resembling the NRM2 measured items that relate to their particular trade or sector of contracting – such as groundworks, cladding, roofing and mechanical and electrical work.

This is straightforward to do because – being a single trade – the number of measured items is relatively small and fairly standardised, and the ensuing calculations are not too complex.

7.2.3 Medium-size Companies

Larger contractors, and general contractors who employ their own trades, will probably use an estimating system based on labour outputs or constants.

The calculations of materials costs will include allowances for cutting and waste, laps and ground loss for stone or concrete, together with detailed calculations for preliminaries items such as site accommodation, supervision costs, vans and transport, cranes and access equipment and so on.

7.2.4 Large-size Companies

Large, and very large, contractors, who carry out little or no work on-site, normally rely on the pricing of their subcontractors or specialist contractors to build up a price and then add their own costs of organising and managing the project. Estimating capability is nevertheless required in such firms to price:

- Risk.
- Temporary works, cranes, site access, traffic management.
- Preliminaries – salaries, health and safety, security and the general costs of running the site.
- Subcontract attendances – such as the provision of piling mats, scaffolding and temporary works.
- Self-delivery items – interfaces which subcontractors are not interested in pricing or work that does not fall within their normal activities.
- Quotations for acceleration of projects during construction.
- Additional elements of work under a framework agreement not covered by the agreed prices.

Large firms may also utilise approximate methods of estimating. This is common for design-build projects or when a client requests an 'order of cost' for an outline design or where they are engaged on an Early Contractor Involvement (ECI) basis.

7.2.5 Estimating Methods

Modern methods of estimating have a long history, and there is clear evidence that the calculation of unit prices for construction work differs little from the methods used in the late nineteenth century. Then, as now, the emphasis was on the calculation of resource costs – labour and materials, etc.

This method of estimating, however, relies upon detailed quantities based on a developed design. Alas, this does not work when calculating construction costs at the strategic definition and early design stage of projects.

It was not until the early 1950s that quantity surveyor James Nisbet recognised the need for a means of measuring and estimating that did not rely on detailed quantities such as those found in bills of quantities. Nisbet developed a system of cost planning that was to not only overcome the difficulties of estimating costs at the early stages of projects but also to revolutionise the quantity surveying profession.

7.2.6 Classification

Cost estimation methods can be classified according to whether they are appropriate for the early and design stages of projects or whether pricing is to be based on detailed information which permits a more granular approach to be taken. Essentially, there are two methods:

- Top-down methods.
- Bottom-up methods.

The distinction between these methods can be explained by the two examples illustrated in Figure 7.1.

7.2.7 Early Cost Advice

Contractors are undeniably best placed to give clients early cost advice for a number of reasons:

1. They have their own database of costs from previous projects that they have undertaken.
2. They have access to their supply chain of subcontractors and works package contractors who can offer accurate and current pricing of design options.
3. They have their own project managers, estimators and planning engineers who can advise on the practicality and cost of alternative design solutions, construction methods and construction sequences which can assist designers in value engineering optimum design solutions.

Professional Quantity Surveyors (PQSs) employed by the client are at a disadvantage:

1. They only have access to the prices (not costs) of previous projects either through their own databases of tender figures or through the RICS Building Cost Information Service (BCIS).
2. Such information is historic and not current which requires statistical adjustment based on predicted, not current, data.
3. They are not builders or civil engineering contractors and thus cannot advise so expertly on construction methods and the costs of alternative methods of construction.

TOP-DOWN ESTIMATING

ORDER OF COST ESTIMATE FOR NEW LEISURE CENTRE

Ref	Constituent	Quantity	Unit	Rate £	Totals £	Sub-totals £	Total £
a	Facilitating works estimate						
	Demolition				50000		
	Site clearance	0.4	ha	200000	80000	130000	
b	Building works estimate	1200	m²	3500		4200000	
						4330000	
c	Main contractor's preliminaries estimate	12	%			519600	
	Sub-total					4849600	
e	Main contractor's overheads and profit estimate	8	%			387968	
f	**Works cost estimate**						**5237568**

BOTTOM-UP ESTIMATING
FACING BRICKWORK

Brickwork in skins of hollow walls, half brick thick. Heather facings in gauged mortar (1:4) pointed one side.

Item	No/Hrs	Unit	£	£	£	£
Facing bricks per 1000 delivered				900.00		
Crane offloading				20.00		
Distribution: 2hrs/1000	2	hrs	20.00	40.00		
				960.00		
Waste	5	%		48.00		
Cost per 1000				1008.00		
Materials						
Bricks per m²	0.059	B/m²			59.47	
Mortar	0.035	m³		140.00	4.90	
					64.37	
Labour						
Foreman	1	No	30.00	30.00		
Bricklayers	6	No	25.00	150.00		
Labourers	3	No	20.00	60.00		
Cost per gang hour	6	÷		240.00		
Cost per Bricklayer hour	6			40.00		
Bricks/m²			59			
Output per bricklayer hour			50			
Multiplier/m²				1.18	47.20	
					111.57	
Overheads and profit	15	%			16.74	
Rate per m²						**128.31**

Figure 7.1 Top-down/bottom-up estimating.

7.3 Measurement

Estimating – especially bottom-up estimating – is the process of building up rates/prices which are then applied to the quantity of work to produce an estimate capable of being converted into a bid. For this to work, a document is required containing descriptions of the work and the associated quantities.

> Traditionally, pricing documents were produced in paper form and priced by hand by each tendering contractor, but modern technology offers a number of much slicker and interactive options for producing quantities.

There is a wide variety of software capable of producing quantities – some using common methods of measurement and others relying on simpler builders' quantities.

7.3.1 Pricing Documents

Quantities are commonly prepared for most projects except where the project scope cannot be defined in sufficient detail. In such cases, schedules of rates are used to measure and value the work done. In the majority of cases, however, quantities are needed to define the price for individual items of work or complete activities which facilitates calculation of an estimated total for bid submission.

The common pricing documents used are bills of quantities and activity schedules which are contrasted in Figure 7.2. However, many subcontractors use spreadsheets for quantities and estimating with the benefit that they are easily understood without much training needed and readily transmittable as all businesses will have access to Excel or to the Google and iOS equivalents.

7.3.2 Quantities

Production of SMM-complaint bills of quantities or builders' quantities using modern software has the great advantage that they can be packaged and sent out for subcontract quotes electronically.

The output from measurement packages such as QSPro can be sent out as full bills of quantities in various formats such as Excel spreadsheets, CITE files and text files which is a great benefit when the recipients do not have the native software available. This is illustrated in Figure 7.3.

As a consequence of the wide choice of software available, there are many ways that quantities can be produced, including:

- Manual data entry into a spreadsheet, mobile app or measurement software package from site measurements or measurements taken from PDF files or paper drawings.
- 2D on-screen measurement using markup and measurement software such as Bluebeam Revu.
- 2D on-screen measurement using an enterprise resource planning (ERP) software such as Procore.

BILL OF QUANTITIES

	Masonry		
	Brick/block walling		
	Walls; half brick thick		
A	Facing brickwork; PC price £750 per 1000; Skins of hollow walls; Stretcher bond; Gauged mortar 1:1:6; Pointed one side [14.1.1.1]	596	m²

ACTIVITY SCHEDULE

Activity reference	Activity description	Programme reference	Price £
A	Form access and set up contractor's compound and site offices	1	
B	Excavate to remove topsoil and store in temporary spoil heaps on site	2	
C	Excavate to reduce levels and remove spoil to tips off site	3	
D	Excavate pad foundations, remove spoil from site and prepare formation for concrete	4	
E	Supply and fix steel reinforcement cages to pad foundations	5	
F	Pour ready-mixed concrete to pad foundations against earth faces	6	
	TOTAL		

Figure 7.2 Pricing documentation.

- 2D or 3D on-screen measurement using PDFs, Building Information Modelling (BIM) models or photographs in an estimating and take-off software package such as Cubit Estimating or Access ConQuest Estimating.
- Using third-party software (such as Bluebeam Revu) to produce quantities and importing them into estimating software such as Bidcon.
- Extracting quantities directly from a BIM model in estimating software such as Cubit or Bidcon.
- SMM-based bill of quantities production in a stand-alone on-screen measurement package such as QSPro.
- SMM-based bills of quantities from PDFs, computer-aided design (CAD) drawings or BIM models inside a high-end suite of fully integrated cost planning, estimating and tendering software using on-screen measurement. This software from Causeway also offers integrated budget, cost and value management capabilities in a single Cloud-based platform.

Subcontract enquiry for Structural Steelwork package

Structural metalwork

	Qty	Unit	Rate £	p
Structural Steelwork				
Framed members, framing and fabrication				
A Lengths over 1.00 but not exceeding 9.00m; Weight 25–50kg/m; Columns	18.00	t		
B Lengths exceeding 9.00m; Weight 25–50kg/m; Beams	8.00	t		
Framed members, permanent erection on site				
C Lengths over 1.00 but not exceeding 9.00m; Weight 25–50kg/m; Columns	18.00	t		
D Lengths exceeding 9.00m; Weight 25–50kg/m; Beams	8.00	t		
Allowance for fittings				
E Percentage allowance; 10%; To framed members	2.60	t		
Holding down bolts or assemblies				
F 4 No M25 x 500mm holding down bolts; 4 No 100mm x 100mm x 20mm Anchor plates; Nut and washer; 4 No cones	70	nr		

Full Bill of Quantities (NRM2) Format

Item	Description	Qty	Unit	Rate £	Total £
	Structural metalwork				
	Structural Steelwork				
	Framed members, framing and fabrication				
A	Lengths over 1.00 but not exceeding 9.00m; Weight 25–50kg/m; Columns	18	t		0.00
B	Lengths exceeding 9.00m; Weight 25–50kg/m; Beams	8	t		0.00
	Framed members, permanent erection on site				
C	Lengths over 1.00 but not exceeding 9.00m; Weight 25–50kg/m; Columns	18	t		0.00
D	Lengths exceeding 9.00m; Weight 25–50kg/m; Beams	8	t		0.00
	Allowance for fittings				
E	Percentage allowance; 10%; To framed members	2.6	t		0.00
	Holding down bolts or assemblies				
F	4 No M25 x 500mm holding down bolts; 4 No 100mm x 100mm x 20mm Anchor plates; Nut and washer; 4 No cones	70	nr		0.00

Excel Format

OUTPUT FROM MEASUREMENT PACKAGE (QSPro)

Figure 7.3 Subcontract enquiry.

222 | 7 Estimating and Bidding

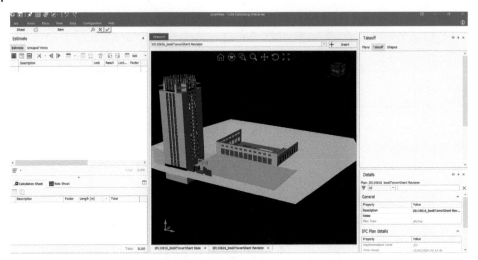

1. Cubit allows BIM models to be imported into the Viewport.
2. From here, quantities can be extracted from the model properties.
3. The quantities can be used for planning and/or estimating.
4. Some interpretation may be necessary as the model might not be complete and some quantities could be missing.

Figure 7.4 BIM measure. BIM model courtesy of Buildsoft.

An example of on-screen measurement is shown in Figure 7.4. This shows a BIM model imported into Buildsoft Cubit Estimating, from which quantities can be extracted and priced.

7.3.3 Link with Planning

There is an important link between measurement documentation, quantities and project planning.

Quantities form the basis of the planner's evaluation of activity durations at both the pre-tender and pre-construction stages. They also inform the choice of plant and equipment for earthmoving, concreting, drainage works and so on, as well as formwork systems, site batching facilities and crushing and recycling plants.

When an activity schedule is the client's preferred pricing document – prepared either by the client or the contractor – there are no quantities provided. An activity schedule is simply a list of activities – quantities for pricing and planning purposes must be generated by the tenderer.

For contracts based on an activity schedule, this forms the basis of the contractor's interim payments. However, unlike bills of quantities contracts – where interim payments are based on the quantity of work completed – interim payments on contracts based on activity schedules are usually geared to completion of each activity on the schedule. Where a number of activities are grouped together, the contractor will not receive interim payments for individual activities but will have to wait until the entire group of activities is complete.

As a result, the contractor's pre-tender and subsequent programmes need to mirror the items on the activity schedule in order to avoid cash flow problems because the programme is the document used to monitor progress and determine interim payments.

7.3.4 Software

Some estimating software includes an integral on-screen measurement facility. In other cases, proprietary measurement packages can be used to provide the quantities, and the estimating database can either be integrated into the measurement software to produce a priced bill of quantities or be used stand-alone with the estimating package.

> BIM models also generate quantities but not yet in accordance with a standard method of measurement and only to the extent that the model has been sufficiently developed to generate the quantities required. Models can only provide quantities for elements in the model but maybe not for ancillary items. Therefore, quantities for an internal wall might be obtained from the model but not skirtings or other ancillaries.

Software packages – including apps – make the estimating process easier and faster and facilitate iteration and the exploration of 'what-if' scenarios. The prerequisites are, therefore:

- A means of measuring, entering or importing quantities.
- A database of costs or the ability to enter rates.

The size of the company or cost consultant looking to introduce computer-based estimating is no guide as to which software packages are used in the industry.

Small companies use high-end packages as do freelance quantity surveyors and estimators. On the other hand, some medium and large companies use spreadsheets for their estimating and quantity take-off. It is more likely to be the size and complexity of the projects undertaken that is the determining factor.

7.4 Estimating Software

Computer-based estimating has been around since the 1980s, pioneered by the likes of John Laing's Elstree Computing, but was never adopted on a widespread basis. However, modern software is far more accessible, and a variety of both proprietary and bespoke systems are used at all levels of the industry.

Spreadsheets are also very popular both for cost planning and for analytical estimating. Despite lacking collaborative sophistication and requiring considerable effort to build a database, spreadsheets are thought to be used by some 80% of the industry for creating estimates.

The software considered below is a cross-section of the products available to the industry but is not intended to be exclusive. Some of the software is designed for trades and small builders whilst more sophisticated software might be attractive to mid-range contractors. High-end packages, perhaps linked to an ERP or to construction management software – are more likely to be used by larger contractors.

7.4.1 Trade Apps

Selco, a leading UK builder's merchant, provides a free app which enables trades people – such as painters, carpenters and joiners and electricians – to price work on a smart 'phone and to send a quote to a client. Features of the app include:

- Pre-populated current materials prices.
- Ability to allow for waste.
- Facility to include relevant labour rates.
- A markup facility.

MKM Building Supplies Ltd provides a mobile app that facilitates invoice management and order tracking. This enables trade customers to view prices, download and pay invoices and keep abreast of orders, including tracking order status, delivery dates and quantities.

7.4.2 Price-a-Job

Price-a-job is Cloud-based estimating software that helps trades and small builders to prepare quotes, invoices and control costs. Its attributes include:

- Quantity and materials take-off.
- Pre-populated materials prices from nationwide suppliers updated every 24 hours.
- A Gantt chart planning facility.
- A database of regional labour rates.
- Facility to insert own labour rates.

7.4.3 ConX

ConX is an Australian Cloud-based estimating package that sells internationally including in the United States and the United Kingdom. It is designed for smaller builders and trades such as plasterers, plumbers and ground workers.

The user defines their own template of items that are usual to their trade and populates the software with prices for labour and materials. The software allows the use of country-specific terminology.

Projects are stored in the Cloud and sophisticated letter-headed quotes with company logos can be prepared. Contract terms and conditions can be added as well as inclusions and exclusions such as scaffolding and waste removal.

Features of the software include:

- Ability to upload PDF drawings.
- Multiple related measurements can be taken at the click of a mouse for areas, lengths, volumes and counts.

7.4.4 Cubit Estimating

Software house Building Software Services (BSS) has developed Cubit – a powerful and flexible mid-range estimating and take-off software package that enables the user to produce top-down and bottom-up estimates and quantity take-off, all on the same screen.

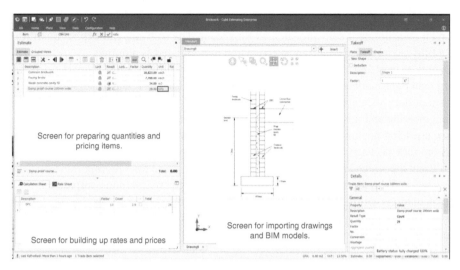

1. Cubit Estimating allows on-screen measurement from imported drawings and BIM models.
2. The Estimate screen permits builders' quantities to be prepared without reliance on any standard method of measurement.
3. Quantities can be prepared for unit rate estimating or operational estimating.

Figure 7.5 Cubit estimating.

The single-screen view has a special window where scanned drawings, PDF, CAD, BIM and image (photograph) files can be imported, and take-offs can be created, priced and seen at the same time without switching views. This is illustrated in Figure 7.5.

The software permits users to create bespoke project-specific or standard templates using their own item descriptions. This is ideal for creating builder's quantities or activity schedules. Cost databases can be created within the software or imported from Excel.

Features of Buildsoft Cubit Estimating include:

- Quantity take-off directly inside the estimate.
- Everything on the same screen.
- Simple or complex quantity take-offs can be performed based on standard methods of measurement if desired.
- No need for CAD software.
- BIM take-off module.
- Build-up project-specific rates and compile pricing templates from previous projects or import supplier price lists.
- Create subcontractor templates for future projects.

7.4.5 Bidcon

Bidcon is a powerful high-end, Cloud-based estimating database that can be used to prepare order of cost estimates, subcontract enquiries and analyses, work package quotes or analytical bottom-up estimates based on full bills of quantities (BQ).

Quantities are provided by importing BIM files, Excel files or quantities produced by on-screen measurement software such as Bluebeam. The software can extract quantities from a BIM model which can then be coded and manipulated as required.

Customised cost libraries can be created within the software and internal prices for labour, materials, plant and subcontractors can be attributed to the database. Alternatively, Spon's cost library databases can be imported into the software including their building, civil engineering and mechanical and electrical services (MEP) price books.

Once a pricing document has been created, tender enquiries can be sent to suppliers, trades or work package contractors for pricing. This information is sent out as a locked macro Excel file which can be priced by the subcontractor and then re-imported into (Bidcon is a registered trademark of Elecosoft AB.) for detailed analysis and comparison.

All calculations within Bidcon can be resolved to a summary sheet categorised as the user requires. This could be a tender summary sheet, for example, or a summary of subcontract or resource costs as desired. Features include:

- Multiple users in different locations at the same time.
- Full integration with Asta Powerproject project planning software.
- Pricing documents, such as bills of quantities, can be assembled from a pre-populated database using NRM1, NRM2 or other measurement conventions such as SMM7.
- Suppliers' price lists can also be imported into the database and Excel price lists may be seamlessly imported and exported.
- Estimates can be compiled in a variety of currencies depending on where the estimator or project is located.
- A variety of templates can be created based on individual trades or work packages or by using NRM1 and NRM2 code libraries.

7.4.6 CATO

Causeway is the largest specialist software provider in the UK construction market and provides a powerful, comprehensive and industry-leading modular suite of construction software. The Causeway CATO Essentials suite includes modules for both top-down and bottom-up estimating together with BQ production capability:

- **QuickEst**: for developing high-level order of cost estimates.
- **Cost planning**: for elemental cost planning.
- **Take-off and billing**: for BQ production.
- **Causeway estimating**: for analytical bottom-up estimating.

Features include:

- Full scalability from single user to Tier 1 contractors.
- On-screen measurement from imported documents such as PDF, CAD and BIM files.
- Bill of quantities production using a large number of standard methods of measurement – both UK-specific and those used in other countries around the world.
- Industry standard cost data such as BCIS included.
- Importation of own cost data including from Excel files.
- Importation of bills of quantities from Excel.
- Estimating and bid management capability.

7.5 Top-down Estimating

Some 50 years ago, top-down methods of estimating were largely the province of PQSs who used these methods to advise clients at the feasibility stage of projects and to assist in the development of designs within the client's budget.

Over this period, however, contractors have become increasingly involved in the early stages of projects and have necessarily adopted top-down methods of estimating to be able to offer early cost advice.

7.5.1 Definition

Top-down estimating is more formally referred to as *cost planning* or *design cost estimating*. This is a two-part process:

1. Preparation of an initial budget
2. Expansion of the budget into a detailed plan of expenditure commensurate with the developing design.

Cost planning is part of a system of *cost control*, the aim of which is to ensure that the out-turn cost of a project does not exceed the original budget or cost limit agreed by the client. The cost control process takes place from the inception of a project to its final completion and commissioning.

7.5.2 Traditional Procurement

In traditional procurement – where the design stage is separated from construction – the process of cost control is undertaken by the client's PQS or cost manager, who will assist the design team to develop the design within the client's cost parameters.

During the design stages, the aim is to control the cost of each part of the design as more and more detail is produced by the architect(s) and other designers.

Once a contract is awarded at an agreed price, cost control focuses on comparing projected monthly or periodic expenditure with the payments made to the contractor and the cost (±) of design changes, variations and the expenditure of provisional sums and contingencies. The aim is to maintain the project budget and to ensure that the projected final account is kept within this figure.

The client's cost reporting stages during the construction period are entirely separate from the contractor's cost control procedures in a traditional procurement setting.

7.5.3 Non-traditional Procurement

In recent years, particularly since the advent of design and build procurement, contractors have had to adopt top-down methods of estimating to cost plan designs in the absence of detailed drawings and quantities.

In such cases, contractors are able to benefit from an extensive database of historic tenders from which top-down estimating data can be extracted.

Top-down estimating is also used when clients require Early Contractor Involvement (ECI) in their projects. Contractors are able to assist the client design team by providing:

- Early cost advice.
- Pricing alternative design solutions.
- Advising on the buildability of design choices.
- Helping with development the project master programme.

7.5.4 Principles

The principles of top-down estimating – cost planning – are the same for contractors and the PQS. The aim is to ensure that the agreed price or target is not exceeded by controlling design outputs and by accommodating unforeseen or compensation events within budgeted allowances. Within the system of project cost control, cost planning consists of:

- Establishing the project budget (order of cost).
- Developing the cost plan.
- Estimating the cost of alternative design solutions.
- Aligning the design and the cost plan within the project budget.

There are essentially four methods of top-down estimating which are used in different circumstances and at different stages of the cost planning process:

- Floor area method.
- Functional unit method.
- Elemental method.
- Hybrid method.

7.5.5 Floor Area Method

This is a method of estimating used for establishing a project budget at RIBA Stage 0 – *Strategic definition* and Stage 1 – *Preparation and briefing*. It provides a 'first cost' or 'high-level' estimate at an early stage in the project when little information is available upon which to base an estimate.

This is a simple but powerful method of estimating which is sometimes referred to as a 'single rate method'. It simply requires the provision, or calculation, of a superficial measurement – gross internal floor area (GIFA) – and a unit rate of currency to arrive at a relatively reliable indication of cost at such an early stage in a project. This method of estimating would be appropriate for calculating the cost of a new warehouse, supermarket or house extension for instance – any building where a superficial floor area can be calculated.

The floor area, or superficial, method of estimating may be used where a specific site or location has yet to be chosen for the project and when the client is simply considering options as part of the business case.

> In the absence of little drawn information or specification – or perhaps none – it is usual practice to calculate a 'cost range' or 'cost bracket' at this stage, giving the client a realistic indication of cost whilst at the same time providing flexibility to balance cost, quality and value for money within affordability limits.

ORDER OF COST ESTIMATE FOR NEW CAR SHOWROOM

Item	Quantity	Unit	Rate £	Sub-total £	Sub-total £	Total £
Sales showroom	1200	m²	2200	2640000		
Servicing and parts	1800	m²	1700	3060000		
Offices	200	m²	1900	380000		
Drainage		say		70000		
External works	2000	m²	200	400000		
Total works cost estimate					6550000	
Main contractor's preliminaries	15	%			982500	
					7532500	
Design fees	8	%			602600	
					8135100	
Risk allowance	10	%			813510	
TOTAL						**8948610**

NB:
1. Unit rates/m² include main contractor's overheads and profit.
2. Risk allowance includes design development and inflation risks.

Figure 7.6 Budget estimate.

The unit rate used for this method of estimating cannot realistically include for site-specific works such as roads, drainage, car parking and landscaping, as this could skew the resulting estimate considerably. Allowances for such items may be calculated by using a variety of methods including the superficial method (e.g. £/m² of car park) or £ per linear metre of drainage (including chambers) or by including a lump sum figure.

Figure 7.6 demonstrates how a budget might be calculated for a design and build car showroom project.

7.5.6 Functional Unit Method

This is another 'single rate' method of estimating but this time using a price per functional unit. The method is particularly suited to projects such as schools, hospitals, prisons, cinemas, sports stadia and multi-storey car parks.

> The unit rate might be £ per place (in a school), £ per bed (in a hospital) or £ per seat in a football stadium which is simply multiplied by the number of functional units required (i.e. number of places, number of beds, number of seats, etc.).

The functional unit rate used in the calculation will depend upon several variables such as:

- The morphology of the building – size, shape, height, etc.
- The level of specification based on a qualitative assessment.
- The extent of mechanical and electrical services required.
- Whether the design is of standard or novel construction.

ORDER OF COST ESTIMATE FOR HOTEL COMPLEX

Item	Quantity	Unit	Rate £	Sub-total £	Sub-total £	Total £
Facilitating works				1300000		
Hotel	165	Beds	130000	21450000		
Conference centre	1000	m²	3250	3250000		
					26000000	
Risk allowance	10	%			2600000	
TOTAL						**28600000**

NB:
1. Functional unit and floor area methods used together.
2. All prices include main contractor's preliminaries and overheads and profit.

Figure 7.7 Order of cost estimate.

As with the floor area method, the characteristics and geology of the site, the type of foundations envisaged and need for enabling works, external works and other site-specific costs are separate considerations that cannot be included in the functional unit rate.

An order of cost estimate for a new hotel and conference centre is shown in Figure 7.7 which illustrates how two single rate methods of estimating may be used together.

7.5.7 Elemental Method

The elemental method of estimating was invented by James Nisbet whilst working on the Hertfordshire Council school building programme. Strictly speaking, this is a method of cost planning as opposed to the floor area and functional unit methods which are used for early cost estimating and for establishing a project budget.

> The elemental method of cost planning is a way of developing a project budget in more detail and in a structured way that reflects the way in which designers work. Cost plans are therefore structured on the basis of building elements and not m³ of concrete or m² of plastering.

Elemental cost planning uses the costs of parts (or elements) of previous similar buildings to predict the cost of a proposed project, making appropriate allowance for design differences, location and market conditions.

The elements used are based on those classified in the New Rules of Measurement (NRM1) and are common to all building types:

1. Substructure.
2. Superstructure.
3. Internal finishes.
4. Fittings, furnishings and equipment.
5. Services.

6. Prefabricated buildings and building units.
7. Work to existing buildings.
8. External works.

Other elements may be added to a cost plan, such as facilitating (enabling) works, main contractor's preliminaries, design fees and risk allowances and so on.

Using this method of estimating, an outline cost plan can be developed from the order of cost estimate (the project budget) which provides a cost framework to assist in design preparation and development.

> The sums of money allocated to each element in the cost plan are not fixed but can be redistributed between individual elements as the design demands albeit that the total spend must remain within the agreed budget. The power behind this method of estimating is that it provides a means of cost checking the design as it develops and a way of controlling expenditure to avoid cost overruns.

Figure 7.8 illustrates how a project budget can be structured into an outline cost plan in two ways:

1. Proportionately by deciding upon a suitable percentage for each element.
2. On an elemental cost per m² of floor area[1] basis should, suitable cost data be available.

As the design develops, a detailed cost plan can be developed with the various elements expanded in more detail by using sub-elements, as illustrated in Figure 7.9.

7.5.8 Hybrid Method

Approximate estimating and cost planning for civil and heavy engineering and other non-building projects provide challenges for the engineer or quantity surveyor due to the wide variety, scale, complexity and scope of work in these sectors of the industry. Engineering structures are also invariably constructed in challenging locations and may require novel construction solutions, extensive temporary works or on-site/near-site materials production facilities.

For this type of work, the superficial area and functional unit methods of estimating might not be appropriate or might only be relevant to a certain degree. They might also be combined with other methods including:

- **Approximate quantities**: measured from existing site plans, early sketches/schematics or measured on site.
- **Inspection**: looking at cost data from comparative projects in a database.
- **Proportion**: based on a similar past project that might be smaller, larger or have dissimilar site characteristics.

> The BCIS does not publish a standard form of cost analysis for civil engineering work, although the RICS has issued consultation documents with this purpose in mind (BCIS 2011).

FORMAL COST PLAN 1 FOR HOTEL COMPLEX

Group element	Element	Proportion %	Order of cost estimate* £	Element cost plan £	Total £
			26000000		
0	Facilitating works	5		1300000	
1	Substructure	20		5200000	
2	Superstructure	15		3900000	
3	Internal finishes	15		3900000	
4	Fittings, furnishings and equipment	5		1300000	
5	Services	25		6500000	
6	Prefabricated buildings and building units	10		2600000	
7	Work to existing buildings	0		0	
8	External works	5		1300000	
	Sub-total	100			26000000
	Risk allowance	10		26000000	2600000
	TOTAL COST PLAN 1				**28600000**

*Includes main contractor's preliminaries and overheads and profit

COST PLAN 1 USING PROPORTION

FORMAL COST PLAN 1 FOR HOTEL COMPLEX

Group element	Element	Gross internal floor area	Element unit rate £	Element cost plan £	Total £
0	Facilitating works	4000	325	1300000	
1	Substructure	4000	1300	5200000	
2	Superstructure	4000	975	3900000	
3	Internal finishes	4000	975	3900000	
4	Fittings, furnishings and equipment	4000	325	1300000	
5	Services	4000	1625	6500000	
6	Prefabricated buildings and building units	4000	650	2600000	
7	Work to existing buildings	4000	0	0	
8	External works	4000	325	1300000	
	Sub-total				26000000
	Risk allowance	10 %		26000000	2600000
	TOTAL COST PLAN 1				**28600000**

COST PLAN 1 USING ELEMENT UNIT RATES

Figure 7.8 Development of cost plan 1.

7.5 Top-down Estimating | 233

FORMAL COST PLAN 2 FOR HOTEL COMPLEX

Group element	Element		Gross internal floor area	Element unit rate £	Sub-element unit rate £	Sub-element total £	Element cost plan £	Total £
0	Facilitating works		4000	325			1300000	
1	Substructure		4000	1300			5200000	
2	Superstructure							
	2.1	Frame	4000		250	1000000		
	2.2	Upper floors	4000		100	400000		
	2.3	Roof	4000		200	800000		
	2.4	Stairs and ramps	4000		50	200000		
	2.5	External walls	4000		160	640000		
	2.6	Windows and external doors	4000		90	360000		
	2.7	Internal walls and partitions	4000		85	340000		
	2.8	Internal doors	4000		40	160000	3900000	
3	Internal finishes		4000	975			3900000	
4	Fittings, furnishings and equipment		4000	325			1300000	
5	Services		4000	1625			6500000	
6	Prefabricated buildings and building units		4000	650			2600000	
7	Work to existing buildings		4000	0			0	
8	External works		4000	325			1300000	
	Sub-total							26000000
	Risk allowance		10 %				26000000	2600000
	TOTAL COST PLAN 2							**28600000**

1. Element 2 (Superstructure) is divided into 8 sub-elements.
2. Other elements will be subdivided in accordance with NRM1 protocols.
3. The gross internal floor area (GIFA) applies to both elements and sub-elements.
4. Each sub-element has its own unit rate.
5. The total of Element 2 sub-element unit rates = £975/m².
6. This is the element unit rate used in Cost Plan 1.

Figure 7.9 Development of detailed cost plan 2.

Part of the difficulty in establishing common construction elements for civil engineering work and the like is that there is no commonality between a tunnel, a bridge, a dam or a sewage treatment works. However, common elements for particular types of civil engineering work, such as roadworks, sea defence work, railway works and so on are clearly possible, as demonstrated by the publication of a *Definition of standard elements for pipelines and ducts* by BCIS in 2011.

Further work to address this problem, and to provide consistency in the understanding of global construction cost issues, has been undertaken by the International Construction Measurement Standards (ICMS) Coalition, members of which are the RIBA and RICS.

ICMS has published a framework of measurement standards (ICMS 2019) for a wide variety of 'projects' including buildings, roads, tunnels, wastewater treatment works, refineries, dams and reservoirs and so on. The term 'projects' is used as opposed to 'work' because the standards provide no classification of the work contained within a project.

Under ICMS, the project quantities for a road, for instance, are measured in km/mi and m² of paved area, whereas the functional units used are 'capacity' in terms of the number of vehicles per hour. For wastewater treatment works, the units of measurement are hectares or acres of site area covered by permanent works and functional units of capacity measured in litres or gallons per day.

In all practicality, engineers and quantity surveyors involved in establishing project budgets for engineering works will use a combination of the estimating methods described above. They may also use approximate quantities if sufficient drawn information, or perhaps a BIM model, is available at RIBA Stages 0 and 1 of a project.

7.6 Cost Information

Top-down estimating relies upon historic data from previous projects. This data is extracted from priced bills of quantities and may be presented in a variety of ways:

- The total cost of the project may simply be divided by its superficial floor area, or its number of functional units, which provides rates that could be used for predicting the cost of future similar projects.
- The data could also be collated on an elemental basis to provide elemental unit rates for cost planning purposes.
- The bills of quantities could be collated into trades or work packages such as groundworks, roofing or concrete frame.

However, the detailed makeup of any tender is privy only to the individual contractor and anyone else using this data must therefore be conscious of the possibility that the contractor has manipulated the data for commercial reasons. Consequently, the prices may not accurately represent the true cost of their labour, plant, materials, overheads and profit elements.

7.6.1 Historic Data

Historic data is collected by client organisations and by their professional advisers – especially by quantity surveying practices. This data is, however, private to these organisations in the same way that the original data, and their detailed makeup, is private to the contractor.

An intermediate source of data, based on historic tender prices, is offered by the BCIS which is an extensive and trusted data source and widely used in the industry.

Contractors are, of course, their own best source of cost information and a contractor with any sort of 'track record' will have carried out many past projects. They should have a database of rates and prices that can be used for estimating construction costs at various points in the project cycle.

Whatever the source of price data, it is important that it contains information that can be used for interpretation of the data and for comparative purposes. Such information includes:

- Description of the structural form of a building (e.g. steel-framed building, cladding type and M&E system).
- GIFA in m^2 and/or number of functional units.
- Description of a structure such as a tunnel (e.g. type of tunnel boring machine (TBM), nature of tunnel lining and length and diameter of bore).

- The level of specification including finishings, fire detection/prevention systems, M&E and air handling, lifts and conveyors.
- Extent of external works and landscaping.
- Size, shape, wall:floor ratio, window:wall ratio, storey height, number of floors of a building.
- Base date of cost information (e.g. date of tender).
- Geographical location of the works.
- Market information at the time of tendering.

7.6.2 Sources

Priced bills of quantities submitted by contractors can be re-sorted in a variety of ways such as elemental or trade/work packages and may also be sorted into activity schedule form by allocating prices to the master programme for the project.

Similar cost information, albeit second-hand, has been available to PQS practices for many years based on the analysis of tender documents received from contractors:

- By creating an internal database of cost information – costs/m^2, functional unit costs and elemental costs – extracted from analysed bills of quantities.
- By subscribing to the Building Cost Information Service (BCIS) of the RICS which offers a hard copy or online database of over 2500 buildings of various types, sizes and locations on a subscription basis.

A further source of price data and related information is available from published price books. Builders' price books have a history dating back to the seventeenth century – the first edition of Spon's was published in 1873. They have been in common use in the industry for many years.

Price books provide average prices for building and engineering work – related to specific standard methods of measurement – approximate estimating data, regional pricing and inflation data and so on. The most well-known are:

- Spon's Architects' and Builders' Price Book.
- Spon's Civil Engineering and Highway Works.
- Spon's Mechanical and Electrical Services.
- Spon's External Works and Landscape.
- Laxton's NRM Building Price Book.
- Griffiths Building and Civil Engineering Price Book.

> The cost data found in price books is historic and 'average' and therefore should be used with caution. This data can be used individually or in conjunction with other data (such as BCIS) but requires considerable judgement and experience to achieve accurate results. The only accurate and reliable cost data is the contractor's or subcontractor's own data and this, of course, is not publicly available.

An example of a price database used for establishing an order of cost estimate or an elemental cost plan is illustrated in Figure 7.10.

BUILDING TYPE	Unit	Range	
		£	£
Administrative			
Offices; low rise	m²	2180	2780
Courts	m²	3725	4725
Prisons	m²	2915	3725
Industrial			
Light industrial factories	m²	1025	1315
Shell and core with heating only	m²	1470	1865
Maintenance workshops	m²	1365	1735
High-tech laboratories, air conditioned	m²	4120	5250
Recreational			
Restaurants	m²	2175	2780
Visitors' centre	m²	2915	3725
Multiplex Cinema	m²	2570	3305
Residential			
Two-storey housing			
Detached	m²	1575	1995
Semi-detached	m²	1470	1865

BUILDING PRICES PER SQUARE METRE

BUILDING TYPE	Unit	Range	
		£	£
Utilities			
Surface car parking	car	2615	3805
Underground parking	car	39900	55650
Offices			
Average spec; non-air conditioned	person	17325	23625
Residential			
Hotels, luxury, city centre	person	588000	798000
Student accommodation, en-suite, <20m² per bedroom	person	37275	51975

BUILDING PRICES PER FUNCTIONAL UNIT

Figure 7.10   Price data bases.

7.6.3   Indices

In order to use historic cost data correctly, account must be taken of the origins of the information, remembering that:

- The data is based on contractor's tender prices (not actual costs or final accounts) which include an unknown element of overheads and profit.
- The prices reflect market conditions at the time of tendering.
- The contractor's rates may have been manipulated for commercial reasons such as anticipation of claims or generating cash flow.
- The tender date is the base date for the data and needs to be updated for inflation/deflation to the current date.
- The data relates to a project in a specific geographical location and may need to be adjusted for regional variations in construction prices.

ORDER OF COST ESTIMATE FOR NEW SECONDARY SCHOOL

COST INFORMATION

GIFA	17000 m²
Project location	North East of England
Location factor	0.85
Comparative cost per m² at base date	£2650 per m²
Building location	Inner London
Location factor	1.05
Tender price index at cost base date	160
Tender price index at contract start date	185

ESTIMATE

Item	Quantity	Rate	£	£
Cost per m²		2650		
Adjust for location				
North East of England	0.85 x	2650	2145.24	
Inner London	1.05			
Adjust for inflation				
Contract start date	185 x	2145.24	2480.43	
Price base date	160			
GIFA	17000 x	2480.43	42167336.31	
Risk allowance	12 %		5060080.36	
TOTAL ORDER OF COST				47227417

NB:
1. A previously completed project in Inner London is found to be a close match for a proposed school in the North-East of England.
2. The two projects are similar in size and specification.
3. This means that the cost per m² at tender stage can be used to predict future costs, making suitable adjustments for location and inflation.
4. Building in the North-East of England is less expensive than Inner London.
5. Inflation is adjusted from the cost base date to contract start date.
6. Cost inflation during construction is included in the base cost per m² as this was a firm price tender.

Figure 7.11 Indices and location factors.

BCIS and price books publish tender price and building cost indices together with regional price adjustment factors. The use of indices and price adjustment factors is illustrated in Figure 7.11.

Indices should be used with caution as some inflation indices are based on general inflation indices such as the retail price index (RPI) or consumer price index (CPI) and these are not construction specific. Construction prices cannot be accurately updated based on indices relating to house mortgages, bananas and packets of crisps!!

7.6.4 Inflation

The cost data upon which early cost estimates and cost plans are based is historic information and may date back some time before project conception. This data – whether from price books, BCIS or internal sources, must therefore be updated from its base date to the current date.

> The current date is usually the date of tender when bids are submitted because the idea of cost planning is to predict what the market price will be at tender stage. However, where contractors are asked to take inflation into account in their bids, the current date may be the mid-point of the construction period (i.e. an average).

Consequently, there are two aspects to inflation:

- Tender inflation.
- Construction inflation.

To calculate tender inflation – inflation from the data base date to the tender date – it is normal to use published indices, as shown in Figure 7.12.

Where the contract requires the contractor to include for inflation in its price, however, the extent to which the cost of resources will increase during the project (construction inflation) will have to be predicted as a risk element of the tender. This is usual practice

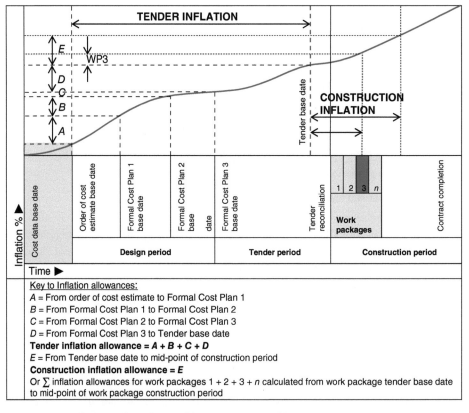

Figure 7.12 Inflation indices. Source: Diagram courtesy of Peter Williams, Managing Measurement Risk in Building and Civil Engineering, Wiley-Blackwell 2016.

for contracts of less than 12 months duration. Being closer to the primary information sources, contractors often use market intelligence to make such allowances rather than indices which are, at best, averages.

In other cases – generally projects over 12 months long – construction inflation will be paid for by the client based on actual cost, where appropriate or based on a recognised index or formula.

It is common to see the CPI or RPI used for this purpose, but these indices are not construction indices – they are concerned with the everyday cost of living. RICS advice is to use a construction specific index to truly represent cost inflation on a fair and equitable basis.

7.7 New Rules of Measurement

The RICS *New Rules of Measurement* is a suite of documents that includes NRM1: *Order of Cost Estimating and cost planning for capital building works.*

NRM1 is the first method of measurement for order of cost estimating and cost planning ever published and has further refined the approach to design cost control.

The principles of design cost control have not changed under NRM1 but the methodology is more formalised and structured and different phraseology is used:

	NRM1 phraseology	Previous phraseology
1.	Order of cost estimate	Cost limit
2.	Formal cost plan 1	Outline cost plan
3.	Formal cost plan 2	Detailed cost plan
4.	Formal cost plan 3	Cost check

NRM1 makes it clear that top-down estimating begins with establishing an order of cost. This would normally inform decisions around the project business case when little or no design information has been prepared. It is also clear that cost planning is an iterative process – cost plan 1, 2 and 3 – and that this requires a system of control to ensure that the cost limit, or order of cost estimate, is not exceeded. There are two elements to this process of control:

- Comparative cost planning.
- Cost checking.

7.7.1 Comparative Cost Planning

Comparative cost planning is a form of control that involves estimating the costs of alternative design solutions for each element of the proposed building.

This provides a guide to the architect or engineer as to the cost consequences of design decisions and facilitates decision making based on value-engineered design solutions. A simple example is illustrated in Figure 7.13.

ELEMENT			COST TARGET			
Internal walls and partitions Aerated concrete block walls 100mm thick Rates taken from analysis 16443			Element Unit Quantity	Unit	Element Unit Rate	Total
Item	£	£			£	£
Element Unit Rate Adjust for: Price level 199 x 177 Quality 15 % New EUR	47.10	52.95 7.94 **60.90**				
			9000	m²	60.90	548076
ELEMENT			COST TARGET			
Internal walls and partitions Metal stud and board partitions Rates taken from analysis 12996			Element Unit Quantity	Unit	Element Unit Rate	Total
Item	£	£			£	£
Element Unit Rate Adjust for: Price level 199 x 186 Quality 10 % New EUR	36.70	39.27 3.93 **43.19**				
			9000	m²	43.19	388724

1. Cost data taken from two different elemental cost analyses.
2. Facilitates comparison between two different forms of construction.
3. Each EUR has a different base index (177/186) but is updated to a common point in time (Index 199).

Figure 7.13 Comparative cost planning.

7.7.2 Cost Checking

Cost checking is a means of determining the current cost of elements or sub-elements of a design to determine whether the original cost plan allowance is still valid.

This process is illustrated in Figure 7.14 which demonstrates the use of approximate quantities and unit rates as a means of detailed cost checking. At the cost checking stage of a design, quantities can be measured because the design is sufficiently advanced to do so.

This technique is especially useful when a contractor is developing a target with a view to establishing a Guaranteed Maximum Price (GMP) and provides greater certainty, and less risk, at a time when the contractor is close to making a definitive commitment to the client in the form of a legally binding price.

ELEMENT			COST TARGET				COST CHECK				
Internal walls and partitions			Element Unit Quantity	Unit	Element Unit Rate						
Rates taken from analysis 16443						Total	Item	Quantity	Unit	Rate	Total
Item		£	£		£	£				£	£
Element Unit Rate		149.70					100mm block walls	1200	m²	40.95	49 140
Adjust for:							Stud partitions dry lined both sides	2000	m²	31.50	63 000
Price level	199 x		168.31								
	177						Hardwood glazed screens	300	m²	1950.0	5 85 000
Quality	15 %		25.25								6 97 140
New EUR			193.55								6 77 435
							Underspend				
			3500	m²	193.55	6 77 435	Overspend				19 705

1. The cost target is set using element unit rates taken from a previous elemental cost analysis.
2. The target is established by the element quantity and EUR.
3. This includes all types of partitions comprised within the element.
4. The cost check is more granular with detailed quantities for each type of partition.
5. The rates used in the cost check are current measured rates.
6. The current cost of the internal partitions is £ 19 705 over the cost target.
7. This indicates that the element must be redesigned, or savings made in other elements, to keep the cost plan on target.

Figure 7.14 Cost checking.

7.7.3 Cost Reconciliation

The key feature of traditional competitive tendering is the separation of design and construction with a tender period in between.

This effectively separates the process of design cost control from the compilation and submission of the contractor's tender. Consequently, tenderers are not involved in the preparation of the cost plan, and the client team has no influence over the value of tenders submitted. In some cases, this results in tenders that are misaligned with the client's budget.

The only control, therefore, is when tenders received are checked or reconciled against the cost plan. This is not an effective control for over-budget tenders, however, as the only means of adjustment is to reduce project scope, reduce specification or otherwise compromise the design to make savings.

7.8 Bottom-up Estimating

Bottom-up estimating is a resource-based method of construction cost estimating. There are various sub-sets to this method that an estimator can employ, but the principles of each are essentially the same:

- Rates and prices are calculated by predicting the cost of labour, materials, plant and equipment.
- Assessments of various risks are made, including inflation (where appropriate), ground conditions, weather, method and time risk and so on.[2]
- An allowance is added for overheads and profit, usually as a percentage.[2]

This method of estimating can only be used where detailed quantities are available or where there is a schedule of rates with full item descriptions. Traditionally, this meant 'formal' bills of quantities prepared by the client's PQS in accordance with a standard method of measurement such as NRM2, CESMM4 or the Highways Method of Measurement.

Formal bills of quantities are now far less commonly seen in practice.

7.8.1 Modern Trends

Changes in procurement methods has meant that contractors, and subcontractors, invariably prepare their own quantities or engage a PQS to do it for them. This means that contractors and subcontractors take the measurement risk – the risk of making mistakes – as well as the price and time risk.

The consequence of this is that the bills of quantities produced may follow a standard method of measurement to a certain degree, but the documents produced are usually far less detailed. There is a tendency towards more composite items, and small items tend not to be measured at all. These quantities are referred to as 'builders' quantities' as they do not use standard method of measurement descriptions.

As Williams (2016) records, less detailed measurement follows the Pareto principle, where 80% of the cost is represented by 20% of the items in a 'normal' bill of quantities, and it is therefore less expensive and faster to produce more abbreviated bills of quantities than is traditional in the industry. This method works provided that the estimator appreciates the Pareto principle and makes allowances for the non-measured items.

It is also common industry practice to ignore standard methods of measurement altogether when preparing quantities. This works especially well at a trade or work package level where relatively few measured items are needed.

Whether abbreviated bills of quantities or builders' quantities are used, it is important to disclaim the non-use of standard methods of measurement in enquiries to subcontractors. The main contractor must make it clear that the descriptions and quantities are not based on a formal method of measurement and that it is up to the subcontractor to make appropriate allowances for the fact that the work is not formally described or measured and that prices quoted should include for all ancillary items that are not measured at all.

7.8.2 Builders' Quantities

Non-standard methods of measuring quantities are frequently referred to as 'builders' quantities' which, traditionally, are 'internal' documents usually unseen by outside parties such as a client or a main contractor (Williams 2016).

Williams (2016) suggests, however, that there is nothing 'rough and ready' about builders' quantities and that they are a widely used, legitimate and accurate way of formulating a tender based on drawings and specification or in a design-build situation.

> The key thing that distinguishes builders' quantities from 'formal' bills of quantities is that the estimator must be careful to price the work that is not measured or described but is nonetheless indispensable to formulating a realistic price.

Builders' quantities should not be confused with the term 'approximate quantities' which has a different meaning. Approximate quantities are prepared when the design is insufficiently detailed, and the work must be admeasured (measured again) on completion. Builders' quantities, on the other hand, should be accurate even though they are not based on a standard method of measurement.

7.8.3 Bottom-up Estimating Methods

The common methods of estimating in this category are:

- The unit rate method:
  - using labour and plant outputs;
  - using labour and plant constants.
- The operational method.

Unit rate methods tend, historically, to be used for pricing building work and the operational method for civil engineering. There is no strict rule however, and much depends upon the estimator and upon the nature of the pricing document.

Composite items, consisting of several work items – such as those found in civil engineering quantities – would tend to favour an operational approach, but there is nothing to prevent an estimator from breaking down a composite item into more detail.

The method of estimating chosen will also depend upon the procurement method and whether there are bills of quantities, an activity schedule, a schedule of rates or simply drawings and specifications.

7.8.4 Unit Rate Estimating

Unit rate methods of estimating rely on the availability of quantities to arrive at a price using the formula:

Quantity × Rate = Price

The items of measured work in bills of quantities are usually defined in the relevant standard method of measurement and are *measured net as fixed in position except where otherwise stated* – or words to that effect.

Effectively, this means that they are items of finished work and, as such, the contractor's rates need to include everything necessary to create that item of work including labour, plant, materials, delivery, handling, hoisting, laps, cutting and waste, fixings and so on.

Conventionally, therefore, a basic unit rate comprises four main items:

1. Labour.
2. Plant.
3. Materials.
4. Overheads and Profit.

The principles of unit rate estimating have remained largely unchanged for centuries although technology and the passage of time have added degrees of sophistication to the process.

The initial task for the estimator is to establish current rates for labour. This might be labour directly employed by the company or subcontract labour.

> Different rates of pay apply in the construction industry to different levels of skill as established by the Working Rule Agreement (WRA) Construction Industry Joint Council (CIJC) (2022). Therefore, general operatives, skill levels 4 to 1 and craft operatives require a separate calculation.

The basic rate of pay is augmented by allowances for incentives, travelling, national insurance contributions and other labour 'on-costs' to establish all-in rates for each category of labour. The CIOB Code of Estimating Practice (CoEP) provides guidance on the calculation of all-in labour rates.

Where direct labour is self-employed, the WRA does not apply, and market rates dictate the level of pay. Added to the base rate of pay are on-costs for employer's and public liability insurances, and perhaps travelling and lodging (where appropriate), but the more considerable on-costs of direct employment are avoided.

As far as materials are concerned, it is usual practice to send enquiries to merchants, suppliers, ready-mixed concrete firms, and the like, in order to obtain up-to-date quotations. The estimator usually analyses the quotes and may choose the lowest quotes or may take an average depending upon a view of market risk.

At one time, many contractors had their own fleets of construction plant, and some even had their own plant yard or separate company which hired plant both internally and to the wider industry. Nowadays, most of the plant comes from the hire market from where current rates for operated and non-operated plant can be determined. In some cases, plant hire firms will quote for a specific site operation – such as demolition or earthworks – and it is common for significant lifting operations to be carried out by specialists.

Once the basic cost information is to hand, the estimator can begin pricing the measured items. This process is sometimes referred to as 'analytical estimating', examples of which are shown in Figure 7.15. These examples demonstrate the use of labour outputs and constants and show how materials, waste and plant allowances are added into the rate calculation.

When the estimator is discussing the pre-tender programme with the planner, one of the topics of conversation will be gang sizes, outputs and methods of working as the programme needs to reflect the estimate and vice versa.

7.8.5 Operational Estimating

The estimator's main job is to determine the prices for the measured items of work given in the bills of quantities or other pricing documents. Unit pricing is the conventional method of doing this, especially for building work.

7.8 Bottom-up Estimating

New Warehouse

		Qty	Unit	Rate	£	p
	'In-situ' concrete works					
	Reinforced in-situ concrete; C35/20					
	Horizontal work					
A	<= 300 thick; In structures; Reinforced > 5%	750	m³	218.12	163,590	00
	Surface finishes to in-situ concrete					
	Power floating					
B	To top surfaces	3000	m²	0.95	2,850	00

Reinforced in situ concrete ground slab; 250mm thick; C35/20; power floated

Method
100m x 30m slab to be laid in alternate bays. Bay size 30m x 4m. 25 No bays.
Concrete to be placed direct from ready-mix truck using roller striker screed.

Materials				£	£	£
	Concrete	per	m³	150.00		
	Waste	3 %		4.50		
	Ground loss	5 %		7.50		162.00
Labour						
	Ganger	1 hr		28.00	28.00	
	Joiner	1 hr		25.00	25.00	
	Labourer	4 hr		20.00	80.00	
		0.27 hrs/m³			133.00	35.91
Plant			per hour			
	Roller striker screed	1		10.00	10.00	
	Poker vibrator	2		2.50	5.00	
		0.27 hrs/m³			15.00	4.05
						201.96
Overheads & Profit		8 %				16.16
				per m³		**218.12**

Power floating top surfaces of slabs

Labour			per hour		
	Labourer	1 No	20.00	20.00	
Plant					
	Power float	1 No	5.00	5.00	
		0.035 hrs/m²		25.00	0.88
Overheads & Profit		8 %			0.07
			per m²		**0.95**

1. This method of estimating is based on the use of labour and plant constants (outputs) per unit of quantity.
2. Each bill of quantities item is priced individually which contrasts with the operational method of estimating where the total quantity measured is priced as a complete activity.
3. Labour and plant constants are based on experience, company data or published price books.
4. The overall cost of the two measured items is £163 590 + £2 850 = £166 440. This compares to a total of £168 558 using the operational method.
5. The variance is explained by the different approaches to estimating the labour and plant elements.
6. For some measured items, method-related and time-related charges may be appropriate which might be included in the main contractor's preliminaries.

Figure 7.15 Unit rate estimating.

> Operational estimating relies on an assessment of time and resources to arrive at a price and has a close synergy with estimating the duration of activities on the contractor's programme. It is more common in civil engineering because the bills of quantities tend to contain much fewer items than in building work and are, consequently, composite in nature and financially significant.

This is not to say that operational pricing is not relevant to building work and, indeed, some estimators prefer this approach. It can be used where there is a composite BQ item or where a series of measured items need to be priced as an operation rather than as individual items.

Operational estimating is used when the method of construction, plant and temporary works requirements outweigh the importance of the measured quantities. For instance, the quantity of excavation, concrete, rebar and formwork required to construct a bridge abutment is relatively small compared to the complexity of the work and the time needed to carry out the work.

The following are examples of situations where operational estimating might be more appropriate than the unit rate method:

- Building work:
  o Excavation of a basement within a sheet-piled cofferdam where the duration of the work is determined by the slowest operation – probably disposal of excavated material. Time will also be needed for temporary propping of the sheet piles.
  o A large continuous concrete pour where the rate of delivery of ready-mixed concrete to the site will determine the time taken for completion of the operation.
  o Slip-forming the lift shaft and stairwells of a multi-storey building – a 24-hour operation where the labour and plant costs will be determined by the rate of slide (e.g. one complete floor level per 24 hours).
- Civil engineering work:
  o A one million m^3 cut and fill earthworks operation for a 10 km highway project where the duration of the work is determined by the haul lengths between cut and fill zones and by obstructions such as bridges, culverts and other structures.
  o A 5 m diameter × 25 m deep segmental tunnel ventilation shaft, sunk by underpinning, would generate approximately 500 m^3 of excavated material. The duration of the work would be dictated by the time needed to move the segmental units to the shaft, to crane them into position and to seal and bolt them together.
  o An incrementally launched bridge deck requires a huge amount of falsework, access scaffolding and temporary works for the launch gantry, far outweighing the volume of concrete in the deck itself. The duration for the operational pricing would be dictated by the time for mobilisation and the rate of launch of the deck.

The essential difference between pricing units of measured work and operational items is that the labour and plant content is priced as if it were an activity on a construction programme as opposed to being based on the physical output of the gang doing the work.

7.8 Bottom-up Estimating

New Warehouse

		Qty	Unit	Rate	£	p
	'In-situ' concrete works					
	Reinforced in-situ concrete; C35/20					
	Horizontal work					
A	<= 300 thick; In structures; Reinforced > 5%	750	m³	224.78	168,585	00
	Surface finishes to in-situ concrete					
	Power floating					
B	To top surfaces	3000	m²	incl		

Reinforced in situ concrete ground slab; 250mm thick; C35/20; power floated

Method

100m x 30m slab to be laid in alternate bays. Bay size 30m x 4m. 25 No bays.
Concrete to be placed direct from ready-mix truck using roller striker screed.
Output = 1 No bay/day including power floating = 25 days.

Materials			£	£	£
Concrete	per m³		150.00		
Waste	3 %		4.50		
Ground loss	5 %		7.50		
	750 m³		162.00		121500
Labour			per day		
Ganger	1 No		224.00	224.00	
Joiner	1 No		200.00	200.00	
Labourer	5 No		160.00	800.00	
		25 days		1224.00	30600
Plant			per day		
Roller striker screed	1		80.00	80	
Poker vibrator	2		20.00	40	
Power float	1		40.00	40	
		25 days		160	4000
					156100
Overheads & Profit		8 %			12488
					168588
					750
			per m³		**224.78**

1. This method of estimating is based on the total quantity measured in the bills of quantities and is priced as a discrete activity.
2. A duration for the activity is calculated based on the output of the gang.
3. Labour and plant costs are priced on the basis of cost per day x duration.
4. The overall cost of the activity is divided by the billed quantity to give a rate per m³.
5. Ancillary items (e.g. power floating) are priced as 'included' as the labour and plant costs are for the complete activity.
6. This is to be contrasted with the unit rate method where the concreting and power floating items are priced individually.
7. For some measured items, method-related and time-related charges may be appropriate which might be included in the main contractor's preliminaries.

Figure 7.16   Operational estimating.

Calculating how long an operation might take depends upon:

- Time to mobilise the operation.
- The rate of progress per day.

Time may be calculated on the basis of 'forward travel' – say for a pipeline or for drainage work on a major highway project – or the rate of continuous progress by estimating the number of units or components that can be placed per day.

An example of an operational estimate for laying a reinforced concrete slab is given in Figure 7.16 which illustrates the synergy of operational estimating with programming construction work.

7.9 Bidding

Contractors, whatever their size, are in business to make money which means that they need to obtain work – to win contracts. The usual process in doing so is to elicit an enquiry, to price the job and to submit a bid for the work. Estimating is the process whereby a contractor calculates the cost of construction, within acceptable confidence limits, which forms the basis of a commercial offer or bid to carry out the work.

The CIOB (2018) and Brook (2017) amplify this understanding by explaining that a bid is an *offer to carry out work for a price* based on an estimate of cost. The CIOB (2018) also explains that tendering is the process of converting the estimate into a bid and that bidding is the preparatory process of formulating a tender figure prior to formal submission.

This implies, therefore, that the words 'bid' and 'tender' are not synonymous and that each has a particular meaning – *bidding is a preparatory process and tendering is* a formal offer to carry out work or to supply goods and/or services.

The bidding process is in two stages:

1. Estimating and bid preparation.
2. Bid submission and formal tender.

7.9.1 Estimating

Despite the reliance that many contractors place upon subcontractors, in some sectors of the industry it is important for contractors to carry their own directly employed workforce. This might be to be more responsive to the needs of clients or to be able to complete rapid turnover projects on time and on budget.

In such cases it is important to have control over the workforce and not to be reliant on subcontractors who may not turn up on site as needed.

Fit-out contractors, for example, need to have a flexible and multi-skilled team capable of turning over projects quickly and maybe working 24-hour shifts to meet demanding deadlines. General contractors may also prefer the flexibility of having their own directly employed workforce for key trades whilst engaging subcontractors for specialist work.

Consequently, there are many contractors and subcontractors who price work from first principles and have their own estimator or estimating team for this purpose. Additionally, subcontractors and specialist contractors, large and small, also price their own work using traditional bottom-up methods of analytical estimating.

The work of the estimator consists of:

- Organising enquiries for materials prices and specialist subcontract quotations; the estimator may be assisted by an estimator's clerk or enquiries clerk.
- Arranging a visit to see the site of the proposed works.
- Liaising with the contractor's planning department in order to prepare to pre-tender programme:
    - To assist with the pricing of preliminaries.
    - To check that the construction period stipulated in the tender documents is realistic.
- Pricing the bills of quantities items that would normally be carried out by the contractor's own workforce.
- Compiling comparison sheets of quotations received and choosing which quotes to use for formulating the estimate.
- Pricing the contractor's preliminaries based on the pre-tender programme or the time stipulated in the tender documents. An example of pricing preliminaries items is shown in Figure 7.17.
- Presenting the completed estimate to management so that commercial decisions can be made and a tender figure determined.
- Arranging for the form of tender to be signed by a company director and arranging for this to be delivered electronically or by hand.

7.9.2 Subcontract Enquiries

For the many contractors who rely almost exclusively on prices from subcontractors, specialist contractors and plant hire firms when compiling tenders, the work of the estimator is largely focused on:

- Sending out enquiries for subcontract prices.
- Pre-tender discussions with preferred subcontractors as to intended methods of working and programme.
- Analysing subcontract bids to decide which prices to use as the basis of the tender.

The choices made by the estimator, and an analysis of subcontract quotes, will be a major focus of discussion at the bid preparation stage.

> The main contractor undertakes a significant degree of risk when tendering on the basis of subcontractor's prices. The subcontract prices included in the tender may no longer be held firm by the subcontractor once the main contract has been awarded. Also, preferred subcontractors at tender stage may not be available when the project starts and alternatives – perhaps at a higher price – may have to be sought.

PRELIMINARIES SUMMARY

Project duration = 39 weeks

Ref	Item	Quantity	Unit	Rate	Time-related charges	Fixed charges	Total
				£	£	£	£
A	**Site management**						
	Site manager	39	weeks	1600	62400		
	Car and expenses	39	weeks	200	7800		
	Setting out engineer	20	weeks	1100	22000		
					92200	0	92200
B	**Site accomodation**						
	Establishment					1000	
	Removal					500	
	Offices	39	weeks	150	5850		
	Welfare	39	weeks	50	1950		
	Security containers	78	weeks	25	1950		
					9750	1500	11250
C	**Fencing and security**						
	Erect/dismantle					2000	
	Hire	39	weeks	250	9750		
					9750	2000	11750
D	**Communications equipment**	39	weeks	100	3900	0	3900
E	**Materials handling**						
	Telehandler	39	weeks	600	23400		
	Site dumper	39	weeks	350	13650		
					37050	0	37050
F	**Temporary services**						
	Installation					3000	
	Removal					1000	
	Water	39	weeks	40	1560		
	Power	39	weeks	250	9750		
	Drainage	39	weeks	50	1950		
					13260	4000	17260
G	**Site transport**	39	weeks	150	5850	0	5850
H	**Access equipment**						
	Erection/dismantling					1000	
	Hoist	26	weeks	120	3120		
	Scaffolding	26	weeks	300	7800		
					10920	1000	11920
I	**Consumables**	39	weeks	150	5850	0	5850
							197030
J	**Overheads & Profit**	8	%				15762
	TOTAL						**212792**

Figure 7.17 Preliminaries.

The contractor may, therefore, take a view on which subcontract prices to use as the basis for the bid and might well decide to employ a parametric method of establishing the rates:

- A beta distribution formula.
- A weighted three-point probability.
- A simple mean of the prices received.

Modern measurement/estimating software and ERP systems – such as construction-specific 4PSConstruct and Access ConQuest Estimating – facilitate the packaging of subcontract enquiries. They can send subcontractor enquiries online, see the real-time response to the enquiry as to which subcontractor is intending to price the job and make immediate access to documents available to the subcontractors.

The software also enables preparation of comparison sheets to compare and rank subcontract bids and perform analyses to assess pricing levels and risk.

7.9.3 Bid Preparation

When the estimator has completed the pricing, a summary sheet will be prepared which will indicate a bottom-line figure. This is the total estimate for the project. Prior to submitting the tender at the appointed time, the directors will have to consider the estimator's calculations and make some decisions, including:

- Is the estimate accurate?
- Have the items been priced sensibly?
- Are the subcontract prices used achievable?
- Could the prices used be bettered if the bid is successful?
- Is the pre-tender programme realistic?
- Can the programme be shortened?
- What are the risk issues related to this job?
- How much does the company need the work?
- What is the level of competition for the work?
- Does the tender documentation offer any commercial opportunities?

When these questions have been answered, the directors will be able to:

- Agree with the estimator's price.
- Add money to the price.
- Reduce the price.

Whatever the decision, this will be the tender sum which will be offered to the employer via the form of tender. If the tender is accepted, the contractor will have to live with the decisions made and get on with building the project for the price offered. The pressure will then be on the site team to work within the budget and return the desired margin (profit).

The CIOB Code of Estimating Practice proposes a model form for the estimate summary, analysis and report.

The tender adjudication process must be thorough to make sure that key issues are not overlooked, and all information generated in the build-up of the tender figure needs to be reviewed at the meeting including all supporting documentation. It may take into account:

- General information:
    - Tender enquiry form including details of parties involved in the contract.
    - Site visit report.
    - The estimator's report.
    - The quantity surveyor's commercial risk assessment.
    - Knowledge of competitors.
    - Details of current and future contractual commitments.

- Pricing data:
  - Net cost analysis including the value of work to be sublet.
  - Analysis of subcontract and materials prices.
  - Schedule of prime cost and provisional sums.
  - Schedule of project overheads.
  - Tender summary.
- Pre-tender planning data:
  - Overview of construction methods allowed in estimate.
  - Planning report from the planner/contracts manager comprising comments on the contract period given by the client and the feasibility of the pre-tender programme.
  - Pre-tender programme (linked bar chart).
  - Pre-tender method statements.
- Risk issues:
  - Ground conditions report.
  - Temporary works allowances.
  - Design liability (if any).
  - Summary of contract conditions including any amendments to the standard form.
  - Details of the damages for late completion.
  - Details of insurance requirements and bonds.
  - Working capital implications for the contractor's borrowings.
  - Any cash funding requirements required.

Finally, the contractor must consider the desired mark-up in the context of the quality of tender information, the technical and contractual risks identified in the tendering process and the construction time and methods allowed. Profit percentages could be anything from 1% to 20%, depending on how keen the contractor is to win the contract.

In a very keen market, the contractor may look for ways of taking money out of the tender by anticipating discounts and buying savings from subcontractors and suppliers or by gambling on possible claims and other commercial opportunities that might arise during the contract.

A percentage addition is normally applied to cover head office overheads. This might be in the range of 5–10% depending on the size of the organisation.

7.9.4 Integrated Software Platforms

4PSConstruct, Eque2, Procore and Oracle Aconex are just some of the many integrated ERP and software solutions that help the contractor to keep estimating files and pricing notes all-in one place.

This facilitates collaboration and means that it is simple for estimators to share their data with the site management team regarding how certain work activities have been priced, what resources and temporary works have been allowed for and what the estimated waste allowances on materials are.

> Not only does this software allow current accessibility to files but also acts as a permanent archive so that files can be accessed for future similar projects and for cost planning purposes.

For contractors, the choice of estimation tool often precedes that of switching to an ERP system and many contractors have been using the same estimating software for years. This explains, to some extent, the reluctance of some contractors to adopt ERP technology, but some ERP providers can recode software so that it can be integrated into an ERP system. This means that estimators can continue to use a familiar and established software and database as normal.

Some ERPs allow the integration of bespoke third-party measurement applications too, which will help contractors to produce estimates based on standard methods of measurement used in the UK industry such as NRM2, CESMM4, the Highways Method of Measurement and, for projects overseas, POMI – the international method of measurement.

7.9.5 Bid Submission and Formal Tender

An invitation to tender from a client to a contractor – or a contractor to a subcontractor – is an offer to treat. The legal effect of this is that there is no obligation to accept the lowest, or any, tender bid or offer received, and the invitation documentation usually contains a provision to this effect.

Once a tender has been received by a client, the contractor may withdraw the offer at any time before the offer is accepted. This might happen, for instance, when the contractor has made a significant mistake and would be unable to stand by the offer. This is a rare event.

> As far as offers from subcontractors are concerned, there is usually a provision within the small print of the offer submitted that the offer is open for acceptance for a limited period – this might be 30 days from the date of the offer, for instance.

Such a limitation might well be problematic for a contractor because the subcontract offer may expire before a contract has been concluded with the client before which, of course, subcontracts cannot be placed. This is a risk issue which the contractor may need to account for when considering the commercial aspects of a bid.

An example of a contractor's final tender summary is shown in Figure 7.18 which also indicates the sort of time and money decisions that might be taken at this stage in the process.

7.9.6 Bid Documentation

Enquiries from clients to contractors, and from contractors to subcontractors, take many forms and vary depending upon the size and complexity of the project and the size and sophistication of the contractor or subcontractor involved. An enquiry may therefore consist of the following:

- An invitation to express an interest in submitting a bid.
- An unsolicited invitation to tender.
- An invitation to tender from a repeat client or main contractor.
- An invitation to tender as part of a framework or as a member of a standing approved list of contractors.

TENDER SUMMARY
T143: Aldergate redevelopment project

Ref		Item			£	£	£
A		Preliminaries				559000	
B	B1	Measured work	Own labour		180000		
	B2		Plant		108000		
	B3		Materials		432000		
	B4		Subcontract		3526000	4246000	
C	C1	Provisional sums				270000	
		Net total					5075000
D	D1	Overheads & profit		8 %			406000
		Total					5481000
E	E1	**Tender adjustment**					
	E2	Commercial opportunity	Subcontract	5 %	3526000	−176300	
			Measured items	3 %	4246000	−127380	−303680
							5177320
	E3	Risk	Price inflation	4 %	4246000	169840	
			Overheads & profit	8 %		13587	
						183427	
			Ground conditions		50000		
			Shuttering		3500		
			Scaffolding		12000	65500	248927
		TENDER TOTAL					**5426247**

NOTES:
1. The majority of work is subcontracted, with an element of self-delivery.
2. Commercial opportunity is a method of reducing the tender price based on perceived savings once the contract is awarded, time-cost savings and potential for contractual claims arising from the tender documentation.
3. Price risk is based on the measured work items (excluding preliminaries and provisional sums) with the addition of main contractor's overheads and profit.
4. Adjustments for other risk issues are added to the tender figure based on the directors' view of the estimator's pricing allowances.

Figure 7.18 Tender summary.

- An inquiry with full documentation including drawings and specifications.
- An inquiry requesting a proposal from a contractor, or subcontractor, in response to outline information regarding the intended project.

The documentation accompanying an invitation to tender can therefore be limited or fully developed, and this can be extremely influential in how the contractor or subcontractor goes about submitting its bid.

Despite the enduring, albeit diminished, popularity of competitive tendering, all is not as it would appear because a significant proportion of these contracts are tendered without traditional bills of quantities to accompany the tender drawings.

Common practice has, therefore, reverted to the late nineteenth century practice of tender enquiries based on drawings and specification with the measurement risk being passed down to contractors, and to subcontractors. Tenderers habitually prepare their own quantities, or engage a quantity surveyor to do it. The pricing documents produced, however, tend to differ from fully detailed SMM-based bills of quantities and are more likely to be:

- Builders' quantities containing composite items with far fewer minor items measured in detail.
- Activity schedules with measured items linked to the contractor's tender programme.

Activity schedules may be prepared by the client or, more usually, the contractor but in either event tenderers must produce their own quantities for each of the activities on the schedule because they are not sufficiently detailed for pricing purposes.

7.9.7 Other Procurement Methods

Increasingly popular are procurement methods that require a variety of contractor submissions other than a simple price or tender written onto a formal 'form of tender' offer. Tender submissions may include:

- A price and contractor's design proposal based on a statement of client's requirements.
- A price and developed design based on an outline design prepared on behalf of the client.
- A price for the contractor's management costs, overheads and profit with the eventual contract sum determined by:
  - Actual recorded costs.
  - Negotiation.
  - A jointly developed target price.

7.9.8 Bid Management

For individual trades such as ground workers, plasterers and plumbers, sole traders employing 1–5 people – or for small limited companies – management of the bid is fairly straightforward. It may be undertaken by the business owner, by an estimator or by a quantity surveyor for instance. Alternatively, there might be a manager or director of the business who occupies themself with marketing and finding work and they may take overall charge of an eventual bid.

In the case of larger contractors and subcontractors, as well as large or major projects, management of the bidding process can be extremely complex and may need to be undertaken by someone appointed to the specific role of bid manager. Depending upon the size of the organisation, the bid manager role may be:

- Combined with other roles such as estimating or marketing.
- A senior project manager.
- An estimating manager leading a team of estimators.
- An individual role as part of a submissions or pre-construction department.

The bid manager may report to a principal or business owner, a director or senior manager, a submissions manager or to the head of a pre-construction department.

Irrespective, the purpose of the role is to take a proactive involvement in the organisation, preparation and submission of the bid.

7.9.9 Bid Manager

Bid manager is a senior position irrespective of the size of company. It is a role that must be outward facing and client orientated whilst at the same time interfacing with the internal organisation and ensuring compliance with corporate governance requirements. This requires the ability to:

- Understand the needs of clients.
- Engage with designers and understand the design process.
- Be commercially aware.
- Understand the importance of the client's budget and programme.
- Understand the work of planners and estimators.
- Project manage the bid and the input of others in the team.

The bid manager is normally responsible for the overall management of the bid from initial enquiry to bid submission which means identifying the risks and opportunities presented by a project enquiry and delivering a comprehensive bid proposal to the client or main contractor that aligns as closely as possible to their requirements.

Depending upon the project and the method of procurement, the duties of the bid manager may include:

- Appraisal and response to pre-qualification documentation (PQQ).
- Response to expressions of interest (EOI) and invitation to tender (ITT) documentation.
- Assembling initial response to a request for proposal (RFP) documentation.
- Early client engagement and understanding of the needs of clients and stakeholders.
- Interface with design, estimating, supply chain and planning teams.
- Engaging external consultants where necessary, including for the preparation of quantities.
- Formulating and developing alternative design proposals.
- The development of alternative construction methodologies (optioneering).
- Value engineering.
- Integrating off-site manufacture and modern methods of construction into the bid proposal.
- Contributing input into the planning and programming of the project.
- Conducting negotiations with clients.
- Management of Stage 2 of a two-stage tender including engaging with the client team when developing a target price.
- Organising and presenting Red Reviews to ensure the completeness, compliance and quality of submissions.
- Writing and presentation of the bid document.
- Organising and leading the bid launch including client presentations.
- Briefing and handover to the project delivery team.

7.9.10 Bid Submission Documents

Bids for construction contracts can be based on a wide variety of documentation supplied either by the client, by the tendering contractors or both.

This documentation will invariably include some sort of pricing document – these are described in detail by Williams (2016). The documentation supplied and exchanged is summarised below:

- Traditional competitive tendering for new work:
  ○ Bill of quantities, drawings and specifications.

- Traditional competitive tendering for repair and maintenance work:
  - Schedule of rates (with no quantities).
  - Schedule of works (containing 'composite' items but no quantities).
- Non-traditional competitive tendering:
  - The contractor will undertake the measurement risk.
- Design and construct – client led:
  - Client's requirements, specification (prescriptive or performance) and a design developed to RIBA Stage 3.
  - Tenderers will submit contractor's proposals, price and programme.
- Design and construct – contractor led:
  - Client's requirements, specifications and outline drawings only.
  - Tenderers will respond with design proposals to RIBA Stage 3 together with a price and programme.
- Two-stage tenders:
  - Similar to contractor-led design and construction.
  - Once the preferred contractor is chosen, it will develop the design to the point where a target price can be established.
  - The target price, any value engineering provisions and a programme will form part of the contractor's bid.

References

BCIS (2011) *Standard Form of Civil Engineering Cost Analysis Consultation*. RICS.
Brook, M. (2017) *Estimating and Tendering for Construction Work*, 5th edn. Routledge.
CIOB (2018) *Code of Estimating Practice*, 7th edn. Wiley Blackwell.
Construction Industry Joint Council (CIJC) (2022) *Working Rule Agreement, July 2022*.
ICMS Coalition (2019) *International Construction Measurement Standards*, 2nd edn.
Williams P (2016) *Managing Measurement Risk in Building and Civil Engineering*, Wiley-Blackwell.

Notes

1 m^2 of GIFA.
2 Some contractors make these adjustments at tender review/adjudication stage.

	Chapter 8 Dashboard	
Key Message		The planning of construction projects can be extremely complex where a large project is concerned.
		The client must usually commit to a schedule for the intended project before contractors are appointed.
Definitions		**Planning**: One of Henri Fayol's six functions of management.
		Delay damages: Recompense for the client due to contractor delay.
		Loss and expense: Recompense for the contractor for no fault delay.
Headings		**Chapter Summary**
8.1	Introduction	Planning provides a means of taking control of events and the ability to react when things go wrong.
		Planning also helps to identify when funds need to be drawn down to pay for professional fees and work in progress, etc.
		Planning helps to determine responsibility for delay and to establish contractual entitlement for damages.
		Planning helps the client's project team to factor in time allowances for design, statutory or parliamentary approvals, financing and procurement as well as the construction and commissioning period.
8.2	Strategic Planning	Well before construction work can begin, a great deal of thinking is needed to plan the steps required to convert the aspirations of the client into the reality of a completed project.
		Planning helps to establish key appointments and key dates.
8.3	Design Planning	The design team's job is to produce a design in accordance with the client's brief and prepare the information necessary for manufacture and construction of the project.
		The Royal Institute of British Architects (RIBA) Plan of Work suggests that the design will develop through three stages – Concept design, Spatial coordination and Technical design.
		It is common practice to appoint a lead designer to manage the design process and to act as Principal Designer under Construction Design and Management Regulations (CDM) 2015.
8.4	Development Case Study	It is proposed to re-develop the site with a mixed commercial and residential development of luxury apartments, restaurants, bars and designer retail units.
		A project management team prepares development appraisals, including the outline planning application.
		A construction team takes the project through the scheme and detail design stages, including detailed planning.
		Development appraisal is based on concept designs and outline planning permission following which land acquisition is completed.
		A hand-down document is a formal handover of project details assembled by the project management team to the developer's construction team.
8.5	Construction Planning	Planning for the construction stage of projects commences during the bid process and develops in more and more detail pre-construction and during the construction stage itself.
		The contractor's estimating team is involved in the pre-tender planning and the site management team deals with the different levels of planning required prior to and during the production stages of the project.
Learning Outcomes		Understand the role of planning in construction projects.
		Appreciate the need for the strategic planning of projects.
		Understand the design planning process and its importance.
		Follow how a development project is planned from initiation to handover.
		Understand how projects are planned from bidding stage to short-term planning on site.
Learn More		Chapter 13 which explains 4D planning, work breakdown structures and how projects are planned in detail.
		Chapter 16 which explains planning in the context of project control, reporting, delay and disruption.
		Chapters 19 and 20 which provide practical project planning examples.

8

The Planning Process

8.1 Introduction

The planning of construction projects is not a straightforward process and can be extremely complex where a large project is concerned. Even for a small project, designs must be prepared, planning approval may be required, building control regulations must be satisfied, utility connections may be required, and the construction work has to be organised. All these activities must be carefully planned and coordinated to have a chance of a successful outcome.

A further complexity is that the client must usually commit to a schedule for the intended project before contractors are appointed or become involved and therefore the projected construction period is not based upon a contractor's view as to how long the work will take.

8.1.1 The Role of Planning

Planning is one of Henri Fayol's six functions of management which starts right at the outset of a project and continues to handover and defects correction.

There are many benefits to be gained from the thorough and realistic planning of a project, not least of which is to establish a realistic time within which the project may be completed. This includes the preparatory and design stages as well as bidding, pre-construction and the construction period itself.

Planning provides a means of taking control of events and having the ability to react when things go wrong, including:

- Accommodating changes in the scope of the contract with least impact on the completion date.
- Controlling progress – both before and during construction – so that action can be taken should the completion date be in danger.
- Monitoring the financial performance of the project for both client and contractor.
- Monitoring site productivity, the balanced use of resources, cash flow and taking corrective action where necessary.

Planning also helps to identify when funds need to be drawn down to pay for professional fees and work in progress, etc. Additionally, contractors need to determine whether their subcontractors' timescales are a good fit with the master programme and, especially, with the contract start and completion dates.

Should the project be delayed, planning helps to determine responsibility for the delay and to establish contractual entitlement for delay damages (client) or additional time and/or loss and expense (contractor) which might need to be robust enough to stand up in court.

8.1.2 Client View

The client's view of the planning process is illustrated in Figure 8.1 which shows that project planning begins with option appraisal and business case confirmation and ends with handover and conclusion of the final account.

During this process, the client's master schedule for the entire project is developed. Additional client-side schedules include design stage programmes, procurement and

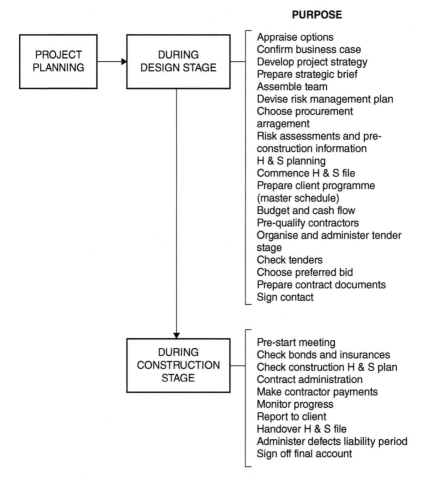

Figure 8.1 Overview of client planning and process.

pre-construction timescales and scrutiny of the contractor's pretender programme (if required) and master programme.

A client's reasons for engaging in a construction project may be based on:

- **Need**: need for more space or improved efficiency, repair or maintenance of buildings, rebuilding following a fire, structural repairs, change of use and so on.
- **Business case**: a speculative development, new hotel, commercial or retail premises, travel or public infrastructure, etc.

The client's view of project planning will invariably be through the lens of completion of the project to meet strategic objectives – occupation and use, income generation, sales or commercial revenues, public utility and so on. This is probably why so many clients push hard – often prematurely – for an early start one site which can lead to poor decision making, incomplete designs, inadequate investigation and information and pressure on the construction phase especially.

From the client's view of the project, the objective is to establish realistic dates for starting work on site and for completing the construction stage. This includes the identification of key dates or gateways at critical stages of the project. Completion might be geared to a specific date – such as a large sporting event or the Christmas shopping period, for instance.

8.1.3 Project Team View

Supporting the client there will normally be a professional team – at least an architect and structural engineer and very often other designers, cost consultants, legal advisers and project managers.

The project team has a narrower perspective than the client, being principally concerned with delivering the client's brief. The design must stand up to scrutiny and the cost plan must be robust because there is the question of liability if they do not. This is why professional advisers carry professional indemnity insurance – to provide for the eventuality of negligent design, poor advice, inaccurate cost advice and unrealistic planning and management of the project.

Planning helps the client's project team to factor in time allowances for design, statutory or parliamentary approvals and financing and procurement as well as the construction and commissioning period. Establishing reliable time estimates for these stages is not easy – especially for statutory consents which can cause considerable delays should there be public consultations or objections to the project.

The organisation, coordination and integration of architectural, engineering and technical endeavour is complex as it involves a variety of heterogeneous professional disciplines often geographically separated. This is the role of project managers.

8.1.4 Contractor View

Contractors engage in construction projects for commercial reasons and planning has the dual purpose of validating the contract period stipulated by the client and, secondly,

8 The Planning Process

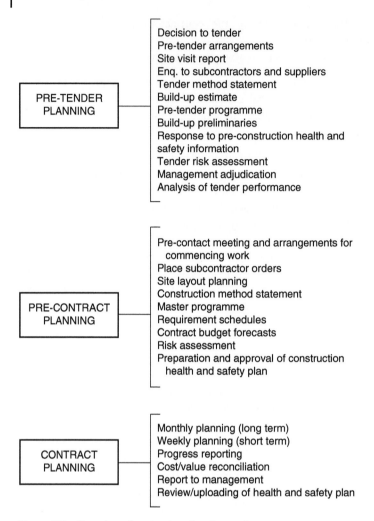

Figure 8.2 Overview of contractor planning and process.

facilitating organisation of the work on site to ensure a safe and profitable contract. This is illustrated in Figure 8.2, where the pretender, pre-construction and construction planning stages are explained.

For contractors, project planning centres around the contract with the client and the availability of resources. The contract states the start and completion dates – or the contract duration – and the contractor must be sure that this is achievable. Even if it is, the contractor will not complete on time without the necessary resources being available as and when needed.

When bidding for a contract, contractors must decide whether the client's timescale is realistic and, if not, overrun costs might have to be factored into the bid. Alternatively, a shorter timeframe might be appropriate enabling the contractor to submit a more competitive bid by reducing time-related costs.

Failure to complete the works on time – without legitimate excuse – might mean that delay damages may be incurred thereby affecting the profitability of the contract.

8.1.5 Types of Programme

The importance of the role of planning is recognised in the RIBA Plan of Work which highlights the need to prepare various programmes from preparation and briefing to manufacturing and construction (RIBA Stages 1–5).

This principle is illustrated in Figure 8.3 which indicates how these programmes relate to the RIBA work stages and when they are required:

- **RIBA Stage 1**: The project programme or project master schedule is prepared as early as realistically possible. This sets the tone for the entire project and provides a framework for professional appointments, for design development and for procurement of the construction works. The intended construction period is also determined at this point.
- **RIBA Stages 2–4**: Construction designs are conventionally developed in stages and may be required to pass through client approval gateways. A design programme for each design stage and a matrix of design responsibilities are needed to establish what needs to be done, by whom and when. This must be developed in line with the project master schedule to avoid slippage and delays to other project stages. The design programme is an essential control tool.
- **RIBA Stage 5**: The construction programme is produced by the contractor following a successful bid for the contract. It will usually be developed from the pretender programme and may form the basis of a target programme should the contractor envisage

RIBA Plan of Work 2020							
Options:							
0	1	2	3	4	5	6	7
Strategic Definition	Preparation and Brief	Concept Design	Spatial Coordination	Technical Design	Manufacturing and Construction	Handover	In Use
	Prepare project programme	Prepare stage design programme			Monitor progress against construction programme		

TYPES OF PROGRAMME/SCHEDULE

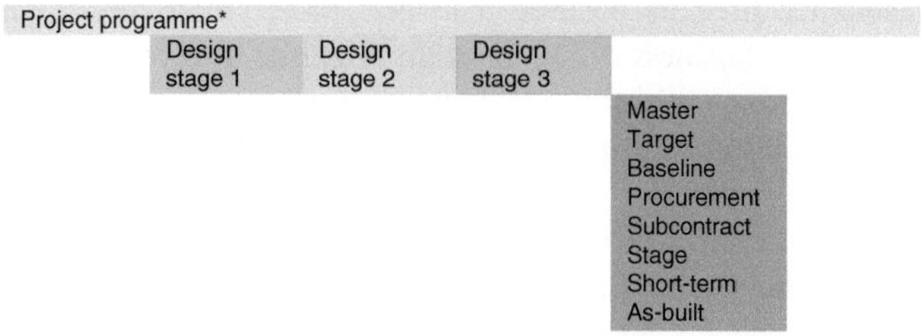

Project programme*
- Design stage 1
- Design stage 2
- Design stage 3
- Master / Target / Baseline / Procurement / Subcontract / Stage / Short-term / As-built

*Often referred to as the Project master schedule

Figure 8.3 Types of programmes.

a shorter/compressed time period for the works. The construction/master programme will become the project baseline which will be the reference point for both client and contractor to control the project.

8.2 Strategic Planning

The planning of construction projects requires consideration of much more than the work activities on site. Well before construction work can begin, a great deal of thinking is needed to plan the steps required to convert the aspirations of the client into the reality of a completed project. This thinking is normally expressed graphically as a programme or schedule.

The RIBA Plan of Work suggests that a **project programme** is prepared during RIBA Stage 1 – *Preparation and Briefing* – whereas the Chartered Institute of Building (CIOB) Code of Practice for Project Management (2002) refers to a **project master schedule**, presumably to avoid confusion with the contractor's master programme.

8.2.1 Key Appointments

The RIBA Plan of Work suggests that clients make three key appointments during a project:

- Appoint client team (Stage 0 – *Strategic Definition*).
- Appoint design team (Stage 1 – *Preparation and Briefing*).
- Appoint contractor (Stages 2, 3 or 4, depending on the procurement method).

The client team is distinguished from the design team as the person or group of professional advisers responsible for preparing the project brief but not for undertaking any design work. The client team's work involves:

- Preparing the client's requirements.
- Developing the business case.
- Preparing the project brief which informs the ensuing design process.

It is likely, therefore, that the client team will be responsible for preparing the project master schedule and, in doing so, will set the 'high-level' strategy for the project.

The make-up of the client team will vary according to the size and importance of the project. For smaller projects, the client team might comprise a project manager, cost consultant and principal designer, whereas for larger projects financial advisers, legal representatives, cost consultants, directors, business managers, project managers or senior personnel involved in the client's business or organisation might be involved. There might also be involvement from representatives of government sponsoring departments or key anchor tenants (where appropriate) and possibly lenders or investors such as banks, insurance companies or venture capitalists.

The client team is also involved with reviewing the design at the end of each RIBA work stage in the light of the client's brief. Such reviews will need to be programmed on the master schedule to enable the design team to proceed to the next design stage.

8.2.2 Key Dates

Once a project has been defined, and its purpose and scope established, the brief can be developed prior to appointing designers. The project master schedule is part of this high-level 'strategic' view of the project and is normally prepared by the client's agent, representative or project manager. It sets out the broad framework and key dates and timings for the project, including:

- Project commencement, phasing and completion.
- Financing, land acquisition and conveyancing.
- Planning and other approvals.
- Design procurement, briefing and concept.
- Design development.
- Site investigation and enabling works.
- Contractor involvement, procurement and bid process.
- Construction period and stage or phased handover dates.

Some projects are driven by an end date – a date when the project must be ready for use or occupation. For large housing projects, phasing is important as sales and market forces will dictate timings. Other projects – such as very big infrastructure projects or programmes – will have vague dates due to complex political, economic, legal and technical considerations.

For any project, it is essential to identify potential risks and to anticipate and avoid possible delays to completion and revenue generation, if any.

A realistic project master schedule can flag up the timing of key appointments – such as designers, project managers, consultants and contractors – and provide a target date for the commissioning and occupancy phase of the scheme.

8.2.3 Statutory Approvals

One of the most unpredictable factors in judging the timescale of projects is the approvals process.

Most projects require planning permission, and some may need parliamentary approval. There is also the possibility of public consultation, or even a public enquiry in some cases and, in others, various stakeholders may need to be consulted before the project details can be firmed up.

The approvals process can consume many months – even years – and seriously impact the viability of some projects. When project financing is reliant on government grants or other finance, delays in the approvals process can result in the finance being withdrawn, unless it has been specifically ring-fenced.

8.2.4 Financing

Often forgotten, but of particular importance to any project, is the time that it takes to negotiate and arrange suitable financing. This might take the form of loans, venture capital, the raising of equity or government grants.

Arranging suitable facilities can take a considerable time as there are complex legalities to be arranged such as the drafting of collateral warranties between funders and designers and contractors.

Most construction projects are based on credit where work and services are provided based on interim payment in arrears. Others, however, may require front-loaded finance where advance orders must be placed for long lead-time materials and components. Without the ability to draw down liquid funds, clients are unable to pay for design and other professional services or make payments to contractors.

8.2.5 Land Purchase

Land purchase is often contingent on whether planning permission has already been granted. This is not always helpful, however, as developers may have different ideas for developing the site to the existing owners.

Where there is no planning – or the plans are unsuitable – developers may wish to discuss the possibility of securing an option to purchase with the current owners, or possibly a conditional contract. This is a means of securing the land pending discussions with the planning authority to discuss possibilities for the development.

Sometimes, planning authorities may wish the developer to purchase derelict and unoccupied property adjacent to the site which it would like to see restored and integrated into the overall development. This is especially the case where the property in question has historical value and development potential.

8.2.6 Design Procurement

Design procurement is a key strategic decision which includes deciding on the terms and conditions of engagement for designers. Consultant design is the traditional choice, but clients may also consider early contractor involvement in the consultant design process to benefit from their expertise.

A further consideration is full or partial contractor design and clients may also wish to think about engaging consultants for the outline design with a novation to a design-build contractor for the detailed design stages. Specialist designs may be undertaken by package contractors, but this is likely to be for the detailed design stage.

Part of the design procurement thinking is that clients must appoint a principal designer under CDM 2015 to manage the health and safety aspects of design development.

8.2.7 Time and Cost

During the initial stages of a project, time and cost are difficult to predict with any certainty and this unpredictability increases significantly for large or complex projects and those with an unusual design.

Time and cost estimates are high-level predictions at the project definition stage, but it is nevertheless vitally important that they are realistic, especially in terms of the business case for a project. Inaccuracies can erode earnings and profit margins and cause delay.

When time – and especially cost – estimates are wide of the mark, the point can be reached where the cost of a project outweighs its benefit, and the business case may no longer be justified. For projects based on finite budgets, the chances are that they may be shelved or suffer scope or quality changes to save money, thereby reducing client satisfaction or utility.

8.2.8 The Construction Period

In view of the considerable lead time for construction and infrastructure projects, clients will be keen to mobilise construction work as soon as the 'green flag' is waved.

If similar projects have been undertaken in the past, the construction period can be predicted fairly accurately, even at the early stages. Repeat clients will have a database of completed projects as a guide, and so will their professional advisers.

If contractors are involved early in a project, their expertise will be invaluable in determining a realistic construction period, even where the design might be novel or complex. They also have a supply chain to rely on for specialist work, lead times and expert guidance.

The Royal Institution of Chartered Surveyors (RICS) Building Cost Information Service (BCIS) contains a wealth of published data about a wide variety of construction projects of different size, cost and location which can provide useful guidance about construction timescales.

8.3 Design Planning

The RIBA Plan of Work provides an invaluable generic matrix of the processes involved in delivering a construction project. It defines the eight stages from Stage 0 – *Strategic definition* through to Stage 7 – *Use* and identifies what is involved in delivering a construction project in terms of:

- **Core tasks**: such as preparing client's requirements, design studies, cost planning and design development.
- **Core statutory processes**: including planning and building regulations applications and making statutory appointments (such as CDM principal designer).
- **Information exchanges**: concerning business case and project brief, procurement strategy, design reviews and sign-off.

8.3.1 The Design Team

The design team's job is to produce a design in accordance with the client's brief and prepare the information necessary for manufacture and construction of the project. For some projects, designers may be involved with the client team to help develop the brief but will not normally be part of the client's decision-making group.

The exact composition of the design team can vary enormously from project to project. There might be a single designer or a multidisciplined team of architects and engineers or there might be several very large professional practices independently appointed by the client.

It is usual to engage members of the design team under a professional services contract (such as JCT or NEC), but the team could be supplemented by a contractor or specialist contractors engaged under a design and construct contract.

8.3.2 Design Development

There are several ways that the design might be developed but, in each case, the design programme should provide for design review periods, or gateways, and stage sign-offs to ensure compliance with the client's brief:

- An architect/engineer design with traditional procurement.
- An architect/engineer design supplemented by partial contractor, or subcontractor, design with traditional procurement.
- An architect/engineer novated design with partial contractor design and build.
- A contractor-led design with full design and build based on agreed employer's requirements and contractor's proposals.

The RIBA Plan of Work suggests that the design will develop through three stages:

- **Stage 2**: *Concept design*
- **Stage 3**: *Spatial coordination*
- **Stage 4**: *Technical design*.

In the case of design and construct procurement, the contractor-designer could be appointed at any one of these stages according to how much control the client wishes to exercise over the design development. The appointment could simply be made at Stage 2, based on a statement of employer's requirements or at Stage 4 or somewhere in between. Choices made should be reflected in the design programme.

8.3.3 Design Management

In common with any plan, it is important to manage the design programme to avoid slippage, duplication of effort, design clashes and, importantly, to avoid delaying the client's master schedule.

It is therefore common practice to appoint a lead designer to manage the design process and to act as Principal Designer under CDM 2015, whose statutory duties include:

- Planning, managing and monitoring the pre-construction phase and coordinating matters relating to health and safety and
- Estimating the period required to complete the design work stages.

The lead designer will often be an architect (building) or engineer (civil engineering) but might be a services engineer for a highly serviced building or part of a building (such as a hospital).

The job of the lead designer is to direct and coordinate the work of other contributors to the design including any specialist designers who may be appointed. This role might include managing the Building Information Modelling (BIM) model (if any) should a BIM manager or coordinator not be separately appointed.

Where contractor design is involved – for either partial or full design – it is usual for the contractor to appoint a design manager or coordinator. This role is similar to that of lead designer, but the composition of the design team might comprise both in-house designers, external consultants and specialist subcontractor-designers.

8.4 Development Case Study

A developer is interested in a brownfield site at Fletchergate in the historical commercial district of a busy East Midlands university city. The site has several existing properties including a Grade II listed building.

It is proposed to redevelop the site with a mixed commercial and residential development of luxury apartments, restaurants, bars and designer retail units. This will consist of new build and refurbishment aimed at preserving the heritage of the site and integrating new buildings with the existing built environment both on and around the site.

8.4.1 Project Master Schedule

The developer/client has a project management team which prepares development appraisals, including the outline planning application, and a construction team which takes projects through the scheme and detail design stages, including detailed planning. All design and construction work is outsourced.

A simplified example of a project master schedule is illustrated in Figure 8.4 where it can be seen that:

- The calendar for the project master schedule is shown in years and months which indicates the degree of approximation in predicting time scales at the early stages of a project.
- Time zero is at the start of the construction phase. Prior to commencement of construction time is negative, i.e. preparation of the project brief commences in month minus (−) 22.

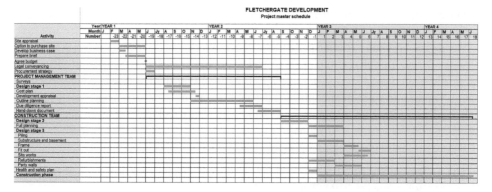

Figure 8.4 Project master schedule.

- The work undertaken by the project management team and by the construction team is shown in detail and within a 'summary' bar. This can be rolled up to make the detail disappear if required.
- The early stages of the project are concerned with acquiring an option to buy the land, arranging finance and ensuring the project is viable before committing to purchasing the land.
- The client's brief is established before committing to the appointment of the design team – this is not always the case for every project.
- The design is undertaken in three stages. Stage one is where the design is developed in concept, sufficient to make an outline planning application. Stage two develops the concept design to facilitate a detailed planning application. Design stage three is where the technical design is completed by specialist work package contractors.
- Activities concerned with the preparatory stages of the project are shown in detail, whereas later stages in the project such as design development, pre-construction and the construction stage are shown as a single bar.

8.4.2 Development Appraisal

Development appraisal is based on concept designs and outline planning permission following which land acquisition is completed.

The client project management team manages the development appraisal process up to the point where a hand-down document can be prepared. The construction team then takes over and manages detailed design development and construction of the works. The construction team acts as principal contractor and design coordinator for stage two of the design, with all site works undertaken by works package contractors.

8.4.3 Design Programme

Preliminary ideas for the Fletchergate scheme indicate that the development will consist of 8500 m^2 of two- and three-bedroom luxury apartments on the 5 upper floors and 2500 m^2 of commercial letting space below. Some of the apartments will be duplex, and 1500 m^2 of the residential development will be within refurbished premises.

The entire scheme will be set around a secure landscaped courtyard with multi-level basement parking for 121 cars.

The design programme follows the RIBA Plan of Work process, commencing at Stage 2 – *Concept Design*. Design consultants may well be appointed at Stages 0 and 1 but this would be to advise on project strategy and design thinking rather than as a design team appointment. There is no point in appointing a design team at the very early stages as this is very much a time for exploring options for development. The developer may even decide not to proceed with the project at this juncture.

Figure 8.5 shows the design programme from which it can be seen that:

- The design process has three main stages – concept, spatial coordination and technical design – following the RIBA Plan of Work model.
- The early concept design work is undertaken by the developer's development team up to the point when an outline planning application can be made.

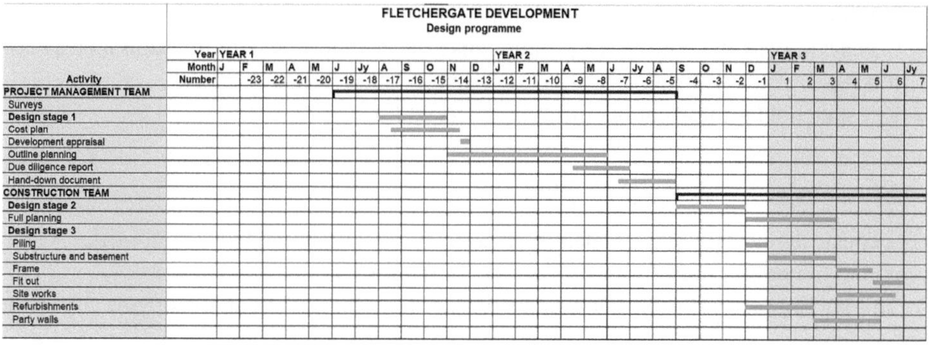

Figure 8.5 Design programme.

- Surveys are undertaken including condition surveys of existing buildings, site contamination, the presence of hazardous substances and environmental and ecological impact investigations.
- The spatial coordination and technical design stages are undertaken by the developer's construction team.
- Consultant architects and engineers provide the *design intent* – including design of the frame and load-bearing brickwork whilst technical design is carried out by specialist work package contractors for elements such as contiguous piling to basements, timber frame for upper floor apartments, bathroom pods, fit out and so on.
- Technical design work overlaps the beginning of the construction phase.

8.4.4 Due Diligence

Development due diligence is concerned with weighing up the risks involved and making sure that all legal requirements for building on a specific parcel of land are satisfied. This is not simply a question of title (legal ownership) but also whether there is any statutory or government interest in the land, as plans for a new motorway or high-speed rail line may affect the viability of the project.

Further considerations include proximity to transport links, site topography, drainage, soil and geology reports, subsidence, flooding and environmental issues. Due diligence also involves discovering details about the property vendor, including previous owners and uses, checks with environment agencies, legal searches and so on.

8.4.5 Hand-Down Document

A hand-down document is a formal handover of project details assembled by the project management team to the developer's construction team which acts as a management contractor, engaging work package contractors to undertake work on site.

The hand-down document contains the following:

- Project details, planning approval and any reserved matters.
- Site plan and geotechnical information.

- Outline design, key dates and deadlines.
- Contact information and access permissions to any Cloud-based information such as BIM models.
- Key stakeholder information, utility contacts and so on.

8.4.6 Pre-construction Programme

The construction stage of the Fletchergate development is planned to take 75 weeks as illustrated in the project master schedule shown in Figure 8.4.

This is a complex project located on the corner of two busy roads on a restricted city centre site involving:

- Demolition of disused buildings.
- Ground remediation due to asbestos contamination.
- Deep basement excavation to accommodate a 4-level car park.
- Refurbishment of existing multi-storey buildings.
- Party wall and right of support issues with adjoining properties.
- Restrictions for the oversailing of tower crane jibs over adjoining property.
- A railway tunnel running under part of the site.
- A tramway passing the main frontage of the site.

The developer/contractor is keen to seek ways of improving operational efficiency on site and has an established research and innovations committee geared to finding new products and construction methods to speed up the construction process and cope with regional labour market shortages.

To this end, the Fletchergate project is designed with a reinforced concrete frame, but the upper-floor apartments are to be constructed using timber frame technology.

The developer's approach to the project involves working closely with subcontractors to reduce waste and prevent pollution from site operations, to minimise dust and noise pollution and to increase waste recovery by recycling materials for reuse or responsible disposal.

8.5 Construction Planning

Planning for the construction stage of projects commences during the bid process and develops in more and more detail pre-construction and during the construction stage itself.

The contractor's estimating team is involved in the pretender planning and the site management team deals with the different levels of planning required prior to and during the production stages of the project. To assist this process, most reasonably sized contractors will employ planning engineers who are skilled in using project management software. They will develop the various programmes required – perhaps using BIM models – and discuss and refine their ideas with those of the estimators and construction managers.

8.5.1 Tender Stage

During the bid preparation stage of a project, it is considered good practice to prepare an outline programme – the pretender programme.

The pretender programme may be developed as an aspect of bid preparation for a traditional tender or could equally be produced during the second stage of a two-stage tender. Either way, this can provide a focus for the estimating team regarding method-related elements of the bid and help with the pricing of time-related aspects of the preliminaries. For early contractor involvement scenarios, or where a guaranteed maximum price bid is required, the pretender programme will provide an invaluable indication as to how realistic the client's expected completion date is and help bring realism to the client's project strategy and planning for occupation and use.

For a traditional tender, an unrealistic contract completion period might require a risk allowance adjustment to the tender figure submitted. In the case of a two-stage tender, the pretender programme can help to produce a more achievable target price as a basis for a contract.

In two-stage tendering especially, the pretender programme can be useful for presenting the contractor's ideas for completing the works and could be used in presentations, pre-contract negotiations or as part of the formal bid submission.

8.5.2 Pre-construction

The contractor's master programme might be developed from the pretender programme but it is quite usual for the construction team to view the project differently to the estimating team. The tender only establishes the contractor's budget for the contract and alternative construction methodologies might well be adopted at the pre-contract stage. This is a perfectly acceptable practice as there can sometimes be better ways to go about the work than that envisaged at tender stage.

The master programme is usually far more detailed than the pretender programme which will normally only indicate the main construction operations to be carried out and, possibly, some basic indications of plant and temporary works to be used.

Before work can start on site, a great deal more detail will be required and should the NEC Engineering and Construction Contract (ECC) be used, this will be a contractual requirement.

Most standard forms of contract require the contractor to submit the master programme to the client/employer or to the project manager and, in some cases, the programme must be approved before work can start on site.

The master programme – the 'accepted' programme under NEC contracts – is an important management tool for the client's contract administrator or project manager. It enables the contractor's physical and financial progress to be monitored, can be used to flag up early warning of potential causes of delay and is helpful in deciding on disruption and prolongation issues.

The master programme will often indicate when information is required by the contractor from the client's design team and can act both as a prompt for the architect/engineer and as justification for prolongation claims/compensation events.

The contractor's master programme is the one that the client team normally sees because it displays the contract start and completion dates and any sectional completion dates stipulated in the contract. This is logical because most construction conditions of contract require the contractor to complete the works on or before the completion date stated in the contract – or words to that effect.

However, contractors quite frequently will take a commercial decision to complete the works earlier than stated in the contract. In other cases, contractors will decide that the contract completion date is unrealistic, and a later completion date would be more achievable.

Consequently, there could effectively be two completion dates – the contract date and the contractor's earlier/later planned completion date. This raises the thorny question as to which completion date should be shown on the contractor's master programme:

- Some clients do not accept programmes that display earlier or later completion than the date stated in the contract.
- The NEC ECC form of contract anticipates that there may be a 'planned completion date' that is different to the contract completion date. This resonates with standard contracts in the United States.
- A master programme may show both a planned completion date and the contract completion date.

The idea of a programme with two completion dates has been tested in the courts, but different legal jurisdictions have different views. In the United Kingdom, the case of *Glenlion Construction Ltd v The Guinness Trust* [1987] 39 BLR 89 clarified that there is nothing to prevent the contractor from finishing early.

8.5.3 Construction Stage

For small projects, it is likely that a master programme will be sufficient to plan, monitor and control site operations.

Very often, small contractors will produce a work schedule using a spreadsheet package such as Microsoft Excel which can be updated with progress quite simply and this will be the only form of formal control necessary to manage the project.

For larger, more complex projects, however, more sophisticated methods will be needed, and project management software will invariably be used to produce the master programme and a variety of other programmes too, including:

- Target programme and baseline for control.
- Procurement and subcontract programmes.
- Short-term stage and look-ahead programmes.
- As-built programmes to reflect what was actually built and when.

The purpose and uses of these programmes are explored in detail in Chapter 13.

Reference

CIOB (2002) *Chartered Institute of Building Code of Practice for Project Management*, 3rd edn. Blackwell Publishing.

	Chapter 9 Dashboard	
Key Message		o Timelines provide managers with an essential control tool. o They facilitate the monitoring of work activities and the management of time slippage.
Definitions		o **Programme**: Refers to several interrelated projects. Also means a representation of the order and sequence of construction work. o **Schedule**: Commonly understood to be a list containing information. Also a synonym for 'programme' meaning a representation of the order and sequence of construction work. o **Critical path**: The longest sequence of dependent activities in a project. o **Float**: Activities in a schedule/programme with free time.
Headings		**Chapter Summary**
9.1	Introduction	o The graphical representation of work versus time is widely acknowledged to have originated in the early 1900s. o The linked bar chart is the most popular display. o Other displays include arrow and precedence diagrams. o They all use the critical path method to determine project duration. o Some project management software incorporates Building Information Modelling (BIM) features that enable the 4D planning of projects.
9.2	Linked Bar Charts	o Linked bar charts are laid out with the top axis as the timescale and a list of tasks or activities down the left-hand axis. o The time required for each activity is shown by a horizontal line (or bar). o Dependency is created in linked bar charts by various relationships.
9.3	Arrow Diagrams	o Distinguished by an arrow representing the activity and circles or nodes between the arrows representing events. o The earliest and latest event times of an activity can be calculated by making forward and backward arithmetic passes through the network. o Dummy activities, which usually have no duration or value, can be introduced to indicate dependencies not shown by the arrow activities.
9.4	Precedence Diagrams	o Precedence diagrams consist of a series of boxes interlinked with lines. o The box or node represents the activity, and the linking arrow indicates dependency.
9.5	Arrow, Precedence and Linked Bar Chart Relationships	o The common feature of all three methods is the logic which links the activities. This logic creates dependencies in the schedule. o Dependencies determine the order and sequence of the work. o Together with the activity durations, dependencies are used to calculate the overall duration of the project. o These calculations are called 'critical path analysis' because they reveal those activities with no free time (i.e. zero float). o The longest route through the schedule is called the 'critical path'.
9.6	Line of Balance	o A line-of-balance diagram is a series of inclined lines representing the rate of working between repetitive operations in a construction sequence. o Also called elemental trend analysis. o Provides a visual display of the rate of working of different activities in a schedule.
9.7	Time-Chainage Diagrams	o Used for linear projects such as pipelines and highways. o This format shows time dependencies between activities together with their order and direction of progress.
Learning Outcomes		o Understand the basics of various scheduling techniques. o Be able to choose the most appropriate technique for a project. o Understand the critical path method. o Be able to make a forward and backward pass through an arrow or precedence diagram and calculate the critical path. o Be able to draw a simple line of balance or time-chainage diagram.
Learn More		o Chapter 16 explains the legal and practical aspects of the contractor's programme including establishing progress, delay and culpability. o It also explores the use of Cloud-based reporting. o Chapter 16 explains the types of delay, extensions of time and delay analysis methods in detail.

9

Scheduling Techniques

9.1 Introduction

It is conventional to represent a construction project – or series of projects – as a timeline depicting the order, sequence and duration of the activities required to complete the work. This could be for an entire project from inception to completion, or a phase of a project – such as the design phase or the testing and handover phase – or it could be a detailed indication of how the contractor intends to carry out the work on site.

Timelines provide managers with an essential control tool. They facilitate the monitoring of work activities and the management of time slippage, should this occur. Timelines also help to establish relationships – dependencies – between different work activities so that the impact of change to related activities and important milestones can be managed to ensure that the desired outcomes are achieved.

9.1.1 Definitions

In the construction industry, a project timeline may be referred to as a 'programme' or a 'schedule' but, whilst the two terms are often used synonymously, they have different meanings. In their broadest sense:

- A **programme** refers to several interrelated projects, such as the various buildings that make up a large hospital development or the many individual projects that make up a large infrastructure scheme.
- A **schedule** is commonly understood to be a list containing information:
  - Drawing schedules list the title, initiator and latest version of the drawings to be used for construction of the works.
  - Manhole and door and window schedules list the reference number, type and size of building components.
  - Rebar schedules list the diameter, shape and type of steel (mild, high-tensile) to be delivered and used for reinforced concrete work.

Schedules are also used in formal contracts to add detail to contractual provisions, and they are to be found in regulations to amplify statutory requirements – such as the welfare requirements in the Construction Design and Management (CDM) Regulations.

In a narrower sense, the distinction between 'programme' and 'schedule' is less evident. In construction, a list of activities or events, including dates when the activities will be carried out, is often referred to as a 'schedule', whereas the term 'programme' might be used to refer to the sequence in which a series of activities or tasks must be carried out to complete a project.

However, the term 'schedule' is not used in standard forms of contract – NEC, JCT, FIDIC and Infrastructure Conditions of Contract (ICC) contracts all refer to the contractor's 'programme' as regards the order, sequence and timing of the construction works. The Royal Institute of British Architects (RIBA) Plan of Work also refers to the project programme, design programme and construction programme.

Conversely, leading software houses tend to favour the term 'schedule' when referring to their products. Oracle Primavera, for instance, states that Primavera P6 can be used to plan, schedule and control projects, whereas Elecosoft markets its construction-specific software – Asta Powerproject – as 'scheduling' software.

This might seem confusing because project management software packages usually comprise both a schedule and a programme. The 'schedule' element is a list of activities with durations and start and finish dates, whereas the 'programme' is a series of interrelated bars depicting the sequence of the activities and their dependencies.

Baldwin and Bardoli (2014) express the view that the term 'scheduling' is more commonly used in the industry and that 'programming' tends to imply broader aspects of construction projects including resource allocation.

In the absence of any definitive definition, the terms 'programme' and 'schedule' are used interchangeably in this book!

9.1.2 Work Versus Time

The graphical representation of work versus time is widely acknowledged to have originated in the early 1900s thanks to Henry Gantt who belonged to the Scientific Management school of thinking of the late nineteenth and early twentieth centuries. It was Gantt who introduced bar charts for ship-building projects which was the first scientific attempt to consider work scheduling against time.

Gantt charts simply consist of a horizontal x-axis representing units of time – hours, days, weeks and so on and a y-axis which is a list of activities to be carried out. They have now become the basis of the modern bar chart, which has several variants.

9.1.3 Gantt Charts

Before the advent of powerful modern computers, Gantt or bar charts were prepared by hand, often using pre-printed blank sheets. The sheets comprised a column on the left-hand side for listing the project activities in approximate order and squares on the right-hand side for drawing the bar lines against each activity.

It was quite common to see the bar chart on the site agent's wall in the site cabin, with progress coloured in with crayon/coloured pencil and a vertical string pinned in place to denote the current date!

A serious limitation with Gantt charts is that they do not show dependency – that is, how preceding, concurrent and succeeding activities relate to each other. Consequently, they do not clearly indicate which operations are directly related to the successful completion of the project – there is no critical path. This makes life difficult for managers to monitor progress, to determine which activities are causing problems and to take action to keep the project on track.

9.1.4 Scheduling Techniques

Since the mid-1950s, a variety of scheduling techniques have been developed to overcome the limitations of Gantt charts, each of which produces a distinctively different type of display. Each relies on a particular form of logic and is chosen according to the type and complexity of the project concerned. The techniques considered in this chapter are:

- Bar charts and linked bar charts.
- Arrow diagrams.
- Precedence diagrams.
- Line-of-balance diagrams.
- Time-chainage diagrams.

Line-of-balance diagrams are used for repetitive work – such as housing – and are based on the rate of progress of activities versus time, whereas time-chainage diagrams enable activities to be related to both time and physical distance.

The relationships between the activities shown on arrow diagrams, precedence diagrams and linked bar charts are based on (1) dependency – preceding, concurrent and succeeding activities or events – and (2) the critical path method (CPM).

9.1.5 The Critical Path Method

The CPM is represented in three main ways using either arrow diagrams, precedence diagrams or linked bar charts. Each type of display is quite distinctive, but they all use the same logic of arithmetic forward and backward pass to determine the critical path.

The calculations may be carried out by hand or algorithmically within the software. This process establishes:

- The longest sequence of dependent activities necessary to complete the project (forward pass).
- The amount of free time available in the sequence of dependent activities (backward pass).
- The critical activities – those with no free time – which must finish on time to avoid delayed project completion.

282 | 9 Scheduling Techniques

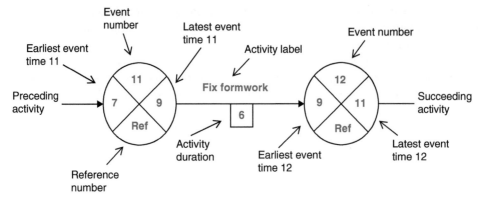

ARROW DIAGRAM NOTATION

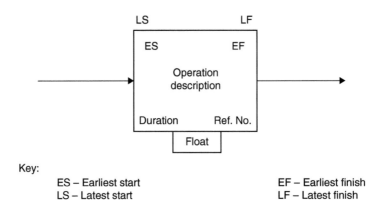

Key:
ES – Earliest start
LS – Latest start
EF – Earliest finish
LF – Latest finish

PRECEDENCE NOTATION

Figure 9.1 Arrow and precedence notations.

For the CPM to work, activities must be linked by logic which is the reason for the notations used for arrow and precedence diagrams, as illustrated in Figure 9.1. The key differences between the arrow and precedence diagrams are that:

- Arrow diagrams represent activities with arrows which also create the logic or dependencies in the diagram. The arrows are joined together by circles or nodes called 'events'.
- Precedence diagrams use boxes to represent activities with arrows solely used to create the logical links between them.

There is no such logic in a traditional Gantt chart, and therefore no dependencies, meaning that a critical path cannot be calculated. This deficiency was resolved in the early 1990s, however, when advances in personal computing facilitated the development of user-friendly project planning software in the form of the linked bar chart.

The linked bar chart is basically a Gantt chart but with logical links added that impose the CPM logic and facilitate calculation of the critical path. The critical path is determined by an algorithm embedded in the software.

The Critical Path Method (CPM) originated in the USA in the mid-1950s when a method of planning and scheduling was developed consisting of a series of arrows linked together with circles or nodes. This form of presentation is often called a Network Analysis or, more graphically, an Arrow Diagram.

In the late 1950s, a method of statistical analysis of the tasks on a network diagram was developed as a result of the Polaris missile programme. The Program Evaluation and Review Technique (PERT) employs three-time estimates of activity durations (optimistic, expected and pessimistic) rather than a single estimate of time.

In the United Kingdom, development work on CPM was undertaken by the Building Research Establishment and in the early 1970s by the Cementation Company which developed the Precedence Diagram to overcome some of the limitations of arrow diagrams. Precedence diagrams consist of a series of boxes linked with lines or arrows where the boxes represent the activities, and the lines represent the logic.

The precedence format can more readily be applied to construction work, and is far more widely used, than arrow networks. They are more flexible, they more easily reflect the way things happen in practice, and they are more software friendly.

Linked bar charts are by far the most popular of all CPM formats used in construction.

9.1.6 CPM Software

Whilst CPM calculations and iterations can be carried out by hand, or by using spreadsheets, it is unthinkable in the twenty-first century not to benefit from the fantastic, cutting-edge software that is now available to the project manager or planner.

Most software is based on the linked Gantt/bar chart, although Micro Planner X-Pert uses arrow and precedence networks. For large, high-end projects and programmes, Primavera P6 is the go-to solution for many clients and contractors. However, there is no integrated BIM interface in P6. Two-way BIM integration requires third-party software such as Bentley Synchro or Zutec to provide the missing digital modelling capability. This can be done by manually interfacing the two applications or automatically using an Application Programming Interface (API).

Asta Powerproject, on the other hand, incorporates BIM features that enable 4D planning using Industry Foundation Classes (IFC) model files in one software solution. This means that the BIM model can be animated to simulate the construction process which populates the Gantt chart directly on the same screen.

High-end packages such as P6 and Asta Powerproject may be cost prohibitive to smaller contractors and subcontractors, but reasonably priced software is nonetheless available:

- Project Commander is a cost-effective and fully functional project management software solution, specifically written for the construction industry.
- Microsoft Project is generic software that can be paid for by annual subscription if desired:
  - For simple projects, *Project for the Web* offers cloud-based Gantt charts showing dependencies, but not resourcing, at reasonable cost.
  - For large or complex projects, *Project Online/Desktop* is more expensive but is fully functional.

Popular project management software solutions are summarised in Table 9.1, some of which are either directly or indirectly BIM enabled.

Table 9.1 Project management software.

Name	Developer	Format	Brief summary
CS Project	Crest Software	Gantt chart	Fully functional and powerful software used extensively in construction allowing task and time management, resourcing, costing, progress tracking and reporting and high-quality printouts.
Easyplan	Elecosoft	Gantt chart	A cut-down version of Asta Powerproject which looks the same and allows resourcing and progress recording but not full critical path analysis, float calculations, baselines or costing.
Micro Planner X-Pert	Micro Planning International	Arrow and precedence	Technically powerful, yet easy to use, CPM software available as a Cloud service. Supports both CPM and PERT (three-time estimates). Incorporates ladder diagrams for repetitive work, earned value analysis and advanced, industry-leading calendars.
Microsoft Project	Microsoft Ltd.	Gantt chart Toggles for precedence	Not designed for construction, but nevertheless easy to learn and with good functionality for all main baseline, progress tracking, resourcing and costing requirements as well as a professional-looking printout.
Asta Powerproject	Elecosoft	Gantt chart Toggles for precedence and time-chainage diagrams	Powerful market leading package with construction specific templates and full functionality for critical path analysis, detailed float calculations, resourcing and costing, baselines, progress recording and high-quality printouts. Asta Powerproject 4D enables models to be imported into the display which interacts with the Gantt chart.
Primavera P6	Oracle	Gantt chart Precedence	High-end software suitable for large-scale programmes and projects. Multiple projects can be opened and scheduled simultaneously. Can be imported into Bentley Synchro to provide BIM functionality.
Project Commander	Project Management Software Centre	Gantt chart	A user-friendly and cost-effective construction industry-specific package ideal for construction managers and students alike which is easy to learn. Produces a colourful and professional linked bar chart with facilities for baselines, progress recording, rudimentary resourcing and costing and good quality printout.

9.2 Linked Bar Charts

Bar charts are similar in layout to spreadsheets. They are laid out with the top axis as the timescale in days/weeks/months/years and a list of tasks or activities down the left-hand axis. The time required for each activity is represented by a horizontal line (or bar), with the length of the line indicating the duration of the activity according to the scale on the top axis.

One of the problems with the traditional Gantt chart is that they are not adapted to the use of the CPM. In other words, they do not show dependency and, consequently, it is not easy to see the interrelationship between activities and how dependent they might be on one another. This is not so bad for a simple project but where a large number of activities are scheduled, real problems can arise for the project manager.

A further issue is that a critical path cannot be calculated using a Gantt chart, and the amount of free time in non-critical activities (float) cannot be determined. This makes life difficult, especially for a site manager, who needs to know what latitude there is in the schedule when things go wrong and what the impact of delay to critical activities will be.

Such problems can be overcome, however, by using the **linked bar chart** which requires the user to enter a logic by creating links between the activities on the schedule, as illustrated in Figure 9.2.

9.2.1 Useful Features

Among the many attractions of linked bar charts are:

- They are readily understood at all levels of management.
- Key milestones, resources and resource histograms can be visually displayed.
- They can be used for 'value–time' financial forecasting by both client and contractor.
- Progress can be easily tracked against the original baseline.
- They can provide an accurate record of progress and delay.
- They can be used as a basis for extensions of time or other claims.
- The effect of late information on programmed operations can be monitored and reported.
- The preparation of management reports is easier.

A useful feature of linked bar charts is that milestone symbols can be introduced to highlight critical dates such as key project stages, information requirements and subcontractor and materials procurement. Milestone are points in time and have no time value. They are often symbolised as a diamond shape (♦).

The ability to show resources on the linked bar chart – such as labour, plant and subcontractors – helps the site manager to relate the rate of working to resource availability. This is useful should activities need to be accelerated and also reveals potential problems with continuity of work and waste.

> Once resources and costs have been added to the project details, labour histograms, value–time forecasts, cumulative labour and plant forecasts and other project budgets can be produced facilitating the monitoring of actual progress against planned progress using several metrics.

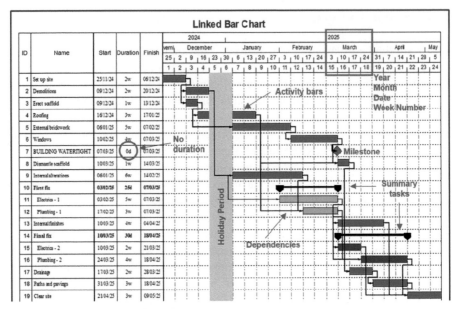

Key components of a linked bar chart

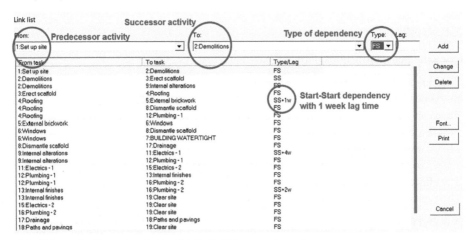

List of dependencies generated by the Project Commander software

Figure 9.2 Linked bar chart.

Linked bar charts can also be readily updated with information such as key site deliveries, progress to date and delays, making it simple to produce an 'as built' programme (or 'programme of the day'). This may prove an asset to the contractor in forming entitlement claims for delay, disruption and loss and expense as can the ability to record the effect of the receipt of late information on scheduled activities.

9.2.2 Basic Principles

Figure 9.3 illustrates the relationships used for linked bar charts. These include:

- **Finish-to-start relationships**: where one activity must be completed before another commences.

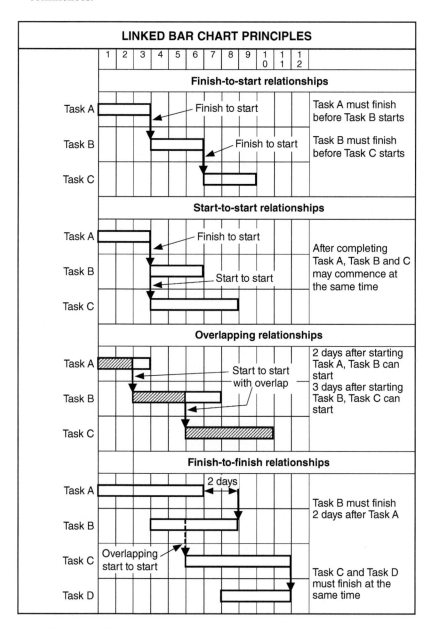

Figure 9.3 Linked bar chart relationships.

- **Start-to-start relationships (overlaps)**: where two or more activities may start at the same time.
- **Overlapping relationships**: where two or more activities may start later than a predecessor.
- **Finish-to-finish relationships**: where a succeeding activity must finish at the same time or after its predecessor.

As with any method of scheduling, it is important to think the process through carefully and to follow logical steps. Adding logic links to a project before the activities have been sorted into approximate order can lead to a tangle and unnecessary complication so patience is a virtue!

There is no single correct way to create a linked bar chart, but following the basic steps below may help the beginner to avoid some of the many pitfalls:

Step 1	a.	It is important to set the project parameters first. All software packages need to know what the proposed start date and project calendar is. Most projects work a 5-day week but in parts of the world Monday is a non-working day.
		Some projects – such as fit out – will work a 7-day week, maybe at night. Others – such as slip-forming – may work a 24-hour shift. Some projects may necessitate the use of several calendars for different parts of the work.
	b.	Decide on the most appropriate timescale for the schedule. Months on the major scale and weeks on the minor scale are usually best for most schedules. For short-term planning and short-duration projects, where a finer level of detail is necessary, weeks and days would be more appropriate.
	c.	Next, list the various tasks/activities that make up the project and sort these into order using the cut and paste facility. Choose the activities carefully. Too few activities will make the schedule of little use, whereas too many will be cumbersome and difficult to print out and read. A schedule for a large project should not exceed 2 000 activities.
		NB: Nearly all project management software initially defaults to a duration of 1 day for each activity.
Step 2	a.	Think about the possibility of using rolled-up tasks (also called summary tasks) which can be expanded into subtasks at a later stage. They can be particularly useful where the project is split into phases.
	b.	Next, add durations to the activities by calculation or from experience. Days or weeks are generally best. Some sources recommend that activities should be no more than $1.5 \times$ the project reporting period (usually monthly/4 weekly). This suggests a maximum duration, therefore, of $4 \times 1.5 = 6$ weeks. There is no strict rule on this, however.
	c.	Make any changes to the number or order of the activities before going on to Step 3.
Step 3	a.	Add logic to the schedule by linking relevant activities to one another.
	b.	Think about which activities must come first, which must follow, and which may happen at the same time.
	c.	Overlaps or delayed starts can be introduced by choosing the appropriate relationship from the menu (e.g. start-to-start) and by adding the necessary lag time.

Further features are easily added such as holiday periods, key milestones (events), resources and cost information.

9.2.3 Worked Example

Creating a useful and professional-looking bar chart using modern software packages is a relatively straightforward task requiring a couple of hours to learn the basics of the package and then it is just a question of adopting a logical and systematic approach. Of course, it does help to know how buildings are put together as well!

Adding complexity to a schedule – such as resources and costs – takes more time and experience to master, as does the use of BIM models and 4D simulation.

One of the common packages which is simple to use, relatively inexpensive and gives good results is Project Commander. Figure 9.4 illustrates how a linked bar chart might be put together using this software.

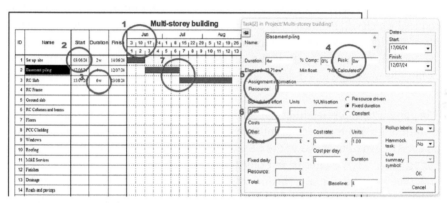

1. Establish project calendar.
2. Decide on start date.
3. Enter activity durations.
4. Make time risk allowances as appropriate.
5. Allocate resources.
6. Allocate costs.
7. Dependencies not yet established.

1. Establish dependencies (logic links).
2. Decide on type of dependency (e.g. FS, SS, SF, FF).

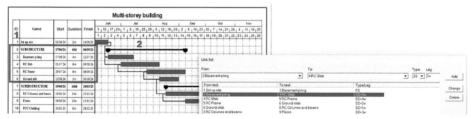

1. Create summary activities and sub-tasks.
2. Summary (roll-up) bar.

Figure 9.4 Developing a linked bar chart.

9.3 Arrow Diagrams

The distinguishing feature of these diagrams is that the arrow represents the activity and the circle or node between the arrows is the event. This is shown in Figure 9.1.

By numbering the events, the arrow activities can be identified – for instance, activity 5–8 Concrete blinding or activity 8–11 Fix formwork. This is usually done in numerical order starting at the beginning of the network and progressing to the end, ensuring that the number at the tail of the arrow is smaller than that at the head. Using this methodology each activity will have a unique reference.

Each activity is given a duration, and the earliest and latest event times of the activity can be calculated by making forward and backward arithmetic passes through the network. These times are recorded in the node or event circles. From this information, a schedule can be produced which will facilitate calculation of the float time (spare time) for each activity.

Critical activities are defined as having no float and, if they are delayed, the entire project will be delayed. The critical path is the longest route through the sequence of activities which must be undertaken in order to complete the project.

> It is important to appreciate that the times at each node are event times and not start and finish times and that each event time is the earliest and latest time that the event can happen – whether this is at the beginning or end of the activity arrow.
>
> For activity $a - b$, the latest event time (b) minus the earliest event time (a) gives the overall time available for undertaking activity $a - b$. If this time is equal to the activity duration, then activity $a - b$ is critical. If not, the balance of time is float.

Dummy activities, which usually have no duration or value, can be introduced to indicate dependencies not shown by the arrow activities.

For straightforward projects, operations can be sorted into priority, the forward and backward pass calculations carried out by hand and the critical path calculated and highlighted. Large or complex projects would require project management software – such as Micro Planner X-Pert – to carry out the time analysis.

9.3.1 Basic Principles

The stages involved in preparing a network based on arrow diagram principles are as follows:

Step 1	Develop a logic diagram based on the sequence of work envisaged and enter the activity descriptions, durations and event numbers.
Step 2	Apply the *forward pass* working from left to right through the diagram and calculate the earliest event times for each activity. At an intersection, always carry the highest number forward. Establish the overall project period and enter on the finish flag.
Step 3	Apply the *backward pass*, working from right to left back through the diagram and establish the latest event times for each activity. At intersections always carry the lowest number back.

| Step 4 | Establish the *float times* for each activity and highlight the *critical path* (the activities with zero float). |
| Step 5 | Convert the arrow diagram to a bar chart format if required for clarity or simplicity. This could be based on either the earliest or latest event times. |

9.3.2 Worked Example

Figure 9.5 shows the plan and cross-section of three bays of a precast concrete retaining wall together with a schedule of the sequence of work and activity durations. In Figure 9.6, a basic logic diagram for constructing section A of the retaining wall foundation is shown.

Below this, an accompanying ladder diagram shows the work sequence for all three sections of the retaining wall – A, B and C – and includes activity descriptions, durations and event numbers. Activity 1–2 is excavation of section A of the retaining wall, activities 2–3 and 3–5 are excavation of sections B and C, respectively, and activity 2–4 is the concrete blinding for section A and so on.

Also included in Figure 9.6 are the forward and backward pass calculations for all three sections of the retaining wall, which establishes an overall duration for the sequence of 21 days.

The critical path is indicated in bold, linking nodes 1, 2, 4, 7, 10, 13, 16 and 18 – wall section (A) – and nodes 19, 20 and 21 which represent wall erection to sections B and C and final backfill of all three sections of the retaining wall.

It will be noted that the critical path is also identified by nodes with the same earliest and latest event times for each of the activities. These activities have no float or free time and are, by definition, critical. Other activities indicate that float is available – the activity for fixing formwork (activity 8–11), for instance, shows a duration of two days and float of two days in a small box.

9.4 Precedence Diagrams

Precedence diagrams follow the same logical procedures as arrow networks except that the activities and their dependencies are drawn differently.

The precedence diagram consists of a series of boxes interlinked with lines. The box or node represents the activity, and the linking arrow indicates the relationships of the activities one to another. The box contains an activity label or name and duration, and there is space for the earliest and latest start and finish times of the activity. A reference number may also be included if required. Each node or precedence box is ascribed a unique activity number.

> In contrast to arrow diagrams, the times in precedence boxes are the earliest and latest times that the activity can start and finish. Therefore, the latest finish minus the earliest start gives the overall time available for an activity. If this time is equal to the activity duration, then the activity is critical. If not, the balance of time is float.

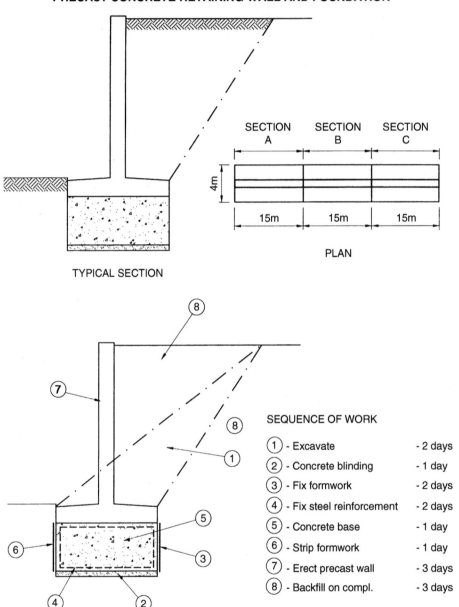

Figure 9.5 Retaining wall plan, cross-section and sequence of work.

Both the boxes and the lines may be given a time value. The time given in the box represents the duration of the activity, whilst any time on the line or arrow adds a dependency. This might be a lead time, if the commencement of a succeeding activity is to be delayed, or a lag if a preceding activity must be completed before completion of a succeeding activity.

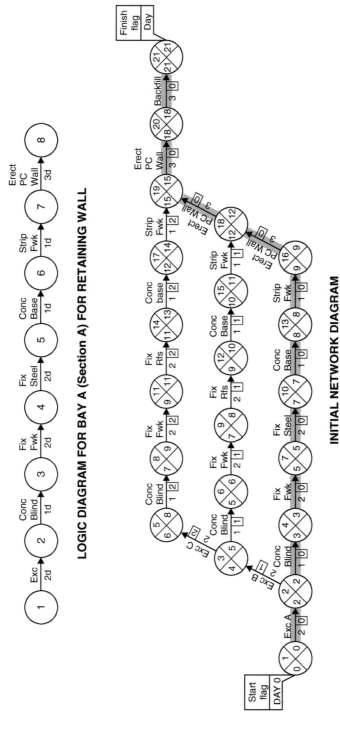

Figure 9.6 Retaining wall logic diagram.

One of the key reasons for the growth of precedence diagrams was the limitation of arrow diagrams when, for instance, one activity is required to start before the preceding activity is completed. This means either dividing the preceding activity into smaller parts or introducing a dummy with a time value. Precedence diagrams do not require dummies to preserve the logic of the relationships as this is achieved with the precedence arrows and any lead or lag times required.

The precedence approach introduced the idea of activity boxes, rather than activity arrows, which permits a number of different relationships to be expressed between activities. This relates more closely to the real situation on a construction project, and this practicality is one of the reasons why precedence diagrams are more popular than arrow diagrams. The relationships which can be included are:

- **Finish-to-start**: where a preceding activity must finish before a succeeding activity can start.
- **Start-to-start**: where one activity can start at the same time as another activity.
- **Finish-to-finish**: where two activities must finish at the same time.
- **Start-to-finish**: where a preceding activity must start before a succeeding activity can finish.

This makes the precedence display easier to follow and permits the introduction of time constraints on the logical links without the need to include dummies or ladders.

9.4.1 Basic Principles

The stages involved in developing a sequence of work in a precedence format are as follows:

Step 1	Develop a logic diagram for the work involved and enter the activity descriptions, durations and reference numbers in the precedence boxes.
Step 2	Enter relationships – finish-to-start, start-to-start, and start-to-finish relationships – using precedence arrows and add any further dependencies such as lead and/or lag times.
Step 3	Analyse the diagram by applying the *forward* and *backward* pass and enter start and finish flags.
Step 4	Calculate the *floats* and establish the operations with zero float – highlighting the *critical* operations.
	NB: The precedence boxes are much easier to analyse than arrow diagrams, as the floats are simply the difference between the latest finish and earliest start numbers in the corner of the precedence boxes, less the activity duration.
	If the difference is zero, then the activity is critical.
Step 5	Convert the precedence diagram to a bar chart, if required.

9.4.2 Worked Example

A precedence diagram for the construction of the foundations and ground floor slab of a small factory building is shown in Figure 9.7. There are nine activities, and the critical path indicates an overall construction period of 34 days for the work.

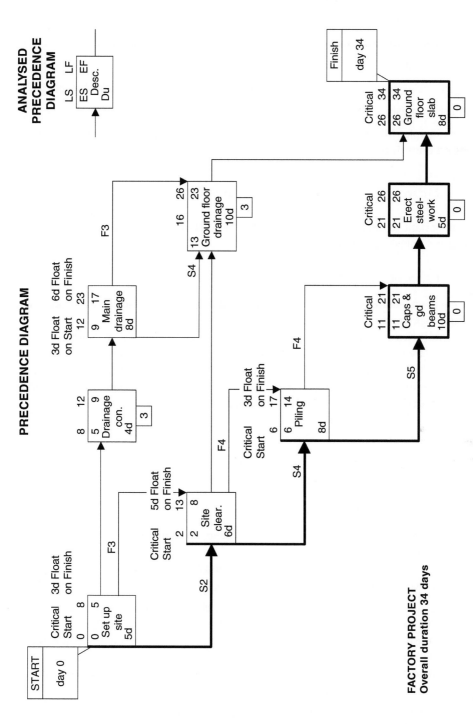

Figure 9.7 Factory project precedence diagram.

A variety of relationships are illustrated:

- Finish-to-start:

  E.g. **Ground floor drainage** must be finished before **Ground floor slab** starts.

- Start-to-start:

  E.g. **Site clearance** can commence two days after the start of the **Set up site** activity.

- Finish-to-finish:

  For example, **Main drainage** must be finished three days before the completion of **Ground floor drainage.**

9.5 Arrow, Precedence and Linked Bar Chart Relationships

It is important to understand the relationship between arrow diagrams, precedence diagrams and linked bar charts.

9.5.1 Common Features

The common feature of all three methods is that there is a logic which links the activities, and this logic creates dependencies in the schedule.

Dependencies are used to determine the order and sequence of the work and, together with the activity durations, may be used to calculate the overall duration of the project.

These calculations are called 'critical path analysis' because they reveal those activities with no free time (i.e. zero float), and it is these activities which determine the longest route through the schedule. This is called the 'critical path'.

9.5.2 Networks

Arrow diagrams and precedence diagrams are known as 'networks' (because that is what they look like), and the analysis of the diagrams is referred to as critical path or network analysis.

There are two ways of presenting networks:

- Using arrow diagrams where the arrow represents the activity, and the circles join the activities together.
- Precedence diagrams where the activities are shown as boxes, and the boxes are linked together with lines or arrows.

Linked bar charts are simply Gantt charts (or bar charts) where the activities are linked with arrows to show dependencies. The various links are illustrated in Figure 9.3, which can be at the beginning or end of the activities or somewhere in between (this is sometimes called 'mid-bar linking').

The different activity relationships for the arrow, precedence and linked bar chart formats are shown in Figures 9.8–9.12:

- Figure 9.8 indicates finish-to-start relationships in all three formats.

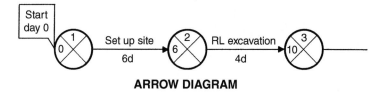

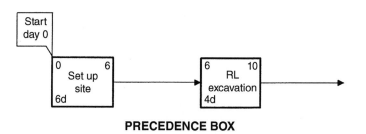

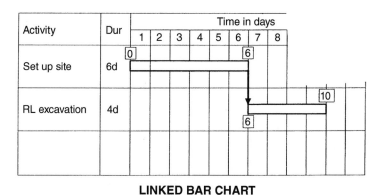

Figure 9.8 Finish-to-start relationships.

- Figures 9.9 and 9.10 indicate start-to-start relationships or methods of introducing overlaps into the sequence of work.
- Figures 9.11 and 9.12 indicate finish-to-finish relationships and start-to-finish relationships, again in all three formats.

9.5.3 Overlapping Activities

On a construction project, activities on site mainly overlap one with another.

In the example shown in Figure 9.9, the setting up site activity starts at the commencement of the project on day 1 (time 0) and takes six days to complete. Reduced level excavation work starts on day 4 (time 3) and takes four days to complete. On the bar chart display a three-day overlap has been introduced between the starting dates of the two operations.

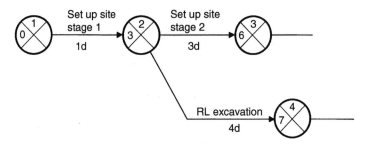

ARROW DIAGRAM

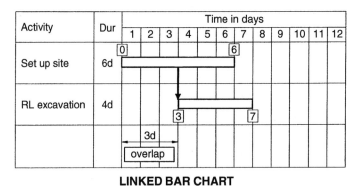

PRECEDENCE BOX

Activity	Dur	Time in days
Set up site	6d	bar from day 0 to day 6
RL excavation	4d	bar from day 3 to day 7
		3d overlap

LINKED BAR CHART

Figure 9.9 Overlapping relationships – 1.

The overlapping of operations is easier to express on either a linked bar chart or in a precedence format. Overlapping on an arrow diagram is much more difficult (see Figure 9.9). This cannot be done by starting an overlapping activity part-way along an arrow but involves breaking activities down into more detailed operations – see 'set up site' activity in Figure 9.9 which has been split in two.

9.5 Arrow, Precedence and Linked Bar Chart Relationships

START-TO-START RELATIONSHIPS

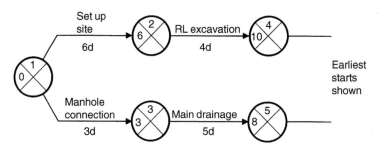

ARROW DIAGRAM

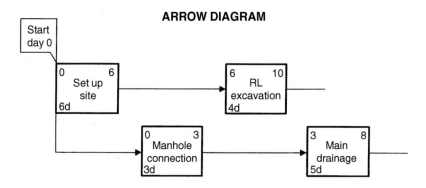

PRECEDENCE BOX

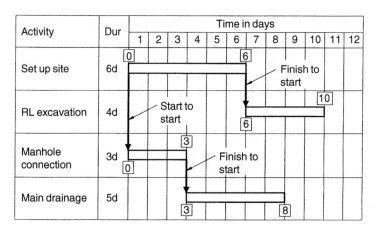

LINKED BAR CHART

Figure 9.10 Overlapping relationships – 2.

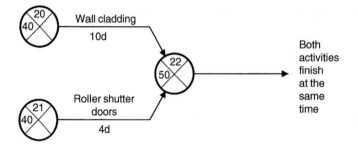

ARROW DIAGRAM

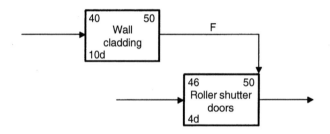

PRECEDENCE BOX

Activity	Dur	Time in days
		40	41	42	43	44	45	46	47	48	49	50	51
Wall cladding	10d	40										50	
Roller shutter doors	4d							46					

LINKED BAR CHART

Figure 9.11 Finish-to-finish relationships.

9.5 Arrow, Precedence and Linked Bar Chart Relationships

START-TO-FINISH RELATIONSHIPS

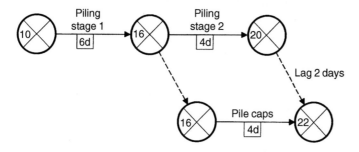

ARROW DIAGRAM

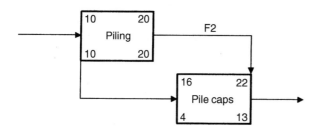

PRECEDENCE BOX

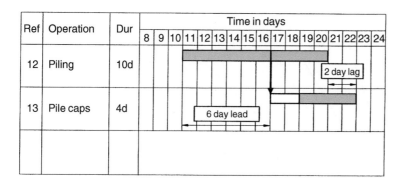

LINKED BAR CHART

Figure 9.12 Start-to-finish relationships.

Alternatively, overlaps can be created by using a 'ladder' diagram which necessitates complex lead and lag terminology (see Figure 9.6 for an example of this).

9.6 Line of Balance

A line-of-balance diagram comprises a series of inclined lines which represent the rate of working between repetitive operations in a construction sequence.

This approach to scheduling was devised in the United States in the early 1940s, and later developed by the UK National Building Agency, which recognised its applicability to house construction.

As well as repetitive new-build housing and apartments, the technique has been widely used for the planning of refurbishment works and in the civil engineering sector. Line-of-balance diagrams can be incorporated into Gantt chart displays for the scheduling of repetitive sections of projects.

Line of balance – also called elemental trend analysis – is a visual display of the rate of working of different activities in a schedule. Ideally, all balance lines should run parallel to each other, but this is often difficult to achieve in practice.

9.6.1 Basic Principles

The principles of line of balance may be illustrated by considering the three sequential construction activities, A, B and C, that are to be repeated several times to complete '*n*' number of housing units.

Figure 9.13 illustrates the line-of-balance diagram for each of these activities. It shows that each activity has a different rate of working and that activities B and C are out of

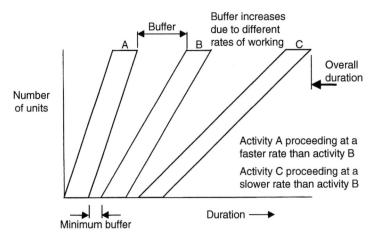

Figure 9.13 Line-of-balance principles – 1.

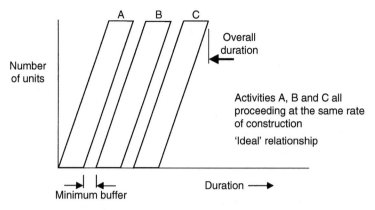

LINE-OF-BALANCE DIAGRAM – IDEAL RELATIONSHIP

Figure 9.14 Line of balance principles – 2.

balance with operation A. As a result of non-parallel working, the initial minimum time buffers (float) shown between the activities increase significantly, thereby increasing the overall duration of the project.

By increasing the number of gangs employed on operations B and C, the line-of-balance diagram shown in Figure 9.14 can be achieved. This results in parallel working and a considerable saving in the overall project period.

The 'buffers' shown in Figure 9.13 are conventionally introduced into line of balance diagrams – either at the start or finish of an activity – to build some degree of flexibility into the schedule. This provides float in case of delays and also allows the labour gang to clear the work area or house unit before the next activity commences.

9.6.2 Worked Example

Table 9.2 identifies the five operations/activities that make up the construction sequence for building a house on a project comprising 10 house units. Also shown are the durations

Table 9.2 Five construction operations.

Activity	Duration per unit in weeks	Number of gangs
Foundations	2	2
External walls	4	3
Roof construction	1	1
Internal finishes	4	3
External works	2	2

Note: No of house units = 10.

(in weeks) of each operation per house unit and the number of gangs available. Nine steps are required to produce the line of balance schedule for the project:

Step 1	Draw the sequence logic for the five operations, as shown in Figure 9.15.
Step 2	Assess the start and finish date of each operation in the construction sequence for the first and last units. This enables the balance lines to be plotted. Use the following formula in each case:

$$\frac{\text{number of units less 1} \times \text{duration of operation}}{\text{number of gangs used}}$$

Operation 1–2 (Foundations)

Start of unit 1 = **week 0**
Finish of unit 1 = **week 2** (i.e. the duration of the operation)

$$\text{Start of unit 10} = 0 + \frac{(10 \text{ units} - 1) \times 2}{2 \text{ gangs}} = \frac{18}{2} = \textbf{week 9}$$

Finish of unit 10 = week 9 + 2 weeks = **week 11**

Step 3	Plot the balance lines on squared paper. Figure 9.16 shows the balance line drawn for the Foundations operation.
Step 4	Assess the start and finish date for the next operation:

Operation 2–3 (External walls)

NB: Allow a minimum buffer of 1 week between one operation and the next. This will be either at the start or finish of the next operation depending on the rate of working.

Start of unit 1 = week 2 + 1 week buffer = **week 3**
Finish of unit 1 = week 3 + duration of 4 weeks = **week 7**

$$\text{Start of unit 10} = \text{week 3} + \frac{(10 \text{ units} - 1) \times 4}{3 \text{ gangs}} = 3 + \frac{36}{3} = \textbf{week 15}$$

Finish of unit 10 = week 15 + 4 weeks = **week 19**

Foundations

$$\text{Slope of balance line} = \frac{\text{duration}}{\text{no. of gangs}} = \frac{2}{2} = 1.00$$

External walls

$$\text{Slope of balance line} = \frac{\text{duration}}{\text{no. of gangs}} = \frac{4}{3} = 1.33$$

NB: Strictly speaking, the numbers do not really mean anything except that the lower number for the foundations operation gives an approximate indication that this operation is progressing at a faster rate than the external walls activity. This quick calculation indicates from where to plot the balance line for the external works operation. If the number is higher than that of the preceding operation, plot from the bottom of the diagram, otherwise the operations will bump into each other. If the number is lower than that of the preceding operation, plot from the top. In either case, be sure to include the minimum buffer allowance.

Step 5	Plot the balance line for External walls as shown in Figure 9.17.

Step 6   Assess the start and finish date for the next operation:
Operation 3–4 (Roof construction)
Start of unit 10 = week 19 + 1 week buffer = **week 20**
Finish of unit 10 = week 20 + 1 week duration = **week 21**
Start of unit 1 = week 20 − $\frac{(10-1) \times 1}{1 \text{ gang}}$ = 20 − 9 = **week 11**
Finish of unit 1 = week 11 + 1 week duration = **week 12**

Step 7   Consider the rate of construction between the External walls and Roof operations.
External walls
Slope of balance line = **1.33**
Roof construction
Slope of balance line = $\frac{1}{1}$ = **1.00**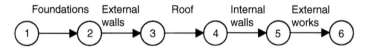
NB: As the roof operation is progressing at a faster rate than the external walls, the relationship between the balance lines will commence at the finish of the external walls operation for house unit 10.

Step 8   Plot the Roof construction balance line starting from the top of the diagram, as shown in Figure 9.18.

Step 9   Complete the calculations for the remaining operations. These have been added to Figure 9.18 to indicate an overall completion at week 32.

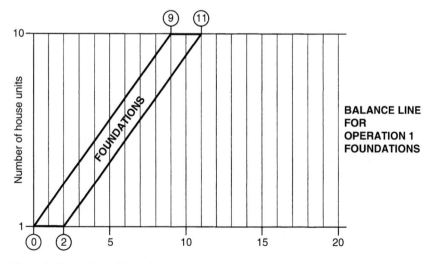

Figure 9.15   Sequence logic.

Figure 9.16   Balance line – foundations.

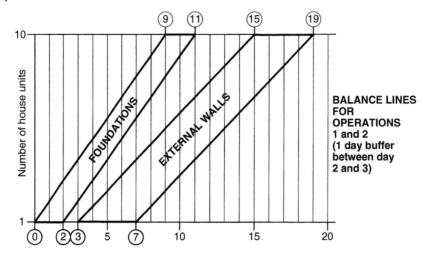

Figure 9.17 Balance line – external walls.

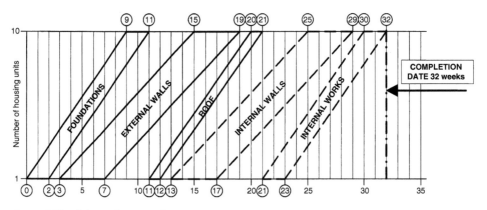

Figure 9.18 Balance line – roof.

9.7 Time-Chainage Diagrams

Line of balance provides a simple and clear means of scheduling and displaying repetitive construction work that is less easily accommodated in Gantt chart format. It is ideal for static, repetitive work such as housing, slip forming, modular building or large-scale repair and maintenance work.

In some cases, however, construction work takes place over long distances – sometimes over hundreds of kilometres – but also includes both repetition and an element of static work. Projects such as motorways and major highway schemes, pipelines, railway track work and tunnelling are repetitive but also include static structures such as bridges, culverts, pumping stations, railways stations and cross-over tunnels.

The time-chainage diagram – sometimes referred to as time–location or time–distance diagram – is a scheduling technique designed to adapt to all of these requirements in an easy-to-follow visual display.

9.7.1 Basic Principles

Some construction projects can be viewed as mainly linear in nature, where work starts at one point and proceeds in an orderly fashion towards another location. This would be typified on a highway project by activities such as fencing, drainage, road surfacing and road markings. In practice, such work does not necessarily start at chainage 0 but might commence at several points along the spread of the job, perhaps going in several directions, but nevertheless working in a repetitive fashion.

> This type of work calls for a different planning technique because bar charts would not be useful in giving locational information, and precedence/arrow diagrams would not reflect the time–location relationship which clearly exists on such projects.

For 'linear' projects, most operations take place on a forward travel basis with the gang starting at one point or chainage and moving along the job. As one activity leaves a particular location, other activities can take its place. This ensures the correct construction sequence and avoids over-intensive activity in one location.

The time-chainage format enables the time dependencies between activities to be shown, together with their order and direction of progress along the job. Most of the lines on the time-chainage diagram have no appreciable thickness. This is because the time spent by each gang at a particular location is relatively small and the gang moves along the site quite quickly. Examples of this are drainage, road surfacing and safety barrier erection on a motorway.

Retaining walls would also constitute a linear activity but would tend to occupy a particular location (or chainage) for a more appreciable time due to the nature and duration of the construction operations involved.

With earthworks cut and fill operations, the situation is different in that earthworks plant will occupy a particular cut or fill zone for some time before moving to another location.

Bridges, culverts and underpasses, on the other hand, are 'static' operations and can be viewed as individual 'sites' in their own right. Such activities act as restrictions and forward travel activities may have to be programmed around them. For instance, on a highway project, drainage work may be interrupted by a bridge site and consequently the contractor will have to return later to finish the drainage once the bridge nears completion.

Various types of time-chainage representations are possible, but basically the diagram comprises two axes – time and distance – with the various activities shown as lines or bars on the chart. Linear activities are represented with a line or bar which is positioned on the chart to show its commencing and completion chainages and is inclined in the direction of progress at an angle consistent with the anticipated duration of the operation.

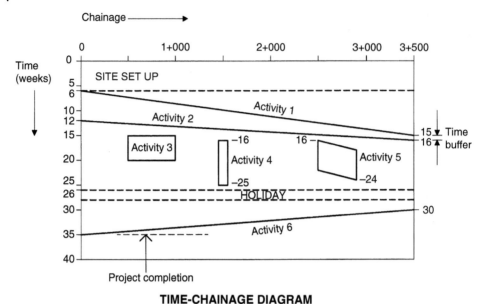

Figure 9.19 Horizontal format.

A static activity, such as a bridge on a motorway, is represented by a line or thin bar positioned at a particular location or chainage, with the duration of the activity expressed by the length of the line or bar. Activity labels are annotated on the respective line or bar to distinguish one operation from another.

Figures 9.19 and 9.20 illustrate two methods for presenting time-chainage diagrams with chainages located on either the vertical or horizontal axes and time – usually in weeks – shown on the other axis.

The advantage of showing chainages along the horizontal axis is that this more readily resembles the way the drawings are laid out and the schedule can therefore more easily be related to what must be constructed. On the other hand, time-chainage diagrams showing time on the horizontal axis may be easier to read by those familiar with Gantt charts. The choice is a matter of personal preference.

9.7.2 Worked Example

Drawing a time-chainage diagram is not as easy as it looks, and a good deal of practice is required to become proficient. Traditionally, time-chainage diagrams are drawn by hand but project management software – such as ChainLink, Turbo Chart or TimeChainage – can take the drudgery out of the task.

Some time-chainage software packages can take an existing Gantt schedule (such as P6 or Asta Powerproject) and convert it to the time-chainage format. Asta Powerproject can also toggle to time-chainage format within the software.

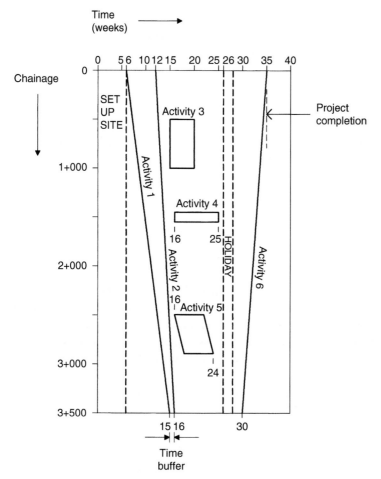

Figure 9.20 Vertical format.

Before looking at the worked example below, consider the following steps:

Step 1	Consult the project layout drawings and note the chainage positions. Main chainages on a highway project are at 1 000 m intervals.
Step 2	Draw an outline time-chainage diagram with time along one axis and distance or chainage along the other, using either the horizontal or vertical format. Add main holiday periods, allowing 2 weeks for Christmas and 1 week for Easter.
Step 3	List main activities or operations in approximate construction sequence. Include activities for site set-up or mobilisation and clear site at the end. Estimate the duration of each activity in weeks.
Step 4	Fill in the site 'set-up' and 'clear site' activities on the time-chainage diagram. Using the preferred format, plot the appropriate number of weeks over the entire length (chainage) of the project.

	Step 5	Decide in turn where and when each activity will take place. For linear activities (e.g. drainage), start at the appropriate location (chainage) and week number and draw a line for the correct distance (chainage) and time (weeks). For static activities (e.g. bridgeworks), draw a thin box at the appropriate chainage with a length representing the activity duration.
	Step 6	Complete all activities on the list. It is sometimes helpful to produce an outline Gantt chart to help clarify the correct time-chainage display.

To put this into a practical context, consider a project to carry out improvements to an existing highway 3.5 km long, which comprises the following activities:

- **Activity 1**: Fencing.
- **Activity 2**: Drainage.
- **Activity 3**: Bulk earthworks.
- **Activity 4**: Footbridge.
- **Activity 5**: Retaining wall.
- **Activity 6**: Road surfacing.

The start of the job is at chainage 0 + 000 and the project finishes at chainage 3 + 500. A footbridge is required to cross the highway at chainage 1 + 500 which would be 1500 m up chainage from the beginning of the job. Distances or locations can be conveniently determined by these chainages and are also useful on site where the contractor will usually position chainage boards along the job to enable anyone to quickly identify exactly where they are.

Referring to either Figure 9.19 (horizontal format) or Figure 9.20 (vertical format), the time-chainage diagram is prepared as follows:

	Activity	Time/Location
		NB: All timings are week commencing.
		Therefore a 5-week activity commences Monday week 15 and is finished before Monday week 20.
	Set up site including offices for the contractor and engineer.	Starts at time 0 with a duration of 6 weeks. The activity occupies the space from chainage 0 to 3 + 500.
1	Fencing	Starts week 6 with a duration of 9 weeks. Forward travel is up chainage finishing week 15. A time buffer (float) of 1 week is added between the completion of activity 1 and completion of the succeeding activity 2.
2	Drainage	Starts week 12 with a duration of 4 weeks. Forward travel is up chainage finishing at week 16.
3	Bulk earthworks	Starts week 15 with a duration of 5 weeks. Finishes week 20. Occupies the zone between chainage 0 + 500 and 1 + 000 for the 5-week duration.

4	Footbridge	Starts week 16 with a duration of 9 weeks. Finishes week 25. This is a static activity taking place at chainage 1 + 500.
5	Retaining wall	Starts week 16 with a duration of 8 weeks. Finishes week 24. Takes place between chainage 2 + 500 and 2 + 900. The sloping box represents a forward travel up chainage at the rate of 50 linear metres per week.
6	Road surfacing	Starts week 30 with a duration of 5 weeks. Finishes week 35. Forward travel is down chainage from 3 + 500 to 0.
	Holiday – Christmas shut down	Starts week 26 with a duration of 2 weeks. Finishes week 28.

Chapter 10 Dashboard

Key Message		○ Developing logical construction sequences is fundamental to creating a realistic and achievable programme. ○ The thought process should consider both the permanent works and any necessary temporary works.
Definitions		○ **Temporary works**: Items that are needed to facilitate construction works but do not usually remain as part of the permanent works. ○ **Sequence study**: Provides a means of determining the time and cost implications of the various construction methods.
Headings		**Chapter Summary**
10.1	Introduction	○ The programme represents a reasonable view of the project based upon the information available at the time. ○ It is most unlikely that the works will be undertaken strictly in accordance with the planned programme. ○ The programme illustrates one of the many possibilities of how the work may progress. ○ A construction programme illustrates the order and sequence of work, dependency, key dates and milestones. ○ A work sequence indicates how each activity will be carried out, which resources will be used and which temporary works will be needed to facilitate each work activity.
10.2	Resources	○ The planner needs to make decisions in relation to the resources available to carry out a specific work sequence. ○ Labour, plant and subcontractors have finite availability. ○ The development of a method statement is important for underpinning the programme and for reducing the risk to health and safety. ○ Ensuring continuity of work is essential to the efficient use of resources.
10.3	Temporary Works	○ Temporary works are an essential element in the construction process. ○ They are fundamental to achieving the correct and safe means of sequencing construction work. ○ In order to comply with legislation and good practice guidance, temporary works must be designed to ensure their safety. ○ It is usual practice for the principal contractor to appoint a temporary works coordinator (TWC) for a project/site.
10.4	Work Sequences	○ Deciding upon suitable work sequences requires detailed and considered thought. ○ It cannot be left to operatives to decide on matters that require engineering expertise. ○ Work sequences explained include demolition, façade retention, basements, core walls, top-down construction and concurrent top-down and bottom-up methods.
10.5	Sequence Study	○ The use of sequence studies is planning at a fine level of detail. ○ They can be very useful when considering alternative methods of construction for a particular operation.
Learning Outcomes		○ Distinguish between a programme/schedule and a work sequence. ○ Appreciate the importance of health and safety and continuity of work when developing work sequences. ○ Understand the importance of temporary works, the applicable legislation and guidance and how temporary works are managed. ○ Understand the logic of work sequences for a wide variety of construction activities. ○ Be able to develop a simple work sequence.
Learn More		○ Chapter 11 explains the role and importance of method statements. ○ Chapters 3 and 12 deal with risk, risk assessment and risk management. ○ Chapter 12 explains how safe systems of work are developed. The use of safety method statements and the role of planning in developing safe work sequences are also explained.

10

Construction Sequences

10.1 Introduction

Developing logical construction sequences is fundamental to creating a realistic and achievable construction programme. The thought process considers not only the permanent works but also the temporary works needed to facilitate construction and completion of the project.

The role of temporary works in construction cannot be overstated, as the installation and removal of such works can often dictate the order and sequencing of the permanent works. In fact, it is arguable that all works on site are temporary if they are in a temporary state of construction and remain so until such time as the permanent works are complete.

The preparation of any pretender or construction programme needs to be structured and logical and is best undertaken collaboratively involving members of the planning, estimating and construction management teams. A thorough knowledge of construction technology and a practical understanding of construction sequences is fundamental, but this is not to say that the knowledge and experience that go into the contractor's programme implies any particular degree of accuracy.

Lowsley and Linnett (2006), for instance, emphasise that the programme is merely a *reasonable view* of the project based upon the information available at the time. They further explain that *it is most unlikely that the works will be undertaken strictly in accordance with the planned programme* but that this simply *illustrates one of the many possibilities of how the work may progress*. The planner's art is to develop an achievable construction sequence, within the contractually required timings, that fully relates to the contractor's available resources.

The implication of this argument is that the contractor's programme is insufficient to provide a basis for planning the day-to-day sequencing of construction operations on site. This is why short-term programming, sequence studies and method statements are crucial to ensure that site activities are rigorously thought through and that safe systems of work are developed that reflect the reality of what happens on site.

10.1.1 Programme versus Work Sequence

It is essential to appreciate the distinction between a construction programme and a work sequence. They both provide a means of planning construction work, but the planning of a work sequence is at a much more granular level of detail than preparing a programme. Durations on a master programme are usually shown in weeks, and those on short-term programmes are in days. Even so, a programme is not the place to plan work sequences in detail.

- A programme simply indicates:
  - The main work activities for completing any given project.
  - The order in which the work will be carried out.
  - The salient dates in the contract and other key dates and milestones.
  - How long each activity is planned to take.
  - The logic or dependencies between an activity and any preceding, succeeding and concurrent activities.
- A work sequence indicates:
  - How the contractor will carry out each activity.
  - Which resources will be used.
  - Which temporary works will be needed to facilitate each work activity.

The distinction between a programme and a work sequence is recognised in the NEC4 Engineering and Construction Contract (ECC). This not only requires the contractor to submit a programme for acceptance to the project manager but also to submit details concerning intended working methods. They include:

- A statement of how the contractor plans to do the work.
- The principal equipment and other resources that the contractor plans to use.

This contractual obligation is fulfilled by considering work activities at a much more detailed level than might be adopted when preparing the programme. The programme will show how long an activity such as piling will take – its duration – and will indicate the activities that precede and follow the piling, whereas a work sequence will show how the activity is to be carried out in detail.

A work sequence for the piling activity on the programme will include considerations of:

- The type of piling rig to be used (CFA, auger, driven, etc.).
- The plant needed to assemble the rig – MEWP, for example.
- The means of access for the piling rig and available headroom.
- The provision of a piling platform or mat for the rig to work on.
- The thickness of the piling mat – 300 mm is usual but they can be up to 1 m deep.
- Material for the pile mat – such recycled 6F2 produced on site or 6F5 imported to site.
- The method and plant required for compaction and subsequent excavation and disposal of the pile mat.
- Plate testing/CBR of piling mats.
- Whether a geogrid is required under the piling mat.
- Whether the piles will be installed from ground level or from a lower level.
- The order in which the piles will be installed.
- How spoil (arisings) from the piling operations will be removed.
- How concrete will be delivered to the rig position.
- Arrangements for the dismantling and removal of the piling rig when the work has been completed.

In this respect, those responsible for preparing a programme must fully understand the link between the sequence of operations to be undertaken, the impact of any temporary work requirements and the practicalities on site. This means planning suitable measures for managing the interfaces between concurrent activities which can influence or even dictate how a particular activity will be conducted or sequenced.

This ability comes with experience, but nowadays, BIM models linked to planning software (such as Asta Powerproject) and virtual reality animations can be used to simulate the build. This can greatly assist the planner and construction team to think through the likely sequence of operations and flag up risks that need to be managed.

> Simple sketches, hand drawn sequence diagrams or temporary works design drawings can all be used to illustrate construction sequences and aid discussion about working methods. Even a simple sketch on the corner of a working drawing can work wonders.

Thinking through and discussing the build sequence helps to create safe systems of work and prevent accidents and ill-health.

10.1.2 Construction Methods

For any given project – be it a demolition, a new building or infrastructure works – there may be several ways of carrying out the works to achieve the desired result. In fact, contractors often consider alternative methods of construction and various temporary works solutions before settling on their preferred option – based on cost, time and risk factors.

Depending on the procurement method chosen for the project, it might also be the case that discussions are held with the permanent works designers to find more buildable solutions whilst, at the same time, satisfying the client's brief.

The construction industry is well known for devising novel and innovative solutions to solve design problems and to shave time off the schedule and money off the budget. When this happens, different construction methods can dictate a different programme of work, different plant and resource requirements and different relationships with other activities. This is equally the case with respect to demolition works.

Take the example of a multi-storey building that is to be demolished. There are several ways that the demolition could be carried out:

- Deliberate collapse
  - **Blow down**: using explosives to induce collapse by weakening structural elements.
  - **Pull down**: using hawsers/chains attached to an excavator to pull over the structure.
- Progressive demolition
  - **Top-down**: using a variety of methods to demolish the building from the top to the bottom.
  - **High reach**: using high-reach excavators and concrete 'nibblers' to gradually pull the structure down.
  - **Deconstruction and dismantling**: removal of structural elements in reverse construction order, perhaps using robotic equipment or hot/cold cutting techniques, etc.
  - A combination of techniques.

Clearly, there are fundamental differences between each of these techniques which would require completely different working methods, expertise and resources. The choice of method would have implications for the programme, and each would create its own problems with regard to danger to the public, crushing and disposal of debris, noise, dust and risk to adjacent properties.

Irrespective of the demolition method chosen, a great deal of detailed thinking will be necessary to plan issues such as:

- Site security.
- Safety exclusion zones.
- How remote plant operation (if used) will work.
- Segregation and resizing demolition waste ready for on-site crushing.
- Choice of plant for handling and loading of debris.
- Location of on-site crushing plant.
- Choice of plant for moving demolition materials to the crushing plant and, subsequently, off-site.

The end result of this planning will be the preparation of a method statement. Method statements have become an art form in bureaucracy in the construction industry, but they should not be, nor were they ever intended to be, anything other than a detailed explanation of how the contractor proposes to carry out the works safely.

10.2 Resources

The construction industry has a well-documented skills shortage, and a great deal of thought needs to be given to planning resource availability for a project rather than simply assuming that everyone will slot into the programme conveniently.

The planner needs to make decisions in relation to the resources available to carry out a specific work sequence because labour, plant and subcontractors have finite availability.

When it comes to the detailed design and planning of temporary works, in-house specialists or consultants need time to develop proposals which then need to be discussed, revised and, importantly, checked for design robustness.

10.2.1 Labour

Labour, and some trades in particular, may be difficult to find 'on demand' and good quality subcontractors invariably have more work than they can handle. Consequently, good 'subbies' can often 'pick and choose' their work and so subcontract lead times and availability are important questions for the planner to ponder over.

In order to overcome resource availability problems, it is becoming common to use multi-skilled labour gangs who are capable of multitasking. For example, a labour gang may be able to undertake excavation work, fix steel reinforcement, position bolt boxes and place concrete. This allows the gang to move from one section to another without having to wait for other trades, such as joiners, when undertaking the construction of a repetitive sequence of foundation bases.

10.2.2 Construction Plant

Apart from 'normal' subcontractors, the construction industry relies heavily on specialists such as:

- Specialist plant hire firms.
- Contractors who specialise in certain types of work:
  - Demolition; heavy lifting; self-propelled modular transporters (SPMTs); marine lifting barges; tower crane erection and dismantling; tunnelling and shafts; piling; diaphragm walling.
- Ground engineering
  - Dynamic compaction; vibro-compaction; vibro-replacement; soil nailing; ground anchors; groundwater control.

All these activities require specialist plant and equipment which represent a substantial investment and are, therefore, in limited supply.

This has implications for the programming of the works and the planning of work sequences, which need to take into account the lead times necessary to secure these resources to fit in with the contractor's programme.

10.2.3 Health and Safety

The development of a method statement is important for underpinning the programme and this will assist the contractor to develop the preferred approach to the work and the resources needed. However, this cannot be prepared in isolation without thinking about the safety of construction operations and the development of safe working methods.

The planner must also think about the statutory requirement to reduce risk to health and safety. This is done by carrying out risk assessments and by devising suitable control measures. These issues must be considered as part of the process of developing the method statement.

10.2.4 Continuity of Work

An important aspect of planning which concerns contractors is the efficient use of resources, as nothing eats away profits more quickly than plant and labour working inefficiently or standing idle. Ensuring continuity of work is essential irrespective of whether the main contractor is doing the work or whether it is sublet.

Subcontractors are just as keen as main contractors to ensure continuity of work, as they normally want to get in and out as quickly as they can. Good subcontractors are usually busy and want to move on to their next job quickly. Additionally, should the main contractor cause them delay or disruption, or not provide attendances such as cranage and scaffolding on time, contractual claims will undoubtedly arise, and this will have a damaging effect on the main contractor's bottom line profits.

All of this is especially true where repetitive work is being carried out, such as housing or multi-storey construction. Following trades or operations cannot afford to be delayed by

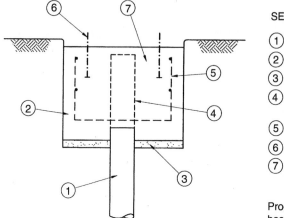

Figure 10.1 Sequence of construction for pile cap formation.

slower preceding activities, and attention must be paid to this in the detailed planning of the project. Whilst perfect continuity of work is not possible in practice, sequence studies enable the contractor to analyse construction activities at the resource level to make the best use of the labour and plant available.

During the development of a programme sequence, consideration must be given to the continuity of work for the gangs involved. This relates to plant, labour and subcontract operations.

Consider the situation where several pile caps are to be constructed, as shown in the cross section in Figure 10.1. In this example, no temporary formwork is required to the sides of the base due to good ground conditions. The sequence of programmed operations will be as follows:

1. Insert bored piles – specialist contractor operation.
2. Excavate pile cap and expose pile.
3. Concrete blinding to pile cap base.
4. Crop pile (cut-off surplus pile shaft) and expose rebar.
5. Fix pile cap rebar and tie it into the pile reinforcement.
6. Fix holding down bolts or column starter bars.
7. Concrete pile cap.

The linked bar chart shown in Figure 10.2 is based on working on pile caps in groups of four and using a multitask labour gang (operatives experienced in undertaking a range of tasks). A 4-day cycle has been developed for each pile cap group, and the overall duration for constructing 16 pile caps is 19 days including piling.

In this situation, the objective should be to move plant and labour from bases A to B to C, etc. Consideration should also be given to the work balance between related operations. Excavation, rebar, formwork (if used) and concreting operations, for instance, progress at

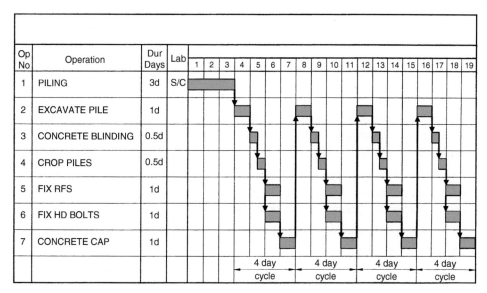

Figure 10.2 Bar chart – Pile cap sequence.

different speeds and therefore the plant, labour and other trade gangs involved need balancing in order to develop cycles of work.

In such a case, it would be uneconomical to keep each operation running at the same rate as the leading operation, and it would be unrealistic to suppose that the rebar and carpentry gangs could be balanced with the concreting gang simply to keep the concrete gang continuously employed. Consequently, the planner would need to look for other activities on the programme where labour could be usefully employed in periods of 'downtime' whilst waiting for other trades or gangs to catch up.

Despite the resultant 4-day cycle for each pile cap group, in practice, it is likely that greater continuity of work will be achieved by excavating the pile caps over a continuous four-day period and then blinding the bases in two continuous days starting part way through the excavation works.

10.3 Temporary Works

Temporary works are an essential element in the construction process and are fundamental to achieving the correct – and safe – means of sequencing construction work.

They are used in both building and civil engineering work:

- In the construction of basements.
- Façade retention.
- Provision of temporary access.
- Provision of support to structures in their temporary condition.
- For supporting liquid concrete in slabs, columns, beams, decks, etc.

10.3.1 Definitions

Temporary works is defined in BS5975:2019 (Clause 5.1.1.1) which explains that it:

- Provides an *engineered solution* that is used to support or protect either an existing structure or the permanent works during construction.
- Supports an item of plant or equipment, or the vertical sides or side-slopes of an excavation during construction operations on site or to provide access.
- Controls stability, strength, deflection, fatigue, geotechnical effects and hydraulic effects within defined limits.

In brief, these are items that are needed to facilitate construction works but do not remain as part of the permanent works. There are exceptions to this where temporary works are left in place or where the permanent works are used as temporary works during construction:

- Earthwork supports left in place such as trench sheeting or steel sheet piling.
- Permanent formwork.
- Permanent works in their temporary condition used as earthwork support systems – such as non-cantilever (propped) contiguous or secant piling or diaphragm walls.
- Profiled metal decking supporting in situ concrete suspended slabs.
- Precast concrete tunnel and shaft linings.
- Sprayed concrete linings used to support tunnel excavations and the like prior to installing the permanent lining.

10.3.2 Legislation

The main legislation that governs temporary works in the United Kingdom is:

- The Health and Safety at Work etc. Act 1974.
- The Management of Health and Safety Regulations 1999.
- The CDM Regulations 2015.

This legislation imposes duties on employers, designers and contractors regarding temporary works in the same way as for the permanent works. Duties include:

- Designing temporary works to avoid risks to health and safety.
- Creating safe places of work that do not endanger workers or others affected by the work (e.g. passers-by).
- The identification, elimination, reduction or management of risk to health and safety.
- The use of safe systems of working based on risk assessment.
- Preparation of method statements and communication with the workforce.
- Planning, controlling and monitoring temporary works during installation, use and removal.

In particular, CDM 2015 regulation 19 imposes a statutory duty to ensure that:

- Steps are taken to ensure that new or existing structures do not become unstable during construction and that:
- Any buttress, temporary support or structure, etc. is designed, installed and maintained so as to withstand any foreseeable loads which may be imposed on it.

10.3.3 Guidance

In common with construction work generally, the legislation regarding temporary works is underpinned by a raft of guidance which the courts would expect duty holders to follow to demonstrate adhesion to the legislation. This guidance includes BS5975:2019.

> Following several major temporary works collapses in the 1960s and early 1970s, when the UK motorway construction boom was in full swing, the government of the day commissioned a report[1] to determine whether the industry was capable of managing falsework – the temporary support structure required during bridge construction. This report – best known as the Bragg Report – was published in 1975.
>
> Amongst other things, the report set out requirements for the erection and stability of temporary structures and its recommendations were substantially incorporated into what is now BS5975:2019.

BS5975 is not legislation, but it provides best practice guidance on how duty holders may comply with health and safety legislation and manage the design, installation, adaption and removal of temporary works on construction projects. Unusually, BS5975 sits alongside the EuroNorm BS EN 12812 – *Performance Requirement and General Design of Falsework* – and practitioners are free to adopt either design methodology. The procedural control measures provided by BS5975 should, however, be followed.

Other guidance available includes CIRIA Guide 517 – *Temporary propping of deep excavations*, the ICE publication *Temporary works – principles of design and construction* and the Temporary Works Forum (TWf).[2]

10.3.4 Management of Temporary Works

The effective and safe use of temporary works on site and the management and procedural control of temporary works are central features of BS5975:2019. This is aimed at ensuring that errors and potential failure of installations are avoided.

Notwithstanding the legislation and guidance around temporary works, failures are not uncommon in the industry – failure of temporary waling beams in basements and shafts, failure of jacks when installing a bridge deck, failure of suspended floor formwork due to removal of supports and lack of redundancy in the formwork design, failure of lifting beams for precast concrete erection, incorrect lifting of wall formwork panels and so on.

The consequences of temporary works failures are not limited to the temporary works themselves – e.g. collapse of excavations, scaffolding failures, collapse of falsework, collapse of unsupported rebar walls, collapse of partially built walls, etc. – but can also lead to failure and collapse of the permanent works whilst in its temporary condition. One of the issues here is that permanent works design does not take account of wind and other loadings when the work is in its temporary condition which is why incomplete permanent works are regarded by many in the industry as temporary works.

Consequently, it is common practice for contractors and subcontractors to appoint competent persons whose job is to ensure that all relevant procedures and checks are carried out and that the work on site is properly supervised by a competent person. Unsurprisingly, responsibility for the management of temporary works begins at the top.

In this regard, BS5975:2019 recommends the appointment of a Designated Individual (DI) to establish and maintain a procedure for the control of temporary works during construction operations. This appointee is usually a director or senior manager.

As part of the hierarchy of temporary works management, BS5975:2019 recommends the appointment of temporary works coordinators and temporary works supervisors who work together with the temporary works designers (TWDs).

10.3.5 Temporary Works Designers

In order to comply with legislation and good practice guidance, temporary works must be designed to ensure their safety and stability. The design effort required will clearly be determined by the level of complexity of the temporary works. TWDs have the same statutory duties as permanent works designers, so it is essential that such works are designed by a competent person to be fit for their intended purpose.

Temporary work design should not be left to site managers or operatives unless they are trained, qualified and experienced, and it will usually be carried out by an engineer.

TWDs may be employed in-house by contractors, or they could be consultants. Alternatively – and commonly in the industry – temporary works are designed by subcontractors or specialist contractors and by companies who specialise in the supply, design, erection and dismantling of proprietary systems. Whoever is undertaking the design, it is essential that designers are fully briefed as to what is required.

On relatively small sites there may be no need for extensive temporary works – just site hoardings, site accommodation, simple excavation supports, scaffolding, etc. Larger projects will need much more sophisticated temporary works which can be extremely complex and influential on the sequencing of the permanent works and upon the contractor's programme.

Temporary works should be checked for:

- Design concept.
- Strength and structural adequacy.
- Compliance with the design brief.

The level of checking required for temporary works designs is defined by BS5975 and is split into four categories 0–3, CAT 3 being the most rigorous. The extent of checking required, and who performs the check, is determined by the complexity and scope of the project. The more complex the project the more scrutiny is required and the greater the organisational separation (i.e. independence) needed between the designer and the checker.

10.3.6 Temporary Works Coordinator

It is usual practice for the principal contractor to appoint a temporary works coordinator (TWC) for a project/site. This person is referred to as the principal contractor's TWC (PCTWC).

The TWC plays an important role in liaising between the site team, the temporary and permanent works designers, subcontractors, temporary works checkers and the client. For large-scale projects, there may be multiple TWCs each dedicated to specific sections of the works or to specific work packages. They may be employed by the principal contractor or, if the subcontract package is large enough, by a subcontractor. In any event, TWCs need to communicate and interact with each other – usually through the PCTWC – who in effect becomes the lead TWC for the project.

The job of the TWC is to prepare a temporary works register of all the temporary works required on the project. This becomes a 'live' document which is changed, adapted and updated as the demands of the project dictate. The TWC must also ensure that suitable temporary works designs are prepared, checked and implemented on site.

The TWC is not responsible for designing temporary works which is the job of a TWD, but acts as liaison between the construction team and TWDs. This ensures that temporary works are designed to suit the needs of the construction team so that the work can be carried out in a safe and controlled manner.

Some projects do not lend themselves to a single lead TWC, and the BS suggests arrangements for large linear projects – such as motorways or railways.

10.3.7 Temporary Works Supervisor

As most construction work is packaged and carried out by subcontractors or specialist contractors, it is recommended that each company appoints a temporary works supervisor (TWS) for their own work to oversee the day-to-day construction, use and safety of temporary works on site. The TWSs will support the PCTWC.

On any particular site, therefore, there may be a number of TWS whose report to TWC and work with the TWDs.

On large sites, there may be several Temporary Works Coordinators in which case the TWC working for the principal contractor will become the lead TWC. Several subcontractors on site may have their own TWC and TWSs in connection with their own work. On major projects there will be a lead TWC and dozens of other TWCs and TWSs.

The TWC and TWS appointments recommended by BS5975:2019 are not necessarily full-time roles and could be undertaken by suitably trained and qualified site managers, foremen or engineers. On large/mega projects, TWCs and TWSs will be fully occupied by their roles.

10.3.8 Types of Temporary Works

Different types of temporary works are described in BS5975:2019 and include measures designed to:

- Support or protect existing structures or permanent works during construction, modification or demolition.
- Provide a means of stabilising permanent structures during construction, pre-weakening or demolition.
- Secure the site, provide access to the site or a workplace on the site and segregate the movement of pedestrians and vehicles around the site.
- Support or restrain plant, materials or equipment.

- Support the sides or roof of excavations during construction or provide earthworks or slopes to excavations including supports or diversions to watercourses during construction operations.
- Provide safe working platforms.
- Undertake measures to control noise, dust, debris, fumes or groundwater during construction.
- Provide protection or support to existing services.
- Facilitate the testing of drains, piles and pre-demolition floor load capacity.

The extent of temporary works in construction is almost endless and includes:

- **Site set up**: hoardings and security gates; accommodation; lay-down areas; aggregate bunkers; materials silos.
- **Earthworks**: access ramps; haul roads; groundwater control (e.g. wellpoint de-watering).
- **Earthwork support**: interlocking trench sheets; sheet piling; drag boxes; manhole boxes; trench boxes.
- **Access**: piling mats and platforms; temporary surfaces/trackways; protective mats and sleepers; scaffolding – tube or proprietary systems.
- **Traffic management**: cones; barriers (e.g. lego blocks, etc.); temporary traffic lights.
- **Propping and supports**: shoring systems; steel/timber needles and beams; dead shores; flying shores; façade retention supports.
- **Formwork**: timber/plywood formwork; proprietary systems; table forms.
- **Falsework**: timber, scaffold tube, proprietary systems or (in some countries) bamboo used to support formwork when casting concrete floors or bridge decks; supports (centres) for forming brick arches; back-propping to cast-in-place concrete decks; temporary piers to support bridge decks during construction.
- **Temporary structures**: footbridges, Bailey bridges.
- **Cranes**: tower crane bases; mobile crane platforms.

10.4 Work Sequences

Deciding upon suitable work sequences requires detailed and considered thought. It cannot be left to operatives – however competent they may be – to decide on matters that require engineering expertise, be it the degree of slope to a battered excavation, the type of trench support system to use for deep excavations, propping systems for floor slabs or the order and sequence of demolition work.

10.4.1 Demolition

Demolition is a very risky business, but no more so when buildings are being adapted or deconstructed. Where such work is contemplated, BS6187:2011 – *Code of practice for full and partial demolition* – provides useful and authoritative guidance.

The usual procedure is to demolish in reverse construction order but, in some cases, the original construction process may be unknown or uncertain. In other cases, buildings may have been previously adapted and load paths may have been changed. Temporary propping may therefore be needed.

> Archive drawings are a useful source of information, but they are not necessarily as-built drawings. Consequently, intrusive site investigations may be needed such as breaking out parts of the existing structure.

Designs for temporary works and intrusive site investigations take time which needs to be factored into the programme and into the project budget.

Where partial demolition is being planned – creating openings in existing brick or concrete walls, for instance – needling and propping may be required. This is a way of safely supporting the structure above whilst the opening is being created, with the loads being transferred to the floor through temporary supports. The work sequence includes:

1. Drill apertures in the existing structure above the opening to be created.
2. Insert horizontal steel or timber 'needles'.
3. Grout or wedge the needles so that they do not move.
4. Insert steel, timber or proprietary props under the needles and brace as necessary.
5. Remove the structure.
6. Insert beam or lintel.
7. Grout or wedge the beam/lintel and remove the temporary works.

Where loads cannot be transferred to the floor, reverse needling may be required where both top and bottom and needles are inserted with dead shores in between.

10.4.2 Façade Retention

Façade retention is used where a building façade is to be preserved but the building behind it is to be demolished or adapted. This often happens when the façade is listed or otherwise of architectural merit.

Once the building is demolished, or the structure behind the façade is removed, a means of preventing collapse of the façade is required. During the temporary unsupported condition, the façade would be propped until such time as it is tied in to the new building. Proprietary propping systems can be used, or alternatively, bespoke steel structures or scaffolding could be erected.

Temporary supports would be installed in sequence with the removal of existing structure. Such work needs to be carefully planned.

10.4.3 Trenches

Deep trenches – for drainage works, for instance – can be supported in various ways according to the ground conditions.

Where the ground is generally stable, drag boxes can be used. Drag boxes are basically moveable shields which are pulled along the excavation by an excavator as the work proceeds. They are heavy pieces of equipment, so the choice of excavator is important.

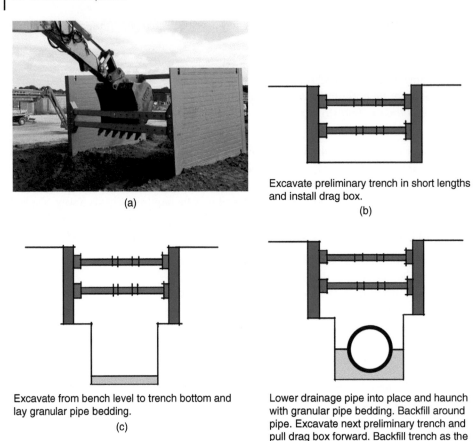

Figure 10.3 Drag boxes. Source: Drag box courtesy of Mabey Hire https://www.mabeyhire.co.uk.

Unlike trench boxes, drag boxes do not provide full-depth support to the ground and usually sit on a 'bench' on either side of a shallow trench excavation below, as illustrated in Figure 10.3.

Drag boxes comprise two rectangular panels separated by fixed-length steel struts. The rear struts are positioned such that the box can be 'dragged' past any pipes laid in the trench. Top boxes can be fitted in order to extend the working depth of the drag box.

The installation sequence is as follows:

1. The excavator digs the first section of trench to the required depth slightly wider than the outside dimensions of the drag box. The ground must be stable enough for the drag box to be inserted.
2. The drag box is lifted into place which protects the workforce in the trench. A shallow, narrower, trench of say 1.2 m depth could be excavated below the base of the drag box if required for drainage work.

3. Pipes and granular fill can then be laid.
4. The next section of trench is then excavated, and the drag box is 'dragged' into its new position by the excavator.
5. The first section of trench can then be backfilled if desired.

Trench boxes look like drag boxes but function differently. They have a cutting edge at the bottom of each panel and incorporate adjustable, not fixed, struts.

The work sequence is as follows:

1. A preliminary trench approximately 1m deep is excavated.
2. The trench box is positioned in the trench.
3. Excavation is then carried out inside the box.
4. The excavator then 'pushes' down on the corners of the box, and further excavation is carried out.
5. The process is repeated until the desired depth is reached.
6. Upper boxes can be added if needed for deep excavations up to approximately 6m.
7. The process is then repeated, and another trench box is installed.
8. For drainage works trench boxes will be installed for the full drain run, with similar but wider manhole boxes installed at each end of the run.
9. Pipework and manhole components are then installed.
10. Trench boxes are progressively removed, and the excavation backfilled.

Unlike drag boxes, trench boxes are not designed to be 'dragged'. Drag boxes and trench boxes are used where there are no utility services present.

Timber trench sheeting and individual steel trench sheets braced with adjustable props can be useful used as earthwork support in situations where excavations are being carried out around existing utility services.

In some cases – where the ground is unstable and is insufficiently self-supporting for drag boxes or trench boxes – there is no alternative to sheet piling. This is a very expensive option and could lead to contractual claims under certain conditions of contract. Installing sheet piles is also a slow process and can seriously impact the programme if needed.

10.4.4 Soil Engineering

A simple example of temporary works in earthworks is the use of sloping (battered) sides to excavations the purpose of which is to remove the ground reposing in the failure zone of the material. The degree of slope will vary according to the nature of the soil.

Such slopes must be engineered and not left to the excavator driver as this is considered a risk due to the potential for ground failure. A suitable slope stability analysis should be undertaken by a competent person.

A popular and cost-effective method of stabilising deep excavations is by using a combination of soil engineering techniques. For instance, the use of soil nails and shotcrete or sprayed concrete can achieve stable sloping or even vertical excavated surfaces. The excavated face must remain self-supporting until stabilised and so the technique is not suitable for all ground conditions.

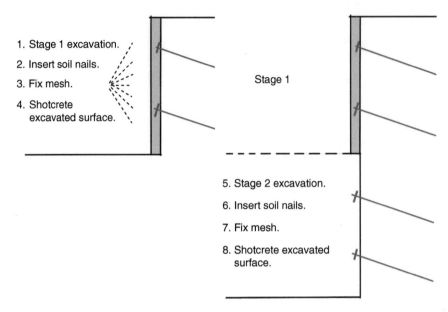

Figure 10.4 Soil nailing/shotcrete.

Figure 10.4 shows the sequence of work which requires the excavation to be carried out in stages, and each stage is stabilised before proceeding with excavation for the next stage until the required depth of excavation is achieved:

1. Excavate to the first soil nail level.
2. Insert soil nails in the ground and grout.
3. Drill for drainage pipes to relieve ground water pressure on the excavated face.
4. The exposed face of the excavation is then covered with a steel mesh.
5. Next, a specially designed concrete mix is sprayed to the required thickness to stabilise the face of the excavation.
6. Install soil nail head plate, nut and washers.
7. The process is repeated until the bottom of the excavation is reached.

Alternative methods for the stabilisation of excavations include the use of rock or ground anchors. In anchored earthworks support systems – where soil nails or ground anchors are used – stability of the system relies upon tensile resistance of the ground beyond the failure zone to provide support to the excavation. This needs an engineered solution.

Temporary works in water-bearing ground includes the use of wellpoint dewatering. Wellpoints may be installed at ground level where the water table is high and where there is the danger of running sand and/or the unintentional collapse of the excavation. Alternatively, reduced-level excavation may be carried out with battered/sloping sides with the wellpoints being installed at the reduced level.

Once the wellpoints have been installed, the second stage excavation can be undertaken safely on dry ground. A work sequence for the installation of a precast concrete box culvert in water-bearing ground is illustrated in Figure 10.5.

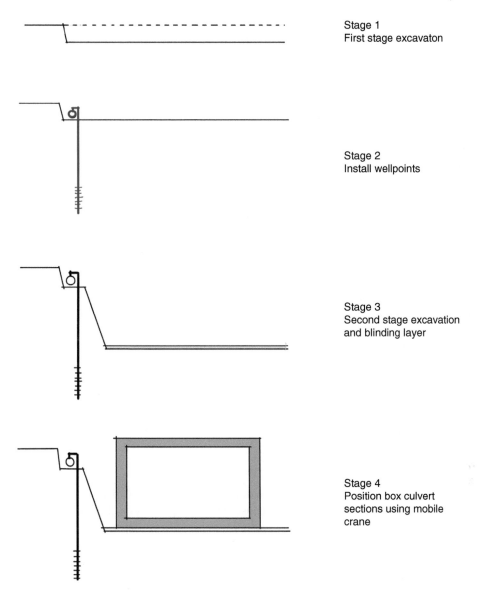

Figure 10.5 Box culvert.

10.4.5 Pile Caps

Pile caps are a type of foundation or base for a structure to be built on. They are constructed on, and are connected to, piles – they can be driven piles, bored or CFA piles, etc. Pile caps can be constructed on an individual pile or on clusters of piles to form the desired base which could be square, rectangular or triangular. Such bases may be separate or linked via a system of ground beams.

Pile caps are frequently constructed on site as the temporary base or foundation for a fixed tower crane. Tower crane bases are quite big – typically 7 or 8 m square on plan and 2 m or so deep – and will incorporate the tower crane legs which are cast into the base. The size and number of piles in a tower crane base will vary according to the ground conditions but may typically comprise 14 No 300 mm diameter CFA piles.

Pile caps are constructed after the piles have been inserted, and this follows a fairly standard procedure:

- Excavate to formation level to expose the pile(s).
- The excavation may be battered (sloping) or benched (stepped).
- The work area should be secured with temporary fencing with access to the excavation via access stairs with handrails or, for larger excavations, a ramp may be constructed.
- The piles are cropped to expose the rebar – normally using a hydraulic pile cropper attached to an excavator.
- The piles are integrity tested to ensure that there are no cracks or defects.
- The formation is then prepared to receive a layer of concrete blinding – usually 75 mm thick.
- The pile rebar is then cut to the desired level or bent according to the engineer's drawing.
- 50 mm rebar spacers are placed on the blinding and the reinforcing cage is either lifted into place or built up in situ.
- It is common practice to use a flexible proprietary formwork for the next stage – concreting the pile cap – such as Correx® polypropylene sheets which are rigid, lightweight and reusable. Alternatively, concrete can be poured into the excavation without formwork. This can save time on the programme and on time-related costs but is wasteful of concrete. Much depends on the cost–benefit of the over-break (over-excavation and disposal) and additional concrete versus time-saving.
- Where steelwork is to be erected on the pile cap, holding down bolts may be cast in during concreting using suspended formwork.

10.4.6 Piled Walls

Where piled walls are used as temporary ground support, they will invariably become part of the permanent works. Such walls may be constructed contiguously (with each pile close to or touching those either side) or using secant piles, where piles overlap to form a continuous 'cut-off' wall.

> Secant piles are often installed using a temporary concrete guide wall – see Chapter 11, Figure 11D. This enables the secant (interlocking) piles to be positioned accurately. Excavation of the shallow trench, erection and striking of formwork and concreting of the guide walls add activities to the schedule and, therefore, take more time.

Normally, the piled walls will be designed to incorporate a reinforced concrete capping beam that enables the piles to be propped whilst in their temporary condition. Props will remain in place until such time that the permanent restraint is in place. This could be a permanent ground floor slab, a basement slab or a tunnel roof slab (for an underground railway station, for instance). Piled walls may also be cantilevered where, for instance, a road is to be constructed in a cutting with buildings on either side and no working space is available.

The planner needs to recognise that there are limitations to horizontal propping of piled walls due to span (say max 50 m) or where there is no opposing capping beam or waling to prop from. Where this is the case, local propping will be needed which can impose limitations on the manoeuvrability of plant working inside the excavation.

10.4.7 Basements

Basements are complex structures and require substantial temporary works. This enables excavation works to proceed safely whilst supporting the adjacent ground.

This could be achieved using:

- Sheet piling.
- King post/soldier walls consisting of vertically embedded H-section steel with precast concrete slabs or timber baulks inserted between the webs of the steelwork.
- Contiguous or secant piling.

Such temporary works are not usually designed to be cantilevered and therefore need propping as the excavation works proceed. An important consideration in the use of propping to excavations is how to sequence the installation and subsequent removal of props.

This propping can be horizontal across the excavation using bespoke steel walings and props, proprietary tubular or hydraulic braces and shores or it may be propped using raking or angled braces locally placed around the earthwork support system.

Whichever propping system is used, this is heavily influential on both excavation methods and the choice of plant. Propping naturally reduces the headroom available and therefore smaller excavators are needed to be able to manoeuvre within the excavation. The same can be said of disposal of excavation arisings, as large dump trucks need headroom for loading. Should smaller site dumpers be needed, this will impact the excavation cycle times as load capacity is reduced, and the excavations will take longer.

In some basement excavations, mini excavators are used to load muck buckets attached to tower cranes for removing spoil. Further plant is thus required at the surface level to both load and carry away the spoil for the excavations.

10.4.8 Core Walls

Core walls are a specialist aspect of concrete forming which create the central shafts of medium-high-rise buildings. The core of such buildings accommodates staircases, lift shafts, mechanical and electrical risers, storerooms, washrooms, etc.

Central cores are usually constructed from a basement floor, ground slab or first-floor podium deck before the surrounding slabs, walls, and columns commence. In civil engineering, a tall hollow or solid core – perhaps for a bridge pier – may start from a pier foundation. There are two main approaches to constructing core walls:

- Traditional timber or proprietary formwork.
- Slip forming.

With traditional methods, formwork needs to be readily strippable by using collapsible forms or methods involving wedges or mitred internal forms. Also, because work takes place in a hollow shaft – which is hazardous for workers and for those below – crash decks are usually installed in the hollow core. This usually comprises scaffolding and scaffold boards that are dismantled and moved up the shaft as the work proceeds. Their purpose is to catch falling debris, tools, equipment, etc.

For really tall buildings, proprietary self-climbing lifts can be installed in the central core shaft to provide working platforms protected by a permanent crash deck above.[3]

Slip forming of central cores is very popular nowadays and enables concrete to be continuously poured into slowly moving formwork. This is not a new idea, having been invented in the early 1900s for constructing grain silos in the United States.

The success of slip forming is heavily dependent on the concrete mix which is normally pumped into place. Concrete must be sufficiently quick setting to provide support to the fresh concrete above but also workable enough to be compacted and allow the formwork to slip by without resistance.

As the formwork rises continuously to avoid horizontal joints in the walls, 24-hour working is required. This will usually mean two 12-hour shifts per day, round-the-clock site management and supervision requirements, uninterrupted and reliable concrete deliveries (usually ready mixed), standby plant (hydraulics, concrete pumps, etc.) and plant fitters in case of breakdowns.

The rate of slip forming varies according to the project but could typically be 300 mm (1 ft) per hour. This means that the core of a 10-storey building could be completed in approximately four days. However, several weeks may be required to initially set up and test the sliding formwork and hydraulic jacking systems and prepare the temporary 'box-outs' for forming openings in the walls.

The slip form has three platforms – upper, middle and lower – as illustrated in Figure 10.6. The middle working area is a confined and dangerous place of work. It is very noisy from the concrete vibrators and physically demanding, with the pump hose continuously moving and delivering concrete at high pressure.

Whether central cores are constructed traditionally or by slip forming, they usually require several openings for doorways, lift shaft doors, mechanical and electrical services, etc. These box-outs must be accommodated within the shuttering system and therefore add complexity to the process and add time to the construction programme.

10.4.9 Top-Down Construction

Apart from some demolition techniques, most building and civil engineering work is conventionally carried out by the bottom-up method. This is where the construction starts at the lowest level (substructure) and proceeds sequentially with superstructure construction.

In some instances – where there are site limitations, for geotechnical reasons or where construction time is limited, for instance – top-down methods might be considered. This method is construction in reverse order, where construction of the upper elements of a structure precede the lower elements. It is commonly used for basement construction which can enable the superstructure to proceed upwards at the same time as the basement is constructed downwards.

The upper platform is a general area used for storage and distribution of materials and small plant. There is space for the provision of basic welfare facilities.

The middle platform is set at concrete placing level and is the main working area.

The lower platform provides access to the fresh concrete surface for finishing works, especially around box-outs.

Figure 10.6 Slip forming.

Other uses of the top-down method include the construction of subterranean railway stations or bridge construction, where the deck is constructed first at ground level, followed by the abutment walls.

Figure 10.7 illustrates a further application of top-down construction where a canal aqueduct is to be constructed at ground level and a new road is to be subsequently constructed beneath. The sequence of work is as follows:

1. Install parallel secant pile walls both sides of proposed road carriageway.
2. Construct capping beam.
3. Erect falsework for aqueduct deck construction.
4. Complete deck and aqueduct walls and spill channels.
5. Carry out first stage excavation below aqueduct to reveal secant piles.
6. Install ground anchors to provide restraint to the piled foundation.
7. Complete second stage excavation to road formation.
8. Construct road.
9. Clean off and face up secant piles with engineering brickwork.

Figure 10.7 shows the installation of the secant piling, construction of the aqueduct deck and installation of the ground anchors following first stage excavation.

An application of top-down construction for simple multi-storey buildings (such as car parks and hotels) is lift slab construction. This briefly consists of casting the roof and floor slabs at ground level, separated by a bond-breaking membrane, and then jacking each one up on pre-installed columns using lifting collars cast into the concrete slabs. Completion of the building can then proceed top-down.

Installation of secant piling. Steel casings provide earthwork support prior to concreting using tremie pipe.

Falsework, formwork and rebar for construction of aqueduct deck and walls.

First stage excavation reveals secant piled abutment wall, capping beam and bridge bearings.

Installation of ground anchors prior to second stage excavation to road formation level.

Figure 10.7   Secant piling/top-down.

10.4.10   Top-down, Bottom-up Construction

The conventional method of constructing basements to buildings is to excavate down to form the basement foundations and slab – using sheet-piling or secant piles to support the ground – and then to build upwards to ground level. The superstructure of the building then follows.

This method of working can be very time-consuming and so, where appropriate, building down to construct the basement whilst concurrently building the superstructure upwards can considerably shorten the overall programme of works. The top-down method of constructing basements is illustrated in Figure 10.8.

10.4 Work Sequences

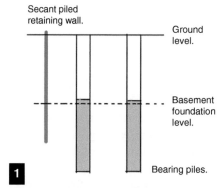

1 Construct retaining wall/cofferdam. Install bearing piles and concrete to underside of basement slab level. Partially withdraw pile casings and leave in place to retain soil prior to completion of plunge piles.

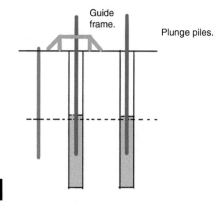

2 Erect guide frame and insert plunge piles. Fill void with gravel and withdraw pile casings.

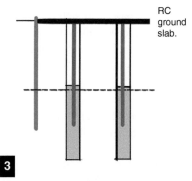

3 Cast reinforced concrete ground slab. Form opening in slab for access to basement.

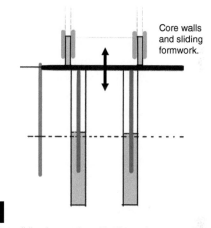

4 Set up sliding formwork and begin casting core walls. Crane in plant to excavate basement and remove spoil. Cast reinforced concrete basement foundation slab. Construct core walls bottom-up using conventional formwork.

Figure 10.8 Top-down construction.

Top-down construction using plunge columns was used for constructing the 4-storey basement of the iconic London Shard (Channel 4TV, 2013), which was completed in 2012. It was the first project in the world to use this technique for constructing a sky-scraper – a 306 m (1016 ft) tall glass-clad tower – which was the tallest building in Western Europe and one the tallest in the world.

Using this technique reduced the construction period by three months.

For the top-down, bottom-up method to work, the superstructure needs a foundation to sit on and, in the absence of the completed basement, this can be provided by:

- Pile columns or
- Plunge columns.

A pile column is a conventional pile designed to act as both a pile foundation and a column. The piles are installed from ground level, extend beyond the lowest basement level and are reinforced to act as a column within the basement once excavation is completed. As the piles extend from bearing level to ground level, this enables the ground slab to be constructed and the superstructure to commence. At the same time, basement construction can begin.

Plunge columns, by contrast, are constructed within the hole created by the piling operation. Piles are drilled from ground, concreted and the rebar cage inserted to the level required to support the eventual basement foundation slab. Whilst the concrete is still wet, steel H-columns or precast concrete columns are lowered into the concrete through the empty bore using a guide frame and are held in place until the concrete is set. The pile hole is then filled with gravel to restrain the column and keep it plumb.

Once all plunge columns are in place, a ground slab or sub-basement slab can be constructed, providing a foundation for superstructure work to proceed. Often, this will be a concrete core constructed using slip forming. Any core walls required below this level would be cast using conventional formwork. As the superstructure work advances, basement construction can proceed concurrently.

Whether using pile columns or plunge columns, piling work is preceded by construction of a basement cofferdam – a vertical structure to retain the ground around the basement. This could be constructed using contiguous or secant piles, diaphragm walling or steel sheet piling.

There are downsides to the pile column method due to the slacker tolerances for piling work and the need to clean and face the rough surface of the piles within the completed basement. Using plunge columns, on the other hand, requires more design effort and the columns may have to be oversized due to their slenderness ratio.

Top-down construction can add direct cost, cause excessive demands on resources and can add logistical and management complexity. Basement construction is also more difficult and slower than the bottom-up method. However, time savings and reduced time-related costs can be significant. The major advantage of the method is concurrent working on the basement and superstructure which is where the time savings come from.

10.4.11 Offline Construction

Offline construction techniques are widely used in civil engineering projects:

- Full or partial highway upgrades.
- Highway bridge replacement.
- New roads under live railways.
- New roads or railways over existing roads.

This method has the benefit of causing the least amount of disruption to existing road or rail services and means that the majority of construction work can be undertaken safely away from live traffic. Various techniques are used including:

- Constructing highways alongside existing roads and completing tie-ins once the new road is ready for use.
- Preparing structural foundations and abutments alongside live railtrack or road carriageways ready to take the new structure once it has been constructed/prefabricated.
- Constructing structures on land adjacent to the site.
- Sliding structures into place using proprietary hydraulic jacking systems.
- Lifting and transporting deck structures onto prepared bearings using SPMTs – this is illustrated in detail in Chapter 20.

10.5 Sequence Study

The use of sequence studies is planning at a fine level of detail and can be very useful when considering alternative methods of construction for a particular operation. A sequence study provides a means of determining the time and cost implications of the choices available. For instance:

- In situ concrete ground-bearing floor slabs can be laid in alternate 'chequer-board' bays, in narrow long-strip alternate bays or in huge single pours utilising laser screed technology.
- Multi-storey concrete-framed buildings can be constructed using traditional formwork panels for the central core walls and slabs. Alternatively, the upper floors could be constructed using in situ concrete on metal decking as permanent formwork, with the central core being slip-formed.

In the following example, the upper floor slabs of an eight-storey in situ concrete-framed building are to be constructed using table forms. Figure 10.9 shows the plan and section of the building, and Figure 10.10 illustrates the construction process:

- The building is of flat slab construction and contains 24 columns per floor.
- A fixed tower crane is to be used for lifting all formwork and rebar and for the column concrete pours.
- The larger concrete floor pours are to be pumped.
- Table forms are to be used for the floor slab construction with traditional formwork used for the columns.
- The central core of the building is to be constructed by slip forming.

10.5.1 Sequence Diagram

Formwork resources are critical to the success of the project.

For the columns, sufficient formwork is to be available for 12 columns. As there are 24 columns per floor this means that the forms will be used twice on each floor level. Once the fifth floor has been completed – i.e. after 10 uses of the forms – it is anticipated that wear

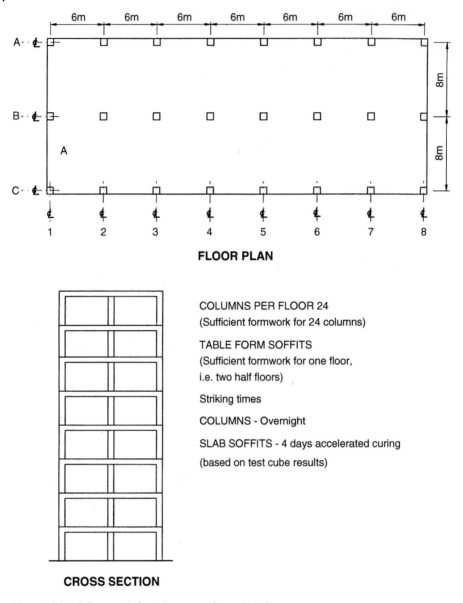

Figure 10.9 8-Storey reinforced concrete frame building.

and tear on the forms will require a degree of remaking which will add some time to the programme.

Sufficient table forms are available for two complete half-floor slabs. As shown in Figure 10.11, however, one set of table forms will be fixed, the slab poured, and the next level of column formwork erected before the table forms to the other half of the floor are fixed in position. Prior to completely removing the table forms following concreting,

Positioning table forms follows on behind slip forming of central core walls.

Table forms to 5th floor are ahead of table forms around the slip formed central core.

Fixing rebar on table form deck and tying in to starter bars cast into slip formed core wall.

Reinforced concrete columns using traditional timber formwork and props.

Figure 10.10 Table forms.

the floor slabs will require propping. The table forms will be moved floor by floor up the building by tower crane as each floor slab is completed.

The sequence diagram shown in Figure 10.11 is for the column and floor construction sequence. The movement of the column formwork and table forms for each floor is also shown.

Formwork striking times are of critical importance to the construction sequence. Vertical column sides can be stripped overnight – horizontal slab soffits, however, must be left in position for a minimum period of seven days. Early stripping times may be achieved by proving high early cube strengths at say four days. A four-day stripping time is therefore to be allowed in the programme.

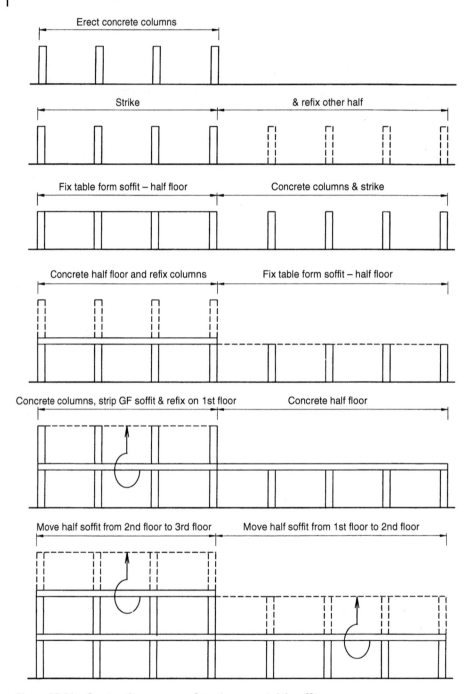

Figure 10.11 Construction sequence for columns and slab soffit.

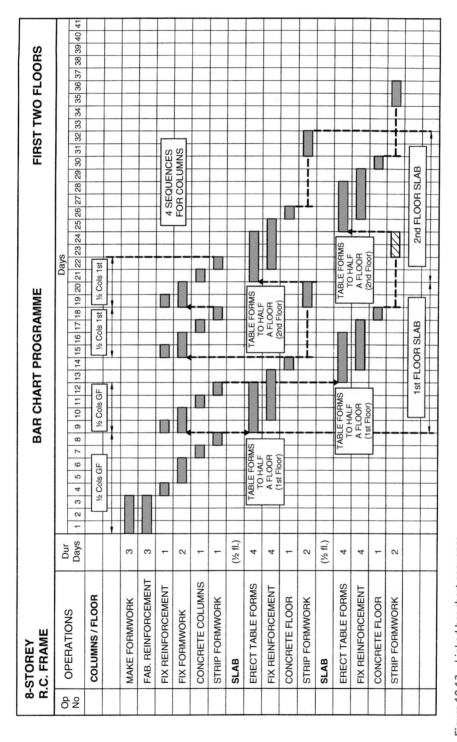

Figure 10.12 Linked bar chart sequence.

10.5.2 Construction Cycles Times

Each floor of the building is to be constructed in two cycles comprising floor slab and 12 No columns.

The cycle times per half floor are:

Operation	Duration
Floor slab construction (durations per half a floor)	
Fix table forms in position	4 days
Fabricate and fix rebar	4 days
Concrete floor slab (curing time = 4 days)	1 day
Strike floor soffit	2 days
Columns (durations per 12 No columns)	
Fabricate column reinforcement	3 days
Fix column rebar	1 day
Fix formwork to columns	2 days
Concrete columns	1 day
Strike formwork to columns	1 day

Figure 10.12 shows a linked bar chart sequence for the construction of two floors of the building. It should be noted that lead times of 3 days, respectively, are allowed for making the column formwork and for fabricating the column rebar cages.

References

Lowsley, S. and Linnett, C. (2006) *About Time – Delay Analysis in Construction*. RICS Books.
Channel 4TV (2013) *The Tallest Tower – Building the Shard*. DCD Publishing.

Notes

1 The Final Report of the Advisory Committee on Falsework (HMSO).
2 https://www.twforum.org.uk.
3 Schindler Climb Lift – https://www.scaffoldmag.com/news/upgraded-schindler-climb-lift-goes-to-work-on-marquee-high-rise-projects/8023520.article.

Chapter 11 Dashboard	
Key Message	o Method statements are widely regarded as an essential component of construction best practice.
Definitions	o Method statements are formal, written documents. o They provide a written explanation of proposed methods of working.
Headings	**Chapter Summary**
11.1 Introduction	o Method statements are commonly used and play a major role in construction planning. o They help to ensure that construction work is carried out safely. o The primary purpose of method statements is to facilitate the planning, monitoring and control of site operations.
11.2 Format of Method Statements	o There are no rules that govern the style, format or content of method statements and there is no standard way of writing them. o There are two major method statements formats – tabular and written.

11.3 Types of Method Statement	o There are three broad categories of method statements commonly used in the industry:
	  o Tender method statements.
	  o Construction or work method statements.
	  o Safety method statements and lift plans.
	o Tender method statements play and important role in the contractor's pricing of preliminaries or general cost items.
	o Construction method statements are frequently referred to as 'work' method statements because they relate solely to specific work activities and to the resources needed to carry them out.
	o At the pre-construction stage a method statement is a great way to formalise a safe system of work.

11.4 Preparing Method Statements	o Method statements can be written on a single page, or they may extend to several pages for a complex operation or activity. o The general layout of method statements is defined by the headings which subdivide the document. o The first consideration is the type of method statement required and who it is for. o The structure of a method statement should ideally mirror the programme to which it relates thus providing a useful level of detail behind each activity on the programme. o Any chosen method of construction cannot be considered in isolation from safety because it is illegal to use a system of work that is not safe. o Whilst safety method statements are not required by law, it is common practice for contractors to prepare them and, together with subcontractors' risk assessments and method statements, they form part of the site safety management system.
11.5 Worked Example	o A city-centre brownfield site is to be re-developed with two 8-storey apartment blocks and a basement car park beneath, accessed by a ramp. o Tender stage, construction stage and safety method statements are developed in detail.
Learning Outcomes	o Understand the purpose of method statements and how they are used in construction. o Understand the types and structure of method statements. o Appreciate the link between work method statements and safe systems of work. o Be able to prepare a simple method statement by following a worked example.
Learn More	o Safety method statements are explained in Chapter 12. o Understanding construction sequences is essential for developing method statements – see Chapter 10.

11

Method Statements

11.1 Introduction

Method statements are commonly used in the construction industry and play a major role in construction planning and in ensuring that construction work is carried out safely. They are formal, written documents and, whilst they are used mainly during the construction stages of a project, method statements may be employed at other times too.

The primary purpose of method statements is to facilitate the planning, monitoring and control of site operations, thereby avoiding confusion and possible accidents. To be successful, however, planned working methods must be conveyed to everyone involved such that they understand what is meant to happen when work commences. Should a planned method need to be changed – a better way of working may have been found or different plant or equipment might be preferred – then the formal document must be amended and communicated to operatives and other site personnel.

Method statements are usually initiated by contractors, subcontractors or specialist contractors prior to work operations but they can be used by clients and consultants especially during the procurement stages of a project. Method statements are widely regarded as an essential component of construction best practice.

11.1.1 Definition

Method statements provide a written explanation of proposed methods of working. This might be for part of the permanent works, such as cladding or roofing or it could be for any temporary works needed to facilitate construction operations, such as the erection of scaffolding or mast climbers.

There is no need for a method statement for every site activity and, on some sites, there may be no method statements at all. On very small projects, for instance, individual trades people or small contractors may instinctively plan their work because they are experienced and have probably done the same tasks many times in the past. Plumbers and electricians, wet trades such as plastering and painting, kitchen fitters and bathroom installers are obvious examples. They do not usually formalise their risk assessments, let alone their programme or working methods although, strictly speaking, they should.

However, where there is any degree of novelty or complexity in the proposed work, planning 'in the head' is simply not good enough. A simple bar-chart programme would be a good starting point, but this is not sufficient to fulfil the statutory duty to *apply the principles of prevention*[1] to the planning process. This is because a programme depicts the 'when' but not the 'how', which is crucial to carrying out the work safely.

Method statements provide a means of formalising the 'how' bit.

11.1.2 Uses of Method Statements

Method statements come in all shapes and sizes. They can contain various levels of detail and can be aimed at different audiences. They are used extensively throughout construction projects for a wide variety of purposes:

- At tender stage to assist contractor selection or during bid preparation.
- At construction stage to support the master programme or to assist short-term planning and safety management on site.
- During the final cleaning and handover stages of projects.

During the pre-construction stages, method statements can be used to give a broad understanding of how a project might be carried out or to identify and evaluate alternative methods of construction. Method statements may also be requested by clients as part of their preferred contractor selection process.

> At contract stage method statements can be used to validate and add further detail to the contractor's programme or they can be used in conjunction with risk assessments to plan safe systems of working and to discuss the intended construction process during task-talks with the workforce.

In the NEC ECC form of contract, contractors are required to integrate details of their intended method of working with their programme. Thus, any programme submitted for acceptance must include a statement of how the contractor plans to carry out each operation on the programme and must identify the main equipment and other resources to be used. Consequently, should the stated method of working need to be changed – perhaps due to ground conditions, a design change, weather conditions or an unforeseen event – this triggers a contractual requirement for the contractor to submit a revised programme.

There are several types, formats and styles of method statements that can be used according to the circumstances or stage of a project; they may be employed by a variety of project participants including clients, consultants, main contractors, subcontractors and specialist contractors.

11.1.3 Clients

Method statements are often used as part of the contractor selection process, especially in two-stage tendering, and will form part of the client's quality assessment of its preferred bidder.

Prior to appointing a contractor, clients may ask for a method statement to appreciate how the contractor intends to undertake the work and to assess whether this raises any issues with regard to:

- Organisation and site management.
- Communications and applications for payment, extensions of time, claims, etc.

- Site access and circulation.
- Use of cranes and their impact on neighbouring properties.
- Neighbour issues where there are shops, schools or housing nearby or where the project raises local sensitivities.
- Environmental concerns such as recycling or removal of waste materials, noise, dust, wagon movements, etc.

Clients may also ask to see details of the contractor's proposed methods of working to supplement the programme. This might consist of notes on the programme indicating the plant and resourcing of each activity or the contractor might produce an additional document explaining planned working methods in more detail.

11.1.4 Consultants

A variety of consultants – including architects, engineers, quantity surveyors and health and safety advisers – are invariably employed by clients and by contractors to provide professional services and advice.

Clients, for instance, may require help to interpret method statements submitted by tendering contractors or to judge the contractor's intended methods of working in conjunction with the submitted programme.

Contractors and subcontractors may need specialist guidance as well. For design and build contracts – or where there is partial contractor design – outside consultants may be required. If so, it is important that they work closely with the contractor's planning or construction department to provide a design which is compatible with the contractor's preferred working methods.

Where contractors and subcontractors have no safety department or in-house health and safety adviser, health and safety consultants can be engaged to provide many services, including the preparation of risk assessments and methods statements (RAMS). They will usually have standard documents which can be adapted to produce method statements for specific activities on a specific site. They may even have software packages which can adapt generic information to produce site-specific method statements for their clients.

11.1.5 Main Contractors

Main contractors use method statements throughout their projects as they are an essential part of the planning process. Method statements are used during the tender/procurement stage, pre-construction, during the works and during the final cleaning, handover and defects correction period.

As such, method statements take many forms:

- Minutes of formal meetings.
- Estimator's pricing notes.
- As part of a contractor's pre-qualification or first-stage tender submission.
- As a PowerPoint presentation.
- As simple strategic statements of intended methods of working.
- As notes on a programme or
- As detailed documents included in the Construction Design and Management Regulations (CDM) Construction Phase Plan.

Consequently, there are several uses for method statements in a contracting firm:

- As a means of identifying and evaluating appropriate working methods and resource requirements at tender stage.
- To explain to clients how they intend to organise and construct the works.
- To underpin the contractor's master programme and to plan site activities in detail.
- To plan safe methods of working in response to risk assessments.
- As a means of communication with the workforce to explain how the work is to be carried out, to take on board their views and experiences and to control activities on site when work commences.

A major element of the contractor's tender price is the general cost items or preliminaries coupled with the fixed and time-related costs of plant and temporary works. It is little wonder, therefore, that the estimator or estimating team will devote a great deal of attention to working out the best way of carrying out the works so that the eventual price and programme accurately represent the intended method of working.

Once the contract has been awarded, the construction team will develop the tender strategy in more and more detail as the work proceeds.

11.1.6 Subcontractors

Subcontractors are major contributors to the main contractor's methods of working as it is they who carry out most of the work on site. There may, however, be several subcontractors on site at any one time and they will all have their own way of working, depending on their trade, the composition of their workforce and the type of plant they use.

Consequently, subcontractors' method statements will have to be studied carefully by the main contractor to avoid clashes with other trades, and pre-contract discussions will be held to iron out the detail.

Subcontractors' resources and methods of working may also have to be tailored to suit the contractor's master programme to ensure a smooth and efficient workflow on site. Subcontractors' method statements will:

- Indicate how they intend to carry out their part of the works.
- Identify attendances to be provided by the main contractor (such as scaffolding, use of cranes, etc.).
- Be used to instruct their workforce and make sure that they understand how they may be impacted by other activities on site.

11.1.7 Specialist Contractors

Specialist contractors are often engaged by main contractors to provide expert skills or special plant, such as access systems, cranes, temporary works, abseiling, robotic equipment, etc.

The work of specialist contractors will invariably involve out of the ordinary expertise and risk control measures that need to be documented in a detailed method statement. This could be done in another document – such as a lift plan – where a method statement might be included alongside other details appertaining to their work.

The function of this documentation is to formalise and communicate the planning, monitoring and control of the installation and operation of specialist work such as scaffolding, common towers, mast climbers, hoists, etc. or provide a lift plan to ensure that lifting operations involving cranes are conducted safely.

11.2 Format of Method Statements

There are no rules that govern the style, format or content of method statements and there is no standard way of writing them. They can be simple documents but, such is the complexity of construction work, they can be lengthy and very detailed. Therefore, a method statement might be:

- A strategic document describing how a contractor aims to go about an entire project
or
- It might be a detailed step-by-step explanation of how an individual site operation is to be carried out.

There are two major method statements formats – **tabular and written**. Whichever format is chosen, method statements are an essential aid to conducting activities on site in accordance with a well thought-out and agreed procedure and, importantly, they need to be easily communicated at site level.

> Method statements should not be too long, or overly complex, but empirical evidence suggests that quite the opposite is true in practice. Many method statements in the industry are far too lengthy and contain extensive, and often generic, information such that the real purpose of the method statement is lost.

Internet downloads provide a plentiful source of method statement templates. They tend to favour the tabular format but there is no right or wrong choice and there may be circumstances where both tabular and written methods are adopted to good advantage.

Some contractors prefer to adopt a standard company-wide method statement format. This can help benchmarking across sites and provide a measure of safety performance when undertaking regular health and safety audits. This can help contractors to satisfy the statutory duty to *monitor construction work carried out*[2] on a consistent basis.

11.2.1 Tabular

The tabular format of method statements is probably simpler and easier to read but, as with all standard forms, there is a space restriction on the content.

Limitations of space can be a plus, however, as this encourages the writer to be succinct and to choose words carefully. Overly verbose method statements should be avoided as they can be difficult to communicate to the workforce. Succinct messages are more likely to be understood and retained.

11.2.2 Written or Prose

Using a written or prose format (rather like a mini report) gets over the difficulties of the tabular format. On the proviso that they are well laid out, with a judicious choice of headings, such method statements can be both useful and comprehensive.

The written format can be particularly valuable where the construction work intended is complex. In such cases a tabular format might prove to be too restrictive and could inhibit the writer's ability to convey the intended method clearly and comprehensively.

11.2.3 Generic

Generic documents are commonly used in the construction industry – specifications, risk assessments and safety method statements being prime examples.

Such documents give the impression that they have been originated for a specific project, but it is usually the case that they have been adapted from a pre-existing document. This might be a document from a previous project or from a standard document produced by a consultant or downloaded from the Internet.

From a practical point of view, generic documents can save a great deal of time and effort, but the quality of the end product depends upon the skill of the editor, and this is invariably not the best.

It is very easy to fall into the trap of thinking that a method statement is just another administrative chore, but this can lead to lazy habits. Simply copying another document and changing the company logo is not good practice and, in the case of risk assessments or safety method statements, would, according to the Health and Safety Executive (HSE), not satisfy the law. Failure to think about site-specific hazards and controls could put the lives of construction operatives at risk.

Whilst there is a place for generic documents for non-site-specific activities – such as handling cement, using abrasive discs, etc. – generic documents can never replace the thoroughness and specificity of a bespoke document.

Method statement templates, on the other hand, can be quite useful as they provide a style and a set of headings that can be a starting point for ideas. A template could then be adapted to a specific situation or business but, essentially, the contents of the method statement would be project specific.

This is crucial because the idea is to devise a safe and practical method of working for a specific task on a specific site. It is far better, therefore, to devise a method statement format and contents that suit the working methods of a specific trade or company than to blindly copy someone else's ideas.

11.3 Types of Method Statement

With method statements, there is no 'norm' or standard practice, and different contractors have their own preferences. However, there are three broad categories of method statements commonly used in the industry:

- Tender method statements.
- Construction or work method statements.
- Safety method statements and lift plans.

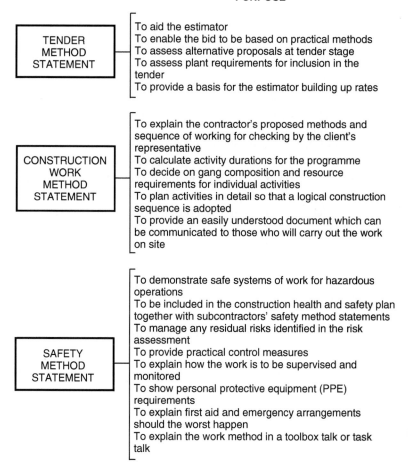

Figure 11.1 Overview of method statement formats.

Which of these might be used in practice depends very much on the circumstances. Figure 11.1 summarises the purpose of each type of method statement and indicates the relationship between them.

11.3.1 Tender Method Statement

Decisions relating to the choice of construction method are an integral part of the tendering process. These decisions directly influence the rates included in the estimator's price build-up and consequently impact the final tender sum. In some cases, the tender documents may require tenderers to submit method statements as part of their bid or this might be requested at the negotiation stage of a project.

Tender method statements will not be at a detailed level but are intended to convey a broad understanding of how the contractor intends to go about the project. They may also be prepared for individual key activities, for high-risk aspects of the project or to identify key plant and temporary works requirements that will impact the tender price.

For a relatively straightforward project, there may be no formal pretender method statement and the estimator's pricing notes may suffice. In this case, individual priced items in the tender document – bills of quantities, activity schedule or schedule of works, etc. – together with the preliminaries items, will be sufficient to convey the estimator's thinking about how the work might be carried out. This can then be explained at the tender adjudication meeting, where directors will take a view as to whether the proposed methods of working are appropriate.

Where a project is more complex, the contractor might invest considerable effort into the tender method statement, and a formal document may be prepared. Estimators and planners often work together at tender stage to decide on methods of construction which can then be priced and included in the pretender programme. This collaboration is especially important in competitive tendering, where a faster or more efficient method of construction might give a contractor a competitive edge.

Tender method statements play an important role in the contractor's pricing of preliminaries or general cost items and typically focus on:

- Site security, layout and organisation.
- Site management and staffing.
- Key/high-value operations or temporary works.
- Strategic issues including cranes, access equipment, materials handling and on-site/off-site concrete batching, etc.

When considering different methods of construction at tender stage, a programme and price might be produced for each alternative to provide the contractor with comparative costs and timings.

11.3.2 Construction or Work Method Statement

Construction method statements are frequently referred to as 'work' method statements because they relate solely to specific work activities and to the resources needed to carry them out.

A contractor's work method statement will not necessarily follow the method statement prepared at tender stage as the construction team may have a preferred method of working or, on reflection, may think that there is a better or more efficient method of carrying out the works.

This commonly happens in practice because the rates and prices included in the tender bid are simply 'allowances' that establish budgets for the construction team to work with. Pre-construction changes in method may be justified in terms of increased efficiency and/or a shorter programme duration on the proviso, of course, that the profitability of the contract is not jeopardised.

Prior to commencing work on site, a method statement provides a means of adding detail to the contractor's programme and helps to identify the plant, temporary works requirements and resource requirements for specific programme activities. They focus on gang sizes and outputs of plant and labour and, as such, help to validate the time allowances on the contractor's programme.

A work method statement is not required for every activity on the programme. They tend to be prepared for resource-intensive activities or for high value, high-risk or complex operations where the working method needs to be carefully considered and well thought out beforehand. This includes subcontract and specialist work such as complex lifting operations, steelwork erection, roofing and so on.

Work method statements can be used to discuss proposed methods with the workforce as part of task talks on site. In this case, however, it is usual to include health and safety provisions in the method statement to ensure safe working methods are understood and implemented.

11.3.3 Safety Method Statement

For any construction project, there is a statutory duty – under regulations 13(1) and 14(2) of CDM 2015 – for the contractor to plan the timing and sequence of the work and to estimate how long activities or stages of work will take. This duty applies no matter how large or small the project is.

There is no statutory duty to formalise the decisions made, but it makes sense to do so if only to prove that the contractor has thought about the health and safety aspects of the planned work beforehand and to provide a means of communicating this thinking to workers and other contractors on site.

At the pre-construction stage a method statement is a great way to formalise a safe system of work as well as to identify the resources needed to carry out specific activities on the programme in a safe manner. They are often referred to as safety method statements but, unlike risk assessments, are not a legal requirement.

The emphasis in a safety method statement is, of course, on the measures required to carry out the activity in a safe manner in the light of a risk assessment. These measures are commonly called 'control measures' because they are the provisions required to control residual risk following the process of hazard identification and initial risk reduction.

Safety method statements place more emphasis on safety and less on the mechanics of the work activity and so they are sometimes combined with work method statements in a single document. This makes sense because safety is an integral factor in the planning process, and considerations of construction methods can never be divorced from planning a safe working environment. A disadvantage of doing this is that the combined method statement may become too complicated or key messages may be lost in a sea of paperwork.

11.3.4 Lift Plans

Lifting operations are everyday events in the construction industry that involve the movement of materials, equipment and people using a wide variety of excavators, telescopic handlers, cranes and hoists. Such equipment – when used for lifting operations – must comply with the provisions of the Lifting Operations and Lifting Equipment Regulations (LOLER) 1998 which defines them as *operations concerned with the lifting or lowering of a load*.

Under LOLER 1998, regulation 8, employers have a duty to ensure that every lifting operation involving lifting equipment is properly planned by a competent person, appropriately supervised and carried out in a safe manner.

> It is common practice to refer to the competent person as the *appointed person* following BS7121 Code of Practice for Safe Use of Cranes. The appointed person could be the lifting equipment operator, a supervisor or someone else who is qualified to plan and supervise lifting operations.

The level of competence expected of the appointed person depends upon the nature of the lifting operation. This could be a suitably trained and experienced slinger or banksman for simple lifts, such as lifting drainage pipes using an excavator but, for very complex operations, specially trained expert personnel might be required. This is especially the case for crane lifts such as tandem lifts using two cranes. Cranes can easily become unstable if overloaded or if they are operating on unstable ground.

The key to planning any lifting operation, and to choosing the right equipment and method of working, is risk assessment. For simple, everyday lifting operations the HSE suggests that trained and competent people may be able to manage risk adequately by carrying out minimal on-the-job planning. However, with more complex lifting operations, a more sophisticated and detailed approach may be required to plan, monitor and supervise the work.[3] In this case, the level of resources and extent of expert input needs to be commensurate with the task in hand.

Major projects – such as a multi-£billion power station or high-speed rail link – can demand many hundreds of lifting operations comprising large numbers of relatively small lifts and several large, heavy or complex operations. Projects of this nature are frequently organised using framework agreements, and Tier 1 or Tier 2 contractors may be able to engage the same specialist contractor from the framework of approved suppliers for pan-site lifting requirements.

On more modest projects it will be either the main contractor or subcontractors who arrange lifting operations. Such operations might be relatively straightforward – such as installing ready-mixed mortar silos for bricklaying or lifting timber roof trusses on housing projects – or they might be complex such as the tandem lifting of complex structural steel roof trusses on a major sports stadium project.

In any event, it is likely that the cranes will be hired, and contractors will have a choice to make:

- Hire a crane and qualified crane operator only with the lift itself to be planned and supervised by the contractor using in-house staff.
- Arrange a contract lift which will be fully managed by the crane hire company.

In the United Kingdom, cranes are normally hired under a Contractor's Plant Association (CPA) Crane Hire contract, but this will only become a CPA Contract Lift Hire contract when the appointed person is provided by the crane owner who then accepts responsibility for the lift in its entirety.

With any lifting operation, the key to a successful and safe outcome is an experienced and competent lift team. For heavy or complex lifts, the operation will have to be designed in detail and all the arrangements will be fully documented in a Lift Plan. The Lift Plan will include:

- Site and pre-lift information.
- Site and crane layout plan.

- Drawings and other information.
- Access, signalling and communications arrangements.
- Crane details, capacities and temporary works requirements (e.g. crane mat).
- Design of the lift and slinging methods.
- Risk assessment
  - **Method statement**
  - Safe system of work.
- Lift team details, including the appointed person.

11.4 Preparing Method Statements

The preparation of method statements needs to be taken seriously because they are open to scrutiny:

- A visiting HSE inspector will certainly seek out method statements for high-risk activities or where an unsafe system of working is spotted.
- The contractor may be required to submit a method statement for scrutiny by the client's representative prior to commencing a major operation.
- The contractor's own safety adviser may also wish to check specific method statements as part of a regular site safety audit.

Method statements can be written on a single page, or they may extend to several pages for a complex operation or activity. They can be hand-written, or word processed from a database or generated from a bespoke software package – such as Risk Assessor, HandsHQ or HBXL Building Software.

On any meaningful construction project, the contractor's CDM Construction Phase Plan will often extend to several lever-arch files. Much of this paperwork will be RAMS, mostly prepared by subcontractors and specialist contractors, who are routinely required to submit them for scrutiny prior to commencing work on site. They will vary greatly in style and content, but crucial thing is that they are clearly written and can be readily understood.

Writing good method statements is challenging because they are employed at various stages of the planning process and can be aimed at clients, consultants, main contractors, subcontractors and operatives. The level of detail required will not be the same for all recipients.

11.4.1 Contents

The contents of method statements vary considerably according to their intended use and target audience. However, they are not simply a list of construction operations with notes written alongside.

Method statements often include drawings, sketches and photographs as well as excerpts from the programme. Related documents may also be referenced within a method statement such as product data sheets, technical details of cranes and other equipment and Control of Substances Hazardous to Health (COSHH) assessments for certain materials. Method statements may also be a subset of another document such as a Lift Plan or RAMS.

In any event, method statements require some thought, especially regarding the order and sequence of the work and how it can be carried out in a safe manner. It is also vital that the person or team preparing a method statement is competent to do so and fully understands the activity or task at hand, together with any associated hazards, both generic and site-specific.

Everyone involved in preparing a method statement will have their personal interpretation of what it should contain, but the main thing is that the job is thought through carefully and a safe and efficient method of working is devised.

11.4.2 Layout

The general layout of method statements is defined by the headings which subdivide the document. This is true irrespective of whether the method statement is in tabular or written/prose format.

In order to determine appropriate headings, the first consideration is the type of method statement required and who it is for:

- Tender method statement for the client/consultants:
  - A written document as part of a formal submission.
  - A PowerPoint presentation at a pre-qualification or pre-selection meeting.
- For the contractor's tender adjudication panel:
  - A condensed extract from the estimator's pricing notes.
  - A summary of the key aspects of the project such as:
    - Site management and accommodation.
    - Key plant, equipment and temporary works.
    - Major elements of work.
    - High-risk activities.
- For the contractor's tender handover meeting at contract award stage:
  - Description of the project.
  - Approach taken as basis for tender.
  - Plant and temporary works allowances in the tender.
  - Key subcontract packages.
- Work method statement for the use of site management:
  - Description of activity.
  - Sequence of work.
  - Plant and equipment.
  - Waste management.
  - Supervision and management.
- Safety method statement:
  - The task in hand.
  - Generic and site-specific hazards.
  - Risk assessment(s).
  - Control measures resulting from risk assessment(s).
  - Residual risks and PPE.
  - Supervision and emergency procedures.

- Combined work/safety method statement:
  - For the use of site management personnel.
  - For briefing operatives or construction teams as part of a task talk.
  - As part of the CDM Construction Phase Plan:
    - Description of activity and sequence of work.
    - Plant and equipment access provisions.
    - Hazards and control measures.
    - Residual risks and PPE.
    - Supervision.
    - First aid arrangements and emergency procedures.

11.4.3 Structure

The structure of a method statement may be conditioned by the procurement method chosen for the project and especially the type of pricing document employed.

A bill of quantities might be classified by work sections or perhaps design elements or there might be an activity schedule, schedule of works or a schedule of rates. These documents will describe the work in different ways which will inform the structure and contents of the method statement.

The contractor's programme will usually relate to the pricing document and will be structured on the basis of a series of operations or activities. The type of operations or activities on a programme vary according to the nature of the project, the procurement arrangements and the preferences of the contractor or client.

These activities are normally linked by some form of logic which gives meaning to the interrelationships between the component parts of the programme. Consequently, a programme might relate to **specific operations** on site, to **elements** of the building or to **stages of work**:

- **Specific operations**
  - Excavation works to foundations.
  - Demolition works.
  - Laying precast concrete floor units.
- **Elements**
  - Substructure work.
  - Superstructure frame.
  - External envelope.
- **Stages of work**
  - Groundworks to damp proof course level.
  - Brickwork to first floor level.
  - Brickwork to eaves level.
  - Site drainage and manholes.

The structure of a method statement should ideally mirror the programme to which it relates thus providing a useful level of detail behind each activity on the programme. It also helps site managers to 'drill down' into the detail of site operations when making decisions and when checking or changing safe systems of work.

An important consideration when thinking about the 'how' part of planning the work is that no individual programme activity can be viewed in isolation without reference to those preceding, succeeding or concurrent activities on the programme. Method statements may well include a section dealing with these interfaces.

In practice, method statements can be lengthy and complex documents and beyond the scope of a book of this nature. The examples that follow in Section 11.5 are, therefore, simplified but they nevertheless follow the principles of good practice explained in this Chapter.

11.4.4 Safe Systems of Work

Any chosen method of construction cannot be considered in isolation from safety because it is illegal to use a system of work that is not safe. Part of the contract planning process, therefore, is to think about the provision of safe systems of work which is a statutory requirement under the Health and Safety at Work etc. Act 1974. Safe systems of work are commonly expressed in the form of safety method statements.

Safety method statements covering the main stages of the works, or high-risk activities, are prepared at the pre-construction stage and will be reviewed and adapted as work proceeds. The method statement will be informed by risk assessments which are a statutory requirement under the Management of Health and Safety at Work Regulations 1999.

Whilst safety method statements are not required by law, it is common practice for contractors to prepare them and, together with subcontractors' risk assessments and method statements, they form part of the site safety management system.

11.4.5 Link with the Programme

A safety method statement usually includes details of the likely hazards that might be encountered during the work activities to which it relates. The hazards will be derived from the risk assessment process.

Some planners include safety measures as specific activities on the construction programme such as the provision of access equipment or the installation of safety nets, but this is not sufficient to be regarded as a method statement. Programme relationships may be included in the descriptive notes on the method statement which helps to clarify both the sequence of working and the working method.

11.4.6 Changing Method Statements

There must be no deviation from a method statement, as this can lead to confusion and possibly accidents. Where a change in working method is required, this should be discussed, and the method statement should be changed formally before going ahead with the work. It is crucial that the workforce is notified of any changes to the agreed method via a toolbox talk or a task talk, and any subcontractors must be informed of the impact on their operations.

On no account must a method statement be changed on an ad hoc basis as there have been several multiple fatal accidents for this reason.

One of these concerned the demolition of a multi-storey reinforced concrete building. The normal site foreman went on holiday, and his stand-in came to site without being briefed. He adopted a method of working which was contrary to the agreed method, and an 18-tonne excavator fell on to the floor below as a result. Two workmen were crushed to death.

11.5 Worked Example

Well-written method statements are logical, concise and to the point. Even so, they may be lengthy documents in practice simply because of the complexity of construction operations. The RAMS for a straightforward activity such as brickwork could run to several pages to explain the order, sequence and safety provisions for the work. Matters to be covered include materials storage and handling, how mortar is to be mixed and delivered to the place of work, scaffolding and access provisions, COSHH assessments for hazardous substances, noise and dust control, PPE, welfare arrangements and so on.

Consequently, this worked example is necessarily simplified, but the same thought processes and principles of good practice apply, nonetheless.

11.5.1 Project Description

A city-centre brownfield site is to be redeveloped with two 8-storey apartment blocks and a basement car park beneath, accessed by a ramp. The £46 million development comprises 210 one and two-bedroom luxury apartments, set either side of a landscaped courtyard, within a nineteenth-century canal basin in an area of rich industrial heritage.

Figure 11.2 provides a site layout plan which shows the location of the apartment blocks and the surrounding canal basin area. It also shows the main site access to the north-east of the site. The road system to the west of the site is restricted to light site traffic only but provides a further limited access to the west of the north canal basin, as shown on the site layout plan.

The basement is approximately 50 m × 38 m × 4.5 m deep and covers most of the west side of the site. Further ground level parking is to be provided on waste land on the east side of the site. The apartment blocks are of in situ reinforced concrete construction clad with facing brickwork and multicoloured feature balconies as shown in Figure 11.3. Each block has an internal reinforced concrete staircase and lift shaft.

11.5.2 Tender Method Statement

The space limitations of the site, and the proximity of bodies of open water to three sides, are key considerations for the contractor at tender stage. Site access, space for site accommodation and suitable lay-down areas for materials are also a primary concern. The space to the east of the site is available for materials storage but this is also the main access to the site which creates limitations.

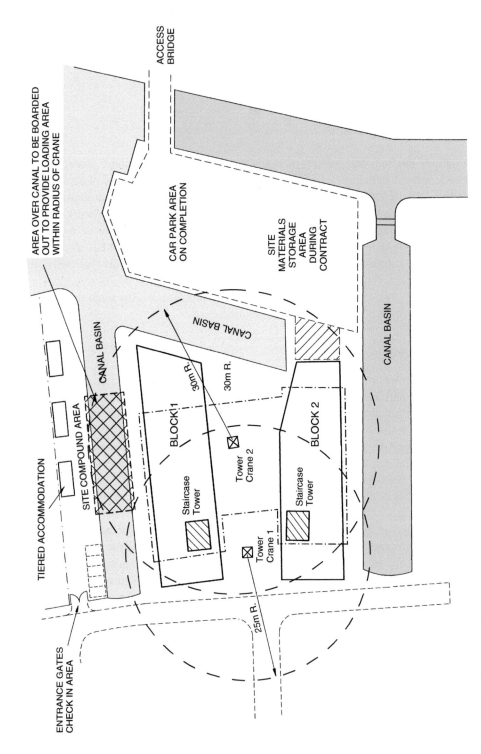

Figure 11.2 Site layout plan.

Completed blocks 1 (left) and 2 with basement parking below. Garden area was location for the two tower cranes.

Completed Block 2 and canal basin.

Figure 11.3 Development photos.

The initial possibilities for materials handling are:

- Two static tower cranes, positioned between the two apartment blocks, as indicated in Figure 11.4.
- Alternatively, rail-mounted tower cranes could be erected between each apartment block and their respective canal basins, with mobile crawler cranes stationed to the east and west servicing the basement construction, as shown in Figure 11.5.

The 50 m × 38 m basement perimeter is to be constructed with 800 mm secant piles – bored interlocking piles used for controlling groundwater. On top of the piles is a reinforced concrete ring beam which forms the foundation for the external frame of each building. Internal columns are constructed on 350 mm diameter continuous flight auger (CFA) piles and pile caps, as shown in Figure 11.6.

Piling is, therefore, a two-phase operation involving two different types of piles and two different piling rigs – one for large-diameter bored piles and the other for CFA piling.

All piling work could be carried out from ground level, but this would require some 5 m of empty boring for the CFA piles. Alternatively, the foundation piling could be carried out after basement excavation which would require a second mobilisation of the piling gang and a second piling mat to be installed and later removed. Lateral support for the secant piles is to comprise permanent ground anchors. Access for the piling rig would be via the permanent access ramp to the basement.

The basement construction sequence is illustrated in Figure 11.7 which shows the construction of the temporary guide walls for the secant piling, the reduced level excavation following construction of the basement ring beam, the ground anchors and basement slab construction and formwork to the basement walls.

With two basement levels and eight floors above ground level, the staircase/lift shafts are ideally suited to using slipform construction, especially as they are a regular shape. This would have to be sequenced around the ground floor slab and reinforced concrete frame construction because the slipform would have to commence from basement slab level.

CRANAGE PROPOSALS AT TENDER

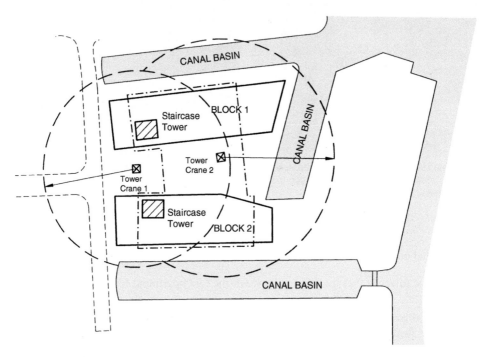

Figure 11.4 Proposal one.

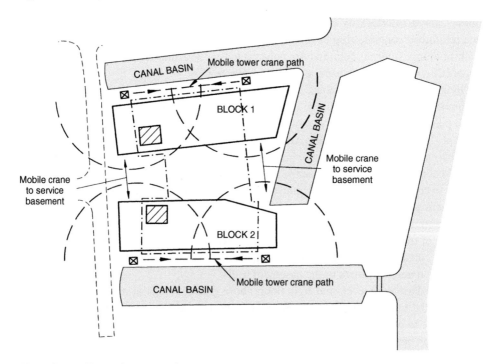

Figure 11.5 Alternative proposal two.

Basement ring beam and starter bars for ground floor columns.

Basement ring beam on secant piles. Pile caps constructed on second stage piling.

Figure 11.6 Basement piling/pile caps.

Despite the high set up costs of slipforming, each lift/stair shaft could be poured continuously with no joints, saving considerable time compared to conventional formwork.

As the frame is reinforced concrete, plate floors constructed using table forms would be economical in terms of formwork costs and would cut down on site labour and time-related costs, thereby benefitting the budget and the programme. This is illustrated in Figure 11.8

Figure 11.9 shows a full tender method statement for the project and identifies key work packages to be procured.

11.5.3 Construction Method Statement

Figure 11.10 shows how a construction method statement for the basement CFA piling operation might be written. It demonstrates the basics of a good method statement:

- A clear description.
- A simple scope of works.
- A clear and logical explanation of the working method.
- Plant and labour requirements for the job.
- An easily understood document which can be communicated to those who will carry out the work on site.

In this example, reference is also made to the main hazards associated with the work and the control measures for each of those hazards derived from risk assessments. This emphasises the importance of health and safety when considering working methods.

Studying Figure 11.10, it can be appreciated that even a method statement for a straightforward and well-understood piling operation requires a fair amount of detail. The level of detail is to be contrasted with the tender method statement shown in Figure 11.9, which is a more strategic document.

From this explanation of the contractor's proposed working method and resourcing for the CFA piling activity – and from the quantities of work indicated – the durations and sequence of work for the master programme can be determined.

Fixing formwork to guide walls for secant piles.

Completed guide walls. Main site access gates in background.

Secant piles. Rebar and formwork to ring beam in progress.

Reduced level excavation following secant piling and ring beam construction.

Secant piles and ground anchors. Waterproofing and formwork to basement walls. Basement ground slab in progress.

Proprietary formwork to basement walls.

Figure 11.7 Basement sequence.

Block 1 RC frame and plate floors. Staircase and lift shaft. Tower crane 2.

Staircase and lift shaft. Moving tableforms into place.

Figure 11.8 Tableforms.

Furthermore, temporary works, working space requirements and subcontractors' attendance requirements – such as piling mats, access and hard standings – can be identified for pricing and inclusion in work package contracts.

11.5.4 Safety Method Statement

The bedrock of a safety management system is provided by risk assessments and method statements.

Risk assessments identify hazards, assess risk and suggest control measures to manage residual risks. A method statement is a safe system of work that builds on risk assessments and specifies how work activities can be carried out in a safe manner.

> If a method statement contains risk assessments – or is linked or referenced to risk assessments – it is commonly referred to as a RAMS (risk assessment and method statement).

The discipline of thinking through the production process step by step should automatically raise the question of hazards. This naturally leads on to how they can be eliminated, or the ensuing risks controlled.

Method statements, on the other hand, provide a convenient means of detailing the 'how', 'when' and 'why' of any control measures that derive from risk assessments. This is illustrated in Figure 11.10 which identifies the hazards and control measures suggested for the CFA piling to the basement and tower crane bases, together with a safe system of work and suitable resources.

In this context, the Contractors Health and Safety Assessment Scheme (CHAS) considers that method statements should be *used alongside risk assessments to make sure certain high-risk tasks and activities involved in a project are safe*.

Method statements are not required under the CDM Regulations, but the HSE does recognise them as one way of satisfying the requirements of the regulations.[4] Method statements also provide a means of assessing and managing risk and facilitate two-way communication with the workforce.

PRE-TENDER METHOD STATEMENT			
Project	PW763	**Tender No.**	T1014

Project description

Two 8-storey apartment blocks comprising 210 one and two-bedroom luxury apartments and a basement car park, 50 m × 38 m on plan × 5 m deep, accessed by a ramp.

Site access

Site access is restricted to the existing access bridge across the eastern canal basin. All deliveries and muck away from site to be via this access only. The existing service road to west of site can only be used for access to site offices.

Site security

Erect temporary 2.4 m high ply hoarding to site perimeter with two sets of double gates. Site security cabin at main access only. CCTV system to be installed with nightly security patrols.

Site management

Two-storey modular site accommodation and welfare facilities. Management comprising site manager, sub-agent for each apartment block, office manager and secretary, general foreman, full-time site engineer, and part-time site QS.

Site accommodation

The site compound is to be located to the north of the site. Site offices and welfare facilities will be two-storey cabins and limited car parking will be available for main contractor's staff only.

Lay-down areas

The proposed ground level car park to the east of the site is the main lay-down area. Part of the canal basin to the north of the site is to be boarded out to provide a loading area serviced by the tower cranes.

Sequence of work

Clear site and construct guide walls for first-stage basement piling. Construct piling mat with 6F2 stone on geotextile layer. Install secant piling to basement perimeter. Following piling, construct reinforced concrete ring beam to basement perimeter. Excavate basement to reduce levels and construct piling mat. Install ground anchors prior to carrying out second-stage CFA piling and pile cap construction. Construct 2 No bases for tower cranes on CFA piled foundations between blocks 1 and 2. Mobilise and erect static tower cranes. Excavate to basement formation level, fill with MOT Type 1 stone and compact. Construct reinforced concrete basement slab and thickenings in 5 m wide alternate bays. Construct reinforced concrete basement walls using single-sided proprietary wall formwork. Set up sliding formwork on basement slab for lift/stair shaft to block 1. Once shaft slide is sufficiently advanced, commence reinforced concrete (RC) frame construction to block 1. Fix rebar, erect formwork and pour ground floor concrete columns. Assemble and install table forms and construct suspended first-floor slab. Construct external scaffolded staircase tower as the work proceeds. External access scaffolding to be erected as RC frame proceeds. Dismantle and move sliding formwork to block 2. Commence lift installation and precast concrete stairs to block 1 tower. Continue with RC frame and floor slabs to block 1 installing temporary propping between floors. Commence shaft slide to block 2 and follow on with RC frame, plate floors and scaffolded staircase tower as before. Erect scaffolding and commence external brick/block cladding to block 1 followed by block 2. Once external envelope is complete, dismantle tower cranes and remove from site. Mobile cranes, external mast climbers and 'Cantidec' or similar transfer decks to be used for further materials handling requirements.

Crane requirements

2 No static saddle jib tower cranes – 30 m operating radius.
2 No 150Te all-terrain mobile cranes for duration of basement construction.

Key work packages

Piling	Main contractor attendances: Guide walls, piling mats, excavator and wagons for clearing piling arisings and muck away.
Ground anchors	Ground anchors to be designed and installed by a specialist contractor. Anchors to be installed following first-stage basement excavation. Priced as a separate subcontract but could be included with the piling package.
Groundworks	Timber formwork made on site with proprietary steel soldiers and Dywidag anchors. Rebar cages to be made up in materials storage area and craned into position. Ready-mixed concrete. Placing by crane and skip.
Slip forming	Three-deck slipform rig to be designed, installed and operated by a specialist contractor. Concrete to be ready mixed. Placing by mobile concrete pump.
Reinforced concrete package	Package to include RC columns and plate floors. Concrete to be ready mixed. Placing by crane and skip for columns and mobile concrete pump for suspended floor pours. Two sets of table forms per block. Each floor level to be poured in two halves.
Brickwork	2 No dry mortar silos and crane-lift mortar tubs. Provide reinforced concrete base slab 8 m × 3.5 m × 125 mm thick for silos. Access from external scaffold. Materials to be craned onto scaffolded transfer decks as the work proceeds. 2 No 5:3 bricklaying gangs per block.

Figure 11.9 Tender method statement.

	CONSTRUCTION METHOD STATEMENT
Operation	350 mm diameter CFA piling to basement car park and tower crane bases.
Quantity	126 No piles average 10.5 m deep including piles for tower crane bases.
Method	• Excavate basement to reduce levels 500 mm above basement formation level. • Form access ramp for piling rig as the work proceeds. • Construct 450 mm thick temporary piling mat and access ramp using 6F2 imported crushed stone laid on 'Terram' or similar geotextile membrane. • Offload and set up concrete agitator and pump with 150Te all-terrain mobile crane. • Commence CFA piling and proceed in accordance with the agreed piling plan. • Concrete piles and insert prefabricated rebar cages. • Remove piling arisings to temporary spoil heap on site as piling work proceeds. • On completion of piling, excavate to remove piling mat. Cart to temporary spoil heaps on site using 9T dumpers. • Pile cropping – see separate method statement. • Excavate temporary spoil heap, load into wagons and cart to tip off site.
Hazards	1. Work at height assembling and dismantling CFA rig. 2. Moving plant and vehicles: Piling rig, excavator, vibrating roller, concrete agitator and pump, site dumper. 3. Noise from plant and vehicles. 4. Hazardous substances: concrete, cement in aqueous solution. 5. Rebar cages: risk of cuts, scratches, abrasions, impaling.
Control measures from risk assessments	1. Access from mobile elevating work platforms (MEWP) wearing fall arrest harness and lanyard. 2. Segregate immediate piling zone with heavy duty interlocking safety barriers and limit access with walk through pedestrian barrier. Limit operative exposure to vibrating machinery. 3. Wear ear defenders. 4. PPE: gloves and eye protection. 5. PPE: gloves, hard hats Fit brightly coloured rebar safety caps to ends of reinforcing bars.
Resources	Soilmec SF-65 CFA Piling Rig Genie Z-34 articulated boom lift (MEWP) Mecbo Betoncap crawler-mounted concrete pump and agitator Takeuchi 16T back acter Bomag BW135 sit-on tandem vibrating roller Thwaites 9T site dumper Liebherr LTM1150-5.3 150Te all-terrain mobile crane 6 No 8-wheeler wagons Piling foreman Site engineer Banksman 2 Labourers

Figure 11.10 Construction method statement.

Notes

1 Management of Health and Safety at Work Regulations 1999, regulations 5 and 4 (schedule 1).
2 CDM Regulations 2015, regulation 15(2).
3 https://www.hse.gov.uk/work-equipment-machinery/loler.htm.
4 https://www.hse.gov.uk/construction/safetytopics/admin.htm#method.

Chapter 12 Dashboard		
Key Message		o Good health and safety practice is part of good management. o Good managers integrate health and safety into their thinking. o Good workers think about the consequences of their actions.
Definitions		o **Hazard**: Something with the potential to cause harm. o **Risk**: The likelihood of potential harm from a hazard. o **Risk assessment**: The formal process of identifying hazards, evaluating risk and recording and reviewing the findings. o **Method statement**: A formal document that defines a safe system of work based on a risk assessment. o **Control measure**: Arrangement to reduce risk to ALARP.
Headings		Chapter Summary
12.1	Introduction	o Construction is dangerous with a high potential for accidents. o Human behaviour in the workplace is an important factor. o Planning is important to manage the risks of site operations.
12.2	Hazard and Risk	o Hazards give rise to risks – if there are no hazards, there are no risks. o Hazards may be generic or site-specific. o Hazard identification is an essential first step in risk control. o The basic idea is to eliminate risk, where possible, and to control any residual risk by effective management.
12.3	Legal Framework	o The Health and Safety at Work etc. Act 1974 (HSWA) is the primary health and safety 'enabling' legislation in the UK. o This permits the enactment of 'subordinate legislation', such as the Management of Health and Safety at Work Regulations and CDM. o The main thread running through all UK health and safety legislation is the **management of risk**. o This is largely achieved through 'goal setting' or 'objective' legislation.
12.4	Managing Health and Safety	o It is essential to create and maintain a positive health and safety culture. o The aim is to provide a safe place of work, to provide safe systems of working and to promote an active safety culture on site.
12.5	Planning the Work	o Planning for safety can be **task-based** or a **hazard-based**. o The NEC ECC specifically requires the contractor to show provisions for health and safety requirements on the accepted programme. o A safe system of work is a method for doing a job in a safe way. o Safety method statements define safe systems of work.
12.6	Industry-Specific Legislation	o Nearly all UK work-place health and safety legislation is generic. o The CDM Regulations are industry-specific. o CDM imposes statutory duties on clients, designers and contractors.
12.7	The CDM Regulations 2015	o Duty holders must plan, organise and manage construction work. o The regulations apply to ALL construction projects, large or small.
12.8	Health and Safety Training	o Training is the responsibility of Construction Skills. o On-the-job training includes site induction, tool-box and task talks.
12.9	Measuring Performance	o Health and safety performance on site is a measure of how well the health and safety management system is working. o Sites should be regularly monitored, audited and reviewed.
12.10	Enforcement of Legislation	o The HSE inspectorate is the enforcement authority, with some exceptions. o HSE inspectors may offer advice or issue legal notices. Notices may lead to prosecution if not heeded. o Many contractors have their own safety advisers who audit sites.
12.11	Accidents and Incidents	o If hazards are not dealt with, they may cause injury, ill health or damage to persons and/or property. o Accidents can have a dramatic impact on individuals and businesses.
Learning Outcomes		o Understand the principles of health and safety management. o Understand the basics of UK H&S legislation. o Identify common hazards and assess risk. o Devise a safe system of work and write a method statement.
Learn More		o Chapter 3.3 – Risk assessment. o Chapter 11.3 – Types of method statement.

12

Planning for Safety

12.1 Introduction

There is no escaping the fact that construction is an inherently dangerous industry, and the potential for accidents to happen is high. After all, the industry relies upon all sorts of heavy plant and equipment and on most sites, large or small, a wide range of hand or machine-mounted power tools can be found.

Construction operatives work in all kinds of weather and conditions – deep excavations, live sewers, tunnels, working at height, etc. – and, more often than not, have to wear awkward personal protective equipment (PPE) to keep themselves safe. The work is dirty and dusty, and operatives often must work with materials that are carcinogenic or otherwise dangerous to health. Additionally, some workers are more vulnerable than others – they may be temporary or may work through an agency or their first language may not be the indigenous language (e.g. English).

> Current statistics show approximately 1 fatality and 1200 non-fatal injuries *per week* in the UK construction sector.

No wonder, then, that construction has had a poor reputation for safety over the years, but the dangers of construction work are only one of the complex interaction factors that can lead to accidents, ill health and fatalities. Not least of such factors is the way that people behave.

12.1.1 Human Factors

Health and safety cannot be effectively managed without appreciating the impact of human behaviour and other related factors in the workplace (HSE 1999). However, whilst behavioural scientists and psychologists help us to understand the human factors operating within complex organisations, it must be acknowledged that there are no perfect explanations as to why people behave the way they do in the work situation.

It is too simplistic, however, to accept *the commonly held belief that incidents and accidents are the result of a "human error" by a worker in the "front line"… which is beyond the control of managers'* (HSE 1999).

Construction Planning, Programming and Control, Fourth Edition. Brian Cooke and Peter Williams.
© 2025 John Wiley & Sons Ltd. Published 2025 by John Wiley & Sons Ltd.

What is more believable is that *a human error problem is actually an organisational problem* (Dekker 2014).

Accidents happen to people at work due to their involvement in work activity and the HSE (1999) estimates that up to 80% of accidents are, at least, partly attributable to the actions or inactions of people. This is not surprising considering the extent to which people 'at the workface' are exposed to danger.

Human failures in health and safety are called 'errors' or 'violations' where:

- Errors are unintended actions.
- Violations are deliberate actions, inactions or deviations from rules.

However, blaming the individual for accidents ignores the less obvious causes of accidents which may lie deeper within the supporting organisation. These are identified in HSG48 (HSE 1999) and include:

- Poor work planning.
- Lack of safety systems.
- Poor communications.
- Poor health and safety culture.

Recognising that organisational failings can contribute significantly to workplace accidents is a starting point, but construction businesses face many pressures that impact on their ability to manage health and safety effectively.

12.1.2 Pressures on Health and Safety

Dekker (2014) views the organisation as having a 'sharp-end' and 'blunt-end'. This view is depicted in Figure 12.1 where:

- The 'sharp-end' represents 'frontline' workers, such as site operatives and subcontract labour.
- The 'blunt end' is the supporting site and head office organisation which underpins the 'sharp-end' activities.

In this model, the blunt end supports and drives the sharp-end activities, and it is the blunt end that provides the necessary technical support, resources, training, etc., which enables the sharp-end to deliver the end product.

However, despite the support that the internal organisation offers, the blunt end also imposes pressures on the sharp-end activities such as keeping on budget and programme and making sure that the desired profit target is achieved.

When these 'internal pressures' are added to the 'external pressures' imposed by legislation, client expectations and the dangers presented by the working environment, etc., it is little wonder that the behaviour of those engaged in physical production sometimes manifests itself as errors.

The HSE (1999) explains that *human factors refer to environmental, organisational and job factors, and human and individual characteristics, which influence behaviour at work in a way which can affect health and safety*. It also suggests that human factors can be viewed simply by considering three issues – the job, the individual and the organisation.

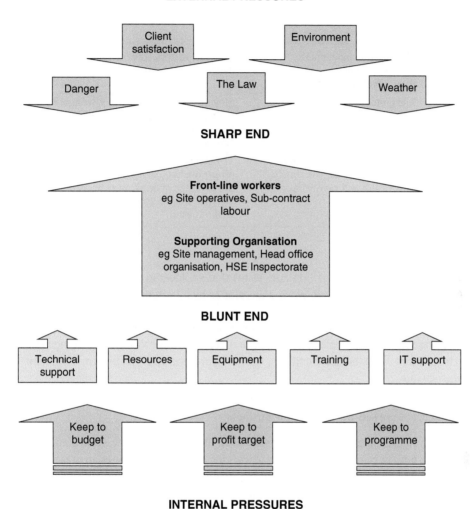

Figure 12.1 Pressures on health and safety. Source: Adapted from: The Field Guide to Understanding Human Error. Sidney Dekker.

Attitudes, risk perception, culture and leadership are some of the factors that impact people's health and safety-related behaviour.

12.1.3 Definitions and Context

Health, safety and welfare considerations – defined in Table 12.1 – begin at the top of the organisation with a positive and active safety culture. In many companies, this is a strategic priority which should be interwoven with ambitions for sustainability, modernising and evolving ways of working and delivering value for clients, the workforce and all stakeholders.

Table 12.1 Definitions.

	Definition	Examples
Health	The physical and mental well-being of persons at work, or those affected by work activities.	Exposure to noise, dust, chemicals or other dangerous materials or processes.
Safety	The protection of workers, passers-by or visitors to site from physical injury from work activities.	Moving plant and machinery, work at height, falling materials or tools, collapse of excavations or scaffolding or overturning of cranes or other lifting equipment.
Welfare	The provision of facilities for those at work so as to maintain their health and well-being.	The provision of site accommodation with proper washing, drying and sanitary facilities, the provision of drinking water, eating and rest facilities and the provision of first-aid facilities, etc.

In its annual report, a major contractor[1] states that engineered safety is central to the way in which the group goes to work, underpinned by its values of placing care at the heart of the business.

Another major contractor[2] states that its consistent culture of health, safety and well-being excellence – embedded in its practices and approach to managing risks and behaviours – has led to a significant reduction in reportable incidents and accident frequency rate.

The safety culture sets the tone for the business:

- At tender stage, the contractor must make sure that the tender price reflects the cost associated with developing safe systems of working on site. This requires an organisation structure suitable for the work in hand and a pre-tender programme that allows sufficient time for the planning, resourcing and control of site safety.
- At the contract planning stage, detailed measures must be put in place to ensure that safety, health, welfare and environmental issues are properly managed and do not pose a risk to those who are working on or affected by the project.
- Managers must make sure that there is legal compliance on all sites with regard to health, safety and welfare.

12.1.4 Planning for Health and Safety

Any construction project, no matter how small, needs careful planning. This is, in fact, a statutory requirement under regulation 13(1) of the CDM Regulations 2015, as illustrated in Figure 12.2. This states that the principal contractor must plan the construction phase and estimate the time required to complete the work.

Duties of a principal contractor in relation to health and safety at the construction phase

13.—(1) The principal contractor must plan, manage and monitor the construction phase and coordinate matters relating to health and safety during the construction phase to ensure that, so far as is reasonably practicable, construction work is carried out without risks to health or safety.

(2) In fulfilling the duties in paragraph (1), and in particular when—

    (a) design, technical and organisational aspects are being decided in order to plan the various items or stages of work which are to take place simultaneously or in succession; and

    (b) estimating the period of time required to complete the work or work stages,

the principal contractor must take into account the general principles of prevention.

Figure 12.2 Planning for health and safety.

Planning is important in managing the risks associated with site operations, avoiding time and cost overruns and making sure that all necessary resources will be available when required. This requires careful thought about matters such as:

- Site layout, organisation and traffic management and site security.
- Provision of welfare facilities for those working on site.
- The safety and well-being of site workers, visitors and passers-by during the works.
- The environmental impact of proposed construction work.
- The order and sequence of the work.
- The choice of plant, equipment and temporary works.

Dealing with such matters requires an effective risk management strategy and an active safety culture driven by top management. The best of intentions will fail, however, if hazards are not identified and correctly evaluated to ensure that they are effectively managed. It should also be recognised that new hazards appear as the work evolves, making the process of hazard identification and risk assessment a continuous theme throughout the project.

It is suggested that site managers will be concerned with five main issues:

1. The identification of **hazards** that might give rise to accidents on site.
2. **Assessing risk** arising from the hazards in question and putting in place appropriate **control measures** to deal with the risks.
3. Devising **safe systems of work** for the project in hand.
4. Making sure that there is an appropriate **safety management system** in place for the particular site in question.
5. Ensuring **compliance** with the law.

12.2 Hazard and Risk

The main emphasis in UK health and safety legislation is the control of risk. Hazard identification is an essential first step in this process.

Hazards give rise to risks and *if there are no hazards, there are no risks* (HSE 2000). However, construction work is, by definition, hazardous and so the best that can be done is to identify the hazards and either eliminate them or control the ensuing risks.

12.2.1 Definitions

A **hazard** is *something with the potential to cause harm* (HSE 2000). This might be the result of a work activity, process, plant or machine or substance, such as:

- Working at height, including working from a ladder.
- Deep excavations.
- Moving plant or vehicles.
- Mixing mortar or plaster, machining wood or working with asbestos.

A **risk** is *the likelihood of potential harm from [a] hazard being realised* (HSE 2000). The extent or level of the risk is determined by:

- The likelihood of harm occurring.
- The potential severity of the harm and
- The number of people who might be exposed to the hazard in question.

12.2.2 Types of Hazard

There are two principal categories of hazards in construction as identified in Figure 12.3:

- **Generic** hazards that are common to most construction sites.
- **Specific** hazards that are peculiar to one individual site.

Hazard identification is the starting point for risk assessment and so it is crucial that this is done correctly, or else the risk assessment will be wrong, and the system of work will be more dangerous than it otherwise would be.

For instance, falls from height is often identified as a hazard, but this is not correct:

- **Working at height** is *something with the potential to cause harm* and is therefore the hazard.
- When someone is working at height, there is a **chance** of falling with the **possible consequence** of a serious injury and this person is therefore **at risk**.
- The extent of the risk will depend on the likelihood or chance of falling and the severity of the fall.
- For example, falling 20 m to the ground is likely to lead to a fatality or, at best, a very severe injury and therefore the risk must be high unless measures are put in place to prevent a fall.

Legally speaking, it is not necessary to identify **every** possible hazard or assess **every** risk because UK health and safety legislation is largely founded on the principle of **so far as is reasonably practicable** or similar words to that effect.

This means (with some exceptions) that, so long as everything reasonably possible has been done to seek out hazards and avoid risks to those who might be exposed, this is enough to satisfy the law. The leading case on this issue is *Edwards* v. *National Coal Board* [1949].

Joyce (2015) reports that this principle has been established as 'not unlawful' in the European Court of Justice and therefore *measures to reduce risks do not have to be grossly disproportionate*. It is a matter of balancing the effort and cost involved. Consequently, risks are reduced to the level of as low as reasonably practicable (ALARP).

HAZARDS

Generic	Specific
DEFINITION – Hazards that are present on virtually every site	**DEFINITION** – hazards that are peculiar to a particular site

EXAMPLES

- Dust and fumes
- Noise
- Vibration
- Harmful substances or materials
- Electricity
- Fire
- Hot work (welding, grinding, cutting)
- Abrasive wheel-type hand tools
- Work at height ⎯⎯⎯⎯⎯⎯⎯⎯⎯⎯⎯⎯⎯⎯⎯⎯⎯⎯⎯⎯⎯⎯
- Heavy weights
- Moving plant and vehicles

EXAMPLES

- Nearby school, pedestrian crossing or bus stop
- Busy road junction
- Railway line nearby
- Presence of water (eg river or canal)
- Overhead high voltage cables
- Underground services (eg gas main, electricity cables)
- Underground railway tunnel
- Existing buildings to be demolished
   - Asbestos
   - Unstable/insecure structural elements
   - Sharp objects such as waste hypodermic needles and syringes
   - Unprotected lift shafts or pits
   - Petro-chemicals, solvents, acids
- Deep excavations
- Contaminated ground
   - Chemicals, solvents, toxic metals
   - Toxic/flammable gases
   - Bacteria/viruses from clinical waste

Preventative measures

- Ladders secured to scaffold with proprietary clamps.
- Full-height scaffold to gable.
- Gated access to work area.
- Transfer deck/loading area above ground level.
- Debris netting to prevent falls of tools and materials.

Figure 12.3 Hazards in construction.

However, the possibility of low likelihood but high-severity risks must be guarded against, as must the possibility of overlooking hazards that might give rise to unmanaged risks.

12.2.3 Common Hazards in Construction

Finding or 'spotting' hazards is not an easy or obvious process and requires considerable knowledge and experience.

As with any aspect of construction planning, a thorough understanding of the technology of construction and the assembly process is fundamental to predicting the presence of hazards. Knowing how items on the work schedule might be constructed on site, and in what sequence, leads to a greater appreciation of the potential for harm to arise and, consequently, what the hazards are.

Figure 12.3 identifies some common hazards in construction where it can be seen that work at height is a generic hazard. However, this must also be risk assessed in terms of the specifics of each site or situation. In this example, specific preventative measures are illustrated for a low-rise housing situation.

12.2.4 Identifying Hazards

The secret to spotting a hazard is to ask the question ***does it have the potential to cause harm?*** If the answer is 'yes', then it is a hazard.

Falling is just one of the risks faced by operatives when working at height, but it is frequently the case that other hazards will be present in any given situation or work activity.

For instance, **chasing out brickwork** for pipes or wiring above ground level is an additional hazard to that of working at height because this activity can create dust, noise, vibration, flying debris, etc., in addition to the attendant risk of falling.

Such hazards should preferably be designed out by assessing the risk and then altering the design in such a way that risks are removed. For instance, the risks associated with chasing out can be eliminated entirely by hiding pipes or wires behind dry lining. If this is not feasible, then the risks have to be managed which – in the case of chasing out brickwork – could be achieved by using full body protection and respiratory apparatus.

However, these are risk reduction (not elimination) measures because the hazard (i.e. chasing out) is still there.

Thinking through a work activity to identify potential hazards is not easy to do and requires experience as well as a good dose of common sense. This is probably why there is considerable anecdotal evidence to suggest that it is common practice in the industry to use standard or 'generic' hazard 'checklists' as the basis for risk assessments rather than trying to establish the hazards that relate to a particular project.

These 'generic' checklists inevitably lead to 'generic' risk assessments that have little practical value other than ticking the 'risk assessment done' box. Such bad practice might well lead to an accident and, perhaps, an appearance in court!

12.2.5 Guidance

Despite the shortcomings of using hazard checklists, there is no shortage of useful guidance for those responsible for preparing risks assessments, including:

- Approved codes of practice (Control of Substances Hazardous to Health, COSHH, for example).
- Health and Safety Executive (HSE) guidance (such as for working in confined spaces).
- Research and publications (such as CIRIA publications).
- Product information and data sheets.
- Industry guidance (such as the Federation of Master Builders).
- Company data – such as accident and ill-health statistics, near-miss reports.
- HSE annual statistics of work-related injuries and ill health in construction.

Referring to approved codes of practice and other guidance illustrates a degree of competence in the awareness of the common causes of accidents and ill health in construction.

In the event of a serious accident, a court would expect to see that such guidance – or something equivalent – had been followed.

12.2.6 Persons at Risk

Part of the risk assessment process is to consider the persons at risk from the various hazards identified. This includes:

- Passers-by.
- Directly employed workers and site staff.
- Subcontractors.
- Official visitors (e.g. architect, engineer and HSE inspector).
- Other visitors (e.g. invited guests and students)
- Unauthorised visitors (e.g. children and people 'up to no good!!')

It can easily be forgotten that construction sites pose hazards for members of the public as well as workers, and typically, around five or six members of the public are killed each year because of construction work – many more are injured. Passers-by can be killed or injured by falling materials and equipment, overturning plant or scaffold collapse, by contact with site vehicles and plant or even by being forced into the road by poorly planned diversions. Children are also at risk because construction sites provide tempting playgrounds.

12.2.7 Risk Evaluation

There are two main types of risk assessment:

1. Quantitative risk assessment
2. Qualitative risk assessment

Quantitative evaluation *produces an objective probability estimate* based on known facts and statistical data (St. John Holt 2001), whereas qualitative risk assessments are *based upon personal judgement backed by generalised data on risk* (Hughes and Ferrett 2011).

Qualitative risk assessments can give the impression that they are quantitative because numbers are sometimes used in the assessment. This methodology is still qualitative because the numbers are based on personal judgement and not on statistical data.

Most risk assessments in practice are qualitative and, normally, they are perfectly adequate and simple to do. Many are also generic – i.e. standard risk assessments for commonplace hazards where the risks are well known, and the control measures are the same each time (e.g. manual handling, COSHH assessments, etc.).

This can make sense, provided the assessments reflect specific circumstances, but generic assessments should never be used for site-specific or high-risk activities.

The HSE suggests a simple 3 × 3 matrix approach, as illustrated in Figure 12.4, whereas the other extreme, offered by St. John Holt (2001), suggests that a qualitative risk

3x3 matrix

Likelihood		Severity	
High	certain/near certain	Major	death or major injury
Medium	often occurs	Serious	causing short term disability
Low	seldom occurs	Slight	other injuries/illness

The resulting matrix is as follows where risk is given by severity of harm x likelihood of occurrence:

		SEVERITY		
		Low	Medium	High
LIKELIHOOD	Low	L	L	M
	Medium	L	M	H
	High	M	H	H

Risk = Likelihood x Severity

Some practitioners like to put numbers to the risk evaluation where:	Example
• low = 1 • medium = 2 • high = 3	If the likelihood is 2 and the potential severity is 3 the resultant risk is 6 out of a maximum possible of 9. This is at the lower boundary of the high risk category. The numbers don't really mean anything – they just indicate a level of risk – and it is sometimes better to think high, medium or low risk.

Many other hazard rating systems have been developed including the 6x6 matrix below:

6x6 matrix

Likelihood		Severity	
1	Remote	1	Minor injury
2	Unlikely	2	Illness
3	Possible	3	Accident
4	Likely	4	Reportable injury*
5	Probable	5	Major injury*
6	Highly probable	6	Fatality*

*RIDDOR category

Figure 12.4 Qualitative risk assessment.

rating can be given by taking consequence × (number of persons exposed + probability of harm).

12.2.8 Control Measures

The Management of Health and Safety at Work Regulations 1999 require employers to carry out risk assessments, implement appropriate measures, appoint competent people and provide suitable information and training. This should include:

- Recording the **preventative and protective** measures in place to control the risks and
- Deciding what further action, if any, needs to be taken to reduce risk sufficiently.

Preventative and protective measures are often referred to as **control measures**, the purpose of which is to reduce the risks to ALARP. This is not always achievable, however, and therefore, further measures may be needed to control any residual risks.

For instance, a control measure might be to provide operatives working at height with a safety harness for a particular task, but there is still the risk that operatives will not use the harnesses correctly or that they will not clip on to a safe anchorage.

This residual risk could be reduced to an acceptable level by appointing a foreman to supervise the task who will ensure that the safe system of work operates properly.

Control measures are precautions to protect those at risk from work activities and are equally relevant to the site set-up as the day-to-day construction operations taking place. Effective site safety management is a matter of good planning, and this includes making sure that the site itself is well-organised, well-run and kept tidy and that unauthorised access is prevented.

Figure 12.5 illustrates some issues that the site manager may consider when planning the site set-up, together with examples of control measures for a variety of work activities.

12.2.9 Risk Assessment

The process of risk assessment is a legal requirement:

- Implied by Section 2 of the Health and Safety at Work etc. Act 1974.
- Specifically required by regulation 3 of the Management of Health and Safety at Work Regulations 1999.

> The basic idea of risk assessment is to look at planned activities to see (a) what hazards (if any) are present and (b) what risk of harm they might pose to workers and others affected by the work activity.
>
> The intention is to eliminate risk, where possible, and to control any residual risk by effective management.

Losses/accidents are caused by exposure to risk. This may be the result of unidentified hazards or residual risks that have been left unmanaged.

The HSE suggests that risk assessment should not be overcomplicated and that there are five basic steps to follow. We have suggested additional measures in step 3, as shown in Table 12.2.

There is no standard way of doing risk assessments but in firms of over five employees, risk assessments must be recorded, and they must also be kept under review. Figure 12.6 illustrates an approach for assessing below-ground risk on a brownfield site for a £50 million contract to develop a mix of low and high-rise housing.

> The best idea is to keep risk assessments simple and understandable so that they are easy to communicate. This sort of approach is illustrated in Figure 12.7.

Risk assessments must be reviewed regularly as work activities develop because risks may change according to changes in circumstances.

SITE LAYOUT PLANNING	
Site perimeter • Plywood hoarding, proprietary interlocking metal panels or mesh fencing • Covered walkways • Protective fans and debris netting • Pedestrian diversions • Clear directional and warning signs	
Access • Pedestrian security gates • Vehicular barriers/gates • Security office • Measures to protect children o Immobilise plant, prevent access to heights, cover or barrier open excavations and drainage systems, store heavy materials safely	
Site accommodation and welfare facilities • Office and restrooms • Sanitary and washing facilities • Drinking water • Drying facilities • Firefighting and first aid equipment • Statutory notices • Materials storage	**Traffic control** • Separation of pedestrians and vehicles • Signed traffic routes • Loading/unloading area • Car parking area • Wheel washers **Overhead power lines** • Disconnection/diversion • Goal post barriers • Height restrictions
Waste disposal • Skips • Refuse chutes or hoists • Hazardous/special waste containers	**Lighting** • Site lighting • Access lighting • Individual task lighting
WORK ACTIVITIES	

Excavations	• Collapse
	  o Battered sides to excavations
	  o Trench boxes, drag boxes, manhole boxes
	  o Hydraulic waling frames
	  o Shoring and sheet piling
	• Excavation guard rails/barriers
	• Plant stop blocks
	  o Use of detectors and 'dial before you dig'

Work at height	• Access
	  o Full scaffold
	  o Mobile scaffold towers
	  o Scaffold stairways
	  o Use of permanent stairways
	  o Mobile elevating work platforms (cheery pickers, scissor lifts etc)
	  o Passenger and materials hoists
	  o Mast climbers
	• Fall prevention
	  o Leading edge protection
	  o Harnesses with 'distance' restraints
	  o Purlin trolleys
	• Fall arrest
	  o Scaffolded and boarded 'crash deck'
	  o Harnesses and running lines
	  o Safety nets
	  o Air bags

Manual handling	• Mechanical lifters eg A-frames, scissor lifters • Hoists • Elevators, conveyors • Rough terrain fork lift trucks

Figure 12.5 Control measures.

Item/Activity	Hazards	Risk H	Risk M	Risk L	Controls	Residual risks	Further controls/Review
The site	Old mine shafts				Shafts to be grouted and capped under separate remediation contract prior to commencement	Damage by excavators during earthworks and drainage operations	Permit to dig system to be operated by Main Contractor
	Below ground contaminants including arsenic and hydrocarbons				Site to be remediated with a hardcore capping layer under enabling works contract prior to commencement	Contaminants below capping layer	Contractor to give site induction and specific task talks where there is a risk of operations penetrating the capping layer
	Existing services within Stuart Street footway including water, gas, telecom and HV cables				Site connections to be made by utilities companies	Drainage connections to combined public sewer in Stuart Street	Contractor to arrange method statement and task talks for drainage works in or near footway
Foundations	CFA piling operations				Piling works to be segregated from other concurrent site operations	Wagons removing spoil from site via Stuart Street	Contractor to arrange traffic management and pedestrian barriers in Stuart Street
	Starter bars projecting from mass concrete foundations for construction of reinforced concrete ring beam and integral floor slab				Groundworks subcontractor to provide suitable plastic caps to cover exposed ends of projecting rebar	Missing plastic caps due to poor housekeeping or vandalism	Contractor to carry out regular site inspections prior to each work shift

Figure 12.6 Risk assessment.

Contract		Contract No	Prepared by	Date of assessment
Canal renovation		C2314	GHM	23 May

Operation	Potential hazards	Risk assessment			Control measures
		H	M	L	
Construct sheet piled cofferdam	Trapping and crushing by moving plant		M		Warning device/banksman
	Work at height	H			Guard rails along lock
	Falls of materials	H			Leave piles proud for edge protection
	Confined space working			L	
	Working over water	H			Safety harness and life jacket to be worn

Programmed for	Training/certification	Action		PPE Required (specify)
16 June	All operatives to be CSCS certified	Method statement	☑	Safety harnesses
	Plant operators to be CTA certified	Work permit	☑	Hard hats
		Assessment :		Gloves
		COSHH	☐	Welding goggles
				HV vests
		Noise	☑	Life jackets
				Ear defenders

Figure 12.7 Contractor's risk assessment.

Table 12.2 Risk assessment.

HSE steps to risk assessment	
1. Identify the hazards	
2. Decide who might be harmed and how	
3. Evaluate the risks and decide on precautions	**Additional steps** (a) Determine the control measures required (b) Evaluate remaining risks (c) Make contingency plans for the residual risks
4. Record your findings and implement them	
5. Review your assessment and update if necessary	

12.3 Legal Framework

UK law consists of 'primary' and 'secondary' legislation. Acts of parliament are 'primary' legislation passed by the legislative body – Parliament in the United Kingdom. This is known as 'enabling legislation' because it gives Ministers of the Crown the power to enact other legislation under it.

'Secondary' legislation – also called Regulations, or more correctly, Statutory Instruments (SIs), are made by an appropriate Minister with the requisite power to do so.

12.3.1 Primary Legislation

The Health and Safety at Work etc. Act 1974 (HSWA) is the primary health and safety legislation in the United Kingdom which was passed by parliament – the legislature.

This Act sets out a general framework of health and safety law and enforcement in all places of work, not just construction. It provides the executive (i.e. the government) with the power to introduce secondary legislation or Regulations which add detailed provisions to the framework of the Act.

The Act also places general duties of care on employers, employees and others and establishes the statutory requirement for a safe place of work. A safe place of work requires thought and planning and no project, large or small, can be properly managed without planning how the work can be carried out safely and without danger to the health and well-being of workers and others.

Prior to the HSWA, the foundation for UK legislation was the Factories Act 1961. This largely consolidated previous factory acts which had a long history dating back to 1802.

12.3.2 Subordinate Legislation

A feature of both the 1961 Factories Act and the current HSWA 1974 is the underlying 'secondary' legislation or Regulations, which provide the detail of the legislative measures required by law.

> Regulations can be introduced without the need for parliamentary approval. In the case of health and safety at work, it is the HSWA 1974 which gives ministers the power to introduce or change or repeal regulations.

Part of the reasoning behind the UK legislative structure is that Acts of Parliament take a long time to pass through the legislative body, whereas Regulations can be made or changed relatively quickly. 'New law' can thus be enacted more responsively to changing circumstances. The CDM Regulations 1994, 2007 and 2015 are good examples of how legislation can be changed in the light of industry consultation to make them more fit for purpose.

Some of the subsidiary or subordinate legislation in the United Kingdom has developed from European Union directives. Prior to Brexit, this meant that the United Kingdom was obliged to incorporate such directives into UK law which resulted in secondary legislation such as the Management of Health and Safety at Work Regulations 1999 and Construction (Design and Management) Regulations (CDM) 1994.

Legislation of particular relevance to construction is summarised in Figure 12.8, although the list is by no means exhaustive.

12.3.3 Goal Setting versus Prescriptive Legislation

The main thread running through all UK health and safety legislation is the **management of risk**.

This is achieved, in the main, through a 'goal setting' or 'objective' style of legislation which emphasises the setting of standards to be achieved rather than a rigid set of rules to follow.

> An example of objective legislation is the requirement to *take suitable and sufficient measures to prevent, so far as is reasonably practicable, any person falling* … under regulation 6(3) of the Work at Height Regulations 2005.

Prior to the HSWA 1974, rule-based or 'prescriptive' legislation was the 'norm' in the United Kingdom (it is still common practice in North America). This type of legislation sets out precisely what has to be done in particular circumstances in order to achieve a safe place of work.

> An example of prescriptive legislation would be regulation 28(1) of The Construction (Working Places) Regulations 1966. This required working platforms above 6 ft 6 in. (198 cm) from the ground to provide a suitable guard rail of adequate strength to a height of between 3 ft (91 m) and 3 ft 6 in. (106 cm) above the platform.

At the time of the 1974 Act, the change from prescriptive to objective legislation was difficult for many people as different thinking was required. It is much easier to follow the rules as opposed to having to think through how a job of work might be carried out safely.

However, it is suggested that placing reliance on prescriptive rules is no substitute for planning the work and assessing the risks which can lead to different conclusions depending upon the individual circumstances of the job in hand.

Statutory Instrument	Brief summary of key requirements
Health and Safety at Work, etc Act 1974 - HSWA	• General duties of care • Safe systems of work • Health and safety policy • Inspection and enforcement
Management of Health and Safety at Work Regulations 1999	• Risk assessment • Principles of prevention • Formal arrangements for planning, organisation, control, monitoring and review of safety measures • Communication, co-operation and co-ordination
Construction (Design and Management) Regulations 2015 - CDM	• Creation of duty holders • Preventative design • Pre-construction information • Construction phase plan • Health and safety file • Notification of project to HSE • Detailed requirements for construction work
Work at Height Regulations 2005	• Avoidance of risk • Organisation, planning, competence • Selection and inspection of work equipment • Scaffolding plans and specific requirements for guard rails
Lifting Operations and Lifting Equipment Regulations 1998 - LOLER	• Strength and stability of lifting equipment • Lifting and lowering of persons • Safe operation of lifting equipment • Planning and supervision of lifts • Thorough examination and regular inspection
Provision and Use of Work Equipment Regulations 1998 - PUWER	• Duty holders responsibilities • Suitability, maintenance and inspection • Information, instruction and training • Conformity of equipment; Roll-over protection
Control of Substances Hazardous to Health Regulations 2002 - COSHH	• Assessment of health risks • Prevention/control of exposure • Exposure limits, monitoring and surveillance • Provision of information
Personal Protective Equipment at Work Regulations 1992 (Amended 2022)	• Provision of PPE by employers • Assessment of risk • Maintenance, cleaning and replacement
Manual Handling Operations Regulations 1992	• Duty of care • Risk assessment
Noise at Work Regulations 2005	• Duty of care • Assessment of exposure • Action levels
Electricity at Work Regulations 1989	• Duties on employers and workers • Prevention of danger • Competent persons
Control of Asbestos Regulations 2012	• Duty to manage • Plans of work • Licensing • Control measures: air monitoring and testing
Confined Spaces Regulations 1997	• Safe systems of work • Competence and training • Emergency arrangements

Reporting of Injuries, Diseases and Dangerous Occurrences Regulations 2013 - RIDDOR	• Reporting of specified events to HSE:
	  o Major injuries
	  o Over 3-day accidents
	  o Work-related disease
	  o Dangerous occurrences

Figure 12.8 Health and safety legislation.

12.3.4 Health and Safety Policy

Section 2(3) of the HSWA requires firms with more than four employees to have a written **health and safety policy** comprising:

- General statement.
- Organisation.
- Arrangements.

The general statement should include:

- **Aims** that are NOT measurable.
- **Objectives** that ARE measurable.

Hughes and Ferrett (2011) say that aims (e.g. for a safe and healthy workplace) will probably remain in place following policy revisions whilst objectives will change each year (e.g. specific targets or campaigns).

Arrangements are concerned with systems and procedures and comprise:

- A policy and procedures manual.
- Standard forms.
- Legislation register.
- Construction phase plan (as required by CDM 2015).
- Other manuals including COSHH.
- Company risk control guidelines.
- Risk assessments and method statements.

Organisation concerns:

- Structure – depicted by an organisation chart – see Figure 12.9.
- Health and safety roles.
- Responsibilities.

Company policy should reflect the genuine aims and objectives of the firm and should not be a generic document.

12.3.5 Approved Codes of Practice

Underpinning primary and secondary legislation are Approved Codes of Practice (ACoPs). These may be industry specific (the CDM ACoP, for instance) or may be pan-industry such as the approved code of practice to the Control of Substances Hazardous to Health Regulations 2002.

It should be noted that ACoPs are intended to provide guidance and practical advice to duty holders on how they might comply with the law. ACoPs are not legislation in their own right, but following the advice given provides reassurance that the relevant law is being complied with.

Notwithstanding this, ACoPs have a special legal status in the event of a prosecution for breach of health and safety law. In this regard, unless it can be proved that the law has been complied with in some other way, a court will find the defendant at fault if the guidance has not been followed.

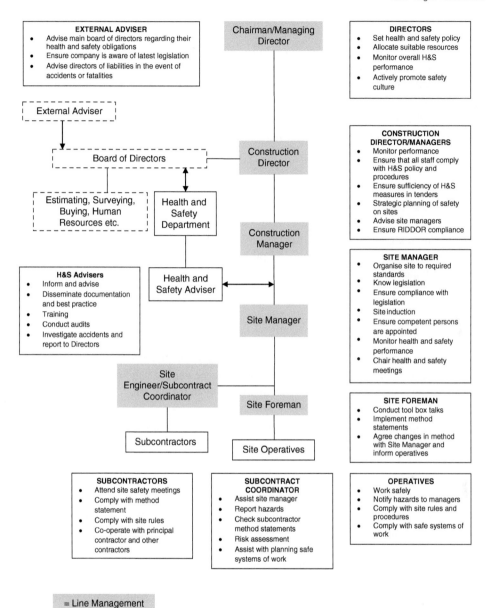

Figure 12.9 Health and safety organisation.

12.3.6 Guidance

As might be imagined, a large body of guidance exists in the field of health and safety, some of which is generic and some being industry specific. Such guidance sits below the level of ACoPs but is nonetheless equally valuable in helping duty holders to comply with their legal obligations.

Some guidance is at a detailed level, such as the use of abrasive wheels or lifting equipment, and some sits at a strategic level in providing guidance for the management of health and safety.

12.4 Managing Health and Safety

HSG65 *Managing for health and safety* (HSE 2013) is widely recognised as the 'go-to' reference point for information on good health and safety principles and practices.

It sets out clear guidance on the key elements of a sound health and safety management system.

12.4.1 HSG65

HSG65 places health and safety management at the heart of good management rather than treating it as a stand-alone system. It also recognises the need to balance management systems with the behavioural aspects of health and safety management.

> The elements of a health and safety management system are 1-**Plan**, 2-**Do**, 3-**Check** and 4-**Act**.

An essential prerequisite for managing health and safety successfully in any organisation is the creation and maintenance of a positive health and safety culture. This requires the active and visible involvement of senior management in the health and safety system, which should encourage health and safety to be an everyday part of work conversations.

A genuine commitment to allocate the requisite resources of time, money and people to the health and safety management system should be part of management's commitment to a positive health and safety culture. However, the pressure of the production schedule, cost and completion date can militate against this aim.

12.4.2 The Four Cs

The '**Four Cs**' are key areas when organising to promote positive health and safety outcomes:

- Control – top-down commitment and collective endeavour.
- Cooperation – participation and ownership, sharing knowledge and experiences.
- Communication – sending out the right signals, talking and listening to each other.
- Competence – awareness, application, skill, knowledge and training.

Table 12.3 Management regulations schedule 1.

Regulation 4, schedule 1	
Requirement	Measure
Avoid risks and evaluate those that cannot be avoided:	Risk assessment.
Combat risks at source:	At the design and planning stages.
Adapt work to the individual:	Including the way the workplace is designed, the type of equipment provided and the choice of production methods.
Give priority to collective measures of protection:	Such as roof edge protection rather than individual safety harnesses.
Take account of technological innovation:	Such as composite roofing panels that provide a safe working platform as well.
Make sure that employees are properly instructed:	Site induction, training, toolbox/task talks, stop-shift interventions.

HSG65 (HSE 2013) explains that organisations need to manage health and safety with the same degree of expertise, and to the same standards, as other core business activities if they are effectively to control risks and prevent harm to people.

Responsibility for this lies with line management, and Figure 12.9 illustrates how this might work within a medium-sized company. Also illustrated are some of the main health and safety responsibilities at each level of management.

12.4.3 Preventative and Protective Measures

Regulation 5 of the Management of Health and Safety at Work Regulations 1999 requires employers to have appropriate arrangements in place for the effective planning, organisation, control, monitoring and review of the **preventative and protective measures** required by regulation 4, schedule 1 (which are shown in Table 12.3).

To help line managers – such as the contractor's project manager/site agent – there should be support from a health and safety department along with in-company safety advisers. Safety advisers are not responsible for health and safety on site but serve in a supportive, consultative and audit role, as illustrated in Figure 12.9.

Small companies may engage external health and safety experts to help them, and many large contractors have external advisers who advise the main board of directors, especially regarding their legal liabilities.

12.5 Planning the Work

Construction sites present many hazards and even the best-managed sites cannot be considered 'safe'.

When planning for safety, it is tempting to list all the hazards you can think of, deal with them and 'tick the box' – job done! – but there are better ways to make sure that risk is properly managed and nothing important is overlooked.

12.5.1 Task-Based Approach

One method of planning for safety is to use a **task-based approach** such as the **job safety analysis** suggested by St. John Holt (2001). This is a logical approach which requires careful consideration of what is involved in the job, and a good understanding of construction methods is needed.

A job safety analysis consists of a number of logical steps:

1. Select the job/task.
2. Break it down into logical steps.
3. Identify the hazards in each step.
4. Devise ways of eliminating the hazard or reducing the risk from it.
5. Record the analysis.
6. Review and update as necessary.

> A lower league professional association football club, with a long and distinguished history dating back to the nineteenth century, has a ground capacity of approximately 15 000. The club's famous Spion Kop stand has stood derelict for some years after its safety licence was revoked, thus reducing ground capacity to 10 000. The club wishes to demolish the old stand and rebuild it as part of a community redevelopment of the area in which the club is located.
>
> The main reason for the club losing its safety licence for this stand is settlement of the mine workings upon which the banking for the terracing is built. The materials used for the banking are also contaminated. Part of the concrete terracing is covered, and this comprises structural steelwork and roof sheeting. Figure 12.10 illustrates a job safety analysis for this work.

One of the disadvantages of this method is that it is possible to miss hazards that may be present at the interface with other tasks. For instance, running concurrently with the demolition activity may be some drainage work to divert an existing sewer away from the new stand.

12.5.2 Hazard-Based Approach

As an alternative to looking at job tasks or activities on the programme, a **hazard-based approach** may be taken. This involves looking at the project and deciding whether specific hazards will arise and, if so, how and where. Common generic hazards are identified in Figure 12.11.

This method involves knowing what the hazards are and recognising them in different circumstances. It is not as easy as it sounds!

An advantage of this approach is that the ensuing risk assessment will be linked to hazards which can be easily recognised by most people from ACoP, published guidance and standard texts, and therefore, the process becomes familiar to those involved.

Task	Task breakdown	Hazards	Risk control measures
Demolish stand	1. Form access	a. Moving traffic	Traffic management plan with pedestrian segregation
		b. Moving plant	Banksman with machine
	2. Break up terracing	a. Moving plant	Banksman with machine
		b. Dust	Hose down regularly with water
		c. Noise	Muffled hydraulic breakers
	3. Remove safety barriers	a. Post-tensioned cables	Relieve stresses before removing
	4. Soft strip roof	a. Work at height	Safety harnesses attached to running line
		b. Wind	Check wind speed with anemometer – no work above Beaufort Scale 5
		c. Falls of materials	Exclusion zone beneath work area
	5. Dismantle steelwork	a. Work at height	Use man baskets and cherry pickers
	6. Remove earth banking	a. Moving plant	Banksman with machine
		b. Dust	Hose down regularly with water
		c. Contaminated ground	Laboratory analysis + PPE for all operatives
		d. Moving vehicles	Traffic management plan
	7. Demolish rear wall	a. Unstable wall	Progressive demolition by hand from MEWP* using temporary shoring system
		b. Dust	Hose down regularly with water

*Mobile elevating work platform (e.g. cherry picker, genie boom or scissors lift)

Figure 12.10 Job safety analysis.

This is likely to ensure a consistent approach to particular hazards across all sites within a contracting organisation. A disadvantage is that individual hazards may appear repeatedly, making the analysis repetitive and overlong and, of course, it is perhaps easier to overlook hazards than the task-based approach.

12.5.3 Utilising the Programme

Some contractors like to include health and safety considerations in their short-term contract planning.

For instance, in a six- or eight-week detailed programme of work activities, health and safety actions may be included on the bar chart, or health and safety-specific plant and equipment may be identified on the programme.

The Engineering and Construction Contract specifically requires the contractor to show provisions for health and safety requirements on the accepted programme (see Core Clause 31.2), and an example of how this might be done is given in Figure 12.12.

Work activity	Hazards
Demolitions	• Work at height • Falling debris • Premature collapse • Asbestos and other hazardous substances • Noise from heavy plant • Vibration from hand-held breakers, etc.
Bulk excavations	• Moving plant, e.g. excavators, dumpers, wagons • Contaminated ground • Underground services • Overhead power lines

Groundworks	• Collapse of excavations
	• Falling materials
	  ○ Into trenches
	  ○ From work above (e.g. roofing)
	• Excavation plant (striking/crushing/trapping, etc.)
	• Confined spaces
	• Water borne diseases (e.g. Weil's disease)
	• Wet concrete (dermatitis)
	• Dust/debris from disc cutters

Steelwork	• Work at height • Temporary instability • Cutting & welding equipment (burns, sparks, toxic fumes) • Manual handling (trapping, cuts) • Weather (over exposure)
Brickwork and blockwork	• Work at height • Fragile materials (roof sheets, roof lights) • Power tools for cutting and nailing • Manual handling (e.g. heavy blocks) • Dust/debris from disc cutters • Cement dust/mortar/plasticisers (dermatitis)
Woodwork	• Power tools and woodworking machinery • Electricity • Sharp hand tools • Wood dusts • Noise and vibration • Glues, paints, solvents, chemicals
Cladding and roofing	• Work at height • Manual handling (heavy rolls of roofing felt, heavy/large sheet materials) • Sharp edged/heavy materials (cuts/abrasions) • Wind and weather

Figure 12.11 Generic construction hazards.

12.5.4 Statutory Duties

Under the HSWA, contractors (as employers) have general duties to their employees – and to others not in their employment – as regards their health and safety. Section 2(1) of the Act

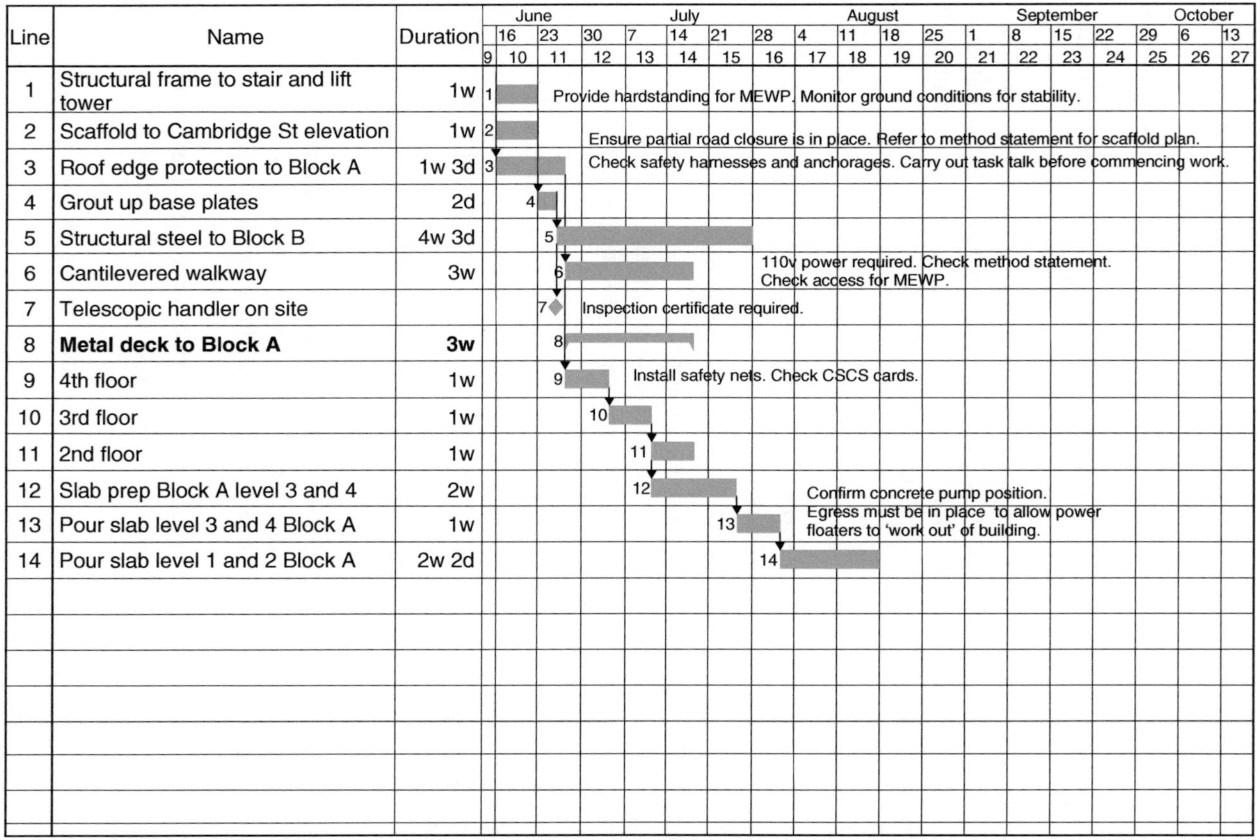

Line	Name	Duration	June			July				August				September				October		Notes
			9 10	16 23	30	7 14	21	28	4	11	18	25	1	8 15	22	29	6 13	26 27		
1	Structural frame to stair and lift tower	1w	1						Provide hardstanding for MEWP. Monitor ground conditions for stability.										
2	Scaffold to Cambridge St elevation	1w	2						Ensure partial road closure is in place. Refer to method statement for scaffold plan.										
3	Roof edge protection to Block A	1w 3d	3						Check safety harnesses and anchorages. Carry out task talk before commencing work.										
4	Grout up base plates	2d		4															
5	Structural steel to Block B	4w 3d		5					110v power required. Check method statement.										
6	Cantilevered walkway	3w		6					Check access for MEWP.										
7	Telescopic handler on site			7					Inspection certificate required.										
8	**Metal deck to Block A**	**3w**			8														
9	4th floor	1w			9				Install safety nets. Check CSCS cards.										
10	3rd floor	1w				10													
11	2nd floor	1w				11													
12	Slab prep Block A level 3 and 4	2w					12												
13	Pour slab level 3 and 4 Block A	1w						13	Confirm concrete pump position. Egress must be in place to allow power floaters to 'work out' of building.										
14	Pour slab level 1 and 2 Block A	2w 2d							14										

Planned by Asta Powerproject (Asta Powerproject is a registered trademark of Eleco UK Ltd. in the UK, the European Union (Community Trademark Office), Norway, the USA, and Australia.)

Figure 12.12 Provision for health and safety on accepted programme.

requires employers to *ensure the health and safety at work of all his employees*, and Section 2(2) extends this in further detail:

(a) The provision and maintenance of safe plant and **safe systems of work**.
(b) Arrangements for ensuring safe means of handling, use, storage and transport of articles and substances.
(c) The provision of information, instruction, training and supervision.
(d) Provision of a **safe place of work** and provision and maintenance of **safe access and egress** to that place of work.
(e) Provision and maintenance of a **safe working environment** and **adequate welfare facilities.**

HSWA Section 3(1) requires employers and the self-employed to *conduct their undertaking in such a way that persons not in their employment are not exposed to risks to their health or safety*. This imposes a duty of care on contractors, for instance, for their subcontractors and for passers-by and visitors to the site, etc.

12.5.5 Safe Place of Work

The idea of a safe place of work is derived from the law and, in particular, Section 2(2)(d) of HSWA referred to above and regulation 17 of CDM 2015 which requires such places of work to:

- Be safe and have properly maintained access and egress.
- Be made and kept safe and without risks to the health of workers.
- Have sufficient working space, taking into account any work equipment present.

Regulation 2 of CDM 2015 states that a **place of work** is *any place which is used by any person at work for the purposes of construction work*. However, a truly safe place of work is very difficult if not impossible to achieve, especially in construction. Even on the best-run projects, there will be hazards to deal with and therefore safe systems of work will have to be developed to deal with them.

12.5.6 Safe Systems of Work

Hughes and Ferrett (2011) and St. John Holt (2001) explain that a **safe system of work** is *a formal procedure that results from a systematic examination of a task in order to identify all the hazards and assess the risks*. More simply, it is a defined method for doing a job in a safe way.

If there are no hazards, there is no risk, but where risks remain, a safe system of work will be required. Wallace (2000) suggests that a safe system of work is made up of five key components:

- **Personnel**: they must be competent to do the job, have suitable training and be properly supervised.
- **Supervision**: the direct supervisor (ganger or foreman) must have knowledge and understanding of the work to be done, must make sure that the defined safe system of work is adhered to, must be capable of motivating the personnel doing the job, and must have the ability to take corrective action when necessary.

- **Working environment**: at the outset, this is defined by the safe place of work standard but, as the work operations unfold, this may well change. Hazards such as noise, vibration, physical strain and moving plant and machinery must be monitored and managed during the work activity. Further safe systems of work may need to be developed as the job progresses.
- **Procedures**: should be well designed and clearly define how the task shall be done correctly, safely and efficiently.
- **Permits**: may be necessary to prevent unauthorised personnel from entering a particular place of work and may be part of a safe system of work where hot working is to be carried out or where entry into a confined space such as a manhole, sewer or tunnel is required.

It is crucial that any **changes** to a safe system of work must be:

- Carefully considered with risks correctly assessed.
- Written down and any changes to the method statement carefully noted in writing.
- Properly communicated to the workforce via a task talk or similar discussion or, if necessary, through additional training.
- Monitored to make sure that they are working correctly.

Some safe systems of work can be verbal, perhaps where the risks are low, and the work is of short duration, so long as the hazards and control measures are explained to the operatives, preferably by their immediate supervisor or manager. St. John Holt (2001) suggests that operatives must not be allowed to devise their own method of working as this is not a safe system of work, and, more usually, safe systems of work are documented in the form of method statements.

12.5.7 Method Statements

Method statements which define a safe system of work are often referred to as **safety method statements** and they are often confused with method statements used in construction planning.

Best practice suggests that they are one and the same thing because safety is both an integral aspect of the planning of work operations and a legal requirement with respect to the provision of a safe place of work. Method statements may be written down in a tabular or prose format but, in both cases, they should include consideration of:

- Time needed to do the job.
- How the job will be done.
- The plant, machinery and temporary works required (e.g. earthwork support, scaffolding, and formwork).
- The order or sequence of work.
- Means of access and egress to and from the workplace.
- Lifting equipment and the means of getting materials in place.
- Means of preventing falls of people and objects.
- Means of preventing unauthorised access to the place of work (e.g. children at night).
- Interface with other trades working in the same place.

Contract	Contract No.	Prepared By	Date	Checked By	Date
Canal renovation	C2314	GHM	26 May	APH	27 May

Operation	Plant
Construct sheet piled cofferdam	22 RB crane BSP 900 pile hammer Komatsu 380 excavator Two 30T wagons 150mm diesel pump

Work sequence	Supervision and monitoring
Construct hardstanding for piling rig in lock mouth using imported quarry waste Erect guide frame and install Frodingham 3N piles Pump out water between hardstanding and piles Fill behind piles for access	Site engineer Piling foreman Operation to be monitored daily by the site agent Banksman to work with mobile plant Daily check on crane equipment, load indicators and operation
	Controls
Remove piling frame Construct top bracing with steel wallings and struts Remove hardstanding and install bottom bracing Erect secure ladder access to bottom of cofferdam	Authorised personnel area only When working over water to remove piling frame, life jackets and safety harness must be worn Area to be fenced off at night with chestnut paled fence Warning notices to be displayed either side of cofferdam 'Danger deep excavation' and 'Danger deep water'

Emergency procedures	First Aid	PPE schedules
Send for site first aider and/or call emergency services where necessary Rescuers must not put themselves in danger Follow first aid drill if appropriate Do not remove evidence Notify site agent	Dinghy to be moored adjacent to cofferdam Two Lifebuoy in wooden locker First aid box in site office 2-way radios	Safety harnesses Hard hats Gloves Welding goggles High visibility vests Life jackets Ear defenders

Figure 12.13 Safety method statement.

A sample safety method statement, using a pro forma style, is illustrated in Figure 12.13. Note that the construction method or work sequence is included together with appropriate risk control measures.

12.6 Industry-Specific Legislation

Nearly all UK workplace health and safety legislation is generic – meaning that the statutory duties apply irrespective of the industry involved. Consequently, legislation dealing with asbestos, hazardous substances, work at height, and the use of work equipment, etc., applies not only to construction but to other industries as well.

Construction does merit its own industry-specific legislation, however, in the form of the Construction (Design and Management) Regulations which impose statutory duties on clients, designers and contractors for the health and safety aspects of qualifying projects.

12.6.1 European Union Directives

The United Kingdom's membership of the European Union meant that EU Directives had to be incorporated into UK law. This resulted in a raft of indigenous secondary legislation – such as the Management of Health and Safety at Work Regulations 1999, the Work at Height Regulations 2005, the Manual Handling Operations Regulations 1992 and so on.

Following Brexit in 2021, this legislation has remained in place including the Construction (Design and Management) Regulations which were first introduced in 1994 as a consequence of the Temporary or Mobile Construction Sites Directive of the European Union (92/57/EEC). Initially, the CDM regulations differed significantly from the EU Directive – and introduced additional duty holders – but subsequent revisions in 2007 and 2015 have brought the regulations ever closer to the format of the directive.

12.6.2 The CDM Regulations

The CDM Regulations are primarily a project management tool placing emphasis on the management of health and safety by imposing statutory requirements on duty holders to plan, organise and manage construction work.

The regulations recognise the impact on health and safety that clients and designers can have because they make important decisions that influence future site operations – decisions such as project scope, design choices and specifications.

The regulations also acknowledge that design and construction are undertaken separately on many projects and so it is important to ensure that safety-critical decisions made at the briefing and design stages are coordinated with the management of health and safety on site.

12.7 The CDM Regulations 2015

CDM 2015 may be briefly summarised:

- The regulations apply to ALL construction projects, large or small.
- CDM applies to all types of work – building, civils, refurbishments, extensions, repairs and maintenance.
- They provide the steps required to ensure the effective management of risk.
- Clients have duties that cannot be delegated.
- The duties of domestic clients may pass to designers or contractors by choice or default.
- A **principal designer** must be appointed at the design stage, where there is to be more than one contractor on site.
- A **principal contractor** must be appointed to plan and manage the construction phase and to prepare a **construction phase plan**.
- Where there is to be more than one contractor on site there must be a **health and safety file**.
- The principal designer takes charge of the health and safety file and hands it to the client.

The regulations are supported by official guidance[3] which explains how duty holders may comply with the law. This guidance is important but does not have the legal status of an ACoP.

The regulations are in five parts, as illustrated in Table 12.4, and there are five schedules at the end of the regulations. The schedules deal with issues such as the particulars to be notified to the HSE on the F10 form (for notifiable projects), legal requirements for the provision of welfare facilities on site, details of work involving particular risks and so on.

12.7.1 Application

The regulations apply to *all* construction work, which is defined in regulation 2(1), irrespective of the size of project – even for small scale domestic jobs such as a new kitchen or bathroom or for external painting or roof work.

'Domestic clients' have duties under CDM but may elect for designers and contractors to act for them.

The wide-ranging applicability of the regulations emphasises the importance of planning how any construction work will be carried out where there is to be more than one contractor on site. This applies equally to a small domestic project requiring, say, a plumber and an electrician or to a multi-million £ project employing many trades or subcontractors.

12.7.2 Notification

Before work commences on site, the HSE must be notified of any significant project above certain thresholds where construction work will:

- Last longer than 30 days AND have more than 20 workers on site at any time
  or
- Exceed 500 person days.

Table 12.4 CDM structure.

	The structure of the CDM Regulations	
Part 1	Introduction	Defines legislation and terms used in the Regulations.
Part 2	Client duties	Stipulates what a client must do to make suitable arrangements for managing a project.
Part 3	Health and safety duties and roles	Sets out the duties of designers, the principal contractor and other contractors and the law regarding the creation of a construction phase plan and health and safety file.
Part 4	General requirements on all construction sites	The law relates to arrangements to ensure the safety of construction operations including safe places of work, site security, traffic management, excavations and the like, emergency procedures, etc.
Part 5	General	Concerns enforcement, transitional arrangements and review of the regulations.

The purpose of the notification thresholds is to limit the number of projects notified but also to capture short duration, labour-intensive projects – even domestic projects. Notification is a client duty but for domestic clients this must be done by the contractor (or principal contractor, if applicable) or by the principal designer should this duty be formally agreed with the client.

Projects may notified to HSE by post (on an F10 form) or, more conveniently, online if planned to last longer than 30 days or exceed 500 person days of construction work. If this is the case, additional duties apply.

12.7.3 Duty Holders

For any legislation to have any chance of being effective, the law must impose statutory duties upon responsible persons. In the case of the CDM regulations such persons, or 'duty holders' are:

- Clients, including domestic clients.
- Designers and principal designers.
- Contractors and principal contractors.

12.7.4 Client Duties

Clients have considerable responsibilities under CDM including making sure that their project will be carried out without risk to the health or safety of anyone, subject to the test of reasonable practicability.

Clients must also be satisfied that the welfare arrangements for the project are suitable. These duties are ongoing, and the client must review arrangements throughout the project.

Where there is likely to be more than one contractor on site, the client must appoint a designer as principal designer and a contractor as principal contractor. Furthermore, any designer or contractor appointed by a client *must have the requisite skills, knowledge and experience* to fulfil the role and this requirement extends to principal designers and principal contractors.

The client has a further duty to give all relevant information to any designers and contractors appointed. This is called the 'pre-construction information' and corresponds to information in the client's possession and information that could reasonably be obtained. Such information could be invaluable especially where it extends to existing buildings or the site and ground conditions of an intended project.

12.7.5 Designers

Designers' duties under CDM apply irrespective of the size of the project or whether the project is notifiable or not. This illustrates the central role that designers play in reducing risks both during construction and when the building or structure is in use.

Consequently, designers must:

- Avoid foreseeable risks to the health and safety of any person.
- Take steps to reduce or control risks through the design process.
- Provide information about such risks to the principal designer.
- Ensure that appropriate information is included in the health and safety file.

Designers must also make sure that clients are aware of their own duties before starting work on the design and share information with the client and other designers and contractors.

12.7.6 Principal Contractors

It is a statutory requirement to appoint a principal contractor where there is more than one contractor on site at a given point in time. Their role is to plan, manage and monitor the construction phase so that there is no risk to health or safety.

This requirement of CDM 2015 extends to small-scale domestic projects as well as large ones, but the scale of the project or the degree of sophistication of the duty holder does not minimise the duties to be fulfilled, subject to the test of 'reasonable practicability'.

On a project of any reasonable size, it is usually the main contractor who undertakes the role of principal contractor, but in some circumstances, it can be another contractor or even the client. In any event, the role of the CDM duty holder is quite separate from the position of main or managing contractor under the civil contract.

Preparation of the construction phase (health and safety) plan is also the principal contractor's responsibility, and this must be developed in sufficient detail before work is started.

The client must make sure that the construction phase does not start until the construction phase plan is drawn up by the principal contractor.

Other duties of the principal contractor include:

- Preventing unauthorised access to the site.
- Providing information to contractors and workers.
- Consulting with others.
- Drawing up site rules.
- Making sure that workers are provided with a site induction.

12.7.7 Other Contractors

Under CDM 2015, contractors (commonly 'subcontractors') are required to comply with directions given by the principal contractor and with such parts of the construction phase plan that are relevant to the work required of the contractor.

Key requirements of contractors are that they:

- Plan, manage and monitor their work.
- Apply the general principles of prevention when estimating how long their work will take.

Contractors are also required to provide information to their workers so that the work can be carried out safely. This includes a site induction and furnishing information concerning risks to health and safety such as will enable workers to comply with the law.

12.7.8 Principles of CDM

In fulfilling their duties under CDM 2015, all duty holders must apply **the principles of prevention** during both design and construction. The principles of prevention are laid down in schedule 1 of the Management of Health and Safety at Work Regulations 1999.

Under CDM 2015, taking into account the principles of prevention in fulfilling their statutory duties is a legal duty imposed on duty holders such as the principal designer and principal contractor. For instance, the principal contractor must plan, manage and monitor the construction phase taking such principles into account.

If there is only one contractor on site then this duty falls on that contractor when *design, technical and organisational aspects of the project are being decided* and when estimating the duration of the work or work stages.

12.7.9 Welfare Facilities

Construction work is dirty and dangerous, and workers have to endure all the elements of nature and more besides. Sanitary and washing facilities, together with the provision of changing and drying rooms are clearly vital, therefore, for the health and well-being of people working on site.

As a consequence, the standard of welfare facilities provided on site is given central importance in CDM 2015, and the various provisions expected by law are specified in some detail in schedule 2 of the regulations.

Contractors must price the required standards of sanitary, washing, drying, changing and resting facilities into their tenders and make sure that allowances have been made for the provision of drinking water and first aid.

In addition:

- The client must be satisfied that welfare arrangements comply with the regulations and that they are maintained throughout the project.
- For notifiable projects, the principal contractor must make sure that schedule 2 is complied with.
- Contractors must ensure that the requirements of schedule 2 are complied with in so much as the provisions apply to its workers.

12.7.10 Pre-construction Information

As far as any contractor is concerned, their ability to plan and manage health and safety on site is very much dependent on the quality of information provided by the client and by designers, as well as pre-construction.

Pre-construction means both during the design and preparatory stages of a project and during the construction phase as well. This is recognised in CDM 2015 which requires the client to provide information to both designers and contractors to help them to manage risk.

Designers are required to add to this pre-construction information by taking appropriate steps in the design process to eliminate, reduce or control risks and provide information about the design to help other duty holders to comply with the regulations.

Pre-construction information should refer to:

- The site and any associated hazards or risk issues.
- Existing services and hazardous substances.
- Existing information such as surveys or a health and safety file.
- Significant risks which may not be obvious to contractors and others.

The information supplied should be provided early in the procurement or tendering process and must be project specific. It may include notes on drawings or suggested construction sequences where this would be helpful to the contractor.

Pre-construction information might be presented to contractors and others in a convenient pre-construction information pack. Figure 12.14 summarises matters that might be included in such a pack.

12.7.11 Construction Phase Plan

Since they first came into force in 1995, the CDM regulations have distinguished between 'the principal contractor' and other 'contractors'. The purpose of this distinction is to elevate one of the contractors on site – usually the main contractor – to the statutory role of principal contractor with overall responsibility for the management of health and safety on the project.

As part of this responsibility, regulation 12(1) of CDM 2015 requires the principal contractor to prepare a construction phase plan which is defined in regulation 2(1) as *a document recording the health and safety arrangements, site rules and any special measures for construction work*.

This plan must be prepared before the start of the construction phase of a project, and it is the client's duty under regulation 16 to make sure that this happens. The purpose of the construction phase plan is to ensure that the construction phase is planned, managed and monitored effectively so that the work is carried out, so far as is reasonably practicable, without risk to health or safety. The plan is intended to be developed over the course of the project.

Regulation 12(2) stipulates that the construction phase plan must:

- Set out the health and safety arrangements and site rules.
- Take account of the industrial activities taking place on the site.
- Make specific provisions for work that falls into one or more of the ten categories listed in schedule 3 of the regulations – work deemed to pose particular risks to workers include the risk of being buried, of drowning, of falling or of working in compressed air (e.g. in tunnels).

Pre-construction information

Topic	Examples
1. **Description of project**	• Project description • Programme and key dates • Details of client, principal designers, designers, etc.
2. **Client's considerations and management requirements**	• Health and safety goals • Communications and liaison • Site planning and security, e.g. hoardings, permits, emergency procedures
3. **Environmental restrictions and existing on-site risks**	• Safety hazards, e.g. existing buildings or services, hazardous materials, unstable structures • Health hazards, e.g. asbestos, contaminated land
4. **Significant design and construction hazards**	• Design assumptions and erection sequences • Design co-ordination during construction • Design risks
5. **The health and safety file**	• Description and format • Requirements regarding content

Construction phase plan

Description of project	• Description of work and site details, programme and key dates • Contact details of duty holders
Management of the work	• Organisation structure and responsibilities • Arrangements for communication and consultation ○ Meetings, site induction, training • Arrangements for dealing with information flow, design changes • Selection and control of subcontractors • Production and approval of risk assessments and method statements • Reporting and investigation of accidents and near misses • Site rules • Fire and emergency procedures
Arrangements for controlling significant site risks	• Delivery, storage and removal of materials and waste • Plant management • Services • Falls, lifting operations, excavations, traffic management, etc. • Health risks including asbestos, hazardous substances, noise and vibration
The health and safety file	• Format and layout • Data collection • Storage

Figure 12.14 Pre-construction/construction phase plan.

Appendix 3 of the CDM 2015 guidance *Managing health and safety in construction* (HSE 2015) explains the role of the various duty holders in the construction phase plan and sets out what it should include. This is summarised in Figure 12.14.

12.7.12 Site Rules

Site rules are part of the principal contractor's safety management system and form the basis of the organisation, planning and control of site activities. They lay down standards for safe working practices, fire and emergency procedures and behaviour on site. They also include issues such as restricted areas, permits-to-work and hot work.

Under CDM 2015, regulation 12(2), it is the duty of the principal contractor to draw up a set of site rules that must be included in the construction phase plan. The site rules should be readily available to anyone working on the site in accordance with regulation 14(c).

12.7.13 Health and Safety File

Regulation 12(2)(e) of CDM 2015 requires the principal designer to prepare a health and safety file during the pre-construction stage, and this must be reviewed, updated and revised from time to time during the project.

The purpose of the health and safety file is to provide information about a project that might be useful to anyone who carries out construction work on the finished structure in the future. It is also applicable to those involved with the cleaning, maintenance, alteration or demolition of the structure.

The health and safety file is a legally required document that goes with the building or structure for which it was intended – the idea being to alert others to risks that only people involved in the original project know about. Residual hazards, design principles, special construction methods, location of services, the presence of hazardous materials, and so on, should be included in the health and safety file.

At the end of a project, the health and safety file must be passed to the client in accordance with regulation 12(10) who must equally pass on the file should someone else acquire the structure according to regulation 4(7). In doing so, the client must explain the nature and purpose of the file to the person acquiring the structure.

12.8 Health and Safety Training

Training in the construction industry is the responsibility of a sector skills council – Construction Skills – which is a partnership between CITB-Construction Skills, the Construction Industry Council (CIC) and Construction Skills-Northern Ireland.

The scheme is employer-led and operates under a levy and grant system. The objective of Construction Skills is to tackle the skills and productivity needs of the industry.

A large number of training courses are available to the industry, for both operatives and management, aimed at improving competence and addressing the huge skills shortage in construction. Training may take place on or off site. Site-based training includes site induction, toolbox talks and task talks.

Industry training includes national vocational qualification (NVQs), experienced worker assessments and card schemes. Different card schemes exist for demolition workers, scaffolders, plant operators, etc. Industry Card schemes include:

- **CSCS**: Construction Skills Certification Scheme which includes a health and safety competence test for over 200 trades, general operatives, supervisors and managers.
- **CTA**: Certificate of Training Achievement for plant operators now replaced by CPCS (Construction Plant Competence Scheme) for plant operators based on a test of professional competence and health and safety awareness.

12.8.1 Site Induction

In common with any employee starting a new job, people arriving at a construction site for the first time should undergo a site induction. This applies equally to workers, site management, client representatives, subcontractors and visitors.

The objective of the site induction is to make sure that the person concerned is fully aware of the rules, responsibilities and arrangements on site. On notifiable projects (lasting more than 30 days), it is the principal contractor's statutory duty to ensure that this is done. Personnel should also be made aware of site-specific hazards and how safe working is being managed. This is in addition to any information and training required by statute.

A site induction agenda will typically cover the following:

- The company health and safety policy, organisation and arrangements.
- The site safety management system and who to contact.
- Site rules.
- Personal responsibilities.
- Accident and unsafe work practices reporting procedures.
- Fire and emergency procedures.
- Welfare facilities.
- Site hazards.
- Risk assessments and safe systems of work.

Induction training will need to be repeated on a regular basis because there is a considerable 'churn' of personnel on construction sites.

12.8.2 Toolbox Talks

Toolbox talks are part of the contractor's safety management system and take place in the site cabin, usually between the supervisor or foreman and the operatives. A toolbox talk affords an opportunity to raise awareness and promote a two-way discussion about day-to-day health and safety matters.

HSG65 (HSE 2013) emphasises that the full participation of the workforce is essential in the management of health and safety. This requires genuine collaboration in order to solve problems and the trust that views will be listened to seriously and without recrimination.

A toolbox talk agenda will cover such items as:

- General site health and safety.
- Traffic management.

- Roofwork and ladders.
- Manual handling.
- Noise and dust.
- Abrasive wheels.
- Cartridge-operated tools.
- Buried services.
- Asbestos and asbestos-cement roof sheeting.

12.8.3 Task Talks

Task talks are similar in principle to toolbox talks, but they deal with specific site operations rather than general matters. Typical task talk topics would include proposed construction methods for:

- Deep excavations.
- The use and handling of drag-boxes for deep drainage.
- Installation of sheet-piled cofferdams.
- Hoisting and fixing external precast concrete cladding panels.
- Demolition work by hand.
- Fixing roof sheeting from a leading edge.

The task talk gives the operatives who will be doing the work the chance to listen to the contractor's method statement and make suggestions for improvements on the basis of their own previous experience.

12.8.4 Walk-through/Talk-through

Prior to undertaking a work activity on site, it may sometimes be beneficial to hear what an experienced person has to say about the job in hand and to see how the task is carried out.

The walk-through/talk-through process involves an experienced person demonstrating each step of the activity, no matter how minor, in order to fully communicate to others exactly what is involved in the task. Part of the procedure is to identify what might go wrong if the system of working is not carried out correctly.

This process is much more demonstrative than task talks and should be carried out in the location of the work to be done using the intended plant or equipment and PPE.

This will provide the workers involved with a step-by-step understanding of the job to be done including the use of PPE and the tools and equipment needed for the work.

12.9 Measuring Performance

It is often said that good health and safety management is good for business and it is certainly true that clients look to a contractor's health and safety record as part of their quality evaluation process.

However, a good health and safety record is not an end in itself as this could equally be achieved by good fortune as well as by good management. It is thus the measure of how well the health and safety management system is working that is the determining factor, as this is the basis for reliably well-managed sites and consistent advancement in learning from experience.

12.9.1 Monitoring

Health and safety performance on site should be regularly monitored in accordance with guidance in HSG65. This can either be done before things go wrong or after:

- Active (or proactive) monitoring involves regular inspections and checks to make sure that the safety management system is working.
- Reactive monitoring looks at what has gone wrong, why and how to prevent a recurrence of the event.

It is common practice to measure accidents by calculating the **incidence rate** or **frequency rate** either on a particular site or in the company as a whole where:

$$\textbf{Incidence rate} = \frac{\text{Total number of accidents}}{\text{Average number of persons employed}} \times 1000$$

and

$$\textbf{Frequency rate} = \frac{\text{Total number of accidents}}{\text{Total number of hours worked}} \times 100\,000$$

It is usual to have a rate for fatalities and major injuries, over 3-day injuries and reportable injuries which can be benchmarked with industry standards or 'best-in-class' for different-sized contractors.

'X-days without an accident' boards are frequently seen on construction sites. This, however, is a measure of failure, and it is much better to measure the success of the organisation in making sure that its safety management systems are working properly before accidents happen.

Active monitoring includes making direct observations of site conditions and peoples' behaviour (or unsafe acts), talking to people to elicit their views and attitudes and looking at documentation, reports and records.

12.9.2 Audit

Audit and review is part of the health and safety 'control loop' aimed at seeing how well the health and safety management system is working and where the defects are. HSG65 (HSE 2013) gives guidance on the subject.

Individual site managers and senior management in the company need to know how sites are performing with regard to management arrangements, risk control systems and workplace precautions.

Audits are often carried out by managers or directors from other parts of the company or other regions or areas where the company works. The idea is to audit without any preconceived notions or prejudices.

The safety audit process collects information about what is happening on the site and makes judgements about the adequacy and performance of the safety management system.

Some companies have long lists of 'compliance' issues and give the site a score out of 100, whilst other companies merely investigate whether, say, 15/20 key aspects of the management system are working or not and score the site 'compliant' or 'non-compliant'.

As part of the audit process, some contractors conduct a 'stop-shift' audit by stopping individual workers and asking them questions about health and safety issues.

12.10 Enforcement of Legislation

For the vast majority of construction work, the HSE inspectorate is the enforcement authority, notable exceptions being nuclear and rail sites which have their own inspectorates.

Generally, therefore, it is the HSE inspector (formerly factory inspector) whom the construction site manager will meet should their site be subject to an inspection.

Even though many contractors have their own safety advisers, either in-house or consultants, a visit from an HSE inspector – either impromptu or resulting from a tip-off – should be regarded as a welcome intervention. This provides an independent and authoritative view of the site safety management arrangements and should result in positive outcomes. Failure to respond to the inspector's suggestions, however, could lead to serious consequences as inspectors have considerable powers, including:

- Rights of entry at reasonable times, without appointments.
- Right to investigate, examine.
- Right to dismantle equipment, take substances/equipment.
- Right to see documents, take copies.
- Right to assistance (from colleagues or police).
- Right to ask questions under caution.
- Right to seize articles/substances in cases of imminent danger.

12.10.1 Enforcement Action

Prior to taking enforcement action, HSE inspectors are keen to offer suggestions, advice and encouragement. However, should unsafe conditions or breaches of health and safety law be found, the HSE inspector has the right to issue legal notices in order to rectify the situation.

Breaches of legislation may be under the HSW Act 1974 or under secondary legislation such as the Management of Health and Safety at Work Regulations 1999 or the CDM Regulations 2015.

Enforcement notices are written documents requiring a person to do or stop doing something which might be dangerous or in contravention of the law. Ignoring notices is a criminal offence. There are two types of notice:

- Improvement notice:
    - This says what is wrong and why.
    - It says how to put it right.
    - It gives a set time to take action.

- Prohibition notice
  - This prohibits the carrying out of unsafe practices or the use of unsafe equipment with immediate effect.
  - A prohibition notice may refer to a particular site operation, for example, excavation work or scaffolding or it might require the immediate suspension of all work on site.

HSE inspectors may also give written or verbal warnings or advice so that potentially unsafe practices may be averted without the need for a formal notice. They may also explain regulations and available guidance, etc., in confidence. This is quite important because formal notices frequently lead to prosecution, and any responsible contractor will wish to avoid this at all costs.

12.10.2 Prosecution

If there is a serious accident or incident on site, this will normally be investigated by the HSE, who will determine the cause of the accident, what needs to be done to prevent a recurrence, what lessons can be learnt and whether a prosecution is relevant.

It would be a misconception, however, to believe that prosecutions only result where there has been an accident or fatality. Consider CDM 2015 regulation 13(1) for instance:

> **Duties of a Principal Contractor in Relation to Health and Safety at the Construction Phase**
>
> 13.—(1) The principal contractor must plan, manage and monitor the construction phase and coordinate matters relating to health and safety during the construction phase to ensure that, so far as is reasonably practicable, construction work is carried out without risks to health or safety.

This is a statutory duty and failure to plan and manage the construction phase could lead to a prosecution, notwithstanding any accident occurring. However, a prosecution would not ensue without a warning or official notice to comply.

The prosecution may be brought by the HSE or by the Crown Prosecution Service (CPS). Where there is a fatality, it is the police who will investigate, in conjunction with the HSE who will provide technical support. Any ensuing prosecution will be made by the CPS.

> A construction company was prosecuted following an unannounced inspection by Health and Safety Executive Inspectors who found numerous uncontrolled high risks. They included poor welfare standards, dangerous electrical systems and inadequate health and safety provision on site.
>
> The company was found to have failed to effectively plan, manage and monitor the works in contravention of regulation 13(1) of CDM Regulations. It had complied with previous written enforcement notices, but poor standards were again allowed to develop.
>
> The case was heard in the Magistrates Court and a large fine was imposed.
>
> Source: *HSE*

Death or injury is not a prerequisite for a prosecution and, where there is an incident that could have resulted in serious harm, but did not, a prosecution may well follow.

> The collapse of tunnels under construction at Heathrow Airport in 1994 is a case in point where multi-million-pound fines were handed out by the courts, but no one was even scratched!

It is not only employers who face the possibility of prosecution. Employees and subcontractors also have statutory duties under health and safety legislation, and this could result in a court appearance.

Court proceedings commence in the lower courts – the Magistrates Court in the United Kingdom. If the defendant is found guilty of a summary offence, the Magistrates' Court has power to impose an unlimited fine (for Level 5 offences), a community order or a jail term of up to 12 months.

More serious offences will be referred to the Crown Court, where the fines are also unlimited, but custodial sentences may be considerably longer.

12.10.3 Corporate Manslaughter

When accidents occur, it is not uncommon for a prosecution to follow. This might be heard in the lower courts or, where the offence is more serious, in the higher courts where sentencing can be more severe. The basis of a prosecution will be founded in health and safety law, which is part of the criminal code, and a guilty verdict will result in an appropriate punishment for what is, after all, a criminal offence.

The decision of the court will reflect society's view as to a fair retribution for the crime committed which is reflected in the prevailing sentencing guidelines open to the court.

In the unfortunate case of a fatal accident, a guilty verdict should result in a custodial sentence. However, where the accident is the result of an organisational failure – such as failure to plan, manage and monitor work activity or to provide adequate resources – it can be difficult to pinpoint the individual or individuals at fault.

In a very small company, the guiding hand of the business will undoubtedly be a single identifiable person but for larger businesses there will be a complex management structure which makes the identification of a guilty individual extremely tricky. This is why countless corporate manslaughter cases – including the Herald of Free Enterprise, Clapham Junction and Piper Alpha – have historically failed in the courts.

Corporate manslaughter is where an organisation – through its senior managers – conducts its activities in such a way as to lead to a person's death. With the advent of the Corporate Manslaughter and Corporate Homicide Act 2007, companies can no longer escape their liabilities for management failures that lead to fatal accidents.

The Act introduces a criminal offence of corporate manslaughter in England, Wales and Northern Ireland (homicide in Scotland). This means that an organisation can be tried for causing a fatal accident without the need to identify an individual person or persons to carry the blame. If found guilty, the organisation is culpable, and it is the organisation that will be punished.

The level of fines under the Act are technically 'unlimited' but a court will consider factors such as the seriousness of the offence and the size of the company involved. The annual turnover of the company is a determining factor in deciding what is a 'proportionate' sentence.

For a medium-sized company with a turnover between £10 million and £50 million, the sentencing guideline for a very high level of culpability is a fine of between £1 million and £4 million.[4] A court may also impose remedial orders and publicity orders.

As a consequence of the Act, a distinction has to be made between corporate liability and that of individual company directors or senior managers. Such people would not be prosecuted under the Corporate Manslaughter and Corporate Homicide Act 2007 but would be subject to prosecution under existing health and safety legislation **and** to the common law offence of gross negligence manslaughter.

In this case, and provided that there was sufficient evidence, the failure of individuals who play a significant role in the making of decisions about how the whole company's activities are to be managed or organised could face a custodial sentence in the event of a fatal accident.

12.10.4 HSE Prosecutions Database

The official website of the HSE is an invaluable learning resource on the subject of health and safety generally but no more so than its register of convictions. This publicly available information records recent and historic health and safety convictions in a variety of industries, including construction. It can be found at: https://www.hse.gov.uk/enforce/convictions.htm.

The database is often referred to as the 'name and shame' list, but this would be to miss the point. By searching the database, managers and others can learn from the mistakes of individuals and firms large and small. They might also take time to reflect on how well they are managing health and safety on their own sites or how well they understand their legal obligations and the essentials of good practice.

A random case taken from the HSE database involves a housing developer who was prosecuted for failing to adequately assess the risk associated with the delivery of materials to site. This led to life-changing injuries to a member of the public and resulted in a considerable fine for breaching CDM Regulation 13(1).

> **Regulation 13(1)**
>
> The principal contractor must plan, manage and monitor the construction phase and coordinate matters relating to health and safety during the construction phase to ensure that, so far as is reasonably practicable, construction work is carried out without risks to health or safety.
>
> £53,000.00 fine £2,892.35 HSE costs

Anyone convicted of an offence will have been punished according to the law and there the matter should end, but health and safety offences are a 'black mark' against the otherwise good name of a contractor, not least because they remain on the contractor's record for five years – a bit like penalty points on your driving licence.

The HSE database can be searched either geographically or by industry and there is a special section for construction which is subdivided into four sectors. The data can be searched either by individual cases or by the particular breach or breaches of the law and can be sorted in A–Z or Z–A order. Of particular interest is the option to search Case Details which provides useful background information.

All prosecutions which have resulted in successful convictions are included except those cases which are either subject to appeal or applicable in Northern Ireland or the Republic of Ireland.

12.10.5 HSE Enforcement Notices Database

The HSE also has a public register of notices, similar to that for prosecutions, which makes the details of improvement notices and prohibition notices issued publicly available on the Internet. This information is kept 'on the record' for five years.

The register can be searched geographically or by industry and particular types of notice can be filtered if required. The notice reference number is hyperlinked which then leads to the details of the particular notice concerned – such as the type of notice, when served and upon whom and the specific breach of the law involved.

12.11 Accidents and Incidents

The reason why it is so important to discover the hazards present in any project is so that the risks arising from them can be properly managed. If hazards are not dealt with, they may manifest themselves in such a way as to cause injury, ill health or damage to persons and/or property.

Should a serious injury or fatality occur, there is a legal obligation to file a report to the HSE. Employers have the additional duty to keep an official record of the incident in an 'accident book'.

Quite often incidents occur that do not result in an accident or fatality but nevertheless might be classed as incidents with the potential to cause harm. An incident of this nature is a 'near miss' and might need to be reported to the authorities so that an investigation can be held.

12.11.1 RIDDOR 2013

Reporting accidents and ill health at work is a legal requirement under the Reporting of Injuries, Diseases and Dangerous Occurrences Regulations (RIDDOR) 2013.

The legal duty is on employers, self-employed people and people who control premises. Employers must report any work-related deaths, injuries, cases of disease, or near misses involving their employees, irrespective of where they are working. This must be done by submitting form F2508 online, but fatal and specified injuries can be reported by calling the HSE Incident Contact Centre (ICC).

Not all accidents are reportable, but the following **must** be reported:

- Deaths and major injuries.
- Over three-day injuries.
- Injuries to members of the public or people not at work who have to be taken to hospital.
- Certain work-related diseases.
- Dangerous occurrences – where something happens that might have resulted in an injury (i.e. 'near miss').

Reportable major injuries include fractures (but not fingers, thumbs or toes), amputation, dislocation of the shoulder, hip, knee or spine, loss of sight, etc., and must be reported without delay. Over three-day injuries are those which are not 'major' but result in an employee or self-employed person being absent from work, or unable to work normally, for more than three consecutive days. They must be reported within 10 days of the incident occurring.

Detailed records of any reportable injury, disease or dangerous occurrence must be kept, including when and how the incident was reported, the date, time and place of the event and details of those involved and a brief description of what happened. Records can be kept in a file, on computer or in the accident book.

12.11.2 Accident Book

Not all accidents are reportable but every employer is required to keep an accident book (where there are more than ten people employed) under the Social Security (Claims and Payments) Regulations 1979. Details of every accident must be recorded, including how it happened.

12.11.3 The Cost of Accidents

Anyone who has suffered an accident, no matter how slight, knows of the pain and trauma that this can cause. Imagine, then, the impact of a serious accident at work which could result in many weeks of suffering or even life-changing injuries, meaning that the injured party may not be able to work again. The consequences of a fatal accident are unimaginable.

There is no metric for human suffering, and therefore the cost of accidents is usually measured by loss of workforce morale, loss of reputation of those responsible and money. Monetary loss includes loss of production, time-related costs, delayed project completion and any resulting liquidated and ascertained damages and reduced productivity when work on site actually resumes. Research indicates that such uninsured costs are 11 times more than the costs covered by insurance, such as civil actions for damages taken by the injured party or their family. Even when civil claims are settled, insurers will pay out less than the full damages awarded as there will be an excess or uninsured loss to pay. With regard to the costs associated with criminal prosecution, it is not possible to insure against fines for breaches of health and safety law, but this does not apply to court costs.

In an industry where typical contract margins are commonly less than 5% of turnover, half of which is normally regarded as risk, it is clear that accidents can severely impact a contractor's 'bottom line'. Perhaps the greatest damage, however, is to the reputation of the contractor which no amount of money can repair.

References

Dekker (2014) *The Field Guide to Understanding Human Error*. CRC Press.

HSE (1999) *Reducing Error and Influencing Behaviour*. HSE Books. Free download https://www.hse.gov.uk/pubns/books/hsg48.htm

HSE (2000) *Management of Health and Safety at Work, Approved Code of Practice*. HSE Books Withdrawn – replaced by guidance.

HSE (2013) *Managing for Health and Safety* HSG65. HSE Books. Free download https://www.hse.gov.uk/pubns/books/hsg65.htm

HSE (2015) *Managing Health and Safety in Construction, CDM 2015, Guidance on Regulations*. HSE Books.

Hughes, P. and Ferrett, E. (2011) *Introduction to Health and Safety at Work*. 5th edn. Routledge.

Joyce, R. (2015) *CDM Regulations 2015 Explained*. ICE Publishing.

St. John Holt, A. (2001) *Principles of Construction Safety*. Wiley-Blackwell.

Wallace, I.G. (2000) *Developing Effective Safety Systems*. Institution of Chemical Engineers.

Web References

http://www.hse.gov.uk/riddor/
https://www.hse.gov.uk/enforce/convictions.htm
https://www.hse.gov.uk/statistics/industry/index.htm

Notes

1 Mace Group plc.
2 Laing O'Rourke.
3 HSE (2015) Managing health and safety in construction (L153).
4 www.sentencingcouncil.org.uk.

\multicolumn{2}{l	}{}	Chapter 13 Dashboard
\multicolumn{2}{l	}{**Key Message**}	○ When planning a project, it is essential to follow a logical thought process to develop a realistic and workable schedule.
\multicolumn{2}{l	}{**Definitions**}	○ **Scope**: Describes what the project will include and what it will not include (i.e. what is in and out of scope). ○ **Work breakdown structure (WBS)**: A key tool in 'scope management' which is used to subdivide the project into manageable packages.
\multicolumn{2}{l	}{**Headings**}	**Chapter Summary**
13.1	Introduction	○ Planning the construction stage of a project starts when a contract has been awarded. ○ Post-bid acceptance is known as the pre-construction period.
13.2	Principles of Project Planning	○ Developing the project logic involves several steps including establishing key project dates, activities or events. ○ Computer software is used for speed, for considering 'what-if' options and for high-quality professional presentation. ○ BIM technology facilitates 3D modelling of buildings and structures combined with time and schedule data to create virtual construction sequences.
13.3	Pre-construction Planning	○ The contractor will put a lot of effort into planning the timing and logistics of the project pre-construction. ○ Due allowance needs to be made for the lead times of all major components and associated trades packages. ○ Site layout planning is an essential part of pre-construction planning. ○ The contract master programme is called the 'as-planned' programme.
13.4	The Baseline Programme	○ The baseline programme represents the first version of the master programme 'frozen' in time. ○ The baseline remains 'frozen' when the programme is updated and a parallel set of bar lines shift in accordance with progress. ○ The baseline makes it easy to compare the original plan with actual progress and forensically scrutinise reasons for any differences.
13.5	Requirement Schedules	○ Internal requirement schedules – key materials, plant and subcontractors. ○ External requirement schedules – key dates for the release of information from the client or client's representative. ○ Contractors invariably need more information from designers which generates requests for information (RFIs).
13.6	The Target Programme	○ A target programme may be developed as well as a master programme. ○ This is a compressed version of the master programme with time taken out of the critical path. ○ Target programmes place added pressure on the site management team and subcontractors. ○ Demands for faster working must be balanced with the health, safety and quality standards expected.
13.7	Contract Planning	○ The purpose of contract planning is to maintain control and ensure that the project is completed on time and within the cost limits. ○ Once work commences, detailed planning is required to determine when and how particular site activities are to be carried out. ○ The as-built programme is a record of actual progress.
\multicolumn{2}{l	}{**Learning Outcomes**}	○ Understand the concept of a work breakdown structure. ○ Understand how project planning evolves as the work proceeds. ○ Be able to recognise the key elements of a master programme. ○ Appreciate the role of BIM and project management software. ○ Understand the idea of a baseline programme. ○ Appreciate how actual progress relates to 'as-planned'.
\multicolumn{2}{l	}{**Learn More**}	○ Chapter 16 builds on the principles established in this Chapter and adds competencies to understanding. ○ Chapter 16 illustrates the use of baselines in delay analysis.

13

Planning the Project

13.1 Introduction

Once a successful bid for a contract has been made, the contractor's processes for planning the construction stage will be mobilised to make sure that the contract proceeds in a timely and efficient fashion.

The period post-bid acceptance is known as the pre-construction period, and this is followed by the construction phase and handover of the completed project – Stages 5 and 6 of the RIBA Plan of Work.

13.2 Principles of Project Planning

When planning a project – irrespective of the stage – it is essential to follow a logical thought process to develop a realistic and workable schedule.

Initial thoughts may be developed on paper, but computer software should be used for speed, for considering 'what-if' options and for high-quality professional presentation. Developing the project logic involves several steps or thought processes:

- Getting a feel for the project.
- Establishing key project dates.
- Establishing key activities or events.
- Assessing how long the activities will take.
- Establishing the sequence.
- Deciding which programming technique to use.
- Choosing schedule calendars and timescales.

13.2.1 Getting a Feel for the Project

It is important for the planner or project manager to get a feel for the project because a clear appreciation of the scale, scope and complexity of the works helps to trigger the natural human instincts of when things look right or look wrong. These instincts are not simply based on hunch but also on experience and familiarity with the construction process.

It is just as important to appreciate the financial scale of the project as well. Experience will often identify the project value or rate of expenditure as being inconsistent with the time allowed. To get a feel for the project:

- Visit the site.
- Read and understand the project brief (if applicable).
- Study the drawings, models and project documentation.
- Assess the scale and scope of the project.
- Assess the approximate value of the project.
- Consider the rate of expenditure (i.e. the relationship between value and time).

13.2.2 Establishing Key Project Dates

The overall parameters of a project will be determined by establishing the key dates. Some of these will be established by the client/client team – the procurement process and the design and construction periods, for instance – and others will be subject to less predictable factors – such as negotiating finance and obtaining statutory approvals.

Some key dates are particularly important because they will be included in the tender documentation and, subsequently, incorporated in the contract. Others will be common sense. They will include:

- Key contractual dates:
  - Signing of the construction contract.
  - Site possession/access date and any sectional possessions/access.
  - Project start and finish dates.
  - Sectional or phased completion dates.
- Other key dates in a project include:
  - Financial approval.
  - Client approval of design brief.
  - Planning approval granted.
  - Tender submission date.
  - Holiday periods.
  - Practical completion date.
  - Commissioning or handover.

13.2.3 Completion

One of the essential purposes of the contractor's master programme is to capture the key dates for the project. Some of these dates derive from the Contract Particulars (JCT) or Works Information (NEC) such as the start date, possession/access date(s), sectional completion dates and contract completion date.

Prior to the commencement of construction work, the contractor must be granted possession of, or access to, the site. There is a distinction between the two:

- Possession means that the contractor controls the site.
- Access means that the contractor is permitted to be on the site for the purpose of carrying out the works.

> There may be legal implications to this distinction because possession means that no one can access the site without the contractor's permission. In some countries – the USA and Canada, for instance – the contractor may be granted a lien (legal charge) over the site. Other implications include responsibility for site security, protection of the site and insurances.

Possession or access dates are usually stated in the contract but sometimes these dates can be deferred, leading to delays commencing the works. There could be several reasons for such delays including the late completion of enabling works by other contractors, problems with legal ownership, party wall issues, planning permission delays and so on.

Once possession/access is granted – either partially or in full – the contractor's obligation is to proceed with the works with due diligence with the aim of completing the works on or before the contract completion date.

There are, however, several concepts of 'completion' in construction, as explained in Table 13.1, which the contractor needs to consider when programming the work or during the construction period itself.

13.2.4 4D BIM

By using BIM technology, the contractor can combine 3D models of buildings and structures with time and schedule data to create virtual construction sequences. Consequently, 3D models + time = 4D BIM.

Table 13.1 Completion.

Contract completion	The date required for completion of the works as stated in the contract. This date may be stipulated as a specific calendar date or might be a stated number of weeks counting from the start date.
Sectional completion	Where a project is to be carried out in stages and handed over to the client for use before the whole of the project is complete.
Planned completion	This is the date on which the contractor plans to complete the works. It could be a date earlier than the contract completion date. If the contractor can complete by this earlier date, then savings could be made on the time-related costs for the project thus enhancing the profitability of the contract.
Extended completion	A date beyond the contract completion date as determined by the contract administrator or project manager. This date will be calculated with reference to the time impact of variations to the contract, delays due to weather and events beyond the contractor's control in accordance with the conditions of contract.
Practical completion	This is the date, as certified by the contract administrator, that the works are effectively complete, barring the rectification of patent defects. It marks the beginning of the defects correction period which is usually 12 months. The concept of practical completion is usual in building contracts.
Substantial completion	This is when the works are largely complete but to a lesser degree than 'practical' completion. This is a common concept in civil engineering where, for example, a highway is complete and can be used by traffic but where side roads are incomplete, landscaping is not finished, final tests have not been carried out or where traffic management measures are still in place.

With the information from a BIM model, the contractor can quickly extract the data needed for scheduling the construction phase. There are many advantages to using this technology including encouraging collaboration, providing a deeper understanding of the master programme through the model, increasing the safety of construction operations, reducing cost and rework and benefitting from greater project coordination.

When using project management software in conjunction with BIM models, activities can be entered into the activity list, thereby creating the work breakdown structure (WBS), and objects from the model can be associated with each activity using drag and drop. Once the activities in the schedule have been linked logically (start to start, finish to start, etc.) a timeline can be run to simulate the construction sequence. This can be used to identify omissions, refine the master programme or to plan site operations in more detail.

In common with other uses of BIM models, the only data that can be extracted is that which resides in the model. Therefore, should data be missing, this will be missing from the contractor's programme and the contractor's planning engineer will have to interpret the designer's intentions to produce a complete picture.

When the design changes, and the model is updated, the information changes can be directly reflected in the contractor's programme.

13.2.5 Establishing Key Activities or Events

Once the main project parameters have been established, the next step is to determine the key activities or tasks to be carried out, together with any important events which should be included in the programme:

- **Activities** are tasks or jobs to be done which have a time value – such as piling, steel erection, drainage works and so on.
- **Events** are points in time by when things happen but have no time value. For example, the start of construction work is an event that triggers a series of activities which do have a time value.

Choosing the key activities/events to include in the schedule depends upon the skill and judgement of the planner. They could be based on:

- Objects in a BIM model – such as walls, slabs and doors.
- NRM2 work sections in a bill of quantities – such as precast concrete, carpentry and suspended ceilings.
- Design elements typically found in a cost plan based on NRM1 – such as facilitating works, substructure, services, prefabricated buildings and building units.
- Personal preference, perhaps loosely based on any/all of the above – such as frame/external envelope, floors, roof structure and cladding.

Other generic activities or events can be added including erection of site hoardings and accommodation (site set up), building watertight, practical or substantial completion and clear site. Time must also be allowed for the erection of tower cranes, installation of temporary works, construction of site access, traffic management and so on.

13.2.6 Work Breakdown Structure

The Association of Project Management (APM) Body of Knowledge (BoK) (2006) suggests that the 'scope' of a project must be identified and defined to *describe what the project will include and what it will not include* (i.e. what is in and out of scope). The BoK also suggests that the project scope needs to be managed effectively to both keep track of the many changes that will inevitably occur and to retain control of the project.

A WBS is a key tool in 'scope management' which is used to subdivide the project into manageable packages that can be individually planned, scheduled and controlled. This ensures that the complete scope of the project is included in the planning and control process.

There are several methods of presenting a WBS, but they all share the same common features – a hierarchy or cascade arrangement together with a multi-level numbering system.

Figure 13.1 illustrates the principles of a WBS, where Level 0 is project level and Levels 1, 2 and so on are subdivisions of the project. This enables unique reference numbers to be given to the elements or work packages that make up the project scope. In this example:

- 1.0.0 denotes a new factory,
- 1.1.0 denotes the substructure and
- 1.1.1 and 1.1.2 denote the piling and foundations, respectively.

Should there be a programme of projects on the same site – such as several similar factory buildings – then each factory could be given a unique reference at Level 0 – Factory 1, Factory 2, Factory 3 and so on. The reference for the foundations for Factory 3 would therefore be 3.1.2, making this distinguishable from the foundations of the other factories.

Most linked bar chart project management software packages facilitate the use of WBS and allow activities on the programme to be 'rolled up' or expanded to an appropriate level of detail depending on the audience (e.g. client, contractor or subcontractor).

13.2.7 Activity Durations

Assessing the duration of activities is not an exact science. This might be based on empirical data or calculation, but experience, and a working knowledge of the design and construction process, is an invaluable ingredient.

Calculations can be performed by considering the relationship between the quantity of work to be done and the output or rate of production anticipated. However, outputs used by estimators are not suitable for planning as these relate to bill of quantities items where the measured quantities consist of separate items for the main items of work and their associated ancillary items.

In brickwork, for instance, the bills of quantities will include an item for facing brickwork and there will also be incidental items for laying the damp-proof course, for closing cavities

13 Planning the Project

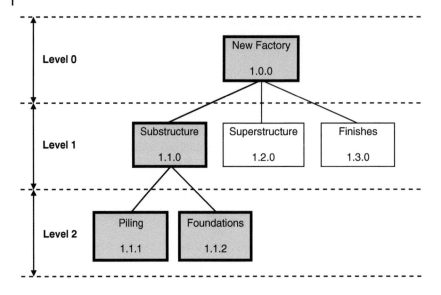

Level 2 elements may be further subdivided as required eg:

1.1.2 Foundations

- 1.1.2.1 Excavate pile cap
- 1.1.2.2 Cut off and trim piles
- 1.1.2.3 Blind base of pile cap
- 1.1.2.4 Fix formwork to base
- 1.1.2.5 Fix steel reinforcement
- 1.1.2.6 Set bolts in base
- 1.1.2.7 Place concrete
- 1.1.2.8 Strip formwork
- 1.1.2.9 Backfill working space

Each element of work is given a unique code number. This may be linked to the master programme or to items in an activity schedule. The referencing system may be used for preparing work packages for subcontracts, for preparing budgets or for costing and linking to the contractor's CVR process.

Figure 13.1 Work breakdown structure.

and for building in lintels. This level of detail is too granular for planning purposes and would make calculations too complex. It is better to:

- Include 'incidental' items in the overall output per m^2 of facing brickwork or, alternatively:
- Calculate the quantity of bricks in thousands and divide this by the output of the gang in bricks laid per day.

Figure 13.2 illustrates how a bill of quantities item for brickwork can be used to decide on labour resources and the planned duration of the activity. The duration of 14 weeks is

13.2 Principles of Project Planning

		Qty	Unit	Rate	£	p
	Masonry					
	Brick/block walling					
	Walls; half brick thick					
A	Facing brickwork; PC price £750 per 1000; Skins of hollow walls; Stretcher bond; Gauged mortar 1:1:6; Pointed one side [14.1.1.1]	3596	m²			

The activity duration is given by:

Bricklaying gang of 5 bricklayers and 3 labourers (5:3 gang).

Output of 600 bricks per day per bricklayer.
Total = 600 x 5 = 3 000 bricks per gang day.

$$\frac{\text{Quantity of brickwork in m}^2 \times \text{No of bricks per m}^2}{\text{Output of gang per day}} = \frac{3\,596\text{ m}^2 \times 59 \text{ bricks/m}^2}{3\,000 \text{ bricks/day}}$$

= 71 days/5 days per week
= **14 weeks**

Figure 13.2 Activity durations.

based on one 5:3 bricklaying gang although, on larger sites, more than one gang may be employed.

The mechanics of this calculation are straightforward, but the difficult bit is to judge what the output should be. This will depend on factors such as:

- The gang make up.
- The type of work.
- Location of the work.
- The standard expected in the specification.

13.2.8 Programming Techniques

There is no strict rule as to which programming technique should be employed. This needs to be considered in the light of the size and complexity of the project, any personal preferences or whether there are any stipulations in the contract documentation.

Bar charts are the easiest to use, but they can give misleading results because there is no strict logic imposed on the schedule. It might be better to use critical path methods such as linked bar charts or arrow or precedence diagrams to overcome this problem. However, for repetitive work such as housing projects, line of balance may be preferred, or for roadworks, tunnelling or repetitive civil engineering work, time-chainage diagrams could be the best application to use.

A further important consideration is whether the software accommodates BIM models. Some software packages – such as Asta Powerproject – allow Revit and other models to be imported and linked directly to the schedule. Others – such as Primavera P6 – rely on third-party software to link the schedule to the Revit model.

When using project management software packages, it is usually advisable to enter activities and events first and arrange them in approximate order. This helps to establish the correct logic and avoid the possibility of getting in a tangle when working at the computer screen – which is particularly likely to happen when making the logical links between activities, as it is not always possible to see the whole picture on screen.

Software packages allow both hard and soft constraints to be included in the schedule:

- Hard constraints are those that cannot be changed and can even override the scheduling logic. Such activities MUST start on a certain date or be completed by a certain date.
- Soft constraints do not change the scheduling logic and are identified as activities that must start on or after a given date or be completed on or before a given date.

13.2.9 Calendars

Computer software packages provide the flexibility to choose a variety of calendars according to the demands of a project.

Calendars normally default to Monday–Friday working but can be changed to suit project requirements, working times adopted in certain countries or cultures, resources available or to individual tasks in the schedule.

Therefore, if a project calendar is using a normal five-day working week but a particular activity requires 24-hour working – such as construction of core walls using sliding formwork – the specific task(s) associated with that activity can be assigned a special 24-hour calendar.

13.2.10 Timescales

Project management software allows the planner to choose the most appropriate timescale for a project, and there is usually a choice of three or four headings.

For projects with a long duration, it might be appropriate to choose a main timescale in years and a secondary timescale in months, whereas a project of shorter duration could have a timescale in months and weeks, respectively.

In either case, it is conventional to give each month or week a number for easy reference. These numbers could be negative to represent the design and pre-construction period with positive numbers for the construction period itself.

13.3 Pre-construction Planning

Pre-construction procedures vary according to the size and importance of the project, but it is nonetheless a busy period for both the client and the contractor.

Following acceptance of the contractor's tender, the client must prepare the contract documents – the form of contract (JCT, NEC, ICC, etc.), specification, drawings and contract pricing document (bills of quantities, activity schedule, etc.) – although, in some cases, work will start on site under a letter of intent.

The Contract Particulars to the form of contract (JCT) – or Contract Data (NEC) – brings into sharp focus the contractor's obligation to complete the works within the contract period and this will drive the planning and programming of the works.

Other legalities to be completed include the provision of a bond or a parent company guarantee by the contractor and a joint venture agreement might also be required for large projects.

At this stage, the contractor will put a lot of effort into planning the timing and logistics of the project and this might include design work where contractor design is part of the contract.

13.3.1 Pre-contract Planning

There is no rule as to how long the pre-construction period is – it may be six weeks or six days or it could be several months for large project – but this is the time when the contract start date is negotiated, unless this is specified in the tender documents. It is also a time to make arrangements for:

- Pre-contract meetings with the client to establish protocols for communications, interim valuations, progress meetings and submission of the contractor's programme.
- Commencement of work on site.
- Procurement of subcontractors and suppliers.
- Site layout planning.
- Preparing the master programme.
- Preparation and approval of the construction phase health and safety plan.
- Preparation of information requirement schedules.
- Preparation of contract budgets.

It must be pointed out that no two companies undertake the same procedures at the pre-construction stage, which will depend on company policy and established standard routines for every new contract. In this chapter, the procedures outlined could be considered normal in a large or medium-sized company.

13.3.2 Pre-contract Meetings

There are no official protocols to determine what happens at the start of construction projects and much will depend on how big the project is or whether there are unusual complexities that need to be discussed before work starts.

Prior to the commencement of the project, however, a series of meetings will certainly be held. Some will involve the client and the contractor and others will be held internally by both the client and construction teams.

Once a contract has been awarded – and while the formalities of the contract and start date are awaited – the contractor will firstly organise a pre-contract meeting. This will normally involve members of the estimating and pretender planning team and senior members of the construction team together with quantity surveying, procurement and planning personnel.

The purpose of the pre-contract meeting is to announce the award of the contract and hand over relevant tender stage documentation to the construction team responsible for undertaking the work.

Following this, more focused meetings will take place concerning risk management, health and safety, procurement, subcontractors, design of permanent and temporary works, site planning and security, mobilisation and, of course, preparation of the master programme.

Prior to the formal pre-contract start-up meeting between the client and contractor, contractors sometimes make a presentation to the client team to explain plans for carrying out the project, to present the programme of works and to introduce key construction team personnel.

Following this, the pre-start meeting – usually chaired by the architect, engineer or project manager – will be organised which enables channels of communication to be set up for the issue and distribution of project information. It is important to establish lines of communication between the client's team and the contractor that are clear and transparent to avoid confusion and disagreement.

Apart from obvious contact details, the contractor needs to know who has authority under the contract to issue instructions, to certify payment and to agree on the value of variations. Other matters to be discussed include organisational and contractual details, dates of progress and valuation meetings, safety management arrangements, the commencement of work on site and so on.

13.3.3 Procurement

Procurement at pre-construction stage is based on quotations received at tender stage for materials and subcontract work.

It is usual to schedule the estimator's tender allowances and to use them as a benchmark for further quotations once the contract has been awarded. More competitive quotes may be obtained in a competitive market by negotiation, re-tender or via reverse auctions but where the construction market is buoyant or resources are in short supply, the opposite may happen.

Subcontract procurement is often handled by quantity surveyors who might also obtain materials quotations if there is no buying department in the company.

A key element of the procurement process will be the negotiation of lead times, as this will be influential in finalising the master programme and the contractor's ability to complete the project by the contract completion date.

Items with very long lead times might be commissioned by the client pretender and included as defined provisional sums in the tender pricing document. The contractor is required to include the work contained in defined provisional sums in the programming of the works.

13.3.4 Procurement Programme

To ensure that subcontract work starts and finishes on time, it is usual for the contractor to produce a procurement programme for each work package. This will be based on the subcontract programme developed from the target programme. Subcontractors will thus be tied to start and completion dates consistent with the target programme and not the longer master programme.

The procurement programme will often show both negative and positive time as the subcontract may well comprise design and fabrication work as well as on-site activities. Negative time is that required to organise design aspects of the workpackage or to pre-order key materials with long lead times for delivery.

Negative time is the lead-in time needed before work starts on site whereas positive time is the time needed to carry out the subcontractor's work on site.

13.3.5 Procurement Programmes

Figure 13.3 illustrates a procurement programme where the bar lines represent the procurement periods for each major component supplier or subcontractor. Each bar line has been flagged with three milestone symbols relating, respectively, to information requirements, placing the order and to commencement on site.

The following examples illustrate how the procurement programme may be developed into specific programmes for the procurement of specific packages – in this case 'steelwork' and 'finishes':

- **Steelwork**: Figure 13.4 shows a detailed bar chart display indicating the procurement requirements for the steelwork activity using a simple linked bar chart and early warning symbols. In practice, the steelwork activity would be shown as a single bar on the main procurement bar chart. However, by using a project management software package, the detailed programme may be accessed simply by clicking on the bar line or, in some packages, by clicking on an icon to expand the activity and show the detail.
- **Finishes**: Figure 13.5 shows a procurement bar chart for the 'finishes' operation depicting lead-in times incorporated within the bar lines. Lead-in times are for the benefit of both the contractor and those providing the information. This might be the project architect or other designer or the design and build designer engaged by the contractor.

Each of the above examples clearly indicates 'must have' or 'must do' information reminders. Management action must be taken when key dates indicated by the milestone symbols are not met or if the data is not available on the specified date. Delays resulting from the late release of information must be confirmed in writing at the appropriate time to have contractual effect.

13.3.6 Lead Times

Before physical construction work or off-site fabrication or assembly works are undertaken on a project, project managers and contractors need to consider the impact of the design, procurement and manufacturing stages required for the various materials and components that make up the project.

It cannot simply be assumed that all such materials are readily available 'off-the-shelf' and available for immediate delivery to site. Whilst 'just-in-time' delivery is common practice in construction, this must be carefully planned and thought out well in advance.

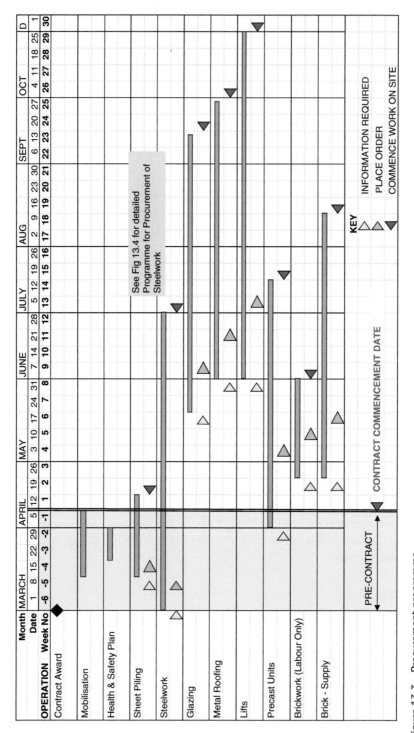

Figure 13.3 Procurement programme.

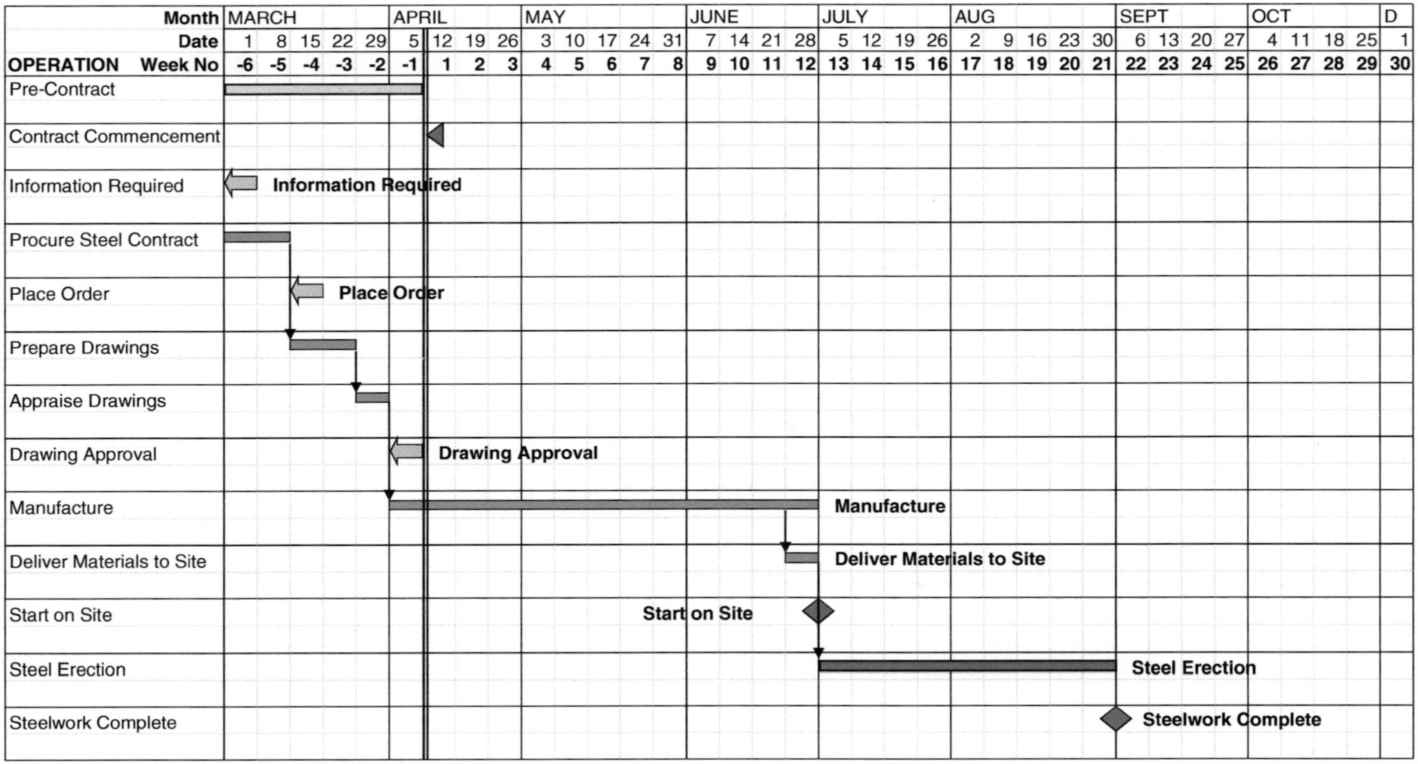

Figure 13.4 Procurement – steelwork.

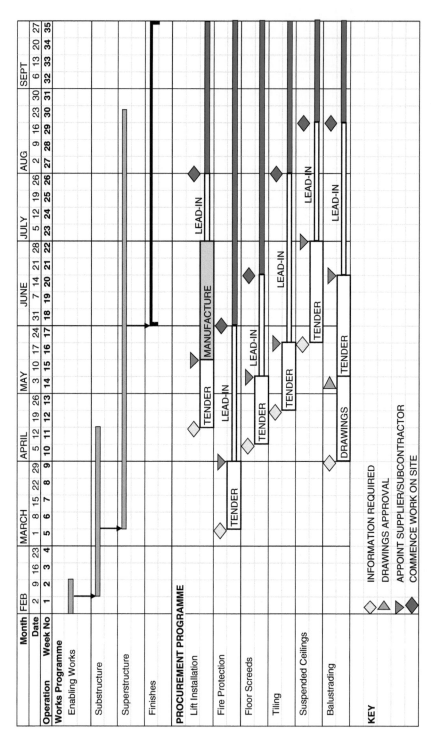

Figure 13.5 Procurement programme – finishes.

Consequently, due allowance needs to be made in the project master schedule for the 'lead times' of all major components and associated trades packages as well as in the contractor's master programme for the works.

Examples of the items that may have to be considered on a project include piling, structural steelwork, cladding, roofing, suspended ceilings and lifts as well as such 'normal' elements as brickwork, blockwork and general joinery components, etc.

Lead times for certain materials can exert considerable influence on a client's procurement strategy – especially post-pandemic. This can be relieved to a certain extent by the early appointment of the main contractor and the key trade packages as well as the early placing of orders for important materials, components and off-site fabrication.

13.3.7 Site Layout Planning

Site layout planning is an essential part of pre-construction planning and forms the platform for the contractor's management of the works. This includes consideration of site security, site accommodation and services, storage areas and hardstandings, access roads and the provision of telephones and internet access where possible.

It is usually the contractor's responsibility to find space on the site for the required facilities such as site cabins, batching plants or materials storage but sometimes it will be necessary to rent land adjacent or close to the site from private landowners. In some cases, the client will be able to provide land to the contractor where there is no space on the worksite itself.

On some large projects – especially in the public sector – the client will require a presence on the site and accommodation and other facilities will be specified in the tender documents and priced in the contractor's preliminaries.

In some instances, the contractor will be required to submit site layout proposals for approval by the client's representative prior to commencing work on the project.

13.3.8 The Master Programme

The contract master programme is referred to in delay analysis as the 'as-planned' programme and it is an important management tool. It is an essential requirement in the coordination and control of the many integrated tasks to be undertaken during a project and is also used by the client's contract administrator to monitor the contractor's progress.

The master programme is prepared during the pre-construction period – there may be several such schedules for a major project or programme. It will show:

- The major phases or sections of work.
- The planned sequence of construction.
- The main subcontract packages, lead times for key material supplies and principal resources such as cranes, access equipment, materials handling equipment and temporary works.
- The key dates for the release of design team information.

The master programme forms the basis of the contractor's budgetary control and financial forecasting procedures and helps the client to assess cash-funding requirements to meet the contractor's interim payments. It also provides a means of recording progress throughout the construction period which helps the client and contractor to take action should the planned completion date be in danger.

Despite its importance, the master programme is not usually a contract document following the case of *Yorkshire Water Authority v. Sir Alfred McAlpine & Son (Northern) Ltd.* (1985). If the master programme was to be included as a contract document, it would somewhat impair its flexibility and usefulness as a management tool. It would involve both parties having to strictly adhere to the programme and the contractor would be obliged to start and finish each activity by the programmed dates, and in the programmed sequence, or risk being in breach of contract.

The master programme may be presented in Gantt chart, linked bar chart, network or precedence format according to preference or contractual requirement. It should show the contract possession and completion dates for the project, together with the main work activities on site. It should also indicate the sequence of operations and relationships or dependencies between them.

> Without dependencies, a critical path cannot be calculated and the programme loses its effectiveness as a control document and a means of demonstrating the cause and effect of delay and disruption, the contractor's entitlement to extensions of time or any loss and expense or compensation entitlement under the contract.

An extract from a master programme is shown in Figure 13.6 which illustrates three stages of a steel-framed building project:

- Site establishment = 7 weeks
- Substructure = 9 weeks
- Superstructure = 14 weeks

A single bar line distinguishes the overall duration of each activity beneath which are its subtasks. Superstructure, for instance, has three subtasks – erect steel frame, erect roof pod and stair installation.

The main activity bar is called a 'summary activity' because it summarises the tasks or activities within it. Using project management software, the subtasks can be 'rolled-up' and hidden to produce a simpler overview of the master programme. Finish-to-start and start-to-start dependency links have been shown connecting the various activities on the schedule.

13.4 The Baseline Programme

Once the master programme has been prepared – and approved, if required under the contract – this is saved as a 'baseline'.

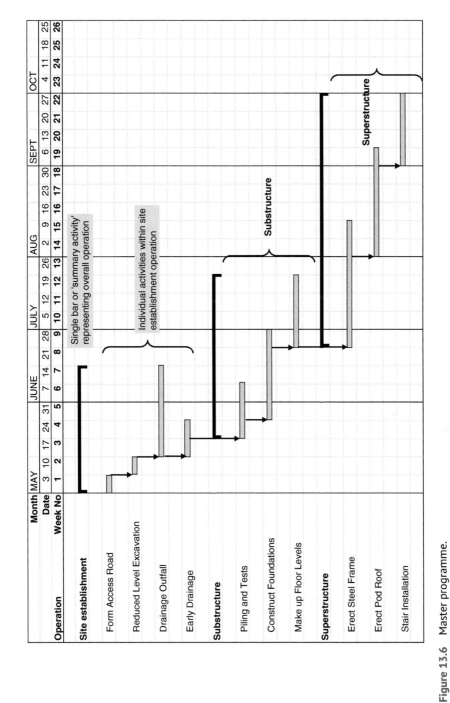

Figure 13.6 Master programme.

> The baseline programme represents the first version of the master programme and may be viewed as the original programme 'frozen' in time. It is often referred to as the 'as-planned programme' especially by forensic delay analysis practitioners.

13.4.1 Managing Change

As the contract works progress, changes invariably occur. This might be because of delays or disruption due to design changes or late information or unforeseen events such as the discovery of bad ground or contamination.

Clearly, there is no point working from the original master programme, or baseline, as this does not reflect reality and will be out of date. The contractor, therefore, needs to accommodate changes in a revised or updated programme. The purpose of the baseline programme is to allow the updated programme to be compared with the original programme.

When the baseline is first created, the project management software will indicate two identical bar lines against each activity on the programme, as indicated in Figure 13.7. Once the impact of changes is applied in the software, however, the baseline remains 'frozen' and the updated programme is seen in parallel with the bar lines shifted in accordance with progress. Figure 13.8 shows the updated programme set against the original baseline.

13.4.2 Progress

Having created a baseline, it is an easy task to compare the original plan with actual progress and forensically scrutinise reasons for any differences. This is crucial in establishing cause and effect when applying for extensions of time or loss and expense/compensation events under the contract.

Comparing the updated programme with the baseline is an important aspect of project control and can have implications regarding the contract completion date, extensions of time, liquidated damages and entitlement to additional payments under the contract – compensation events under NEC contracts or loss and expense where JCT contacts are used.

> By using a baseline, the impact of change on the remaining activities and, especially critical tasks, can be identified and instances where sub-critical tasks have become critical can be highlighted. The contractor's order of working could also be affected as could the use and availability of resources.

It should be noted that the tracking facility provided by project management software is imprecise and not an acceptable foundation for entitlement claims. This is because progress is normally recorded as a percentage completion rather than by using precise dates. This is illustrated in Figure 13.8 which shows the project baseline and progress alongside expressed in percentages.

This method of project control is a matter of good practice for contractors, but it can also be a contract condition that the contractor is required to submit an updated

13.4 The Baseline Programme

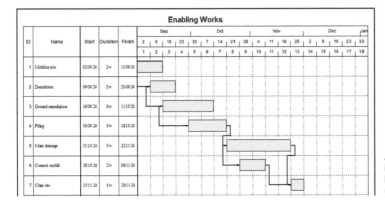

A series of 7 activities is shown together with logic links (dependencies).

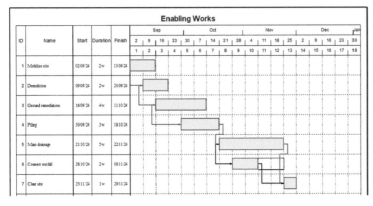

A critical path analysis (CPA) of the linked bar chart is shown.
Activity 6 has 2 weeks float.
All other activities are critical.

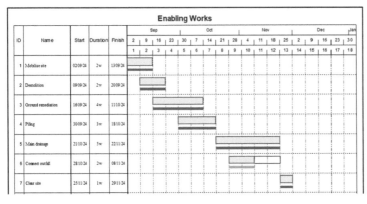

Logic links are removed for clarity. The project baseline (red/green lines) is shown.

Figure 13.7 Baseline programme.

programme to the client/employer, especially if change is likely to impact contract completion.

In some cases, several baselines may be saved on a single project. This might be done, for instance, when a significant instruction has been received by the contractor and the impact of the instruction needs to be monitored independently as the work proceeds.

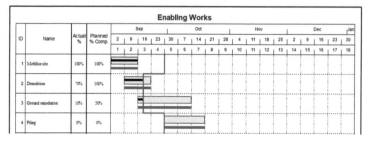

Progress is shown at week 4 (red line). Activities 2 and 3 are behind schedule.

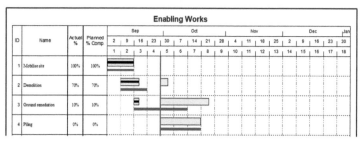

The project is rescheduled at week 4. The (red) baseline remains static.

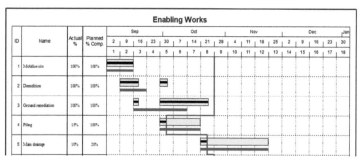

Progress is shown at week 8 (red line). Activity 4 is behind schedule. Activity 5 is ahead of schedule.

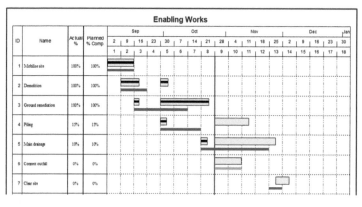

The project is rescheduled at week 8. Activities 5 and 7 indicate that the project is in delay compared to the baseline.

Figure 13.8 Updated baseline.

13.5 Requirement Schedules

Schedules are an important aspect of pre-construction planning and during the construction stage. Conventionally they are produced on standard forms or by using spreadsheets but nowadays many contractors and project teams harness the power of the Internet. This could mean using Cloud-based collaborative technology – such as OneDrive and Dropbox – or, on larger projects, common data environments such as Oracle Aconex or Zutec, as described in Chapter 4.

Requirement schedules can be simply classified:

- Schedules of the contractor's internal requirements.
- Schedules of key dates for the release of information from the client or client's representative.

The contractor's internal schedules include those concerning key materials, plant and subcontractors, whereas those relating to client-based information requirements can include nominated/named subcontractors and suppliers (where permitted under the contract), drawings and information release and requests for information from the contractor.

13.5.1 Key Materials Schedules

A materials schedule should be prepared in conjunction with the contractor's planning or buying department showing the materials required to meet key operational programme dates. The schedule should show ordering information referenced to the bills of quantities or other pricing documents, delivery required-by dates and supplier contact details. The material requirements should be assessed from the contract drawings and not from pricing documents as these may not be accurate or up to date.

Phased deliveries may be arranged for materials such as bricks and precast floor units. On sites with restricted space, or inner-city refurbishment projects, just-in-time deliveries to suit the contract programme may be preferred.

It is important that the material schedule contains full information regarding what is expected of the supplier. Additionally, site management personnel must be encouraged to use and update the schedules as it is important that material suppliers are kept informed of the progress position and any amendments to planned deliveries.

13.5.2 Plant Schedules

The plant schedule indicates key dates for the major items of plant required to carry out the contract works. This could be presented in a tabular format or included on the master programme. Most linked bar chart software packages allow resources to be attached to activities for such purposes and to satisfy contractual requirements under NEC contracts.

Most construction plant is hired and this needs to be ordered in good time as some equipment – such as large mobile cranes, tunnel boring machines, SPMTs, etc. – is not available off-the-shelf.

Scheduling plant requirements is a good discipline as a careful check should be kept on plant expenditure compared to the allowances within the contract preliminaries. Overspending on plant is common at site level mainly because site managers can always find an excuse to keep plant on-hire for a bit longer than intended.

13.5.3 Subcontract Schedules

The coordination of subcontractors and suppliers is an essential part of project control.

A suitable schedule can enable the site manager to overview key contract dates with respect to progress, to chase up the architect for information outstanding and to make sure that subcontractors are 'on message' as regards their start date.

Details may also be shown on the schedule relating to the subcontract order, contact address and notification dates. Reference should be made to the subcontractor's programme prepared at the contract stage. The late release of key information relative to subcontractors and suppliers may lead to an extension of time claim when the delay is caused by the contract administrator.

13.5.4 Information Requirement Schedules

It is essential to monitor and record the receipt of contract drawings and other information issued by the architect or engineer and it is the contractor's responsibility under the contract to give adequate notice of information requirements such as:

- Setting out dimensions and measurements to site boundaries.
- Reinforcement details for pile caps, foundation beams, etc.
- Details of ground floor services and pockets for fixing bolts.
- Fixing details for cladding panels.
- Door, window and ironmongery schedules.
- Colour schedules for internal decoration.
- Service layout details.

Should the contract administrator fail to release information on time, the contractor may be entitled to an extension of time and loss and expense. In this respect, the schedule shown in Figure 13.9 allows the contractor to record the date of the information request and compare this with the information release date.

Key information requirements may also be highlighted in the form of milestone events on the master programme. Alternatively, an early warning system, such as that described in Chapter 16, could be employed to highlight project requirements. On NEC contracts, early warning is a contractual prerequisite in any event.

The purpose of requirement schedules is to aid the smooth running of the contract by providing a programme for the preparation and release of information by the contract administrator to the contractor throughout the project. However, late information has been a long-standing bone of contention between contractors and contract administrators and has been the cause of many claims and disputes. The problem is especially prominent with conventional contracts where the design is incomplete at the tender stage, and information is still missing long after work starts on site.

Figure 13.9 illustrates information requirements regarding several work packages. The date when information is required from the client's design team is indicated on the

DETAILS REQUIRED	Contract Newcastle						Contract No		
	Design team						Workpackage requisition		
Work element/package	Arch	Str. eng	Serv. eng	Date required	Lead-in period	Start on site	Requisition required by	Last date for order	Start on site
Demolitions	✓	✓		15 Dec	3 weeks	19 Jan			19 Jan
Foundations and lift pit	✓	✓	✓	5 Jan	4 weeks	2 Feb			2 Feb
Precast units	✓	✓	✓						
Raised access floor	✓		✓						
Suspended ceilings	✓		✓						

Figure 13.9 Information requirements schedule.

schedule. Common sense is needed when specifying such dates, as it is pointless asking for information relating to painting schedules at week 1 of a 70-week project and expecting to formulate a claim at week 20 simply because the information has not been received.

13.5.5 Requests for Information (RFIs)

Designers do not produce fully detailed drawings and other information for projects – they produce what is known as a 'design intent'. This is sufficient to indicate what is to be built but leaves the contractor to 'fill in the blanks' to produce detailed construction information.

As part of this process, documents issued by the design team will need to be clarified, information gaps will have to be filled and ambiguities between documents will have to be resolved.

This is the role of requests for information (RFIs) which may be made using standard forms submitted by email – as illustrated in Figure 13.10 – or accommodated within a common data environment system.

Contract	Wigan	Request date	15 Jan
Contract No	84/1	Prepared by	AB

Issued by	Issued to	Distribution
C. Wooton (Site Agent)	Architect – DBS Associates	Quantity surveyor Contracts manager File

INFORMATION REQUIRED	INFORMATION SUPPLIED
Drainage layout for ground floor slab to include: 1. Gully positions 2. Main service ducts in laboratory unit 3. Connections to toilets 4. Floor duct for waste water from machine hall	
CONTRACTOR'S COMMENTS	

Information required by	6 Feb
Date of response	
Further action date	

Signed	Signed
Date	Date

Figure 13.10 Request for information sheet.

The RFI illustrated shows a formal request for information of a detailed nature – in this case concerning drainage layout details. This information is needed by the contractor to determine the location of holes, pockets and ducts prior to pouring a ground floor slab area. Using this format, the information requested can be compared with the information received on the same request form.

13.6 The Target Programme

During the pre-construction period, it is common practice for contractors to develop a target programme as well as a master programme. This is effectively a compressed version of the master programme with time taken out of the critical path.

This can be done because whilst critical activities have no spare time (float), they can contain time risk allowances. These allowances represent the time difference between the most optimistic duration for an activity and the most realistic time, considering variables such as productivity, the weather, resource problems, supply chain issues and so on.

These are intuitive allowances – time contingencies – and not float, which is derived by critical path analysis, i.e. by calculation. Removing these time contingencies shortens the critical path and therefore the overall duration of the project.

> A master programme shows completion in 22 weeks as shown in Figure 13.6. The target programme shown in Figure 13.11 indicates completion in 17 weeks.
> This saving of 5 weeks is achieved by:
>
> - Reducing the duration of the drainage outfall (1w), construct foundations (2w), make up floor levels (1w), erect steel frame (1w) and erect pod roof activities (2w).
> - Increasing concurrency by making the early drainage and piling and tests activities start–start relationships.

To achieve the target programme, time savings must be realistic and not just hopeful. They therefore need to be based on the informed opinion of the construction team – especially the contracts manager and site agent – as well as discussions with subcontractors. Increasing resources, changing methods of working or submitting contractor-led design change proposals can enable the project to be realistically completed faster than the time shown on the master programme.

Target programmes place added pressure on the site management team and subcontractors alike and a careful balance needs to be struck between the demands for faster working and the health, safety and quality standards expected.

13.6.1 Legal Implications

The legal implications of compressed programmes can be far reaching in terms of the contractor's own supervision and resources, the procurement and management of subcontractors and the ability or willingness of the design team to provide the necessary drawings and other information in time.

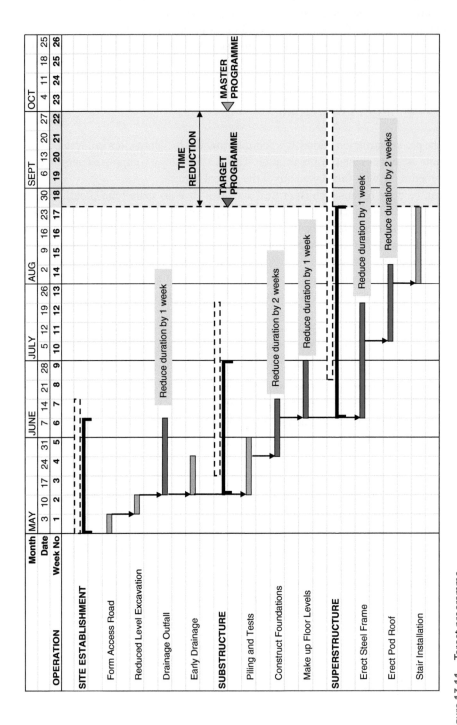

Figure 13.11 Target programme.

Following the judgment in the case of *Glenlion Construction* v. *The Guinness Trust* (1987), a programme showing a shorter completion period than that required under the contract – even though it may be annotated with information deadlines – will not be successful in putting pressure on the architect or engineer to produce drawings and other information to a tight timescale.

Any claim by the contractor based on the failure of the contract administrator to provide information, thereby causing delay, will not be valid unless the contract is delayed beyond the **contractual** completion date.

Furthermore, the architect or engineer is obliged only to furnish drawings and instructions 'within a reasonable time of the conclusion of the contract', and legal opinion takes the view that the employer is not under an implied obligation 'to enable the contractor to complete by the earlier date'.

Therefore, provided that the contractor can still complete within the contract period, prolongation expenses cannot be recovered, and the employer is under no obligation to pay compensation if the contractor is unable to achieve an accelerated programme.

13.6.2 Practical Implications

A master programme with an early planned completion date differs from a target programme in that the employer can see that the contractor plans to finish the contract early. The consequence of this is that the time difference between the planned completion and contract completion dates might be regarded as terminal float. This is incorrect, however, as this time is a risk allowance and not float calculated by critical path analysis.

> A major risk issue in having both a master programme and an internal target programme, is that the time risk allowances in the master programme could be absorbed by employer delay and/or disruption to the programme that does not result in an extension to the completion date. This could become an issue between the employer and the contractor and could even lead to a dispute.

A further consequence of compressed critical activities is that sub-critical activities can become critical thereby opening up the possibility that these activities too could be compressed. Compressing a programme can be a commercial gamble which might not work out as planned, but the reward is higher profits if it does. Contracting is a risk-reward business!

13.7 Contract Planning

Once the contractor has commenced work on site, planning is required at regular intervals to determine when and how particular site activities are to be carried out. This type of planning is called 'contract planning' and is carried out monthly or weekly (short-term planning).

Contract planning is planning at a fine level of detail and takes place during the contract period. This is when the master programme is developed into stage programmes showing part of the master programme in more detail. Alternatively, the contractor might produce a series of short-term programmes at weekly or fortnightly intervals to plan day-to-day work in detail.

Contract planning involves monitoring the master programme and updating it 'as built', reporting progress to management and making sure that the health and safety plan and all safety method statements for specific activities are up to date.

Planning at this stage of the project aims to flesh out the master programme and provide a basis for the detailed day-to-day arrangement of work on site. This helps to prevent inefficient working or duplication of effort on site and enables the contractor to make the best use of resources.

Contract planning is carried out by the main contractor to maintain control and ensure that the project is completed on time and within the cost limits established at the tender stage. Subcontractors contribute to the process either by submitting their work programme for approval or through discussion with the main contractor.

There are several reasons for contract planning, including:

- To monitor the master programme – monthly, weekly and daily.
- To plan site operations in detail in the short term.
- To communicate safe working methods to the workforce and to discuss their views before work starts.
- To optimise and review resources and productivity.
- To integrate subcontractors into the works programme and deal with interfaces between the various work packages.
- To raise early warnings with the contract administrator, to keep progress under review and to identify reasons for slippages in the programme measured against the baseline programme.

13.7.1 Subcontract Programmes

The master or target programme frequently shows subcontract packages as a single bar line simply because it is impossible to show every site activity in detail – otherwise, the programme would be too complex and unwieldy.

This principle is demonstrated in Figure 13.11 which shows, for example, 'piling and tests' as a single bar on a target programme with a duration of three weeks.

In reality, however, 'piling and tests' is effectively a subproject with its own list of activities and events and the contractor needs to exercise control over them.

Contractors, therefore, commonly prepare subcontractor programmes for this purpose and an example is given in Figure 13.12. In this example, the piling and tests package is shown in detail with several activities and events happening within the overall duration of three weeks indicated on the target programme.

Subcontract programmes are usually developed from the contractor's target programme which is a compressed or shortened version of the master programme. This means that the work package contractors are under pressure to complete the works earlier than the date stated in the master programme. Subcontract activities therefore become sub-critical or critical with little or no spare time to absorb delay or disruption to the works.

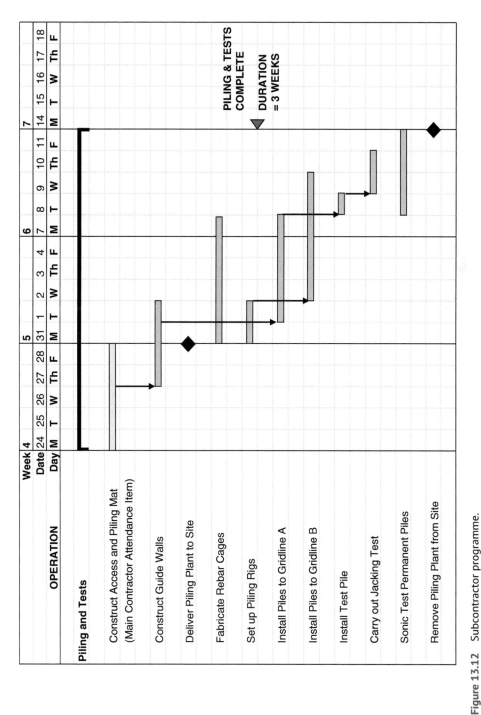

Figure 13.12 Subcontractor programme.

13.7.2 Stage Programme

The contractor's master programme is frequently developed as the project progresses to show a finer level of detail that could not in all practicality be done on the master programme.

This is the purpose of a stage programme which is illustrated in Figure 13.13. This shows part of the substructure activity shown on the master programme which comprises three sub-activities:

- Piling and tests.
- Construct foundations.
- Make up floor levels.

The construct foundations activity is shown in expanded form with logic arrows indicating dependency. Some activities (e.g. fix rebar) are continuous and others (e.g. blind bases) are intermittent.

Stage programmes are useful where alternative construction methods are being considered. By preparing a stage programme, detailed cost versus time comparisons can be made between alternatives.

The outcome could reveal that the time savings made by choosing a more expensive construction method could reduce preliminaries costs significantly and possibly avoid the risk of liquidated and ascertained damages being charged should a cheaper but slower method be chosen.

13.7.3 Short-term Programme

When the contract stage is reached, the level of planning becomes more detailed, and many contractors prepare short-term programmes on a monthly or weekly basis. On refurbishment-type projects of a relatively short duration, planning schedules may even be prepared daily for each trade gang employed on the project.

For example, a programme covering, say, two weeks' work might be produced, and at the end of the first week of this programme, another two-week programme would be prepared for the following fortnight reflecting progress, problems and any changes.

Short-term programmes afford the contractor a much better means of controlling day-to-day operations on site and act as a useful method of communication between the site manager and the foreman, gangers or work package contractors. They can also be used in toolbox talks to discuss the health and safety implications of the programme with the workforce.

The principles of short-term planning based on a two-week review period are illustrated in Figure 13.14 which shows the 'construct foundations' activity from the stage programme in Figure 13.13. Figure 13.14 shows:

- The programme for weeks 5 and 6.
- The short-term programme for the next two-weekly period (weeks 6 and 7).

The short-term programme might be prepared by the planning engineer or by the construction manager and will be used at coordination meetings with trades foremen and subcontractors to plan the work for the forthcoming two-weekly period.

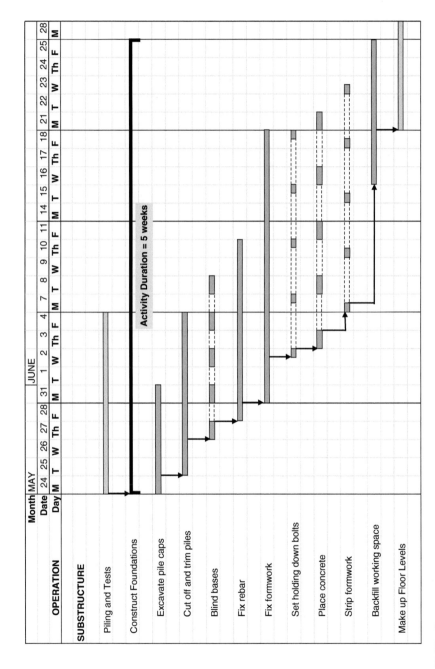

Figure 13.13 Stage programme.

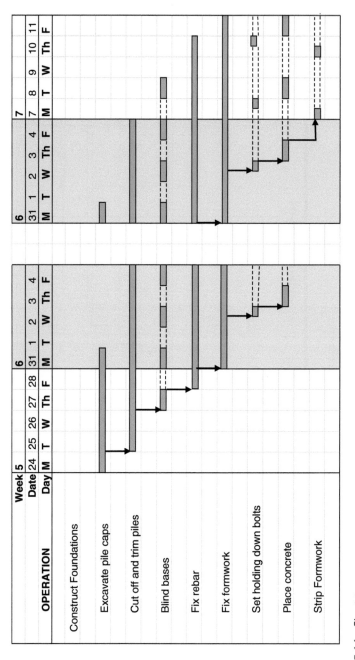

Figure 13.14 Short-term programme.

13.7.4 As-built Programme

The as-built programme is a record of actual progress. It differs significantly from the as-planned master programme as this shows planned activities and not those actually carried out. The as-built programme may also differ from the record of progress recorded against the project baseline because this will invariably be based on percentage completion which lacks precision.

Actual start and completion dates for every activity that actually took place on the project are recorded on the as-built programme and not the planned activities shown on the master programme. The as-planned programme is a 'best-guess' view of the project at its outset and may differ significantly from the truth of what actually took place on site. As-built programmes can be very detailed.

An as-built programme or 'programme of the day' should be developed and agreed (if possible) as the work proceeds but it requires considerable time, effort and cost. Consequently, unless entitlement claims based on the programme are contemplated, progress recording against the as-planned programme is usually sufficient for project control purposes or for simple extension of time applications. For serious claims to be successful, they must be capable of forensic scrutiny; otherwise, they will fail.

	Chapter 14 Dashboard	
Key Message		o There must be cash flow in the building trades – it is the very lifeblood of the enterprise [Lord Denning]. o When the cash dries up, there must be sufficient working capital in the business to keep going and pay the bills.
Definitions		o **Cash flow**: The movement of money in and out of a business. o **Cash flow forecast**: A forecast of income and expenditure – usually on a cumulative monthly basis. o **Working capital**: Capital available to support the day-to-day running of a business. o **Work in progress**: The value of work undertaken on a contract that has yet to be valued or has been valued but not paid for.
Headings		**Chapter Summary**
14.1	Introduction	o Construction firms get their 'cash' from monies received on contracts. o Payments made for contract work represent negative cash flow (money out) for the client and positive cash flow (money in) for the contractor. o Working capital helps to pay the bills before normal revenues are received and helps the business to survive when the cash dries up. o The valuation of work in progress is a well-known problem in construction that can lead to cash flow problems. o Cash flow forecasts are usually prepared on a contract-by-contract basis and accumulated for the company as a whole.
14.2	Earned Value Analysis	o Income from projects is called 'earned value'. o A popular method of forecasting value is earned value analysis. o This relies on S (or ogee) curves to predict cumulative value. o Common earned value forecasting methods include the quarter–third rule, the cumulative percentage method and the bar chart programme.
14.3	Cash Flow Forecasting	o The first step when preparing a cash flow forecast is to determine what the contractor's income will be. o This involves assessing the likely income from the various projects that the contractor has on its books. o Negative cash flow indicates potential problems which need to be investigated and illustrates the need for working capital. o Cash flow forecasts should reflect credit terms and retentions.
14.4	Improving Cash Flow	o Negative cash flow requires working capital. o Interest must be paid on any borrowings. o A number of techniques can be employed to improve cash flow.
14.5	Movement of Money	o In the first two months of a typical project, the contractor will be paying out money for wages, materials, subcontractors, etc. o This represents negative cash for the contractor. o Work in progress represents negative cash until certified and paid for.
14.6	Working Capital	o Working capital is the money a firm needs to fund its day-to-day operations and to pay bills on time. o To fund the negative cash situation, the contractor needs access to working capital in the form of shareholders' funds, bank borrowings or overdraft facilities.
Learning Outcomes		o Understand why cash flow is a problem in construction. o Understand the importance of working capital. o Understand how the valuation of work in progress can impact cash flow. o Be able to prepare an earned value forecast using a variety of methods. o Be able to prepare a simple cash flow forecast reflecting industry credit terms and contract retentions.
Learn More		o Chapter 17 expands on earned value analysis and explains how cost and value should be reconciled to show the true financial position on a contract.

14

Planning Cash Flow

14.1 Introduction

Cash flow is important to any business as it represents the movement of money in and out of the firm. When the cash dries up, there must be sufficient working capital in the business to keep going and pay the bills; otherwise, the threat of insolvency will not be far away.

Cash flow in construction is especially important because it is largely a credit-based industry. This means that firms rely on money coming in to pay for the deferred amounts owed to suppliers and subcontractors as and when they become due.

In the case of *Dawnays v Minter (FG) and Trollope & Colls* (1971), Lord Denning, Master of the Rolls, famously said that *there must be cash flow in the building trades – it is the very lifeblood of the enterprise.* He also added that *one of the greatest threats to cash flow is the incidence of disputes* and that resolving them *is frequently lengthy and expensive.*

14.1.1 Insolvency

The construction industry has historically been one of the biggest contributors to the UK's insolvency figures and currently represents 17% of all insolvent firms. At the time of writing, over 4000 construction firms collapsed in the previous 12 months, owing £millions to the wider supply chain.

> A top-100 contractor, with an annual turnover of £200 million, recently entered administration following £85 million losses on a single contract. The company owed £43 million in outstanding debt.
>
> Another top-100 contracting firm entered administration owing £108 million according to its administrators,[1] costing 500 jobs in a business with an annual turnover of £700 million. This was the biggest collapse since Carillion imploded in 2018 owing huge amounts of money.

It is always sad to see once thriving businesses fail, and many people lose their livelihoods as a result. The wider supply chain also suffers as the suppliers and subcontractors owed money are usually unsecured creditors who recover little or nothing from an insolvency. This can have a devastating impact on cash flow as unpaid subcontractors still have to pay their employees and suppliers nonetheless.

Construction Planning, Programming and Control, Fourth Edition. Brian Cooke and Peter Williams.
© 2025 John Wiley & Sons Ltd. Published 2025 by John Wiley & Sons Ltd.

Surprisingly, many of the companies and sole traders that get into financial difficulties are generating satisfactory profits but fail because they simply run out of cash. An insolvent company is unable to pay its debts as and when they fall due, and if cash is not available at the right time, the company ceases trading.

14.1.2 Cash Flow

Construction firms get their 'cash' from monies received on contracts. Consequently, payments made for contract work represent negative cash flow (money out) for the client and positive cash flow (money in) for the contractor. For house builders, sales revenue is their positive cash flow, but considerable working capital is needed to pay for the construction works before sales start coming through.

Cash flow is a problem not only for contractors but also for developers and construction client organisations. Clients can also run into financial difficulties and therefore contractors need to be as sure as possible that they will be paid in full and on time by their clients if they are not to find themselves with 'cash flow problems'.

Notwithstanding this, reliance on 'money coming in' is not enough to run a business and remain solvent. To do this, 'working capital' is required.

14.1.3 Working Capital

Working capital helps to pay the bills before normal revenues are received and helps the business to survive when the cash dries up for some reason (e.g. bad debts on a contract or insolvency of a debtor).

This working capital is normally obtained through a bank overdraft, bank loans or from funds invested in the company by its shareholders.

14.1.4 Work in Progress

The valuation of work in progress is a well-known problem in construction that can lead to cash flow problems.

When work is carried out on site, it is usual for the client to pay the contractor monthly in arrears or in stages as each part of the project is reached. The same process applies to subcontracts.

This work is valued and an application for payment made but, being an imprecise valuation, the amount applied for might not be agreed. Imprecision in the valuation stems from the fact that:

- The quantity of work done is usually expressed as a percentage of the contract sum.
- Work may have been carried out that has not been officially instructed or signed for.
- Some work might be deemed substandard and subject to re-work.
- There might be outstanding claims for loss and expense or prolongation that have not yet been agreed.

The consequence of this is that the money applied for will be less than expected and this reduces the 'money-in' side of the cash flow equation.

14.1.5 Client Cash Flow

Developers and other clients in the industry provide the capital investment required for construction projects to go ahead. This money may be borrowed from banks or other institutions, invested by shareholders or generated from profits – usually a combination of all three.

The developer or client has a different view of cash flow compared to a contractor because their cash position is always negative until sales income or revenue from the completed building is forthcoming.

Examples of client cash flow are given in Table 14.1 which shows several types of project and their respective sources of income together with the outgoings associated with undertaking a construction project.

The cash flow position of two types of clients is shown in Figure 14.1. This contrasts the cash requirements of a developer-client building three houses with that of a client undertaking a speculative office building project with a six-month construction period.

14.1.6 Contractor Cash Flow

The contractor's cash flow position is somewhat different to that of the client.

The contractor relies on interim or stage payments from the client to provide 'money in' and this helps to pay for the 'money out' payments for wages, materials, subcontractors, etc.

However, because the contractor must wait for perhaps two or three months for the money to come in, reliance must also be placed on the credit provided by suppliers and subcontractors in order to reduce the negative cash flow effect of contract payments.

An example of a contractor's cash flow is given in Table 14.2.

14.1.7 Cash Flow Forecasts

Cash flow forecasts are an indispensable part of running a construction business and are useful for several purposes:

- Negotiating banking facilities, such as an overdraft.

Table 14.1 Client cash flow.

Type of project	Money in	Money out
Housing development	Deposits	• Land purchase
	Sales completions	• Interest on borrowings
		• Planning and legal fees
Office/commercial buildings	Rental income	• Professional fees
Factory buildings	Production revenues	• Infrastructure costs (e.g. roads and sewers)
Speculative offices and factories	Sales of completed buildings	• Site remediation (e.g. removal of contamination)
Bridge or road tunnel	Tolls charged over a concession period – say 25 years.	• Building costs (monthly or stage payments)

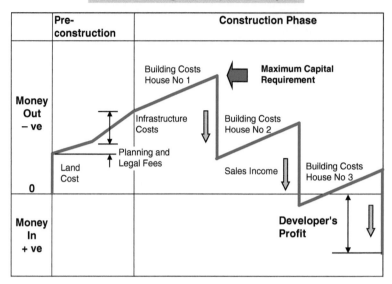

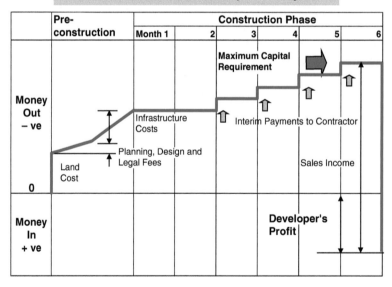

Figure 14.1 Client cash flow.

- For anticipating cash shortages and working capital requirements.
- To aid the financial control of contracts by providing an indication of profit or loss.
- To avoid 'overtrading' – taking on too much work with too little working capital.

Cash flow forecasts are normally prepared annually at monthly time intervals thereby enabling actual income to be compared to that anticipated. They are not easy to do because

Table 14.2 Contractor's cash flow.

Money in	Money out
Finance	**Overheads**
• Medium–long-term bank borrowings	• Head office running costs
• Shareholders' funds invested in the business	• Staff salaries
• Returns on investments (e.g. land and property)	• Company cars and expenses
Contracts	
• Monthly payments	• Payments to suppliers
• Final account payments	• Payments for plant hire
• Retentions released on practical completion	• Contract payments to subcontractors
• Retentions released on issue of the final certificate	• Wages and labour-only payments

the contractor is never sure exactly how much money will be received from its portfolio of contracts.

The cash flow forecast needs to be constantly updated within the contractor's reporting system to record actual value received and actual cost from invoices received. This gives the true cash flow position at any point in time.

Cash flow forecasts are usually prepared on a contract-by-contract basis and accumulated for the company as a whole to give a complete picture of what is happening. By doing the calculations, a company can predict the minimum and maximum cash required over the trading year and thereby arrange a comfortable working capital facility.

In order to prepare a cash flow forecast, a forecast of value and cost is required as cash flow is defined as income minus expenditure i.e. the money flowing into the business less the related costs.

A common basis for forecasting value is to use earned value analysis (EVA).

14.2 Earned Value Analysis

EVA relies on S (or ogee) curves to predict cumulative value.

The basis for this method of forecasting is the presumption that the cumulative expenditure on any construction project normally approximates to an S-shaped curve. Research has established that different types of construction work generate different revenue S-profiles, and these can be used as a basis for forecasting value (income) from contracts.

Consequently, if each monthly payment on a project is added to the total of previous payments, a cumulative S-curve will be described.

The S-curve is best applied to projects where any provisional sums are evenly spread through the contract, as the technique does not allow for high-expenditure provisional items early in the project period.

Figure 14.2 shows the principles of preparing a valuation forecast for a six-month contract. This shows five interim payments – not six, due to the initial payment delay at the

14 Planning Cash Flow

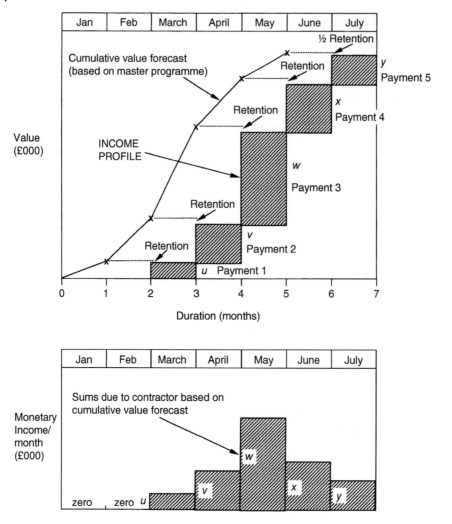

Figure 14.2 Principles of forecasting monthly payments due to contractor.

start of the contract – and release of 50% of the retention money once practical completion is reached. The remaining retention money – which will be released at the end of the defects correction period (usually twelve months) – is not shown.

The histogram shown below the line graph enables the monthly payment sums to be more clearly visualised.

14.2.1 Earned Value

Most of a contractor's income comes from contracts, with payments being made normally on a monthly basis. This income is called 'earned value'. Although it is somewhat of an over-simplification, this value represents the contractors resource costs plus a margin for overheads and profit.

The cost side of the cash flow forecast is, therefore, value minus margin.

Monthly income or value provides useful data for control purposes as it can be:

- Compared to a value forecast to determine a variance which can indicate the financial health and progress of a contract (see Chapter 17).
- Used as a means of forecasting cash flow from contracts, i.e. income (value) less cost.

The earned value data used for cash flow forecasting comes from the contractor's tender sum or, at the pre-construction stage of a project, from an order of cost estimate or cost plan. This data can be used in various ways according to the forecasting technique used, the choice of which depends upon the level of sophistication required.

Common earned value forecasting methods include:

- Quarter–third rule.
- Cumulative percentage method.
- Bar chart programme.

14.2.2 Quarter–third Rule

The quarter–third approximation is a geometric curve which has been found to give a reasonable assessment of value, while other approximations based on the ogee curve give a similar range of cumulative values.

The $¼:⅓$ rule gives a good approximation of this curve, the basis of the rule being that:

- $¼$ of total expenditure will be made during the first $⅓$ of the project timescale.
- A further $½$ of total expenditure will be made during the middle $⅓$ of the project.
- The final $¼$ of total expenditure will be made during the final $⅓$ of the project timescale.

The method works by plotting a graph as shown in Figure 14.3 which illustrates the basic principles of the quarter–third approximation.

Figure 14.4 gives a cumulative value forecast using this method for a contract lasting six months with a value of £160 000. From this forecast, monthly income can be derived – £7500 for month 1, £32 500 for month 2 and so on. By deducting retention at the contract percentage, net monthly income can be established.

This method of forecasting value is particularly useful at the early stages of projects. It can be used where there is only the client's budget to work on or where an order of cost has been established. If the order of cost is given as a value range, then the $¼:⅓$ rule could be used to provide a value envelope rather than a single S-curve.

14.2.3 Cumulative Percentage Value

The cumulative percentage method involves preparing a forecast of the cumulative percentage value per month on the basis of data analysed from similar types of projects. The purpose is to produce a reasonable S-curve approximation of cumulative value.

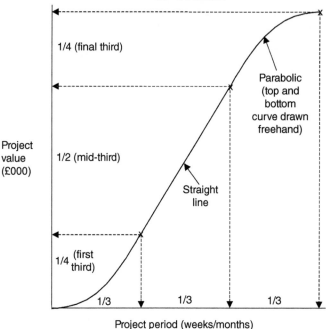

Figure 14.3 Quarter–third rule.

This method – which is particularly useful at the feasibility or tender stage of the project when little or no actual data is available – is illustrated in Figure 14.5. This shows a tabular display of the cumulative percentage value forecasts per month for projects lasting up to 12 months.

The table in Figure 14.6 shows data abstracted from Figure 14.5 for a contract with a value of £160 000 and lasting six months. The appropriate contract duration is chosen along the horizontal scale and the estimated percentage of the total contract value for that month of the project is read off vertically.

This approximation gives a slightly lower cumulative value forecast than the quarter–third method, as shown by the comparison of values and curve profiles in Figure 14.6.

The tabular display in Figure 14.5 is ideally suited to spreadsheets as this allows the analysis to be speedily undertaken and displayed in line graph form.

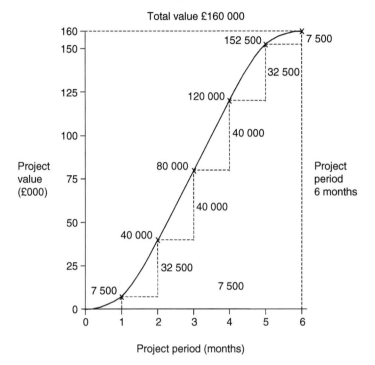

VALUE FORECAST (Cumulative)			
Month	Cumulative value	Month	Cumulative value
1	7 500	4	120 000
2	40 000	5	152 500
3	80 000	6	160 000

Figure 14.4 Quarter–third rule: Monthly/cumulative value forecast.

The idea of percentage value analysis relative to time may be developed by clients and contractors by collecting empirical data for different types of projects – schools, housing, factories, office developments, etc.

In this way a series of different S-curve profiles may be produced for forecasting the cumulative values of specific project types, each of which may have different cumulative expenditure profiles.

PROJECT PERIOD (months)

MONTH	1	2	3	4	5	6	7	8	9	10	11	12
0	0	0	0	0	0	0	0	0	0	0	0	0
1	100	45	24	18	12	9	7	6	5	5	4	3
2		100	76	45	30	24	18	14	12	11	9	7
3			100	79	60	45	35	27	23	20	17	13
4				100	82	67	55	45	37	31	27	22
5					100	85	72	63	53	45	38	33
6						100	88	77	67	59	51	45
7							100	90	80	72	64	56
8								100	92	83	75	66
9									100	92	85	76
10										100	93	85
11											100	93
12												100

Cumulative % value complete forecast

HOW TO USE

Read off cumulative % value relative to contract duration, i.e. 6 month contract

CUMULATIVE VALUE

Month	Cumulative %
1	9
2	24
3	45
4	67
5	85
6	100

Figure 14.5 Cumulative Percentage Value method.

CUMULATIVE VALUE FORECAST

Contract value £160 000
Contract period 6 months

Cumulative value forecast (from % graph)

Month	Cumulative %	Cumulative value	Compared with 1/4–1/3
1	9	14 400	7 500
2	24	38 400	40 000
3	45	72 000	80 000
4	67	107 200	120 000
5	85	136 000	152 500
6	100	160 000	160 000

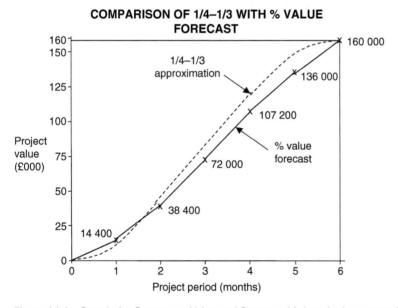

Figure 14.6 Cumulative Percentage Value and Quarter–third methods compared.

14.2.4 The Bar Chart Programme

A bar chart programme – it could be the client's master schedule, the contractor's pretender programme or the master programme – may be used to determine the rate of value generation or accumulative expenditure on a project.

To do this, the bills of quantities or other pricing document must be analysed to produce a total value for each activity on the programme. By dividing the value of each activity by its duration, a forecast can be established. This method gives a straight-line forecast which is slightly unrealistic as expenditure tends not to follow a linear timescale. It nevertheless provides a reasonable approximation and a value forecast directly linked to the sequence of construction operations.

The cumulative value forecast is a useful tool for controlling the project once work starts on site. At monthly intervals, the actual cumulative and forecast values may be compared as part of the company's monthly cost-reporting procedures. Cumulative value forecasts may also be used for assessing the client's and contractor's cash funding requirements. It is often a requirement at tender submission that a cumulative value forecast is provided to help the client to assess funding requirements for contract payments.

Figure 14.7 shows a bar chart for a factory project with the weekly and cumulative forecast values at each of the planned valuation dates. A cumulative value–time forecast based on the programme is shown in Figure 14.8.

If the relationship between value and time is represented by a nearly straight line, as in this case, then so be it. Even though cumulative value forecasts tend to follow the basic S-curve, curves can be distorted where high values are expended on items such as steel erection early in the project period.

It should be noted that the cumulative value forecast is only as good as the accuracy of the planning forecast. If the planning is too ambitious, then so will the resulting value forecast. This could lead to lower than expected contract revenues and cash flow problems and so it is a management responsibility to overview the budget forecasts and ensure that they are attainable.

14.3 Cash Flow Forecasting

The first step in preparing a cash flow forecast is to determine what the contractor's income will be. This is a matter of assessing the likely income from the various projects that the contractor has on its books using one of the forecasting methods described above.

The total income of the company will be the accumulated income from each of the contracts undertaken. A bit of 'crystal ball gazing' is required here as some of these projects may not have started and some will be finished but are awaiting the outcome of the final account, which might be in dispute.

The income from other ongoing projects will be derived from the contractor's internal reporting system.

14.3.1 Preparing a Forecast

Figure 14.9 is a simple illustration of how a cash flow forecast may be developed for a small contractor with an annual turnover of around £3 million. The forecast and graph are prepared using Microsoft Excel.

The net cash flow for each month of trading is given simply by deducting the cash out from the cash in for that month. For example, the net cash flow for February is determined by cash in (£150 000) less cash out (£160 500) = minus £10 500.

The cumulative cash flow is derived by adding the current month's figure to the previous month. Consequently, the cumulative position at the end of February is January net cash

FACTORY PROJECT – BAR CHART PROGRAMME – CUMULATIVE VALUE FORECAST

OPERATION	BUDGET	\multicolumn{20}{c}{Time in weeks}

OPERATION	BUDGET	1	2	3	4	5	6	7	8	9	10	11	12	13	14	15	16	17	18	19	20
Establish	5 000	5																			
Piling	10 000		10																		
Caps & g.beams	21 000			7	7	7															
Drainage	6 000		2	2	2																
Erect frame	20 000						10	10													
Roof cladding	14 000								7	7											
Ext. brickwork	30 000									6	6	6	6	6							
Floor slab	12 000													4	4	4					
Internal services	20 000															5	5	5	5		
External works	16 000																	4	4	4	4
Preliminaries	10 000	0.5	0.5	0.5	0.5	0.5	0.5	0.5	0.5	0.5	0.5	0.5	0.5	0.5	0.5	0.5	0.5	0.5	0.5	0.5	0.5
Weekly value (£000)		5.5	0.5	9.5	9.5	7.5	10.5	10.5	7.5	13.5	6.5	6.5	6.5	10.5	4.5	9.5	5.5	9.5	9.5	4.5	4.5
Cumulative weekly value		5.5	18.0	27.5	37.0	44.5	55.0	65.5	73.0	86.5	93.0	99.5	106.0	116.5	121.0	130.5	136.0	145.5	155.0	159.5	164.0
Monthly value					37.0				73.0				106.0				136.0				164.0
Valuation periods					↑1				↑2				↑3			↑4					↑5

20-week project period

Project value £164 000

Figure 14.7 Value forecasts based on bar chart.

(+ £39 000) plus February net cash (− £10 500). As February net cash is a negative figure, this gives a cumulative cash flow for February of £39 000 − £10 500 = £28 500 as shown.

Negative figures each month illustrate potential problems which need to be investigated and demonstrate the need for working capital to support the business when cash flow is negative.

The graph beneath the spreadsheet illustrates the 'roller coaster' of cash flow forecasting. This is not necessarily a problem but if, as in this case, the contractor's overdraft facility is £50 000, the directors would be getting a little 'twitchy'!

Cash flow forecasts must necessarily reflect the realities of life which are that the money does not flow in as might be expected and not all the money will be received. The first issue is a factor of the industry credit system and the second reflects the retention of monies by clients or main contractors as a hedge against possible defects in the work carried out.

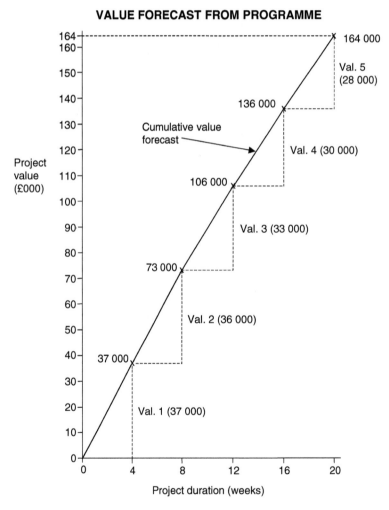

Figure 14.8 Cumulative value–time forecast based on bar chart.

CASH FLOW FORECAST

	JAN	FEB	MAR	APRIL	MAY	JUNE
CASH IN						
CONTRACT A	100000	100000	50000			
CONTRACT B	50000	50000	60000	100000	120000	100000
CONTRACT C			50000	100000	150000	150000
CONTRACT D				50000	100000	200000
TOTAL CASH IN	150000	150000	160000	250000	370000	450000
CASH OUT						
SALARIES	20000	20000	20000	20000	20000	20000
RENT	1000			1000		
TELEPHONE		500			500	
MATERIALS	40000	60000	90000	150000	200000	200000
SUBCONTRACTORS	50000	80000	100000	100000	150000	200000
TOTAL CASH OUT	111000	160500	210000	271000	370500	420000
NET CASH FLOW	39000	−10500	−50000	−21000	−500	30000
CUM CASH FLOW	39000	28500	−21500	−42500	−43000	−13000

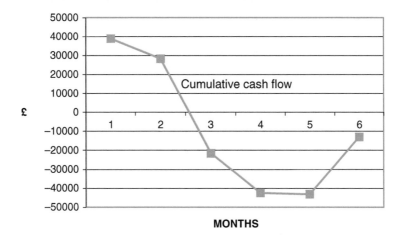

Figure 14.9 Cash flow forecast.

14.3.2 Credit Terms

Construction is largely a credit-based industry where customers expect a period of deferment before having to pay for goods or services received. Traditionally, the construction client expects the contractor to carry out a month's work which is then valued and certified, and the certificate is later honoured by cheque or, more conventionally bank transfer.

The average period of credit varies according to the form of contract used. A period of 14 days from certificate is usual but this can stretch to 28 days under some forms of contract.

For subcontracts the period of credit can be much longer whereas suppliers usually must wait for 30 days on monthly accounts.

It is a feature of the industry that payments are slow to arrive, and many suppliers have a credit control department to chase late debts. They sometimes factor their invoices – sell them to a debt collection firm at a discount.

14.3.3 Retentions

The construction industry operates a system of retentions where a percentage of the payments made for work in progress is held back by the client from the contractor (or by the contractor from subcontractors). Typically, this is 3% or 5% but can be 10% or more where non-standard forms of contract are used.

The retention held is normally reduced by one-half on issue of the practical completion certificate and the balance of the retention fund is released on issue of the certificate of making good defects at the end of the defects rectification period.

The defects rectification period can vary between 6 and 12 months unless otherwise stated in the contract, but 12 months is usual on reasonably sized contracts, especially where there are significant mechanical engineering installations.

The money held back is effectively capital lock-up as far as the contractor (or subcontractor) is concerned, and this has a cost effect in terms of interest payments on working capital requirements. This adds to the cost of building and is counterproductive from a client value point of view.

The practice of retention has been widely criticised over the years, including by Banwell (1964) and Latham (1994).

14.3.4 Payment

The industry culture of late payment of subcontractors and suppliers was a central theme in the Latham Interim Report *Trust and Money*. It was also the reasoning behind Lord Denning's judgment to allow a number of cases to be brought before the Court of Appeal, starting with *Dawnays v Minter* (1971) and concluding with *Modern Engineering v Gilbert Ash* (1973).

Although the House of Lords overruled the Court of Appeal decision in Gilbert Ash – and rejected Lord Denning's view that there was anything unique about building contracts – issues of payment delays and fairness have been addressed to some extent by the 1996 Construction Act[2] that followed Latham's Report.

Subcontract payment terms vary according to the form of subcontract used, which might be a standard form or bespoke main contractor form. Joint contracts tribunal (JCT), New Engineering Contract (NEC) and other families of contracts have their own subcontract forms with conditions 'back-to-back' with the main contract. Applications for payment, however, are invariably disputed and settled late in some cases and in other cases subcontractors never see their retention monies despite years of waiting and chasing payment.

Suppliers tend to invoice monthly and normal credit terms are usually 30 days from the end of the month. However, some contractors keep suppliers waiting for 60 or 90 days, sometimes more. In such cases, suppliers may be forced to suspend supplies or even withdraw credit facilities from persistent offenders. Recovery action through the courts may well ensue.

14.3.5 Payment Terms

Cumulative value forecasts can be an extremely useful means of assessing the payments due to the contractor at the end of each payment period – per week, per month or at agreed stages of the work. To be realistic, however, they need to take into account the deferral of payments made according to the applicable conditions of contract and the retention monies held back until the defects correction period is reached.

The timing of forecast income from interim valuations, and the release of the interim payment certificate to the contractor, depend on the payment terms contained in the various forms of contract – and subcontract – which can differ significantly.

14.3.6 JCT SBC

Under JCT conditions, the payment terms are stated as 14 days from the date of issue of the interim certificate. This is certified by the contract administrator, but it is the contractor who submits the payment application. The Construction Act requires notices to be given should the amount claimed be changed.

The overall period, however, between the date of the interim valuation and the payment of the money into the contractor's bank account, may cover some 28 days as illustrated in Figure 14.10. This time delay is made up of:

- Seven days between the valuation date and the issue of the interim certificate.
- Fourteen days' payment period.
- Seven days to receive and clear the payment.

14.3.7 Other Contracts

Payment terms under the JCT Design and Build Contract are different to the standard form and may be either stage payments (Alternative A) or periodic payments (Alternative B).

Periodic payments are normally monthly, whereas stage payments are based on agreed cumulative values linked to stages or milestones as stated in the contract particulars. The payment period from the payment request date is specified in the contract as 14 days and similar terms to the JCT Standard Building Contract apply to the issues of retention and defects liability.

Payment terms under other forms of contract will be similar in principle but may vary in detail. Under NEC forms, for instance, retention is a secondary option and payment is assessed by the project manager.

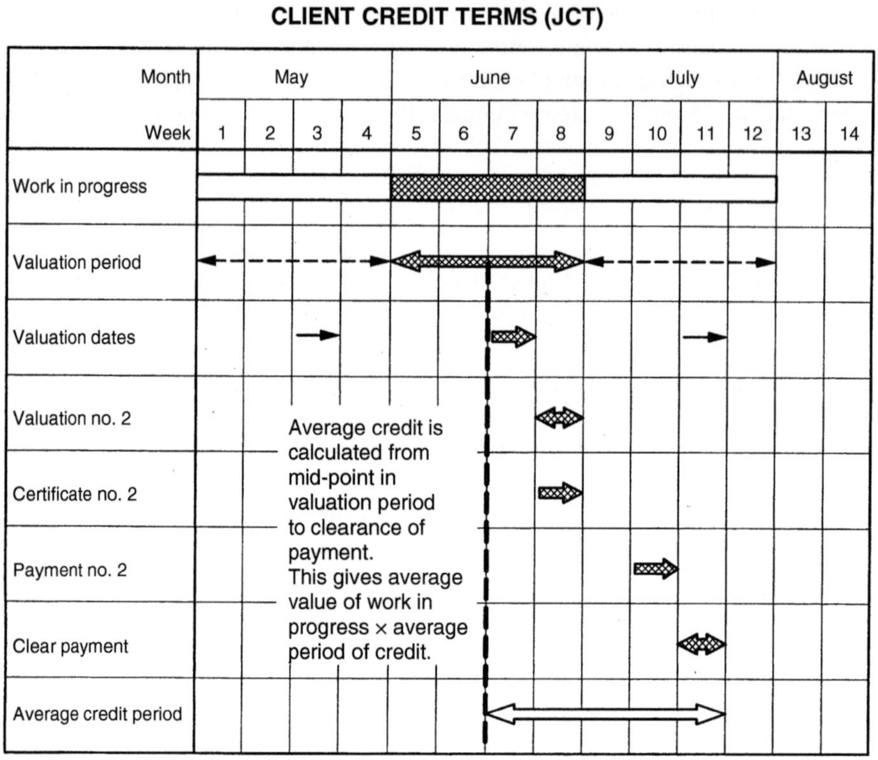

CLIENT CREDIT TERMS (JCT)

Month		May				June				July			August	
Week	1	2	3	4	5	6	7	8	9	10	11	12	13	14
Work in progress														
Valuation period														
Valuation dates														
Valuation no. 2														
Certificate no. 2														
Payment no. 2														
Clear payment														
Average credit period														

Average credit is calculated from mid-point in valuation period to clearance of payment. This gives average value of work in progress × average period of credit.

Figure 14.10 Average credit period (JCT SBC).

14.3.8 Reflecting Payment Terms in Forecasts

The principles of preparing a valuation forecast for a six-month contract reflecting both payment delay and retentions are shown in Figure 14.2.

This shows five interim payments – not six, due to the initial payment delay at the start of the contract – and release of 50% of the retention money once practical completion is reached. The remaining retention money – which will be released at the end of the defects correction period (usually twelve months) – is not shown.

The histogram shown below the line graph enables the monthly payment sums to be more clearly visualised.

To put money to these principles, a cumulative value forecast obtained from a master programme is shown in Table 14.3. The contract period is six months and retention is set at 3%. The sums shown are before deduction of retention.

Figure 14.11 shows the cumulative value forecast derived from Table 14.3 as an S-curve together with the forecast cumulative and monthly income profile for the contract, taking into account payment delay and retention. The contractor's monthly income is shown in histogram form below the graphical display.

Table 14.3 Cumulative value forecast.

Month	Cumulative value forecast (£)
1	45 000
2	124 000
3	198 000
4	265 000
5	320 000
6	380 000

Payments = monthly
Retention = 3%

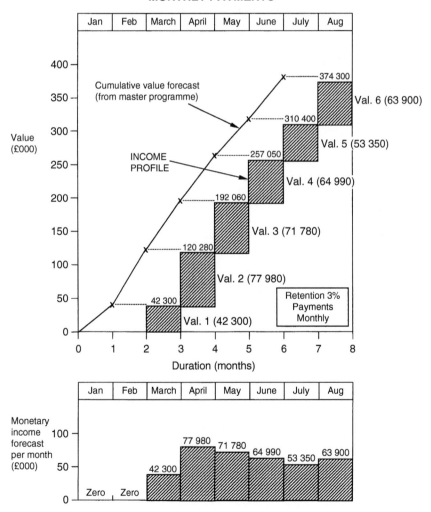

Figure 14.11 Cumulative value forecast.

14.4 Improving Cash Flow

Despite relatively low interest rates at the time of writing, the cost of borrowing is still a matter of concern for contractors because profit margins are generally low in the industry and banks lend at a premium over the base rate.

Consider the situation for a contractor with an annual turnover of £20 million, annual profits of £600 000 (3%) and bank borrowings of £3 million with a lending rate of 2% over base:

Actual cost of money = Base rate (say) 5% + 2% = 7%
Cost of borrowings = 7% × £3 000 000 = £210 000

$$\text{Cost of borrowing expressed as a percentage of turnover} = \frac{£210\,000 \times 100}{£20\,000\,000} = 1.05\%$$

With profits at only 3% of turnover, the contractor is vulnerable should interest rates rise or profits fall. Profit in contracting is always at risk because of the uncertain nature of the work and the likelihood of mistakes in tenders and inaccurate forecasts on contracts. Loss making contracts was one of several reasons why Carillion failed.

For these reasons, and because contractors invariably make more money by keeping other people's money for as long as possible, finding ways of reducing negative cash flows are always attractive. Some common methods are listed below.

14.4.1 At Tender Stage

These methods will improve cash flow but must be done before submitting the priced bills of quantities or other pricing document such as activity schedule or contract sum analysis (design and build):

- Load money into undermeasured items.
- Load money into early items such as excavation and substructures.
- Load money into mobilisation items in the preliminaries.

14.4.2 During the Contract

These methods will reduce working capital requirements:

- Submit interim applications on time.
- Overmeasure the work in progress.
- Overvalue materials on site.
- Agree on the value of variations as soon as possible.
- Keep good records and submit claims early.
- Deal with defective work quickly to avoid delayed payment.
- Make maximum use of trade credit facilities.

14.4.3 Post-contract

These methods will increase profit levels:

- Submit all documentation as soon as possible.
- Ensure timely release of retentions by submitting health and safety file information on time.
- Agree on final account as soon as possible.
- Collect outstanding retentions on time.

14.5 Movement of Money

In the first two months of a typical project, the contractor will be paying out money for wages, materials, subcontractors, etc., and this represents negative cash for the business.

Some of this money will be recovered when the client pays the first interim certificate at the end of month 2. However, this is unlikely to be sufficient to put the contractor in a positive cash position because, by this time, another month's work will have been done, and more costs will have been incurred.

This illustrates the relationship between cash flow and work in progress which represents negative cash flow until certified and paid for. When work in progress payments are less than expected, this has an adverse effect on the contractor's cash position.

14.5.1 Money In and Money Out

The pattern of money in and money out on a contract may be illustrated as a saw tooth diagram as shown in the top part of Figure 14.12.

This shows that the contractor is in a negative cash position until month 6, which means that the project will effectively be financed at the contractor's own expense. As working capital will normally be borrowed money – perhaps from loan or overdraft facility – profit will be impacted by the cost of interest payments.

Consequently, even though cash is coming in from the contract, more money is going out due to the delay in contract payments. The timing and management of cash flows between receiving payment and paying creditors can mean the difference between insolvency and survival.

14.5.2 Retention

A further negative cash flow consideration is that the client will keep some retention money back from the contractor to cover for possible defects in the work. It can therefore be seen that the contractor is in a negative cash position for a considerable period of the contract, until such time as the contract payments begin to outweigh the monies being spent.

14 Planning Cash Flow

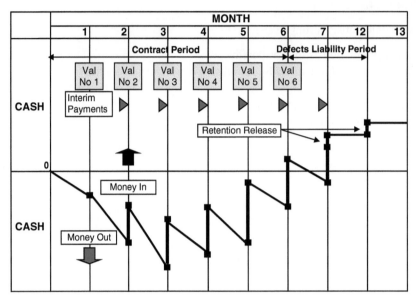

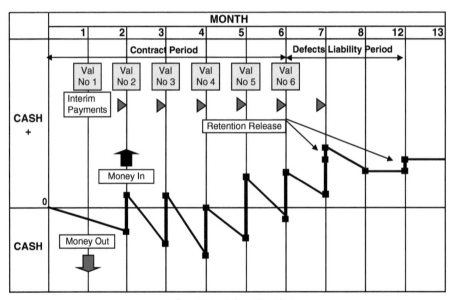

Figure 14.12   Cash flow principles – 1.

14.5.3 Payment Delay

Figure 14.12 illustrates the contractor's cash position, assuming that there is no delay in paying for the wages, materials and services required for a contract. Realistically, however, the contractor will use the credit facilities offered by suppliers and subcontractors to offset the negative cash flow effect of having to wait for the client's payments.

The payment delay situation is shown in the bottom part of Figure 14.12. Here, the contractor makes no payments out until money is due in from the contract. Consequently, the contract will be cash positive more quickly and the contractor will have to find less working capital for the job. This means lower interest payments and more profit for the contractor.

In order to demonstrate the effect of using credit facilities to improve cash flow, a weighted average payment delay may be calculated. This is illustrated in Table 14.4 which calculates the payment delay weighting of various resources by multiplying the credit period for each resource by its percentage of the contract value.

14.5.4 Credit

Because most construction work is carried out on credit, a delicate balance exists in the supply chain between income and expenditure. For instance, under the standard forms of contract – such as JCT, NEC, ICC and FIDIC – the contractor works for a month, but the client is not required to pay for work in progress until a valuation has been carried out and a certificate issued by the contract administrator.

This means that the contractor will have to wait for a month or more before getting paid and will have to find the money to pay wages and salaries in the meantime. On the other hand, the contractor does not pay straightaway for subcontract services, plant hire or materials supplied until a period of credit has expired.

Subcontractors will be paid according to the terms of their contract (around two to four weeks) and suppliers perhaps 30 days from the end of the month when the materials were delivered. For short-term plant hire, invoices may be submitted when the plant has been off-hired from site whereas long-term contracts may be invoiced weekly or monthly with payment normally due a month from invoice.

Table 14.4 Weighted average payment delay.

	Percentage of contract value	Payment delay or credit period (weeks)	Weighted average payment delay (weeks)
Labour	10	1	0.1
Materials	20	6	1.2
Plant	20	6	1.2
Subcontractors	50	3	1.5
Total	**100**		**4.0**

14.5.5 Late Payments

The industry credit system works well unless the chain is broken and one party or the other fails to pay on time. When this happens, cash flow problems can arise for contractors or subcontractors with inadequate working capital.

The longer the contractor can defer paying creditors, the better it will be for the cash position, but care needs to be exercised to avoid withdrawal of credit facilities or possibly court action for recovery of monies outstanding.

Creditors cannot be paid out of profits if the cash is not available and so delaying payment to creditors, while awaiting payment on contracts, has become an art form in construction. However, when small firms are squeezed between a large debtor (e.g. a main contractor) and a creditor (e.g. a supplier wanting payment) – with one controlling the money and the other controlling credit facilities for materials – serious problems can occur.

The Housing Grants, Construction and Regeneration Act 1996 was enacted to address these problems but in practice payment problems are still common in the industry.

14.6 Working Capital

Working capital is the money a firm needs to fund its day-to-day operations and to pay bills on time.

To fund the negative cash situation, the contractor needs access to working capital in the form of shareholders' funds, bank borrowings or overdraft facilities. Contractors who rely solely on overdrafts rather than long-term funding, however, are vulnerable to insolvency because overdrafts can be withdrawn without notice. This usually happens just when the contractor has reached the overdraft limit!

14.6.1 Forecasting Working Capital

Figure 14.13 illustrates the contractor's cash position and working capital requirements (i) when there is no payment delay and (ii) when there is.

In the top part of Figure 14.13, the area between the horizontal zero cash line and the minimum cash line indicates the least amount of working capital needed by the contractor, assuming that contract payments are received on time. If this is not the case, the maximum cash line would apply and the area between the zero cash position and maximum cash line would represent the working capital needed to fund the project.

The areas between the zero cash line and the minimum and maximum cash lines represent the contractor's minimum and maximum capital lock-up on the contract. It is this area which needs to be funded out of working capital.

In the bottom part of Figure 14.13, a payment delay of one month (four weeks) is indicated which means that the contractor is in positive cash by month 2 rather than month 6. As a rule of thumb, a one-month payment delay improves cash flow by approximately 50% with a consequent reduction in the contractor's working capital requirements.

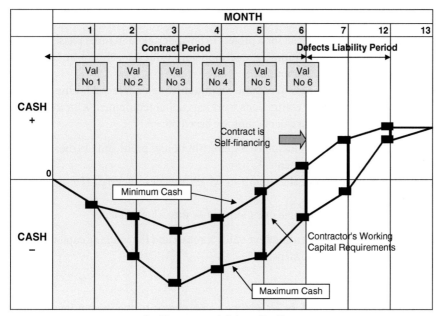

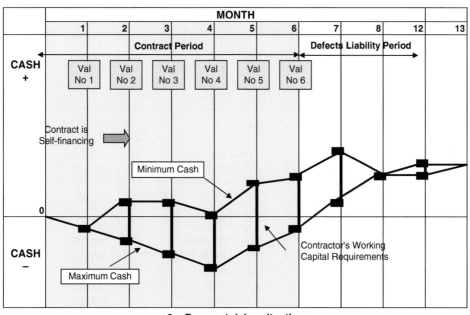

Figure 14.13 Cash flow principles – 2.

14.6.2 Worked Example

A contractor has obtained a contract for a project to be undertaken in three phases. Details of the phases with approximate durations and monetary values are shown in Table 14.5. The overall project duration is 24 months.

This worked example shows how to prepare a cumulative value forecast for the complete project and an assessment of the contractor's working capital requirements for the first six months of the project period. The calculations are based on:

- The values in Table 14.5 which include a 5% contribution to profit and overheads.
- 5% retention applied to the payments.
- Costs to be paid at the end of the month in which they are incurred (i.e. no delay in meeting the cost situation).
- Monthly interim payments payable one month after the valuation date.

The approach to assessing the cumulative value forecast and the working capital requirements involves the following five steps:

Step 1

- Assess the cumulative value forecast for the three phases of the project by allocating project values to a bar chart programme, as shown in Figure 14.14.
- Allocate the value of contract preliminaries separately throughout the contract period with an extra sum of £10 000 for establishing the site at week 1 of the project.
- Calculate the cumulative value forecast along the bottom of the bar chart. This is displayed as a value–time graph in Figure 14.15.

Table 14.5 Tender analysis.

Phase		Item	Cost (£)	Total (£)
Phase 1	Two-storey extension Value = £1 500 000 Duration = 10 months	Operations • Substructure • Superstructure • Finishings and services	150 000 600 000 750 000	1 500 000
Phase 2	Refurbished canteen Value = £750 000 Duration = 8 months	Operations • Demolitions • Superstructure • Finishings and services	37 000 225 000 488 000	750 000
Phase 3	Office refurbishment Value = £1 500 000 Duration = 10 months	Operations • Demolitions • Superstructure • Finishings and services	75 000 450 000 975 000	1 500 000
Contract preliminaries				250 000
Total project value				**4 000 000**

CASH FUNDING FOR THREE-PHASE PROJECT																											
OPERATION		VALUE (£'000)	Time in months																								
			1	2	3	4	5	6	7	8	9	10	11	12	13	14	15	16	17	18	19	20	21	22	23	24	
			PHASE I ←									→															
													PHASE II ←							→							
																	PHASE III ←									→	
Subst.	PH.I	150	75	75																							
Superst.		600		100	100	100	100	100	100																		
Finishings		750						150	150	150	150	150															Phase I must be complete before Phase II and Phase III start
Demolition	PH.II	37											37														
Superst.		225												45	45	45	45	45									
Finishings		488															122	122	122	122							
Demolition	PH.III	78															78										
Superst.		450																90	90	90	90	90					
Finishings		972																			162	162	162	162	162	162	
Prelims	ALL	250	20	10	10	10	10	10	10	10	10	10	10	10	10	10	10	10	10	10	10	10	10	10	10	10	
Monthly value			95	185	110	110	110	260	260	160	160	160	47	55	55	55	255	267	222	222	262	262	172	172	172	172	
Cumulative monthly value			95	280	390	500	610	870	1130	1290	1450	1610	1657	1712	1767	1822	2077	2344	2566	2788	3050	3312	3484	3656	3828	4000	

Figure 14.14 Cumulative value forecast.

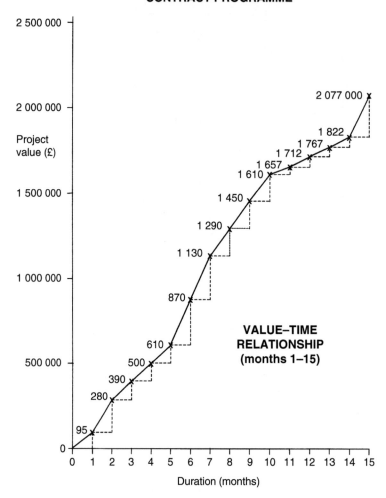

Figure 14.15 Value–time graph.

Step 2

- Establish the cumulative cost forecast, remembering that value is cost plus margin:

$$\text{Cost} = \text{value} \times \frac{100}{(100 + \text{margin})}$$

- Therefore, where cumulative value is £390 000 and margin = 5%:

$$\text{Cost} = £390\,000 \times \frac{100}{(100 + 5)} = £371\,428$$

- Calculate remaining costs and plot the cumulative cost–time relationship on the graphical display (see Figure 14.16).

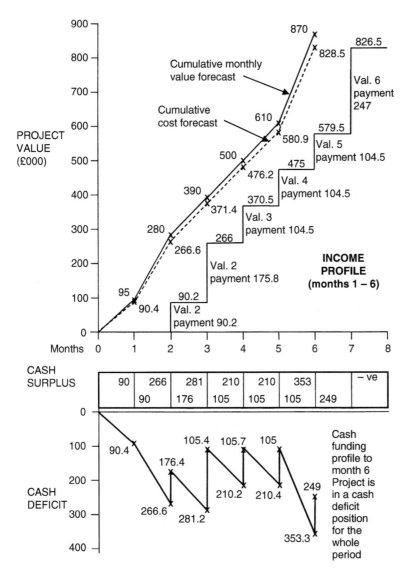

Figure 14.16 Cost, value and cash forecast – months 1–6.

Step 3

- Calculate the contractor's actual income, allowing for a one-month payment delay and 5% retention:

Interim payment no. 1	
Forecast value	£95 000
Less 5% retention	£4 750
	£90 250

Table 14.6 Working capital requirements.

Month	Maximum cash requirement (£)
1	90 400
2	226 600
3	281 200
4	210 200
5	210 400
6	353 300

- Calculate remaining payments and plot them on the display making sure to allow a one-month delay (see Figure 14.16).

Step 4

- Plot a saw tooth diagram for the first six months of the project using the cumulative cost and payment figures calculated above. Money out (cost) is plotted downwards at an angle and money in (payment) is plotted upwards (see bottom of Figure 14.16).

Step 5

- Display the maximum and minimum working capital requirements in the form of a table.

Figure 14.16 indicates the relationship between value–time, cost–time and income–time for the first six months of the project period. The cash funding profile is shown in the form of a saw tooth diagram under the graphical display. This is based on the no delay situation in meeting costs.

The cash funding profile indicates that during the first six months of the project period, cash requirements peak at the end of months 3 and 6, respectively. The maximum working capital requirements at the end of months 1–6 are shown in Table 14.6.

References

Banwell, H. (1964) Report of the Committee (Chairman Sir Harold Banwell). *The Placing and Management of Contracts for Building and Civil Engineering Work*. HMSO.
Dawnays v Minter (FG) and Trollope & Colls (1971) *1 WLR 1205*.
Latham, M. (1994) *Constructing the Team*. HMSO.

Notes

1 Construction News 20 September 2003.
2 Housing Grants, Construction and Regeneration Act 1996.

Chapter 15 Dashboard		
Key Message		o The art of good management is to anticipate a problem before it becomes a crisis. o The ability of managers to do so relies on information.
Definitions		o **Budget**: A financial plan. o **KPI**: Key performance indicator.
Headings		**Chapter Summary**
15.1	Introduction	o Managers can have too little information or too much. o The secret is to have the right information at the right time. o The collapse of Carillion raised many questions about corporate ethics and behaviour and the misreporting of the financial state of contracts.
15.2	Budgetary Control	o The process of making, monitoring and adjusting budgets is called **budgetary control**. o Underpinning budgetary control should be a system of financial reporting that alerts managers to variances. o Mangers need information in sufficient time to make corrective decisions.
15.3	Establishing Budgets	o Budgets can be presented by plotting money or resource allocations against time on a bar chart. o Actual expenditure or resource consumption can be plotted and used to monitor variances. o This would be part of the monthly cost control and reporting system.
15.4	Site Records	o The keeping of site records during a project is imperative if a contractor is to receive the true entitlement prescribed by the conditions of contract. o Records include site diaries, time sheets and daywork records, records of resources on site and progress photographs. o The use of construction management software systems and common data environments facilitate the collection, exchange and validation of all sorts of project data. o Keeping an up-to-date programme of the day represents a very useful record of what has happened on site.
15.5	Meetings	o Meetings are an important means of communication throughout the various stages of any construction project. o They are an integral part of the control process necessary to keep a project on time and on budget. o Meetings may be formal or informal and may be conducted face-to-face or remotely.
15.6	Key Performance Indicators	o Measuring performance is an essential feature of business management. o KPIs are used to measure performance at both corporate and project level using well-established industry-wide metrics. o Benchmarking can be internal project to project or may be measured against national averages or against best-in-class industry standards. o KPIs help businesses to identify strengths and weaknesses and to improve performance year on year.
Learning Outcomes		o Understand the need for the right information to manage a business. o Understand the role of budgetary control. o Be able to explain various types of budgets. o Be able to prepare a variety of simple budgets. o Understand the need to keep good records. o Appreciate the importance of meetings as part of the control process. o Understand the types of key performance indicators and how they are used.
Learn More		o This chapter provides a foundation for the subject matter of Chapters 17 and 18. o Reference might also be made to Chapter 7 which explains how the contractor's tender establishes a project budget.

15

Project Control

15.1 Introduction

In common with other businesses, builders, contractors and developers have to plan and organise their day-to-day activities in order to manage them effectively. The art of good management is to anticipate a problem before it becomes a crisis, but the ability of managers to do so relies on information. There can be too little information or too much, and so the secret is to have the right information at the right time.

The fallout of the Carillion collapse in 2018 raised many questions about corporate ethics and behaviour, not least regarding creative accounting, the misreporting of the financial state of contracts and even the external auditing process (Wylie 2020) and served as a timely reminder of the need to follow accepted accounting practice.

All businesses – large and small, and not just in construction – are required to submit annual accounts to the State to establish tax liability and to ensure that the outside view of the company's affairs given by the accounts are true, fair and not misleading. Achieving this objective requires reliable and accurate reporting systems that give managers an undistorted view of the performance of the business so that informed decisions can be made. This relies on:

- A robust system of budgetary control.
- Effective means of establishing budgets and measuring variances.
- Reliable, accurate and verifiable site records.
- Regular meetings.
- The use of key performance indicators (KPIs).

15.2 Budgetary Control

Financial plans are called budgets and the process of making, monitoring and adjusting them is called **budgetary control**.

Budgetary control is an important management technique used for the purpose of controlling income and expenditure. Control is achieved by preparing budgets relating to the various activities of the business, which provide a basis for comparison with actual performance.

15.2.1 Budgets

A budget is an estimate of income and expenditure over a given period of time. This could be for a business tax year or for a construction project.

Budgets can be for a single contract or grouped for a portfolio of projects or accumulated for all projects undertaken by a company. To an extent, they can be viewed as a financial version of the contractor's programme.

Whilst there are many types of budgets, project budgets are derived from the contractor's price for the job and are closely related to the programme of work.

15.2.2 Reporting

Underpinning budgetary control should be a system of financial reporting of contracts and construction developments that alerts managers to variances in budgets in sufficient time to make corrective decisions and ensure – as far as is possible – that a profitable outcome is achieved.

The basis of effective reporting of contract performance is an accurate estimating system, as no amount of corrective action can compensate for a badly priced contract, but the reporting system can nevertheless help to stem the losses and maybe achieve breakeven.

Effective management requires control, but all firms have different ideas on the degree of control necessary for the projects they undertake. Many factors need to be considered, including the size and organisation of the business and the scale and complexity of the projects in hand.

In any event, to monitor performance, information needs to be collected within a structured reporting system, however simple, so that appropriate action can be taken if and when things start to go wrong.

15.2.3 The Reason for Budgets

It is an essential function of management to prepare forecasts and to monitor performance. Budgets provide the information necessary to keep managers informed so that decisions can be made about how to react to current circumstances.

> Businesses are living entities (going concerns) and therefore money is coming in and going out all the time. There are wages and suppliers' and subcontractors' bills to pay, and monies are being received for contract work. This activity cannot be allowed to happen willy-nilly because the business will soon be in a mess.

Contractors, therefore, must prepare forecasts and these are required for many aspects of their business as well as the contract work undertaken. For instance, the company must plan for the next year's trading to make sure that sufficient work is obtained to enable the company to meet its commitments and keep going. This is very much tied into the estimating, planning and quantity surveying functions in a construction company.

Budgets may also be developed for the labour, plant and preliminaries expenditure on contracts. These will be based on an analysis of the contractor's net estimate – in other words the tender sum excluding profit and overheads. The estimator's allowances in the priced bills of quantities or activity schedule will be extracted into a total for each budget category.

15.2.4 Types of Budgets

Businesses will create various budgets according to their operations and construction firms are no exception. These include:

- **Revenue**: This is the annual sales budget. For contractors, the forecast is anticipated turnover or income from contracts, including incomplete work on existing contracts. For developers sales relate to income generated from reservation fees and completions – such as houses, apartments or commercial/industrial space.
- **Capital expenditure**: For contractors, the capital expenditure budget represents anticipated spending on plant and equipment, off-site production facilities or a new head office building. Capital expenditure for developers relates to the purchase of development land.
- **Operating expenditure**: The operating budget is a forecast of expenditure on staffing, overheads, labour, materials, subcontractors and so on.
- **Cash flow**: The cash flow budget is a forecast of the movement of monies in and out of the business. It is used to monitor income from contracts and to determine working capital and borrowing requirements.
- **Annual accounts**: Larger companies will prepare a forecast balance sheet and profit and loss account several times in a single tax year as well as at the end. The balance sheet will indicate the financial stability of the business in terms of assets and liabilities and the profit and loss account will forecast profits from the company's activities (such as contracts) and give an indication of the annual dividend to shareholders.

In publicly quoted companies, forecast accounts form the basis of regular announcements to the Stock Exchange regarding the company's financial position and any potential profit warnings that may need to be issued.

15.2.5 Presenting Budgets

Budgets may be presented in tabular or graphical format – such as S curves, line graphs or histograms. Many managers tend to favour graphs as a way of expressing data as they are easier to read.

Conventionally, budgets are prepared using spreadsheets. Spreadsheets are useful because they provide the facility to express money over time – such as an annual sales or cash flow forecast. They are ideal for SMEs because powerful budgets can be prepared at minimum cost.

For larger and more sophisticated companies, construction management software is popular. Procore and Trimble e-Builder, for example, provide digital solutions to project delivery and facilitate not only the setting and management of budgets but also cost management, document management and connected workflows – both within and external to the business.

This means that a contractor can grant controlled access to its management system to designers, quantity surveyors, contract administrators and other members of the client team. Collaborative software enables members of the client and construction teams to communicate effectively irrespective of their location and to share information such as costs, drawings and requests for information.

15.2.6 Construction Management Software

Being Cloud-based, construction management software – such as Zutec and Oracle Aconex – can be accessed from anywhere, but some products can also function without an Internet connection which is useful on remote construction sites.

Once logged in, the user can use the menu to view the project portfolio, to navigate between projects, to see the progress status of individual projects and to access the project-level dashboard for functions such as documents, drawings, schedules, instructions, change orders/variations, request for information (RFIs) and budgets. Access can be restricted according to the user.

Whilst construction management software can be used to mark up drawings, embed technical queries into the drawings and to manage project documents, most communications are email based. This simplifies communications – requests for documents, queries about the budget and RFIs, for instance.

An RFI can be raised within the software which generates an email, and the reply comes back into the software and is emailed to the request initiator. This is simple and provides a clear audit trail of communications.

This software is useful for budgetary control purposes both within a contracting organisation and by allowing the client team access to budgets, costs, invoices and other records on cost reimbursement and gain-pain projects.

15.2.7 The Basis of Budgets

In the world of contracting, the basis for all budgets is the contractor's tender – the accepted bid for the project. This can take a number of forms:

- A conventional tender based on drawings, specifications and, perhaps, a pricing document such as a bill of quantities or activity schedule.
- A lump sum price based on drawings and specifications only, with the contractor taking responsibility for the quantities.
- A bid and contract sum analysis for a design and build project.
- An elemental cost plan when work starts early in the design process.
- A cost plan, or other form of price analysis, where the contract is based on a guaranteed maximum price arrangement and is paid for on the basis of actual cost, subject to a gain-pain mechanism.

From this information, budgets can be created to control the costs of labour, materials, plant, subcontractors, preliminaries and general overheads.

15.3 Establishing Budgets

Budgets can be presented in a useful and interesting way by plotting money or resource allocations against time on a bar chart. This can be extended into a graphical display in the form of a value–time curve forecast, for instance.

During a project, the contractor may plot actual expenditure or resource consumption and use the analysis to monitor variances as part of the monthly cost control and reporting procedures.

15.3.1 Labour Budget

Figure 15.1 shows a bar chart display for labour demand expressed in man-hours/man-weeks. It shows the number of man weeks allocated to each site activity.

External brickwork, for instance, has an allocation of 24 man weeks, and this is represented on the bar chart by spreading the number of man weeks over the activity duration of five weeks. Total man weeks are shown at the bottom, along with the cumulative totals.

The graphical display in Figure 15.2 is a cumulative forecast of man-weeks plotted against time. This allows the actual man-weeks used on the project to be matched with the forecast and any variance monitored as the project proceeds. The bar chart and cumulative graph could equally well be expressed in money.

The reason for the variance of two man-weeks could be due to the contract being behind schedule. This could have resulted from insufficient labour being available to the contractor or due to unexpected delays on site. Management should investigate such variances.

This simple approach is highly effective as an aid to resource and cost control but needs to be reconciled with overall progress to be an effective management tool.

15.3.2 Plant Budget

Figure 15.3 shows a plant budget for a project in bar chart form. This is derived from plant cost allocations in the estimator's pricing notes and is shown on the bar chart display under the heading 'budget'.

The total cost of plant for the excavate foundations activity, for instance, is £5000 spread over the five-week duration. The weekly expenditure forecast is totalled at the bottom of the spreadsheet, along with the cumulative totals.

Cumulative expenditure is then plotted on a weekly cumulative graph for comparison with actual expenditure, as shown in Figure 15.4.

Plant costs on site frequently exceed expenditure allowances in the tender, and site managers are notorious for keeping plant on site for too long just 'for convenience'. Whether plant is hired or owned by the contractor, strict control over budget variances needs to be exercised, and plant and equipment must be off-hired as soon as it is not required.

15.3.3 Preliminaries Budget

Preliminaries are the contractor's general cost items for running the site. They include site security, site supervision, accommodation and welfare, plant and equipment not allocated to specific activities, transport, access roads and so on. Some preliminaries are

Labour Expenditure Budget

Operation	Labour budget man hours	Man weeks	1	2	3	4	5	6	7	8	9	10	11	12	13	14	15	16	17
Foundations	800	20	4	4	4	4	4												
Brickwork to DPC	360	9					4	5											
Ground floor slab	200	5							5										
External brickwork	960	24							2	6	6	6	4						
Roof		S/C													2				
1st fix	240	6														2	2		
Plaster		S/C																	
Weekly man weeks			4	4	4	4	8	5	7	6	6	6	4	0	2	2	2	0	0
Cumulative man weeks			4	8	12	16	24	29	36	42	48	54	58	58	60	62	64	64	64

Figure 15.1 Labour budget.

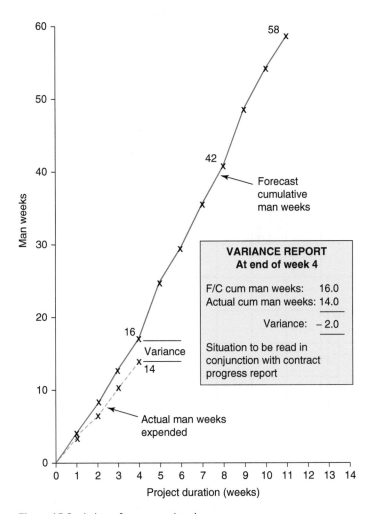

Figure 15.2 Labour forecast and variance.

fixed cost items – such as erecting site hoardings, establishing site cabins and erecting mortar silos – whereas others are time-related including hire of cabins, supervision costs and telecommunications.

The preparation of a contract preliminaries budget follows similar principles to other resources by allocating money to time. The preliminaries allowances in the estimate are analysed, and money is simply allocated to the master programme, being careful to distinguish between fixed and time-related costs.

Figure 15.5 shows a spreadsheet display for a preliminaries budget with forecast preliminaries expenditure allocated to time in bar chart format. The fixed costs of establishing and removing site accommodation are indicated. Some items – such as hardstandings – appear to be time-related as the cost is spread over time but are, in fact, fixed costs allocated to the time taken to carry out the work.

PLANT EXPENDITURE BUDGET

Operation	Plant allocated	Budget £	1	2	3	4	5	6	7	8	9	10	11	12	13	14	15	16
Site clearance	P.Shovel 4 Lorries	2000	400	800	800													
Excavate foundations	Hyd. B/A 2 Lorries	5000			1000	1000	1000	1000	1000									
Concrete foundations	Concrete pump	1000								1000								
Brickwork DPC	Mixer	200									100	100						
Ground floor slab	Concrete pump	1000											1000					
External walls	Scaffold mixer	4000												1000	1000	1000	1000	
1st fix	Crane	1000																1000
Weekly expenditure forecast			400	800	1800	1000	1000	1000	1000	1000	100	100	1000	1000	1000	1000	1000	1000
Cumulative expenditure forecast			400	1200	3000	4000	5000	6000	7000	8000	8100	8200	9200	10200	11200	12200	13200	14200

Figure 15.3 Plant budget.

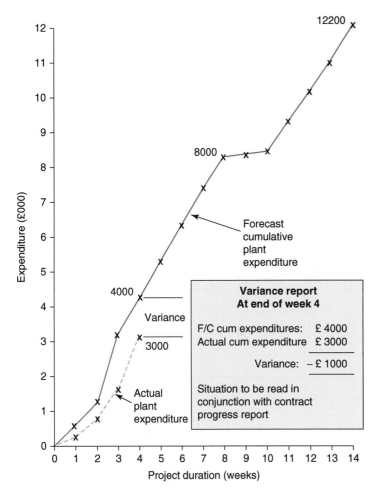

Figure 15.4 Plant forecast and variance.

Figure 15.6 shows the cumulative preliminaries expenditure forecast presented as a line graph which allows actual cost and variances to be plotted. Actual preliminaries costs may be matched with the forecast expenditure in a spreadsheet or graphical format, as both methods permit variances to be monitored continuously as time goes by. However, plotting the variance as a ± in a graphical display somehow focuses attention on the issues.

Preliminaries are one of the most common areas of overspending on construction projects and need to be closely monitored. Overspending the preliminaries budget might be due to management failing to allow sufficient monies in the tender or to overelaborate site organisation. Preliminaries are also the area most subject to adjustment at the tender adjudication stage, either to make the tender more competitive or because the contractor decides to tender on a shorter contract period.

The total of preliminaries typically represents between 8% and 15% of the contract sum, but on small contracts, the percentage could be 25% or even higher.

CONTRACT – BOLTON FACTORY
CONTRACT VALUE – £600 000

PRELIMINARIES BUDGET

BUDGET FIGURES ARE NET PRELIMINARIES SUMS IN THE BILL

Operation	Budget £	1	2	3	4	5	6	7	8	9	10	11	12	13	14	15	16	17	18	19	20
Establish site	8 000	4	4																		
Site accommodation																					
Establish	1 000		1																		
Hire	200/w			0.2	0.2	0.2	0.2	0.2	0.2	0.2	0.2	0.2	0.2	0.2	0.2	0.2	0.2	0.2	0.2	0.2	
Dismantle	1 000																				1
Site management																					
Site manager	1 000/w	1	1	1	1	1	1	1	1	1	1	1	1	1	1	1	1	1	1	1	1
General foreman	500/w								0.5	0.5	0.5	0.5	0.5	0.5	0.5	0.5					
Site engineer	500/w	0.5	0.5												0.5	0.5					
Temporary roads	2 000	0.5	0.5					0.5	0.5												
Hardstandings	2 000			1	1																
Site hoarding																					
Erect/dismantle				1	1															1	1
Weekly expenditure forecast		6	7	3.2	3.2	1.2	1.2	1.7	2.2	1.7	1.7	1.7	1.7	1.7	2.2	2.2	1.2	1.2	1.2	2.2	3
Cumulative expenditure forecast		6	13	16.2	19.4	20.6	21.8	23.5	25.7	27.4	29.1	30.8	32.5	34.2	36.4	38.6	39.8	41	42.2	44.4	47.4
Actual weekly cost		5	5	2	2	2	2	2	2												
Cumulative weekly cost		5	10	12	14	16	18	20	22												

Time in weeks

Total prelims £47 400

Figure 15.5 Preliminaries budget.

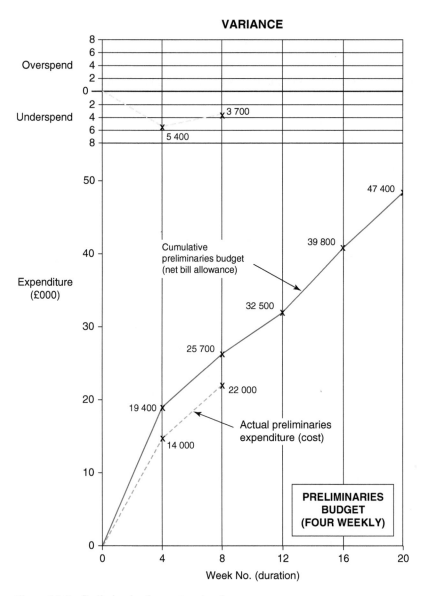

Figure 15.6 Preliminaries forecast and variance.

15.4 Site Records

The keeping of site records during a project is imperative if a contractor is to receive the true entitlement prescribed by the conditions of contract and ensure that subcontractors are not successful with spurious claims for extra payment. Records are necessary for:

- Establishing the basis of a claim for loss and expense.
- Defending a counterclaim from a client.

- Substantiating applications for extensions of time.
- Claiming payment for extra work not included in the contract.
- Reporting to progress meetings.

Contractors keep – or should keep – many types of records. They include:

- Site diaries.
- Timesheets and daywork records.
- Records of resources on site.
- Progress photographs.
- Records of design clashes and errors.
- Changes to the master programme baseline.
- As-built records and programmes.

> Records normally assume greater importance after the event than during it and contractors often fail to keep adequate records during a project unless some formal procedures are adopted and maintained.

The use of common data environments and construction management software systems – such as Oracle Aconex, Zutec and Procore – facilitate the collection, exchange and validation of all sorts of project data and provide a thorough audit trail of site records for future reference.

The contractor should delegate responsibility to the site management team for keeping, circulating and maintaining records, and this includes site supervisors, planning engineers and quantity surveyors.

15.4.1 Site Diary

Contemporaneous records in the form of a daily site diary can prove invaluable to support contractual entitlement for an extension of time or loss and expense, for insurance claims and to record health and safety incidents. They may also be used to settle disputes with subcontractors and for party wall issues.

The site diary will be invaluable when preparing the as-built programme and should record daily events on site, including visitors, and time lost per day due to exceptional rain, cold or heat, late deliveries, subcontractor delays and so on. Site diary details may also be recorded in the formal construction management reporting system and should minuted at monthly site meetings.

15.4.2 Labour Records

Records should be kept of the contractor's own labour force and that of subcontractors. Details relating to site labour are often noted in the daily site diary as well as in the formal reporting system.

Labour records are conventionally made on labour allocation sheets, which record the names, trade and time spent on work tasks as a daily log of site activity. Each work activity is briefly described.

Site labour records provide useful background information on what has happened on site on a day-to-day basis. They are often referred to as substantiation for contractual claims or to prove/disprove reasons for delayed completion of the works. Labour allocation sheets may also be used as a basis for a counterclaim against a subcontractor where delay or non-performance is an issue.

15.4.3 Instructions

Strictly speaking, verbal instructions should not be acted upon, but contractors invariably do so to mitigate any potential delay. Consequently, a system for recording verbal instructions from the contract administrator or architect should be in place, as this is usually a pre-requisite for payment.

Using enterprise resource planning software (ERP) or a common data environment (CDE) can help to automate this process and provide a single auditable version of the truth. Construction problems requiring instructions can be photographed or identified on drawings and sent direct from the site to be actioned by the appropriate individual, with decisions emailed to all concerned with a minimum of delay.

15.4.4 Programme

Keeping an up-to-date programme of the day represents a very useful record of what has happened on site. This can be supported by a variety of short-term programmes including look-ahead programmes and subcontractor programmes.

Other time-related records include weekly and monthly progress reports for submission to head office and records of late instructions or information.

15.4.5 Site Measurements

Keeping records of all site measurements may not seem important at the time, but they could later be vital when concluding the final account. This could include records of foundations and drainage remeasurement.

It may prove invaluable to photograph certain works such as complex drainage connections as this might warrant a new rate for the work.

15.4.6 Variations

Variations are changes to the scope of works and details and measurements should be recorded at the time work is carried out as a permanent record of what took place. This is especially important if the work is to be covered up – such as backfilling of drainage trenches and foundations.

Site measurements do not take precedence over measurements from drawings, but not all variations are detailed on a drawing or sketch or in a BIM model. Dimensions recorded on site are the next best thing.

If in doubt, take the measurements to avoid problems later.

15.4.7 Delay and Disruption

Recording any delays and/or disruptions resulting from undertaking work instructions or variations may be invaluable when formulating the as-built programme.

Claims for extensions of time or additional expense can become very complex and so records taken at the time can provide compelling evidence, especially as memories fade with time.

15.4.8 Daywork

The pricing and agreement of work undertaken on the basis of time and resources is called 'daywork'.

Daywork sheets are normally used to record the names, trade and time spent on work items that the contractor believes are not included in the contract and cannot be measured quantitatively. They also record plant and other resources used for the work.

The site daywork sheets must be completed and submitted each week or recorded in the data management system, and appropriate signatures or approval obtained. They are signed by the contract administrator – if agreed – and priced by the quantity surveyor using rates established in the contract documents.

15.5 Meetings

Meetings are an important means of communication throughout the various stages of any construction project and form an integral part of the control process necessary to keep a project on time and on budget and make sure that the build quality is according to specification.

The most effective meetings are those conducted face-to-face but the pandemic and the need to travel long distances have encouraged the use of Zoom or Teams-based meetings as well.

15.5.1 Purpose

Meetings have a variety of purposes, including:

- To establish communication links between the parties to a construction project.
- To discuss problem areas, suggest alternative solutions and decide on appropriate action.
- To request and distribute project information.
- To review the current progress and reschedule or accelerate the programme as necessary.
- To respond to early warning notices of potential problems and to take the opinion of all interested parties.
- To develop a collaborative approach to problem-solving.

It is important that meetings are short and focused and have the right people in attendance so that the correct facts can be tabled, and decisions made, without constant referring back to superiors.

15.5.2 Types

Examples of the various meetings which may take place during the construction process are included in Figure 15.7, which explains the purpose of the meetings and the main people involved.

Other meetings that might take place include informal meetings with the architect to discuss defects, discussions between the quantity surveyors regarding progress with the final account, weekly/daily meetings to problem-solve contract emergencies and toolbox or task talks with operatives to discuss work methods and health and safety issues.

Project stage	Type of meeting	Purpose	Who is involved
Tender	Pre-qualification	• Determine the suitability and competence of contractors • Short-list contractors	• Client • Design team • Tendering contractors
	Pre-tender	• Discuss bidding strategy • Consider risk issues • Allocate responsibilities	• Director or estimating manager • Estimator • Planning engineer
	Tender coordination	• Ensure bid will be ready on time • Discuss further risk issues • Resolve problems and 'loose-ends'	• Estimating manager • Estimator • Planning engineer
	Tender adjudication	• Scrutinise the estimate • Consider risk and profit • Consider programme and construction period • Decide tender sum	• Director • Estimating manager • Estimator
Pre-contract	Pre-contract (internal)	• Hand over tender documents and pricing notes • Discuss materials and subcontract quotes • Discuss programme and key dates • Discuss risk issues	• Estimator • Planning engineer • Contracts director or Contracts manager • Site manager • QS • Procurement manager
	Pre-contract (external)	• Introduce participants • Hand over contract documents • Establish means of communication • Agree valuation dates	• Client • Architect • Client's QS • Contractor
	Pre-start (external)	• Finalise details e.g. setting out, statutory approvals, existing services, etc. • Agree handover dates • Handover contractor's programme	• Client • Architect and design team • Contractor

Figure 15.7 Project meetings.

Project stage	Type of meeting	Purpose	Who is involved
Contract	Monthly/ progress (external)	• Discuss progress • Receive site manager's report • Receive QSs report • Table problem areas • Request/receive information • Discuss variations/changes	• Architect • Site manager • Client QS/Contractor QS • Other design team members
	Monthly valuation (external)	• Agree value of work in progress • Agree value of materials on site • Discuss physical progress • Agree preliminaries • Discuss variation instructions	• Client's QS • Contractor's QS
	Health and safety	• Discuss construction phase plan • Consider accident reports • Discuss non-compliance issues • Receive information and updates • Discuss risk assessments and method statements	• Site manager • Site foreman • Health and safety adviser (head office) • Planning engineer • Subcontract manager

Figure 15.7   (Continued)

Managers should try to strike the correct balance between working and meeting, as too many meetings may impede or disrupt the work pattern on a project.

15.5.3   Formal and Informal Meetings

The value of informal meetings should be recognised as well as the formal ones.

Formal meetings, such as the client's pre-contract meeting and monthly site meetings, have procedures and rules, whereas informal meetings such as tender coordination meetings and weekly site meetings with the trades foreman may be of a more casual and relaxed nature.

Meetings should normally have an agenda and minutes should be recorded and circulated, but this is sometimes impractical. For instance, when the site manager and site foreman meet for 10 minutes to discuss a site problem, this would not be formalised.

However, where an impromptu meeting like this leads to a change of construction method, a risk assessment should be carried out and recorded and the original method statement should be formally changed to reflect what was agreed. A further meeting to inform the operatives (i.e. a toolbox/task talk) would also be necessary.

15.5.4 Monthly Site Meetings

The meeting agenda for site meetings is the responsibility of the meeting chairman – usually the client's architect, engineer or project manager – and it should be circulated before the meeting in order that the personnel attending are fully aware of the items to be discussed.

Minutes of previous meetings should also be circulated beforehand in order that they can be formally accepted as a true record.

A typical agenda for a monthly site meeting to be chaired by the contract administrator is shown in Table 15.1.

15.5.5 Weekly Progress Meetings

Weekly progress meetings are held at site management level to review the short-term programme from the previous week and to coordinate the activities of subcontractors and the contractor's own trade gangs.

Table 15.1 Meeting agenda.

	Typical meeting agenda
Minute reference[a]	Headings for the order of the meeting
1.1	Personnel attending and apologies for absence
1.2	Confirmation of minutes of previous meeting
1.3	Matters arising from minutes
2.1	Confirmation of matters raised at intermediate site visits by architect
2.2	Weather Report – record of inclement weather to date
2.3	Progress Report – normally given by the contractor or clerk of works
3.1	Drawings and information requirements
3.2	Construction queries – relating to materials or design
3.3	Design issues – anticipated future requirements
3.4	Subcontractor and specialist supplier information requirements
3.5	Liaison with statutory undertakings/utilities
4.1	Health and safety report
5.1	Variation orders – review of outstanding architect's instructions and instructions requiring confirmation
5.2	Dayworks – overview of outstanding dayworks
6.1	Financial review – quantity surveying matters for discussion
7.1	Project completion date – review of projected completion
8.1	Any other business
9.1	Date and time of next meeting

a) To be referred to in all site minutes.

Coordination is essential for the success of a project and the weekly site meetings are used to establish the short-term or 'look-ahead' programme for the following one- or two-week period and to discuss interfaces between the various subcontractors and other site activities.

On larger projects, the meeting may be chaired by the assistant project manager assisted by the site-based planning engineer. A meeting agenda will be prepared, and minutes recorded.

Suggested headings for the order of the meeting are shown in Table 15.2.

15.5.6 Subcontract Coordination Meetings

Most projects these days rely on subcontractors to carry out the bulk of the work. A fairly straightforward project may have 30 or more subcontractors on site, and on large jobs, there may be over 100 subcontractors and utility companies employed.

The work of all these firms needs to be carefully organised and coordinated and many contractors employ a subcontract manager or coordinator to deal with these specialists.

Normally, there will be a pre-start meeting with each of the subcontractors at which the subcontract manager, quantity surveyor/procurement manager and the subcontractors' manager/director will be present. Agenda items will include:

- Subcontract details and price.
- Subcontract programme and start date.
- Resources required.

Table 15.2 Weekly meetings.

	Typical agenda for weekly site meeting
Minute reference[a]	Headings for the order of the meeting
1.1	Personnel attending
1.2	Overview of progress to date
1.3	Overview of the programme requirements for the next period
1.4	Reasons for delays – critical review
1.5	Resources to meet proposed short-term programme:
	Labour review
	Plant review
	Subcontractor resources
	Key material requirements
1.6	Architect's instructions/information requirements
1.7	Interface between proposed operations – coordination of operations
1.8	Date of next meeting

a) To be referred to in all site minutes.

- Common facilities to be provided or charged for by the main contractor – such as water and power, site accommodation and storage, waste removal, scaffolding, cranage and so on.

As work proceeds on site, further meetings will be held with the subcontractors on a regular basis – perhaps weekly for the more complex work packages. These meetings may be held jointly, or there may be separate meetings for each of the larger subcontracts.

Present at the meetings will be the contractor's subcontract manager and planning engineer together with subcontractors' representatives. Items for discussion will include:

- Subcontract progress.
- Problems and delays.
- Resource problems.
- Interfaces with other subcontractors including shared work zones, shared facilities such as scaffolding and cranage, health and safety risks and work overlaps.

15.6 Key Performance Indicators

Measuring performance is an essential feature of business management. This can be done by measuring internal performance – profitability, accident rate, productivity, waste management and so on – or by benchmarking against national averages or best-in-class industry standards.

KPIs originate from the Movement for Innovation, which followed the Egan Report *Rethinking Construction* in 1998. The aim is to measure performance at both corporate and project levels using well-established industry-wide metrics.

15.6.1 Types of KPIs

KPIs can be 'soft', based on subjective assessment or 'hard' using factual data. Soft metrics include:

- Client satisfaction.
- Time and cost predictability.
- Quality.
- Employee satisfaction.

Hard metrics include:

- Financial performance such as profit margin, cash flow, working capital or cost variance.
- Health and safety statistics such as accident rate, incident rate, reportable accidents and near-misses.
- Defects including the number of defects, the time taken to rectify them, the cost of rework and so on.

15.6.2 Uses of KPIs

Using KPIs is justifiable at several levels as they help businesses to identify strengths and weaknesses and to improve performance year on year.

Their use might be identified in the tender documentation so that the client can receive reports regarding safety standards, subcontractor performance or value-engineered solutions and so on.

KPIs can also be used where bonus gain-pain payments are to be based on whether the contractor achieves particular targets – such as cost versus budget or progress in relation to key project milestones.

Within the contracting organisation, several KPIs might be appropriate including financial performance, turnover or sales growth, creditor and debtor days (metrics relating to cash flow), shareholder dividend growth and so on.

In a wider context, KPIs enable businesses to compare their performance with other similar companies as a measure of management performance and corporate efficiency.

15.6.3 Practical Applications

The *CCI KPI Engine* is a software development between the Centre for Construction Innovation (CCI) and Burr IST. Burr IST is a software house, and CCI is the UK Northwest Centre of Excellence for the construction industry.

The software may be used by large and small contractors and can be specified by construction clients to manage their construction and consultant frameworks.

This KPI engine allows KPIs to be calculated, recorded, reported and benchmarked on a project or corporate basis. There are over 200 KPIs and bespoke versions are available too. The software can report KPI scores using tables, graphs and action plans and facilitates benchmarking company to company or project to project.

SmartSite KPIs is an online tool that facilitates the benchmarking of project performance against the rest of the industry by using established standards. This tool has been developed in conjunction with Constructing Excellence.

Reference

Wylie, B. (2020) *Bandit Capitalism – Carillion and the Corruption of the British State*, Birlinn Ltd.

Chapter 16 Dashboard		
Key Message		o Planning is one of the functions of management. o A programme/schedule is a key tool in the management process.
Definitions		o **Delay**: The act or state of being late – such as failing to complete the works by the contract completion date. o **Disruption**: The loss of efficiency due to lower-than-expected productivity or interference with normal progress. o **Acceleration**: Speeding up the work to finish earlier or mitigate delay.
Headings		**Chapter Summary**
16.1	Introduction	o Few plans turn out as expected due to unforeseen ground conditions, delays due to weather and waiting for instructions, etc. o The plan should be monitored in the light of prevailing circumstances. o Normally the contractor's duty is to proceed in a regular and diligence fashion and complete the works on or before the date for completion.
16.2	The Contractor's Programme	o The programme shows how the project will be organised/scheduled. o The precise style and format is largely at the contractor's discretion.
16.3	Progress and Delay	o The contractor may make changes to the plan as the works proceed. o Regular updates to the programme are normally required. o Types of delay include culpable, non-culpable and neutral.
16.4	Recording Progress	o Progress should be recorded and critically analysed. o Progress should be reported at least monthly and preferably weekly. o Progress reporting can be almost instant via the Cloud. o Progress can be assessed by time elapsed or by earned value.
16.5	Delay and Disruption	o Delay and disruption are common features of construction projects. o Fault may lie with the employer, with the contractor or with no one.
16.6	Extensions of Time	o Standard contracts contain clauses dealing with extensions of time. o They are granted for specific events beyond the contractor's control. o The contract administrator monitors progress and makes decisions. o Such decisions establish entitlement to extra time and/or money.
16.7	The 'As-Planned' Programme	o The as-planned programme is the contractor's view of the project at the outset of the construction phase. o From this a 'baseline' programme is created which is used as a fixed reference against which actual progress can be measured.
16.8	The 'As-Built' Programme	o The as-built programme is a representation of what happened on the project so that this can be compared to what was planned. o This should be based on the complete project record database including site diaries, records, returns and minutes of meetings etc.
16.9	Delay Analysis	o Delay analysis seeks to establish the cause(s) of delay and who bears the time and money risk. o The Society of Construction Law Delay and Disruption Protocol is a non-mandatory guide to good practice.
16.10	Delay Analysis in Practice	o Project management software is helpful for speeding up iterations and developing 'what if' scenarios. o Delay analysis can be carried out during a project or after completion.
16.11	Project Acceleration	o Where a project has been delayed, the contractor may wish to bring forward the completion date or may be asked to do so by the client. o Acceleration can be achieved by reorganising the work, increasing resources or both. o Time cost optimisation balances the time and cost of acceleration.
Learning Outcomes		o Distinguish between delay and disruption. o Understand the role of the contractor's programme for monitoring progress and determining culpability for delay/disruption. o Understand the concept of 'as-planned' and 'as-built'. o Understand the basic principles of delay analysis. o Be able to follow a simple practical example of delay analysis.
Learn More		o Chapter 18 emphasises the importance of resource management. It explains the complexity of the site manager's job in meeting the project schedule. o Resource problems are frequently behind delay and disruption of progress and must be accounted for when seeking to speed up the programme or carry out a delay analysis.

16

Controlling Time

16.1 Introduction

Planning is one of the functions of management, and a programme is a key tool in the management process. However, very few plans turn out as expected and this is especially the case in construction work. Unforeseen ground conditions, delays due to weather and waiting for instructions are just some of the many reasons why this may happen.

It is therefore crucial to ensure that the planning process includes an element of control so that the plan can be monitored in the light of prevailing circumstances. Additionally, activities may have to be rearranged or additional resources may be required to cope with difficulties encountered, and a forecast of their likely impact may also be required. This is the role of the contract master programme which both the contractor and the client's contract administrator can use to monitor what is happening.

16.1.1 Duty to Complete

The contractor will normally have a contractual duty to proceed in a regular and diligent fashion and complete the works on or before the date for completion stated in the contract. Failure to do so, without a valid reason, may lead to legal sanctions.

Slow progress, for instance, may invoke a contractual notice that could result in determination of the contract if taken to the extreme. In cases of late completion, the client has a right under many construction contracts to deduct liquidated and ascertained damages from contract payments, unless the basis of the contract is that time is of the essence. If this is an express contract condition, the client will have the right to terminate the contract.

Legal sanctions, however, are of little benefit to the client because the project will finish late – or very late – in any event, and it is much better to work collaboratively to keep the project on track as far as possible. This involves the client's project manager or contract administrator using the master programme to check that the contractor is progressing satisfactorily and to take proactive action if it is not.

If things are not going according to plan, it is conventional for the contract administrator to point this out at the monthly meeting and ask the contractor what action is being taken. Under New Engineering Contract (NEC) contracts, such situations would be covered by the early warning protocols, and action would be taken much sooner.

Construction Planning, Programming and Control, Fourth Edition. Brian Cooke and Peter Williams.
© 2025 John Wiley & Sons Ltd. Published 2025 by John Wiley & Sons Ltd.

Similar obligations fall on subcontractors who, in the worst-case scenario, may have to be replaced and the subcontract terminated.

16.1.2 Time for Completion

Most standard conditions of contract have similar provisions as regards time for completion, and somewhere in the contract will be stated:

- The date for possession of the site.
- The date(s) for possession of specific sections of the site (where appropriate).
- The date for completion of the works.
- The date(s) for completion of section(s) of the works.

The above information is to be found in the Contract Particulars in the joint contracts tribunal (JCT) SBC, in the Appendix in the Infrastructure Conditions of Contract (ICC) form and in the Contract Data in NEC contracts.

Time periods for deferment of possession may also be stated so that the employer may delay handing over the site to the contractor without breaching the contract. The contract will usually state the agreed rate of liquidated and ascertained damages (LADs) for the whole, or where applicable, sections of the works, so that the employer may charge the contractor in the event of late completion of the contract.

Within the main body of common conditions of contract will be found clauses enabling the employer to extend the contract completion date where the contractor has been delayed for reasons beyond its control – provided that the reasons for delay are the same as those specified in the contract.

The purpose of such clauses is to preserve the employer's right to charge LADs for delayed completion where the contractor is at fault.

16.2 The Contractor's Programme

Under most standard forms of contract, the contractor is required to produce a programme of some sort to show how the project is to be organised and scheduled. For example:

- **JCT Standard Building Contract**:
  - Two copies of the programme for execution of the works as soon as possible after the contract is signed.
  - Two copies of programme updates following an extension of time.
- **ICC Form**:
  - Within 14 days of contract award, a programme with a critical path network.
  - At the same time, a general description of the intended arrangements and methods of construction which the contractor proposes to adopt.
  - Revised programmes where actual progress varies from the accepted programme.
- **NEC ECC**:
  - A programme showing key dates including starting, possession, partial possession and completion dates.

- The order and timing of the operations.
- Provisions for float, time-risk allowances, health and safety requirements and contractual procedures.
- A [method] statement for each operation showing planned equipment and resources.
- Revised programmes showing actual progress for each operation and its effect on remaining work.
- The effects of compensation events (such as employer risk delays) and matters giving rise to early warnings (such as design problems or discovery of bad ground conditions) and how the contractor plans to deal with these issues.

JCT, ICC and NEC conditions vary as to the detail that the contractor must include in the programme, but the expectation is nevertheless that the contractor will show sufficient detail so that the employer's contract administrator or project manager may monitor progress and make informed judgements about extensions of time and so on as the work proceeds.

Even in the NEC contract – which is much more explicit than others about the contractor's programme – the extent of detail required is very much at the contractor's discretion, as is its precise style (e.g. bar chart or network) and format (paper-based, computer-based, specific software package to use, etc.).

Consequently, the contractor's master programme may be in any of the following formats:

- Gantt (bar) chart with no logic links.
- Linked bar chart showing dependency logic.
- Precedence network.
- Critical path analysis (arrow) diagram.
- Line of balance.
- Time-chainage diagram.
- A combination of the above – such as a linked bar chart for substructure and envelope in conjunction with a line of balance for frame and fit out.

16.2.1 Milestones

Milestones are symbols shown on the contractor's programme – normally indicated by using a small diamond shape (♦) – that are especially useful in programming and delay analysis:

1. To denote key events on the contractor's programme such as:
   - When the building is planned to be weathertight.
   - When a section of the work is due for completion.
   - Partial handover dates when the employer hopes to take occupation of part of the building or project.
   - Commissioning dates for power and other utilities.
2. To show when something actually happened such as:
   - A drawing issue or revision.
   - A site instruction or architect's approval.
   - The placing of an order.

Milestones have a duration of zero but can still be linked to the programme logic in the usual way (e.g. finish–start, start–start, etc.). In some software packages, such as Asta Powerproject and Microsoft Project, milestones are created by entering a duration of zero alongside the activity description. Project Commander uses 'drag-and-drop'.

> Milestones behave in all respects just like any other activity on the programme, but they denote a point in time as opposed to an activity which has a defined start, duration and finish.

In delay analysis, milestones can be linked to other activities to show the effect of a late instruction or drawing issue on the rest of the programme.

16.2.2 Early Warning Systems

The success of a project is often related to the links developed between the main suppliers, subcontractors and the contractor.

Linked bar chart displays that represent the connection between procurement and the commencement of operations on site have now become an integral part of planning. This is to ensure that key dates are met with respect to design requirements, information flow and the delivery or commencement of works on site.

Early warning systems were developed in the mid-1960s by John Laing and extensively used. The system worked because of its simplicity in application and ease of monitoring; it allowed users to develop their own symbols and letters to denote information requirements.

> The NEC Engineering and Construction Contract highlights requirements for early warning very prominently. This places the responsibility on both the contractor and project manager to give an early warning on any matter which could delay completion or impair performance of the works in use.

Figure 16.1 shows a range of early warning symbols or milestones that may be used to denote an occurrence which affects a supply chain component supplier or subcontractor. The example indicates that actual data were released by the client's representative one week later than planned with respect to information and nomination. This resulted in the order being placed one week later than scheduled, with a possible delay in the commencement of steel erection on site.

Information requirements relative to a project may also be presented in a tabular format, which meets the requirements of the JCT SBC. Delays in the release of information by the architect or other consultants may result in the contractor applying for an extension of time.

16.3 Progress and Delay

Irrespective of the format of the programme, the contractor is free to complete the works on or before the contract completion date, making necessary changes to the plan as the works proceed:

- Provided that the contractor submits regular updates to the programme.

EARLY WARNING SYSTEM

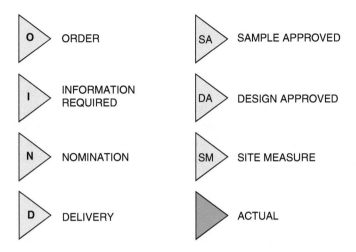

EARLY WARNING SYMBOLS FOR USE ON BAR CHARTS

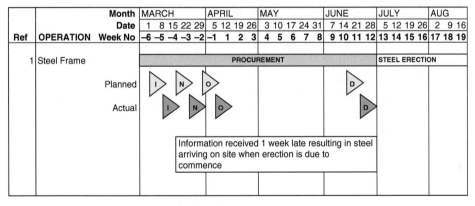

EXAMPLE BAR CHART DISPLAY

Figure 16.1 Early warning system.

- Provided that the contractor proceeds regularly and diligently with the works.
- Provided the contractor does not suffer any prevention or hindrance from the employer or contract administrator.

However, it is frequently the case that the contractor is delayed for one reason or another either due to:

- A contractor risk event, for example:
  o Delay by a subcontractor.
  o Weather delays (other than exceptional conditions).
  o Labour or plant shortages.

- An employer risk event such as:
  - Variations and design changes.
  - Late possession of site.
  - Exceptionally adverse weather.
- Some other reasons.

This does not necessarily mean that completion will be late as the contractor may make up lost time either by working faster or by rearranging the work sequence more efficiently.

Where, however, delayed completion is inevitable or likely, and this is due to an employer risk event, contracts commonly provide a mechanism for the contractor to be granted an extension of time. The reason for this arrangement is not out of kindness to the contractor but to preserve the employer's right to LADs should the contractor otherwise be late finishing the work.

16.3.1 Delay

The usual practice in circumstances where a project is not proceeding as originally intended is as follows:

- The contractor notifies the contract administrator that the works have been or are likely to be delayed beyond the contract completion date.
- The contract administrator then has a set amount of time to consider the situation.
- If the delay is the employer's fault:
  - The contract administrator will decide to grant the contractor an extension of time.
- If the delay is the contractor's fault:
  - The contract administrator will remind the contractor of the contract completion date and its contractual obligation to proceed regularly and diligently with the works.
- In either case, the contractor's master programme will be an important reference point to assist the contract administrator in coming to a decision.

16.3.2 Delay Events

It is frequently the case that matters are not straightforward, and there will often be a number of causes of delay – delay events – some at the employer's risk and some at the risk of the contractor. There will also be issues to untangle such as:

- Whether the contractor's resources were adequate or working efficiently.
- Whether or not certain delays were concurrent.
- Whether the activities affected were on the critical path or not.
- Whether or not the contractor had endeavoured to mitigate the delay – an obligation under English common law.

16.3.3 Change Control

A further problem is that most contractual obligations to provide a programme are insufficiently precise to produce a programme capable of being used effectively as a change control document. For instance:

- Is the programme based on a form of logic which links activities together in such a way as to enable the effects of delay and/or disruption to be determined? This would normally mean making a choice between a linked bar chart and precedence diagram, as arrow diagrams are largely obsolete these days due to their inflexibility and unnecessary complexity.
- Has a method statement been provided which forms the basis of the programme?
- Does the programme or method statement show details of how the contractor intends to carry out the various site operations together with the resources to be used?

Additional problems may include:

- The contractor's programme may be too simplistic.
- The programme may contain imprecise logic.
- There may be open-ended or hanging activities in the programme – in other words, where the logic is incomplete.
- The programme may not be based on a software package capable of modelling the delay/disruption scenario.

16.4 Recording Progress

It is of little benefit to produce a bar chart programme without making best use of it. For the bar chart to be an effective management tool, it is necessary to record progress on it and critically analyse the operations that are ahead of or behind schedule.

The purpose of progress recording is to provide a record of progress to help the construction director, contracts manager and site manager to make informed operational decisions and to report to the monthly site progress meeting with the contract administrator.

16.4.1 Progress Reporting

As a minimum requirement, site progress should be reported at least monthly and preferably on a weekly basis. It is conventional to hold site progress review meetings with the site manager, contracts manager or director and planning engineer present, along with key subcontractors.

The meeting will receive a formal report of progress and an updated programme – preferably comparing actual progress to the baseline – which will assist the participants to discuss issues and look ahead to forthcoming activities.

The progress report should highlight activities of concern, i.e. those which are out of sequence or behind programme. Comments should also be added relating to the action to be taken to get the contract back on schedule. Reasons for the delays may also be indicated, especially where the delays have been caused by variations to the contract, late receipt of information or delays by specific subcontractors. The current position in relation to key subcontractors, materials and orders that have yet to be placed is often highlighted too.

An extract from a monthly site report is given in Figure 16.2.

MONTHLY PROGRESS REPORT			
Contract	City Road	Contract No	1714
Report No	5	Name	A Smith
Date	16 Feb	Contract Period	70
Weeks completed	20	Weeks remaining	50
Contract Completion Date		22 May	
Target Completion Date		4 Feb	
Anticipated completion		4 Feb	

MAIN PHASES OF WORK	PROGRESS COMMENTARY
Site establishment	Commenced on time, now 100% complete
Sheet piling	Started on time, completed 2 weeks later than planned due to work in connection with the adjacent river retaining wall. Work now 100% complete
Riverside retaining wall	Started 4 weeks later than planned due to variable ground conditions and changes in the foundation design. Currently 5 weeks behinds programme. Work now 60% complete
Basement A	Excavation works completed on time. Delays to the wall construction of Basement B have delayed commencement of the basement slab construction. Basement works are currently 4 weeks behind programme. Work now 20% complete
Basement B	Basement concrete and formwork started 3 weeks late due to design delays. Work currently progressing to basement walls at second level. Basement works currently 4 weeks behind programme. Work currently 80% complete

PROCUREMENT	
Subcontracts	Orders placed for steelwork which is due for erection on Block 1 at week 26. The commencement may have to be delayed by 3 weeks
Materials	Brick samples approved and orders placed

COPIES TO   Site manager ☐   Project manager ☐   Chief planning eng ☐

Figure 16.2   Monthly progress report.

16.4.2 Cloud Based Reporting

Whilst conventional weekly or monthly progress reporting and on-site project review meetings are still common – especially on smaller or less complex projects – the computing power available in the twenty-first century – and access to the Cloud – has moved site progress recording to another level. Reports can now be almost instant, can be activity rather than project-specific and can be shared collaboratively with any or all members of the project and construction teams.

By using construction management software (such as Oracle Aconex or Zutec), progress data from site can be integrated into the project dashboard and used for planned versus actual progress tracking, analytical reporting and for reporting key performance metrics. Additionally:

- The 3D Building Information Management (BIM) model can be used to visualise planned versus actual progress.
- Mobile devices on site can capture photographic evidence of progress or problems with live activities.

Alternatively, requests for progress updates from office-based staff could be sent to site-based staff via a mobile progress application such as Elecosoft Site Progress Mobile – which is illustrated in Figure 16.3. Requests can be sent to more than one person each of whom can report on the site activities under their control. Site-based staff can then record progress and upload the information to the Cloud.

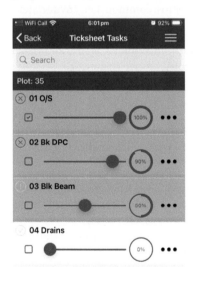
Click on 04 Drains circle to enter progress to date.

Enter progress percentage complete.

Figure 16.3 Elecosoft site progress mobile.

Once received, the progress information can be integrated into Asta Powerproject and the master programme resynchronised. Progress photographs and notes can be added to the report.

A further technological development is the use of simultaneous localisation and mapping (SLAM) technology such as that provided by GeoSLAM – a member of the UK BIM Alliance and the Construction Progress Coalition.

SLAM technology enables a digital model of the site to be created using a hand-held mobile scanner and pre-positioned reference plates. The scanners emit laser beams and work in conjunction with the permanent control points to form a digital map of the site. This is done by simply walking around the site.

Once the scan is complete this is automatically stored at head office and is accessible to authorised personnel who can see progress without having to physically go to the site. Progress can be scanned – and PDF reports automatically generated – on a monthly, weekly or daily basis as required. Key features of the system include:

- Allowing current scans to be compared with previous scans to see what has happened in the intervening period.
- Viewing 3D data and accessing voice notes individually or collaboratively enabling teams to work together on problems or early warning notices.
- Syncing progress results with collaboration tools such as Autodesk Build project management software (formerly BIM360), the Dalux BIM Viewer or Trimble's Viewpoint 4projects documentation management system.
- Data export to computer-aided design (CAD) packages or to a common data environment (CDE).

Using these technologies provides a single source of truth, an auditable record and visual evidence of what is actually happening on site within a collaborative framework.

16.4.3 Using the Bar Chart

There are various ways of recording progress on the bar chart which can be done by the site-based planning engineer on larger projects or the site manager on smaller projects:

- **Colour Coding**:
    - Progress can be recorded by colour coding the bar chart – green, for example, could represent the actual time elapsed on an activity, and red could be used to denote the percentage completion.
    - Alternatively, different colours could be allocated to each week of the project, and progress could be recorded by shading the work undertaken in a particular week with the appropriate colour. This idea is illustrated in Figure 16.4.
- **Progress Tracking**:
    - Project management software – Primavera P6, Asta Powerproject, Project Commander, MS Project, and so on – allows progress recording by entering the percentage complete for each activity.
    - The tracking Gantt is then shown which can be compared with the baseline.
    - A 'jagged' line progress indicator appears as shown in Figure 16.5, where the vertical line shows the progress review date.

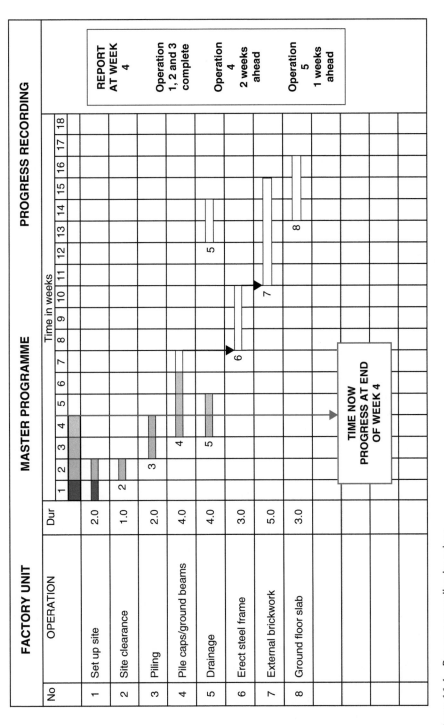

Figure 16.4 Progress recording by colour.

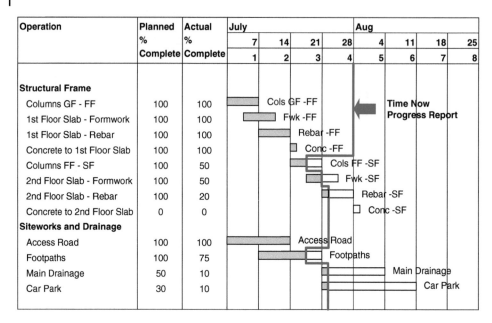

Figure 16.5 Progress recording by computer.

- Should activities be ahead of or behind programme, the line moves to the left or right accordingly. In this example, structural frame activities 1–4 are complete, activities 5–7 are behind and activity 8 is yet to start. Siteworks and drainage activity 1 is complete, and activities 2–4 are behind programme.
- **Earned Value**:
  - The physical progress on a contract can also be measured according to the earned value of work carried out using a value envelope. This is established by plotting the forecast value for the contract based on both the earliest and latest times that work activities can be started. This creates the envelope.
  - The idea is to use the float or free time in non-critical activities to create the control envelope whilst still keeping to the overall programme period as determined by the critical activities (i.e. the critical path). The boundaries of the envelope delineate the intended programme in money terms. The principles of earned value forecasting are explained in Chapter 14.
  - To report progress, the monthly valuation is plotted on the envelope and, where earned value is within the envelope, the contract is progressing according to programme. Should earned value fall outside the envelope, this signifies that the contract is either ahead or behind programme.
  - Figure 16.6 shows an example where the earned value is:
    - Above the envelope in month 1 – generating value faster and, therefore, probably ahead of schedule.
    - Below the envelope in month 2 – behind schedule.
    - Inside the envelope in month 3 – on schedule.

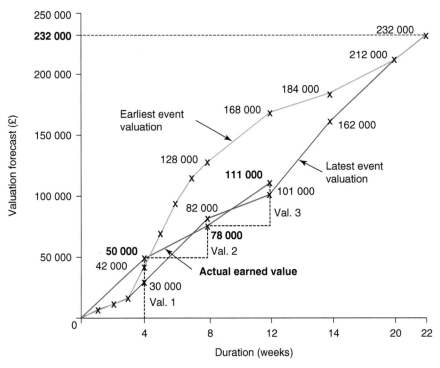

Figure 16.6 Progress by earned value.

16.5 Delay and Disruption

Delay and disruption are common features of construction projects, and this was highlighted in both the Latham and Egan reports. Rethinking Construction (1998) said that *Projects are widely seen as unpredictable in terms of delivery on time.* Whilst improvements have been made since those reports were written, Constructing Excellence maintains that around 40% of construction projects still finish late.

16.5.1 Reasons for Delay and Disruption

Most projects that finish late are likely to be delayed by a few weeks or months, but there have been a number of notable high-profile examples over recent years, including the

Scottish Parliament building and the new Wembley Stadium, which were handed over years later than planned.

Finding plausible reasons for such lengthy delays on major projects is not easy, and all the participants will probably have a different view of things. However, delays in the flow of design information, over-optimistic scheduling, contractual disputes and unusual procurement methods are some of the many reasons cited in the official enquiries into these projects which are also explored in Chapter 1.

On 'normal' projects, the reasons for delays and disruption may include:

- Where the employer/client is at fault:
  - Incomplete design at commencement on site.
  - Delays in the flow of design information.
  - Variations and changes in the scope of the works.
  - Discrepancies in contract documents.
- Where the contractor is at fault:
  - Poor quality of workmanship.
  - Inadequate planning.
  - Under resourcing of site operations.
  - Accidents and incidents on site.
- Where no one is at fault:
  - Unforeseen physical conditions.
  - Strikes affecting site labour or the supply of materials.
  - Fire, explosion or act of God (*force majeure*).
  - Bad weather.

Projects that finish late have implications for both the client and the contractor. For instance, the client will not be able to enjoy a new facility when anticipated, and this may lead to loss of income or utility.

Alternatively, the contractor will not be able to release key managers and other resources for new projects as planned, leading to delays on other projects, or there may be extra costs to bear for running the site establishment for longer than initially planned.

16.5.2 Definitions

'Delay and disruption' is a phrase which is commonly used in construction just like 'loss and expense' or 'health and safety'. The words go together quite naturally but, in each case, the individual words have quite separate and distinct meanings.

As far as 'delay and disruption' is concerned, both words relate to issues of time and completion of the works, but 'delay' is quite distinct from 'disruption'. For instance, a project might have been delayed without any disruption taking place, and activities on site may be disrupted without causing any delay to completion of the works:

- **Delay** relates to the act or state of being late – failing to complete the works by the contract completion date for instance.
- **Disruption** is concerned with loss of efficiency due to lower-than-expected productivity or some interference with normal progress.

Admittedly, delay may become a consequence of disruption, although not necessarily, whereas delay to progress may have nothing to do with the disruption of work on site.

16.5.3 Delay

Delays are caused by events or circumstances which lead to failure to start or complete on time. However, delays to progress and delays to contract completion must be carefully distinguished.

Sometimes, delay occurs when programmed activities do not start or finish on time, or both, but this may not necessarily affect the overall completion date of the project. For instance, the employer may defer the site possession date, causing a delayed start to a project, but the project may still complete on time because there may be spare time in the programme, or the contractor may speed up construction so as to make up for lost time. Consequently, if the contractor is delayed, this will not automatically result in an extension of time.

On the other hand, should there be a delay which impacts on the contract completion date, the cause of the delay must be determined in order that an extension of time can be considered or, where the contractor is at fault, LADs charged.

Delays do not always lead to entitlement to financial reimbursement, but conversely, delays to non-critical activities might result in a claim for loss and expense with no entitlement to an extension of time because the completion date is not affected.

16.5.4 Disruption

Disruption is where the contractor is involved in duplication of effort, out-of-sequence working or where subcontractors or other resources have to be brought back to site to do more work.

Disruption to site activities may not lead to a delay but will invariably result in a claim for loss and expense under the contract due to the extra effort required to do the work.

16.5.5 Types of Delay

There are several meanings to the word 'delay':

- **Delay event**: This refers to something that happens, or a set of circumstances that arise, that lead to a delay. The delay event may be the responsibility of the employer or the contractor or may be beyond the power of either party.
- **Culpable delay**: where the delay is the fault of the contractor. This is also referred to as **non-excusable delay** or **contractor delay**.
- **Non-culpable delay**: Where the contractor is delayed through no fault of its own. The delay may be due to the fault of the client (or agents) in which case it is commonly called **excusable delay** or **employer delay**.
- **Neutral delay**: This is where the delay is caused by factors beyond the control of either of the contracting parties but does not mean that neither party bears the risk:

- Neutral delay at contractor's risk:
  - Bad (but not exceptional) weather.
  - Late delivery of materials.
  - Delay by subcontractors.
- Neutral delay at employer's risk:
  - Exceptionally bad weather.
  - Discovery of ancient artefacts on the site.
  - Strikes affecting site work, off-site manufacture or transport.
- **Concurrent delay**: Where two or more delay events arise at the same time and at least one of the delays has been caused by the client and another by the contractor.

16.6 Extensions of Time

The issue of delay is expressly dealt with in standard construction contracts through specific clauses dealing with:

- Extensions of time for specific events beyond the contractor's control.
- Procedures for the contractor to follow when making an application for an extension of time.
- Procedures for the contract administrator to follow when giving consideration to the contractor's application.

There are usually no similar provisions dealing with disruption and therefore the contractor may have to rely for entitlement upon terms likely to be implied into the contract by a court. Such implied terms may include the employer's undertaking not to hinder or disturb the contractor's execution of the works.

The contractor may also seek to rely upon express contract conditions entitling it to loss and expense by applying for additional payment should the regular progress of the works be affected by specified relevant events.

16.6.1 Contract Provisions

The job of the contract administrator is to monitor progress and make decisions if there are delays. For example, under the JCT Standard Building Contract, the architect is required to:

- Grant the contractor an extension of time where completion of the works is likely to be delayed due to a 'relevant event'.
- State the extension of time granted for each relevant event – there may be several.
- Review the situation later in the light of any further relevant events or instructions which affect the completion date.
- Where the contractor is not proceeding regularly and diligently or if the works are suspended, give notice to the contractor.
- Finally, determine the completion date once the works are complete taking everything relevant into account.

Under JCT contracts, all the contract administrator's decisions must be *fair and reasonable* and subject to the contractor endeavouring to prevent delay and do all that may be necessary to proceed with the works.

Under the ICC form, the engineer is to act impartially and consider whether the delay *fairly* entitles the contractor to an extension of time.

The NEC ECC contract is silent on this matter, but precedence in *Sutcliffe* v. *Thackrah* [1974] is that *the architect* [and the like] *will act in a fair and unbiased manner*.

16.6.2 Establishing Entitlement

Whilst there would seem to be no real science to extensions of time, in most cases, the architect/engineer and contractor will see 'eye-to-eye', and there will be no need for a dispute, or any complex analysis of cause and effect, should there be any delays. However, agreement may not be so easy in some cases where, perhaps, the causes of delay may be complex and, especially where large sums of money are involved.

> In the case of *John Barker Construction Ltd* v. *London Portman Hotel Ltd* [1996], the Judge concluded that the architect did not carry out a logical analysis in a methodical way of the impact which the relevant matters had, or were likely to have had, on [the contractor's] planned programme.
>
> The architect was also criticised for making an impressionistic, rather than a calculated, assessment of the time which he thought was reasonable.

This seems to be a somewhat unfair criticism, especially as most contracts do not spell out how extensions of time are to be decided, nor do they stipulate the information to be supplied by the contractor to support the application:

- **JCT SBC**: The architect/contract administrator *estimates* what is fair and reasonable.
- **ICC form**: The engineer makes an *assessment* of the delay.
- **NEC ECC**: The project manager *assesses* the compensation event (delay).

In practice, it is probable that the contract administrator will base the extension of time decision on the following:

- The contractor's latest programme.
- The information submitted by the contractor.
- Awareness of progress to date.
- Careful consideration of available facts at the time.
- Experience.

16.6.3 Time and Money

Often, there is a tendency to assume that financial recompense automatically follows an extension of time. However, whilst the extension of time and loss and expense clauses are clearly related, it is not necessarily the case that time and money claims are linked. This often comes as disappointing news for contractors, leading to lengthy and costly disputes.

To avoid this happening, it makes sense to employ a fair and equitable means of determining who is responsible for the effects of any delay events that might arise. To do this, it is necessary to establish what happened and why. From a practical standpoint, this is likely to be a complex matter because, unless the project is a simple one, there is unlikely to be a single cause of delay – indeed, there may be concurrent delays, and some delay events may be more significant than others (i.e. a dominant cause of delay).

> Any delay analysis will be dependent for its success on the quality of information and records available and upon the skill, experience and objectivity of the delay analyst.

Standard forms of contract allow for extensions of time to be granted for delay which is not the fault of the contractor. However, individual contract forms treat the issue of extensions of time differently. For instance, under the JCT SBC, the contractor is entitled to an extension of time for relevant events where the contract completion date is *likely to be delayed*, whereas under the ICC form, an assessment is made *of the delay (if any) that has occurred*.

The contractual provisions for extensions of time vary from contract to contract. JCT contracts refer to *relevant events,* whereas NEC contracts use the term *compensation event*. Grounds for extension of time also vary, and the specific contract should be studied carefully to establish entitlement.

16.6.4 Concurrent Delay

Where there is a single cause of delay, there is usually little problem in dealing with prolongation claims from the contractor. However, in practice, the situation is usually far more complex as several causes of delay frequently occur at the same time.

Where there is concurrent delay (i.e. where the contractor is also in delay at the same time as an employer's risk event), the contractor's entitlement will be affected.

Therefore, extensions of time for instructions issued by the architect after the contract completion date when the contractor is in culpable delay must be calculated on a net basis following the judgment in *Balfour Beatty Building Ltd* v. *Chestermount Properties Ltd* (1993).

16.6.5 Mitigation

An important consideration regarding both extensions of time and delay and disruption is the long-standing legal principle of mitigation. Consequently, where there is delay and/or disruption, an implied term of the contract will require the contractor to mitigate its loss.

This means that the contractor in non-culpable delay cannot just sit back, rub its hands and wait for the money to pour in from the client but must do something positive, within reason, to reduce the impact of the circumstances in question.

Some standard contracts also include the express term that the contractor will *constantly make his best endeavours* to prevent delay.

16.6.6 Float

Float – also called 'slack' – may be defined as the amount of 'free time' available in the contractor's schedule. This free time may arise for a number of reasons:

- It may be the difference between the time for completion stated in the contract and the contractor's target programme. This is considered to be 'free' time because the contractor's obligation is to complete the works on or before the contract completion date. The question as to whom this time 'belongs to' has long been debated in the industry.
- It may simply be a contingency allowance made by the contractor where the duration of a particular activity is longer than actually necessary to complete the work required.
- It may be shown on the contractor's programme as a contingency allowance for the entire project.
- It may be the latitude available for carrying out an activity which is not on the critical path – meaning that the activity may be started or completed earlier or later than shown on the programme without affecting other activities or overall completion (i.e. activity float).

Activity float is the most common type of float where total float represents the amount of time which can be used up by delaying either the start or completion of an activity, or both, without delaying project completion. The different types of activity float are illustrated diagrammatically in Figure 16.7 and are explained below:

- **Total float**: This indicates the extent to which an activity may be delayed without compromising the contract completion date. Most modern project management software programmes will calculate total float, and this is often shown on the bar chart display as an extension to the activity bar line. Some software packages will also show free float.

Total float, however, gives a false impression of how much spare time there may be in the programme because there are a number of components to total float:

- **Free float**: This is the part of total float that can be used without affecting the start of any succeeding activities.
- **Independent float**: This is the part of total float that can be used without affecting any preceding or following activities.
- **Interfering float**: This is the part of total float that, if used, will 'interfere' with either the start or completion of dependent activities.

An example of total float may be seen in Figure 16.7 which is given by the equation:
Latest start time of activity B *less* earliest start of activity A *less* duration of activity A: i.e.: $21 - 10 - 5 = 6$ days.

The critical path links all critical activities together and is the longest route through the programme. By definition, therefore, critical activities have no float because they must be started and completed on time so as not to delay completion of the project. In the delay

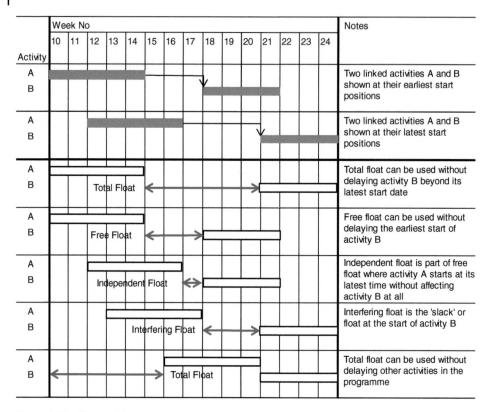

Figure 16.7 Types of float.

situation, critical activities can become sub-critical or even non-critical and other activities can assume critical status in their place.

> It should be noted that there may be more than one critical path in a programme.

A further point to bear in mind about float is that it is possible that critical activities could have float 'built-in' even though they appear on the critical path. This may be because the contractor has included a contingency allowance in the time calculations, or it may simply be the case that estimates of activity durations are not scientifically calculated and are therefore bound to be imprecise.

However, any such 'float' would not be evident when undertaking an analysis of the critical path, as this would reveal only those activities with calculable float.

16.7 The 'As-Planned' Programme

The as-planned programme represents the contractor's view of the project at the outset of the construction phase. It is the contractor's master programme.

16.7 The 'As-Planned' Programme

Project management software – such as Project Commander, Microsoft Project and Asta Powerproject – calls this the 'baseline' programme because the programme is 'frozen' and used as a fixed reference against which actual progress can be measured. The **baseline programme** usually shows:

- Broad bars representing the original planned sequence with logic links hidden.
- Thin bars underneath which will remain in place when the programme is updated later (this is the original 'baseline') – see Figure 16.8 for an example.

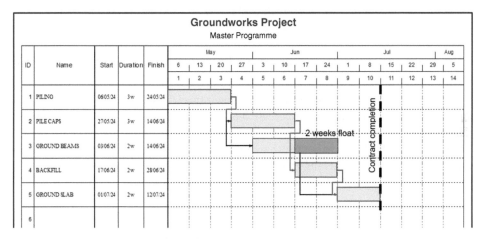

Master programme showing critical activities 1, 2, 4 and 5.
Activity 3 (Ground beams) has 2 days float.
Logic links (dependencies) and contract completion date shown.

Baseline bars shown beneath the as-planned programme.
Logic links hidden.

Figure 16.8 Groundworks project 1.

Note: The baseline is created by saving the original plan as a 'baseline'. Future changes to the schedule will not affect the baseline which will remain in place for change control purposes.

16.7.1 Shortcomings

In delay analysis terms, the as-planned programme may have shortcomings such as:

- It may not show all activities included in the project, either by mistake or because, at the time the programme was prepared, such activities were not highly significant.
- It may not show some activities in enough detail – e.g. a single bar may be shown to represent a subcontract package which contains several sub-activities within it.
- There may be no actual logic, or the logic links may be incorrect – there may be activities which have hanging or incomplete logic or they may simply be wrong.
- The critical path(s) may be incorrect due to either wrong calculation or because the links are wrong.

This may seem inconceivable but programming (especially where the project is large or complex) is not easy and every planner will have a different view of how to put the project plan together.

Consequently, the delay analyst may have to 'tidy-up' the as-planned programme and this may lead to accusations that bias or hindsight may have influenced the analysis.

16.8 The 'As-Built' Programme

The as-built programme is a representation of what actually happened on the project so that this can be compared to what was planned. This programme is not simply the contractor's updated progress programme but is based on the complete project record database including:

- Site diary entries.
- Progress records – usually based on percentages complete.
- Interim valuations – i.e. monthly financial progress (but this may prove to be unreliable data due to under or over-valuation).
- Labour and plant returns such as allocation sheets, daywork sheets and plant hire sheets.
- Subcontractors' accounts and claims.
- Progress photographs (especially useful when memories begin to fade, or management staff move on).
- Correspondence – hopefully, all significant events and instructions will have been confirmed in writing and all contractual notices sent in on time.
- Minutes of site meetings and the contractor's internal progress meetings.

16.9 Delay Analysis

Under the common standard forms of contract, it falls to the contract administrator or project manager to determine whether or not the contractor is entitled to an extension of time under specific circumstances.

For instance, under the JCT Standard Building Contract, a decision has to be made whether a 'relevant event' has happened and what the contractor's entitlement will be. Such decisions need to be made when it becomes apparent or likely that the contract completion date will not be met.

> When delays occur on a project, the contracting parties will be keen to ensure that they recover their due entitlement under the contract – such as an extension of time, loss and expense or liquidated and ascertained damages. It is frequently the case that agreement cannot be reached and adjudication, arbitration or even litigation may follow.

Not surprisingly, each of the parties will want the best settlement and to this end will 'argue their corner' as best they can. This often results in a jaundiced view based more on fiction than fact, which does not stand up to independent scrutiny. The question is a simple one:

- 'What was the cause of the delay and
- Who bears the time and money risk?'

Finding the right way to get the answer is not so easy, though, because, whilst there are several appropriate methodologies available, much depends on the rigour of the investigations undertaken and the skill of the analyst who must determine why a project was delayed.

According to Farrow (2006), the techniques used may be categorised as either **theoretical** or **actual**, although it is recognised that a degree of theory is present in both theoretical and factual techniques due to the extent of personal judgement and interpretation of facts required in most analyses.

When delay or disruption occurs, the conditions of contract for the project must be referred to in order to determine whether any relief is available for the event in question.

16.9.1 Delay and Disruption Protocol

The Society of Construction Law (SCL) Delay and Disruption Protocol (2017) is a non-mandatory guide to good practice for dealing with delay and disruption issues. It provides guidance on compensation and extensions of time, the preparation and maintenance of programmes and records and proposes a number of methodologies for analysing delays and their effects on project completion.

The Protocol recommends that, for all but the simplest of projects, the programme should be based on a critical path network using commercially available software where the activities are linked with logic.

It is further suggested that where the contractor is proposing to complete the project in a faster time than stated in the contract, the employer should be made aware of this intention so that the contractor may be granted fair and reasonable extensions of time and/or costs where delayed by the employer – thereby avoiding arguments such as those which arose in the *Guinness* v. *Glenlion* case.

The Protocol has its critics and some delay analysts – including Lowsley and Linnett (2006) – feel that its overly prescriptive insistence on the use of critical path analysis is costly and places too much emphasis on delay analysis and claims to the detriment of a commonsense approach to project planning.

16.9.2 Delay Analysis Methodologies

In an excellent series of articles, Farrow (2002) summarises the various methods as follows:

(1) **Theoretical methods** – a model of what was planned is produced, ignoring how the project was actually constructed and actual events are imposed on the model to see how they may have influenced the end date.
(2) The **'but-for' methods** (i.e. as-planned but-for/as-built but-for delay events) – create models of either planned intentions or the as-built project and seek to address causation on the basis of assumption.
(3) The **as-planned versus as-built** approach, which overlays two models of the project in order to explain the causes of variances – an approach seemingly favoured by lawyers.
(4) **Update methodologies**, which relive the project in incremental stages, consider events at the time they arose and offer the most extensive analyses.

Farrow maintains that none of the methodologies are perfect because they all include an element of assumption, subjective assessment and theoretical projection. He also explains that there is no preferred delay analysis methodology, that the more theoretical the methodology, the weaker the analysis, and that it is always better to demonstrate what actually occurred.

From a delay analysis point of view, the SCL Delay and Disruption Protocol suggests that a distinction needs to be made between current and completed contracts and that different delay analysis techniques may be applicable during the project and after completion of the project. Consequently, the Protocol categorises the available methodologies as:

- **Prospective**: Looking forward or anticipating future events.
- **Retrospective**: Looking back at past events.

The above methodologies are fully explained by Lowsley and Linnett (2006) in their excellent book.

A further delay analysis methodology called 'windows analysis' is referred to by Farrow (2004), Lowsley and Linnett (2006) and others. It is also called 'time-slice analysis', or the 'snapshot' or 'update' method, because it views the project in time slices (e.g. monthly or weekly intervals).

In time-slice analysis, the time slice is updated for progress, then the delay events are inserted one by one, and the impact of delay events is analysed sequentially. The time delay (if any) at the end of one window is used at the beginning of the next window, so the delay (if any) is accumulated.

16.10 Delay Analysis in Practice

At a practical level, delay analysis is not easy and is a professional discipline in its own right. There are accepted techniques available to help in the process, and modern project management software is helpful for speeding up iterations, developing 'what if?' scenarios and producing colourful and professional displays.

16.10 Delay Analysis in Practice

> Lowsley and Linnett (2006) maintain that all the various techniques have *difficulties and shortcomings*, and it is difficult to find *an approach that can withstand robust examination*.

Where a project is currently ongoing, extension of time judgements will be based on the known facts to date (i.e. the latest updated programme and current site records) together with assessments about what might happen later on in the project.

When a project is completed, on the other hand, a 'forensic' assessment of what actually happened might be more appropriate. However, the reliability and acceptability of this approach will be heavily dependent on the quality of the records available, such as site diary entries and progress records, to build up the 'as-built' view of the project.

16.10.1 Showing Delay on the Programme

From a programming point of view, the difficulty is how to show the effect of delays on the contractor's programme. This is not as easy as it sounds because:

- Additional activities and/or events have to be introduced into the original programme.
- Where a delay event is complex a sub-net or frag-net may have to be introduced into the programme.
- New logic links have to be made and these need to be correct and without bias.
- Existing activities may have to be 'split' so as to introduce the delay events and associated logic links (not all software allows this).

Figure 16.9 shows three bar chart displays:

- An original programme comprising three activities: A, B and C.
- A delayed programme where a two-week delay is introduced into the programme by increasing the duration of the affected activity (activity B).
- A delayed programme which shows activity B split at the point where the delay event occurred and an intervening two-week delay event.
Note: All logic links are retained.

Delays are rarely straightforward, and more often than not, there are several events or activities making up the delay as a whole. The sequence of delaying events could therefore be put together in a mini-programme or sub-network to build up a picture of the delay. The sub-net would have its own logic links, and this can then be incorporated into the main programme.

16.10.2 Software

For delay analysis, and in order to successfully determine the effects of delay and disruption, project management software must be capable of:

- Activity linking, e.g. start–finish, start–start or finish–finish.
- Producing a baseline of the contractor's original master programme.
- Regular updating with actual progress, activity start and finish dates and percentage completion for incomplete activities.
- Rescheduling the programme to see the effects of delays.

Delay Analysis 1
Original Programme

ID	Name	Start	Duration	Finish
1	Activity A	03/06/24	3w	21/06/24
2	Activity B	24/06/24	4w	19/07/24
3	Activity C	22/07/24	2w	02/08/24
4				

A project comprises activities A, B and C each with Finish-Start dependencies. The contract period is 9 weeks.

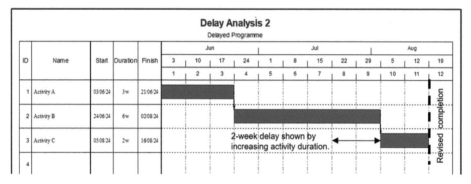

A 2-week delay is shown by increasing the duration of Activity B.

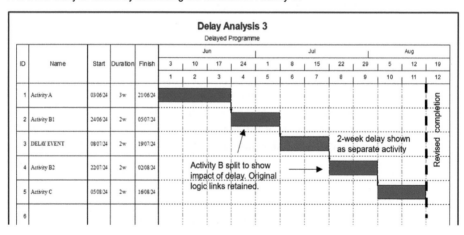

The delay event is shown as a separate activity.
Activity B is split (B1 and B2) at the point where the delay event occurs.

Figure 16.9   Showing delay on the programme.

Farrow argues that, ideally, the software should also have an integrated database for forensic analysis of project records.

He also suggests that, whilst some of the more popular packages are critical path analysis tools, they are weak as regards data processing and vary enormously in terms of cost and the extent of training and experience required to use them effectively for delay analysis purposes.

16.10.3   Worked Example

The following worked example is a very much simplified example of a time-impact analysis.

- Figure 16.8 shows two bar chart displays for a groundworks project:

Master programme	• Critical activities (activities 1, 2, 4 and 5).
	• Non-critical activities (activity 3).
	• Available float.
	• Logic links showing dependencies.
	• Contract completion date.
Baseline programme	• As-planned programme.
	• Baseline bars beneath activity bars.
	• Logic links hidden (a facility on most software packages).

Both schedules are effectively the contractor's as-planned programme, but the baseline programme is more likely to be the contractor's internal control document.

- Figure 16.10 shows the same project with schedule status indicated at week 4:

Progress at week 4	• The baseline programme is shown.
	• Progress is shown by a 'jagged' line at week 4.
	• Piling is 70% complete (behind programme).
	• Pile caps are 10% complete (behind programme).
	• Other activities have not yet commenced.
Reschedule at week 4	• The programme is updated for progress.
	• Incomplete and yet-to-be-started activities are moved to week 4 progress line.
	• The baseline remains in place.
	• A delay of 4 days is indicated.

Note: The delays shown in Figure 16.10 are **contractor risk events** where the cause of delay is the contractor's responsibility under the contract, e.g. bad weather, slow or inefficient working, late delivery of materials or lateness on the part of subcontractors.

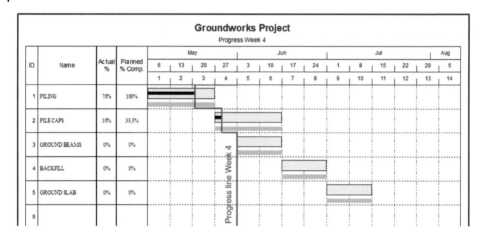

Progress on Piling and Pile Caps at Week 4 showing progress line.

Programme rescheduled at Week 4 showing reporting line.
Original contract completion and revised completion dates shown.
Delay of 4 days indicated.

Figure 16.10 Groundworks project 2.

- Figure 16.11 shows two further displays for the groundworks project, this time at week 8 of the project:

Revised programme at week 4:	• Updated with progress to week 4.
	• Logic links unhidden to show dependencies.
	• Includes two employer risk events impacted after week 4 update:
	   ○ 10 days delay for the re-design of the rebar to the ground beams.
	   ○ A further 8 days delay for off-site fabrication of the new rebar.

Progress at week 8	• Updated for progress to week 8. • Shows revised completion to week 12 + 3 days (i.e. mid-week). • Overall delay to date = 8 days.

Note: The delays in Figure 16.11 are partly contractor risk events and partly employer risk event.

Employer risk events are where the cause of delay is the employer's responsibility under the contract, e.g. delayed instructions, exceptionally inclement weather, unforeseen ground conditions or, as in this case, design variations to the ground beams.

- The contractor's extension of time entitlement would therefore be:

Overall delay to completion	= 8 days	Contract completion delayed from week 10 to week 12 + 3 days.
Less contractor delay events	= 4 days	Contractor was in culpable delay for the piling and pile caps activities.
Therefore, employer risk events	= 4 days	The employer risk events followed the contractor's culpable delay so there was no concurrency of delay.

Even though the employer risk events amounted to some 18 days, the contractor's extension of time entitlement is only four days. This is because the overall delay was only 8 days because the ground beams activity was non-critical and had 10 days float. The contractor was also in culpable delay which amounted to four days.

16.11 Project Acceleration

Where a project has been delayed for one reason or another, the contractor may wish to bring forward the completion date or may be asked to do so by the client.

If the contractor is not at fault in these circumstances, the trouble and effort of accelerating the works should be paid for through a separate agreement with the client. Some standard conditions of contract, such as the NEC ECC and ICC forms, provide for this eventuality.

There are several types of acceleration, which have different legal meanings:

- **Pure acceleration**: Where the contractor speeds up work on site so as to finish earlier than scheduled at the request of the client.
- **Constructive acceleration**: Where the contractor is effectively forced to work at a faster rate because the contract administrator has delayed or refused a legitimate application for an extension of time.
- **Expedite**: Where there is culpable delay and the pace of work on site must be speeded up so as to get back on schedule.

Acceleration, or speeding up the work, can be achieved in three main ways:

- Re-organising the work more efficiently:
  - The work might be re-organised by increasing the concurrency of site operations. This might be successful or could introduce other problems – inefficiencies or accidents for instance – where operatives and package contractors may be working on top of each other.

16 Controlling Time

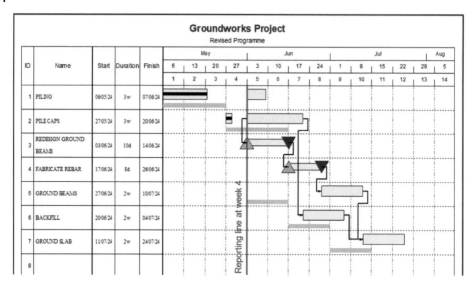

Revised programme at Week 4.
Employer risk events 3 and 4 included.

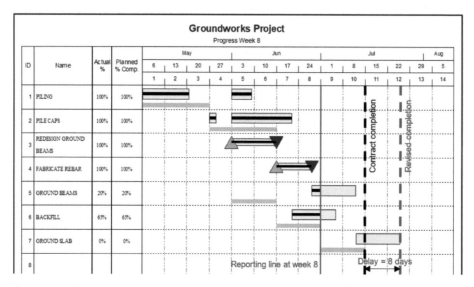

Progress at Week 8.
Revised completion date indicated.
Overall delay = 8 days.

Figure 16.11 Groundworks project 3.

- Increasing the resources on site:
    - By increasing resources, work may be speeded up initially but here again problems may arise. For example, inefficient gang sizes may be introduced, or the site may become congested with plant and operatives. There may also be problems with quality and supervision, and extra costs may be incurred due to the need for additional site management personnel.
- Both:
    - The contractor needs to balance the possible measures and decide to what extent it is physically and financially viable to speed up the work. There is a break-even point here beyond which diminishing returns may set in.

One method of seeking out this break-even point is to use the management technique of time–cost optimisation. This attempts to balance the direct cost of doing the work with the indirect costs of managing the process. The resultant analysis provides an optimum time and cost solution.

16.11.1 Time–Cost Optimisation

There are many circumstances in which the contractor may wish to speed up work on a contract:

- Being behind programme and having to increase production to minimise extensive liquidated damages.
- The client might have requested the contractor to indicate the additional costs of completing the project earlier than the contract completion date for some reason.
- A department store client, for instance, may request an earlier occupation date for the building to take advantage of the winter or summer sales. The additional profit created by the earlier opening may well exceed the contractor's additional costs.

To balance the time savings against the costs of speeding up the work, optimisation studies can be undertaken in order to consider the various options available. These studies allow the client and/or contractor to assess the effect on the direct and indirect costs of reducing the overall project period, and this can then be compared with the potential profits or savings in liquidated damages due to earlier completion.

This method of analysis is variously referred to as 'time–cost optimisation', 'least cost optimisation' or 'crash costing', but perhaps a more appropriate term would be 'project acceleration' as this is really what it is all about.

16.11.2 Terminology

An activity on a contractor's master programme is for the electrical services work. It has a duration of 12 days. To accelerate this activity, a few terms need to be considered:

Normal time, normal cost	This is the usual time that would be needed to carry out the electrical services work under normal circumstances, estimated at 12 days with a cost of £10 000.

Crash time, crash cost	The crash time is the maximum time the operation can be compressed by increasing the resources. This reduction in time leads to an increase in the direct cost. The revised cost is called the crash cost. In this example, the accelerated time or crash time is to be 8 days at a total cost or crash cost of £18 000.
Cost slope	The cost slope represents the cost of accelerating any of the project activities by one unit of time (in this case, 1 day).
Activity ranking	Once the arrow or precedence relationship has been analysed and the critical activities identified, each of the activities on the critical path is ranked in order of their cost slopes, starting with the least expensive.
Direct costs	These are the costs associated with carrying out activities on the programme including labour, plant, materials, subcontractors (if applicable) and overheads and profit.
Indirect costs	These are the time-related costs of the project which change as the project duration changes. They are normally included in the contract preliminaries and will include project supervision, site hutting and accommodation, site office telephones, heating and lighting, vans and site transport.
Total project cost	This is the summation of the direct and indirect costs. The total cost is usually expressed at the normal time and at the optimum project duration.
Optimum project duration	The optimum project duration occurs at the point where the most beneficial least-cost situation occurs, taking into account both direct and indirect costs. In order to establish the least cost situation, the cost increase for each unit of time reduction must be considered.

In order to achieve a reduction in the overall project duration at the least possible cost, the activities on the critical path of the programme must be compressed as much as possible. This is done by first considering the activities with the least cost slope.

The cost slope of the electrical services activity is expressed as:

$$\frac{\text{increase in cost}}{\text{reduction in time}} \text{ or } \frac{\text{crash cost less normal cost}}{\text{normal time less crash time}}$$

$$= \frac{£18\,000 - £10\,000}{12-8 \text{ days}} = \frac{£8000}{4 \text{ days}}$$

$$= \textbf{£2000 cost slope}$$

It is obviously more economical to apply reductions in the project time to the less expensive activities first to achieve the required reduction in the overall project period. As the ranking is applied to the network sequence, the float times and cost slopes of non-critical activities must be considered as at some point these may become critical.

When an activity is accelerated, the corresponding direct costs will increase. This may be due to the need to supply additional resources in the form of increased labour, plant and materials requirements. Accelerating an operation may also involve overtime or weekend work. Also, incentives in the form of bonus payments may have to be made to ensure that the task is completed on time.

Other direct costs may be incurred, such as additional formwork for a concreting activity. This will reduce the number of formwork uses originally envisaged by the estimator and thus add to the direct cost of acceleration.

Acceleration may require extra direct supervision, such as foremen and gangers, to cover weekend work and the supervision of additional labour gangs.

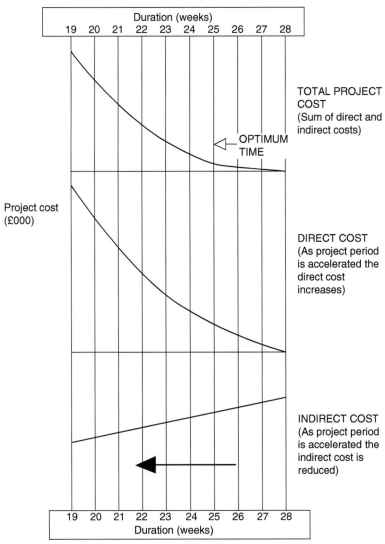

Figure 16.12 Time-cost relationships.

When an activity is accelerated, the corresponding indirect costs will decrease. This is due to the reduction in project duration which directly affects the contract preliminaries. The contractor will, in principle, be on site for a shorter period and therefore the client will expect some reduction in the site administration costs or preliminaries.

Figure 16.12 illustrates the relationship between direct cost, indirect cost and project duration. The summation of the direct and indirect costs (i.e. the total project cost) is also indicated. The optimum duration is shown on the total cost graph, and this is the date at which the costs rise most significantly.

16.11.3 Worked Example

Figure 16.13 shows a construction sequence involving activities A to I in arrow diagram format. The tabular data indicates the normal times, normal costs, crash times and crash costs. The indirect costs (or time-related preliminaries) of the project amount to £2000 per week.

Figure 16.14 illustrates the analysed arrow diagram based on the normal time situation which gives an overall project duration of 28 weeks. The total project cost, based on normal time, has been calculated at £254 000 which is made up of direct costs of £198 000, plus £56 000 (28 weeks × £2000) of indirect costs.

In order to assess the least cost situation, the effect on project cost for each week's reduction in time must be calculated taking into account the change in the direct and indirect costs. This is shown in Figure 16.15 and illustrated in graphical form in Figure 16.16.

The relationship between the indirect and direct costs for each week of the project from weeks 28 to 18 can be observed in Figure 16.16. The summation of the direct and indirect costs is displayed, and the point on the graph where the cost suddenly increases is the position of the optimum time and least cost situation. This occurs at the end of week 24. At week 25, the change in cost per week alters from £2000 to £7000 per week.

The project costs have been analysed back to week 18 to provide an overall assessment of the project cost situation at various points in time.

> Should it be desired to complete the project by week 23 – a reduction of five weeks – the effect on the direct and indirect costs must be considered. To do so, it will be necessary to consider the cost slopes of all activities. The assessment of the cost slopes and their appropriate ranking are indicated in Figure 16.17.
>
> From Figures 16.14 and 16.17, it can be seen that to reduce the project duration by 5 weeks, it will be necessary to reduce the duration of activities on the critical path in rank order (from the least expensive to the more expensive cost slopes).
>
> In Figure 16.14, the critical path follows activities D, A and E in rank order. By using the crash durations for these activities, a five-week reduction in the overall project period can be achieved, as shown in Table 16.1. This gives a crash cost of £267 000, as shown in Table 16.2.
>
> Figure 16.18 shows the revised arrow diagram analysis using the crash times on activities A, E and D. This analysis also indicates the revised float times on the non-critical activities.

INITIAL ARROW DIAGRAM

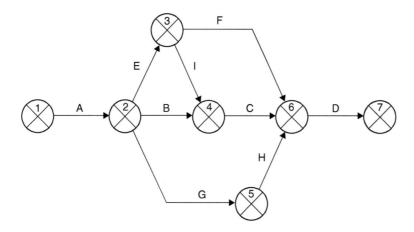

PROJECT DATA				
Activity	Normal time	Normal cost	Crash time	Crash cost
A	6	12 000	4	20 000
B	8	24 000	4	48 000
C	3	18 000	2	20 000
D	6	18 000	4	24 000
E	6	36 000	4	54 000
F	10	10 000	6	50 000
G	5	20 000	3	30 000
H	8	40 000	6	50 000
I	2	20 000	2	20 000
Summation		£198 000		
Indirect costs £2 000 per week				

Figure 16.13 Time-cost optimisation 1.

16 Controlling Time

Normal time = 28 weeks
Normal cost = Direct cost plus indirect cost
= £198 000 + 28 (£2 000)
= £254 000

Total project cost at normal time £254 000

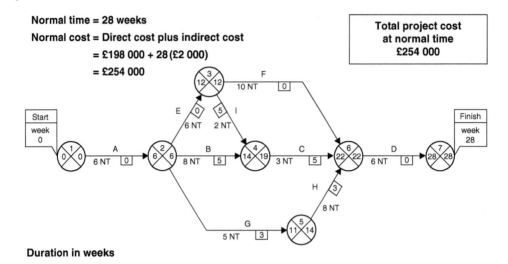

Duration in weeks

NORMAL TIME/NORMAL COST ANALYSIS

Figure 16.14 Time-cost optimisation 2.

Week No. start	Direct cost	Indirect cost +ve	Indirect cost −ve	Aggregate	Total project cost (£000)	Activity ranking
28	198		56		254	
27		+3	−2	+1	255	Activity D
26		+3	−2	+1	256	
25		+4	−2	+2	258	Activity A
24		+4	−2	+2	260	
23		+9	−2	+7	267	Activity E
22		+9	−2	+7	274	
21		+10	−2	+8	282	Activity F
20		+10	−2	+8	290	
19		+10	−2	+8	298	
18		+10	−2	+8	306	

CALCULATION OF CHANGE IN DIRECT AND INDIRECT COST PER WEEK AFTER APPLYING RANKING

Figure 16.15 Time-cost optimisation 3.

16.11 Project Acceleration

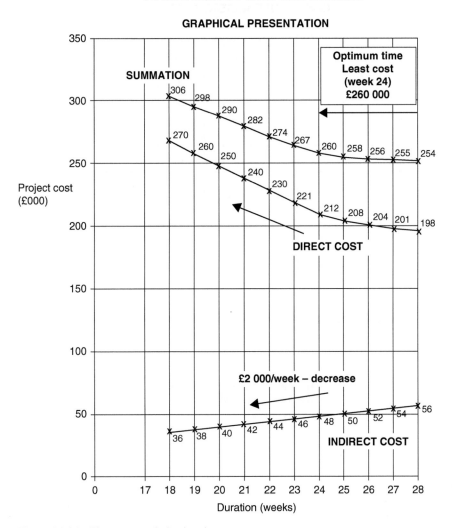

Figure 16.16 Time-cost optimisation 4.

COST SLOPE ASSESSMENT

Activity	Normal time	Crash time	Saving in time	Normal cost (£000)	Crash cost (£000)	Increase in cost (£000)	Cost slope	Float	Order/Ranking
A	6	4	2	12	20	8	4000	Zero	2nd
E	6	4	2	36	54	18	9000	Zero	3rd
F	10	6	4	10	50	40	10000	Zero	4th
D	6	4	2	18	24	6	3000	Zero	1st
B	8	4	4	24	48	24	6000	5	
G	5	3	2	20	30	10	5000	3	Non
H	8	6	2	40	50	10	5000	3	critical
I	2	2	0	20	20	0	0	5	operations
C	3	2	1	18	20	2	2000	5	

Figure 16.17 Time-cost optimisation 5.

Table 16.1 Reduction in project period.

Activity	Normal time (wk)	Crash time (wk)	Reduction in time (wk)	Cost slope (£)	Increase in direct cost (£)
D	6	4	2	3 000	6 000
A	6	4	2	4 000	8 000
E	6	5	1	9 000	9 000
Reduction in overall time =			5		
Increase in direct cost due to this reduction in time =					23 000

Analysis based on accelerated project period of 23 weeks

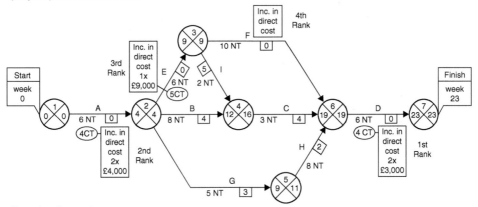

Duration in weeks

REVISED ARROW DIAGRAM

Figure 16.18   Time-cost optimisation 6.

Table 16.2   Total project costs at week 23.

Total project costs at week 23

Direct costs =	£198 000 + £23 000 =	£221 000
Indirect costs =	23 weeks @ £2 000 =	£46 000
Total project cost =		**£267 000**

Therefore, in order to achieve an acceleration of 5 wk, the project cost will be increased by £13 000:

Crash cost	= £267 000
Less normal cost	= £254 000
Acceleration cost	= **£13 000**

References

Farrow, A. (2002) *Delay analysis – methodology and mythology – part 1*. Trett Consulting Digest, Issue 27, Article 1.

Farrow, A. (2004) *Delay analysis – methodology and mythology – part 2*. Trett Consulting Digest, Issue 28, Article 1.

Farrow, A. (2006) *Assessing extensions of time*. Trett Consulting Digest, Issue 31, Article 5.

Lowsley, S. & Linnett, C. (2006) *About Time – Delay Analysis in Construction*. RICS Books.

The Society of Construction Law (SCL) Delay and Disruption Protocol (2017) 2nd edn, The Society of Construction Law.

Web References

National Audit Office (2003), The English national stadium project at Wembley. http://www.nao.org.uk/.

The Scottish Parliament, The Holyrood Inquiry (2004) http://www.scottish.parliament.uk/.

	Chapter 17 Dashboard	
Key Message		o The collapse of Carillion emphasised the need for prudent financial management and robust corporate governance.
Definitions		o **Corporate governance**: A strategy to deal with the principal risks and uncertainties facing a business. o **Earned value analysis (EVA)**: A means of forecasting the cumulative monthly value of work to be carried out on a project. o **Cost-value reconciliation (CVR)**: A means of determining the true financial position of a contract based on prudence and accepted accounting standards.
Headings		**Chapter Summary**
17.1	Introduction	o The turnover of construction companies is largely made-up of income from contracts. o The tender price or agreed contract sum represents the contractor's budget for the project. o Budgetary control includes forecasting the revenues from contracts and comparing the actual revenues received. o Variances indicate that the contract is either on or behind schedule and provide a basis for corrective action. o Profit or loss is uncertain until the contract is complete. o This is when the true revenues from a contract can be compared to the costs which is gives the true level of profit or loss.
17.2	Reporting Procedures	o Procedures for reporting the financial position of contracts must facilitate comparison of what is happening with what was planned. o Comparing actual figures with the forecast is an essential part of the project control process. o Cost and value are compared at a common reconciliation date which enables all projects undertaken to be compared equally.
17.3	Earned Value Analysis	o A method of allocating the contract sum to the various activities on the programme and then plotting the cumulative figures on a graph. o Actual figures can then be plotted to determine variances. o Variances indicate financial progress – positive or negative. o EVA is a crude method of comparing value and cost and has some limitations.
17.4	Cost-Value Reconciliation	o CVR is a method for comparing value and cost at predetermined intervals during a contract. o Costs and values are brought to a common date each month. o Value is determined by adjusting certified contract payments appropriately. o Costs are established from accounting records.
17.5	Cost-Value Reports	o Cost-value reports bring together the value and cost sides of the CVR process. o Reconciled value is the contract value assessed at the cut-off date. o Reconciled cost is the money spent adjusted to the cut-off date. o The CVR process provides the opportunity to add accruals and provisions for future liabilities. o The reporting date should be as close to the cut-off date as possible.
Learning Outcomes		o Appreciate the need for sound corporate governance. o Understand the distinction between cost and value. o Understand the need for reporting procedures which compare cost and value as reliably as possible. o Realise that true profit or loss can only be determined at the conclusion of a contract. o Be able to prepare an earned value forecast and an estimate of cost. o Understand the CVR process and follow a worked example.
Learn More		o Chapter 7.8 explains that the contractor's tender is a forecast of cost and profit. o Chapter 3.5 explains how tender risk is managed. Tender risk strategies can impact the CVR process. o Chapter 14.2 explains the basics of earned value analysis.

17

Controlling Money

17.1 Introduction

The collapse of Carillion in 2018, with losses and liabilities of over £7 billion (Wylie 2020) brought the need for prudent financial management and robust corporate governance into sharp focus.

Reasons for this industry-wide disaster were cited in a House of Commons Library Briefing Paper[1] as:

- Too much borrowing and paying shareholder dividends out of debt rather than profits.
- Under investment in the Company.
- Declining revenues (turnover) and **optimistic forecasting of revenues and profits**.
- Aggressive bidding and accounting.
- **Declaring profits before they are earned**.

This book is primarily concerned with the planning and control of construction projects and so, in this context, two aspects of the Carillion case are highlighted:

- Optimistic forecasting of revenues and profits.
- Declaring profits before they are earned.

17.1.1 Forecasting Revenues and Profits

The turnover of construction companies is largely made-up of income from contracts. Contracts are normally won in competition and the tender price or agreed contract sum represents the contractor's budget for the project.

Part of the budgetary control system within a construction firm consists of forecasting the revenues from contracts so that this can be compared with the actual revenues received once the contract gets underway. In simple terms, any variance from the expected revenues indicates that the contract is either on or behind schedule and this provides a basis for corrective action by management.

Construction Planning, Programming and Control, Fourth Edition. Brian Cooke and Peter Williams.
© 2025 John Wiley & Sons Ltd. Published 2025 by John Wiley & Sons Ltd.

When actual revenue it compared to actual costs, this gives an indication of the profitability of the contract on a month-by-month basis. However, the monthly revenues received from contracts is necessarily approximate for a number of reasons:

- The valuation of work carried out to date (the external valuation) is approximate – usually based on a percentage, e.g. groundworks 90% complete, brickwork 15% complete etc.
- The contract pricing document may have been priced in such a way as to generate early positive cash flow and therefore the basis of the valuation is not true.
- Work in progress may have been overvalued.
- Additional work may have been undertaken which has not been agreed or paid for.
- Some work undertaken may have to be corrected because it is either defective or not in accordance with the specification.
- The contractor may have submitted contractual claims which are in dispute.
- Subcontractors may have contractual claims against the main contractor.

Strictly speaking therefore, contractors should not be expecting revenues from contracts unless there is certainty in the amount that will be received. Furthermore, a realistic valuation of the work to date should be carried out – this is called an internal valuation – and it is this valuation that should be compared with the actual costs to date.

With this information, the contractor can compare true contract revenues with the initial budget. If this process is not carried out, or if the figures are misreported by mistake or to make the picture look rosier than it is, the true position on a contract will not be known and, at the end of the financial year, the balance sheet will not represent a true and fair view of the revenues that the company has earned.

The process of comparing true revenues with actual costs is called cost-value reconciliation (CVR) and is explored in this chapter. Ross and Williams (2013) explain CVRs, the reporting of contracts and corporate governance in detail.

17.1.2 Declaring Profits

When monthly revenues are compared to the initial budget, any variance represents either profit or loss.

However, because monthly external valuations are somewhat unreliable – and internal valuations rely to a degree on forecasting future revenues and liabilities – the profitability of a contract is uncertain until the contract is complete.

It is only at this point that the true revenues from a contract can be compared to the costs which is when the true level of profit or loss on the contract can be determined.

As a consequence, accounting practice requires profits to be declared prudently and they should only be reported in the company accounts when either the profits have been earned or, on long-term contracts, where there is a strong possibility that profits will be earned.

Even so, profits on long-term contracts should only be included in the accounts on a prudent basis which usually means reporting only a percentage of the profits likely to be earned.

In any event, losses should be reported immediately and taken into account in the reporting system and in the annual declaration of accounts.

17.1.3 Corporate Governance

Corporate governance is the system by which businesses are directed and controlled. This is the responsibility of the directors who are appointed by shareholders along with auditors whose job is to verify that the accounts represent a true and fair view of the business – especially to investors.

An important aspect of corporate governance is the implementation of a strategy to deal with the principal risks and uncertainties facing the business including procurement, credit, liquidity, cash flow and creditor risk.

> Management is to be distinguished from corporate governance as it concerns the day-to-day operational running of the business by its full-time executives and managers in the pursuance of company objectives. In larger companies, corporate governance is the province of executive and non-executive directors appointed for the purpose.

Directors are also responsible for preparing the annual report and financial statements required by company and taxation law and selecting suitable accounting standards and practices to be adopted in the preparation of the accounts.

> In construction, projects are the primary source of revenue and so the annual accounts relate directly to the financial reporting of individual contracts undertaken by the business.

The key aspects of the financial reporting of projects that translate directly into the annual accounts are:

- Turnover and profit – which appears in the profit and loss account and
- Debtors and work in progress – which appear in the company balance sheet.

Whilst this chapter explains the basics of the financial reporting of contracts, this a very complex subject that is explained in detail by Ross and Williams (2013).

17.2 Reporting Procedures

Procedures for reporting the financial position of contracts must facilitate comparison of what is happening during the project with what was planned. It is essential, therefore, that realistic budgets are prepared at the pre-contract stage and that they are monitored as the work proceeds.

Comparing actual figures with those forecast is an essential part of the project control process. This includes monitoring turnover – revenues from the contract – and profit but especially entails comparing cost and value (work in progress) as illustrated in Figure 17.1.

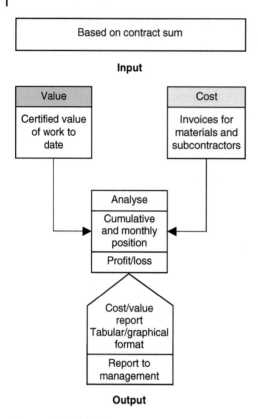

Figure 17.1 Principles of cost-value reporting.

Preferably cost and value are compared at a common reconciliation date which enables all projects undertaken to be compared equally and provide a company-wide perspective of the cost/value and profit situation.

CVR is a process that enables the true picture of a contract to be seen in financial terms and provides output which can be directly translated into the account accounts. Many companies, however, do not have, cannot afford or may not need sophisticated control measures. They should be fit for purpose.

17.2.1 Small Firms

Small contracting firms include sole traders, partnerships and limited companies. They are usually micro-firms with 10 or fewer direct employees and a turnover less than £1 million.

All such companies must submit accounts to His Majesty's Revenue and Customs (HMRC) to establish tax liability, but these accounts are usually simple. They will consist of a balance sheet and profit and loss account but, in the case of limited companies, the profit and loss account is not publicly available at Companies House.

> Small companies are not required to comply with the UK Corporate Governance code that applies to large firms and, therefore, cost and value reporting is often rudimentary at best. The focus is more about the cash flow position, keeping the bank overdraft under control and paying creditors (such as subcontractors and suppliers) as late as possible.

Cash flow is vital for the survival of the small firm because working capital is invariably in short supply and there are pressures from the bank and from creditors – including the regular HMRC PAYE and National Insurance payments.

These pressures – and the need to manage contracts and find new work – means that comparison of value with cost is probably only made at defined stages of the contract – such as substructure, superstructure and on completion – if at all. Even if this is done, it is unlikely that true cost and true value will be compared because the calculations are complex.

Most small contractors operate simple cost control systems at best and staff – and even the accountant – will not be familiar with CVR procedures unless a quantity surveyor is employed. Profit is often regarded as the difference between the final account and the tender figure, but this can be wholly misleading.

An ad-hoc approach to financial management might be satisfactory for the small contractor but it is not a sound basis for control. As a business expands, more sophisticated and formal reporting procedures will be needed.

17.2.2 Medium-Sized Firms

Medium-sized businesses – SMEs – have between 14 and 299 employees with an annual turnover of £2–50 million. This is a broad category of firms which might undertake quite large projects in the upper quartile.

Clearly, when undertaking contracts of significant value, the reporting of financial information takes on more importance. Projects upwards of £250 000 cannot be allowed to drift along with management unaware of the financial position, and procedures must be implemented for reporting on contract profitability as the work proceeds.

Consequently, many medium-sized contractors undertake some degree of cost reporting of current contracts, and some will have CVR procedures as an integral part of the monthly valuation procedure. They will employ quantity surveyors to manage the CVR process.

17.2.3 Large Firms

Emphasis on the analysis of contract performance becomes more important in large firms as corporate governance rules must be adhered to and there are shareholders to satisfy. Reporting procedures must be much more reliable in large firms and compliance with complex accounting standards and legislation requires greater attention to detail.

CVR procedures in large firms may be quite sophisticated and quantity surveyors will probably report to a managing surveyor before financial reports are allowed to percolate to the upper echelons of management.

In the 1970s, a number of large construction firms operated reporting systems which collected data on every single site activity, but this practice appears to be out of favour these days. Collecting data for data's sake is an expensive luxury and the emphasis should be on the quality of data rather than the amount.

The CVR procedures followed in large companies tend to be based on the principles first propounded by Barrett (1981). This guidance was based around SSAP9 (Financial Reporting Council (FRC) 1975) – the accounting standard at the time for stocks and work in progress – which has been superseded by other similar financial reporting standards, such as FRS 102.

17.2.4 Financial Control Methods

It is evident that a construction firm needs some form of financial control over its contracts appropriate to the size of the business and complexity of the work undertaken.

For micro firms, this might simply consist of a system of cash flow forecasting – as described in Chapter 14 – and a simple ledger system to control debtors – the firm's customers – and creditors – the firms that goods and services are purchased from, with a petty cash book to record day-to-day spending.

Such methods will never reveal the profitability of the contracts undertaken until the annual profit and loss account is prepared by the accountant. Even then, profitability will only be reported for the business as a whole and not for individual contracts.

For many small companies, however, keeping the firm's head above water, keeping the bank quiet, having enough liquidity to pay the wages and creditors relatively promptly and making a profit at the end of the year might be considered enough. For more sophistication, two methods of financial control could be considered:

- Earned value analysis.
- CVR.

17.3 Earned Value Analysis

Earned value analysis (EVA) is a forecasting method that enables the amount (value) of work carried out on a project to be compared to a forecast. This is done by allocating the contract sum to the various activities on the programme and then plotting the cumulative figures on a graph. Actual figures can then be plotted to determine variances.

Variances indicate financial progress – positive variances at a given point in time generally indicate progress ahead of schedule whilst negative variances may identify that progress on site is behind schedule.

In this method, cost is derived from value less margin and actual figures are plotted to determine cost and value variance. The method – and various techniques available – is explained in detail in Chapter 14.

17.3.1 EVA Process

The basic principles of EVA are illustrated in Figure 17.2 which shows the key elements of the process:

- **Preparation of a contract budget/forecast**: This is based on the contractor's tender. The figure is broken down and allocated to activities on the contractor's master programme to produce monthly and cumulative totals. This can be done on a spreadsheet or by using a cumulative value graph or S-curve.

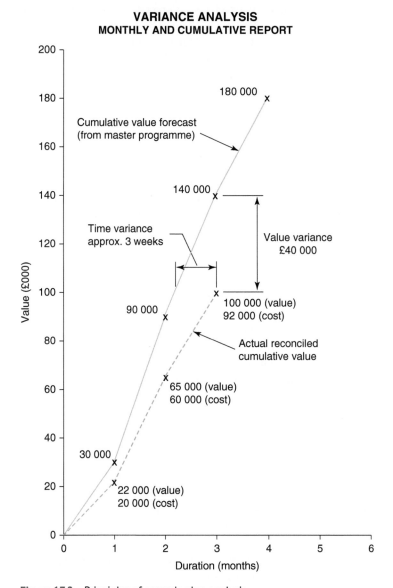

Figure 17.2 Principles of earned value analysis.

- **Determining value and cost**: Value is identified from the contractor's tender figure by multiplying the rate per unit of work by the quantity measured in the bill of quantities. Cost is determined by deducting margin (overheads and profit) from value.
- **Analysis of both the cumulative and monthly position on a contract**: This is the process of analysing value and cost at a point in time – usually the end of a month. This date is specific to the particular contract and is to be contrasted to the more complex cut-off date used in the CVR process. Because the EVA method is simple, there is no adjustment of the external valuation for overmeasure items or for tender loadings, nor will there be an internal valuation to give the true value of work to date.
- **Report to management**: The site quantity surveyor's EVA report will show the approximate cost/value position on an individual contract but cannot be accumulated with other contracts because the reporting dates are dissimilar. Management will see a view of the cost/value position on the contract and an approximate indication of contract profitability. As the figures are not thorough enough to be reliable, this simply provides an indication of where the contract stands as a basis for taking corrective action if necessary.

17.3.2 Limitations

EVA is a crude method of comparing value and cost because:

- Values are based on programme activities and are therefore imprecise.
- The values used are based on the contractor's tender figure which will undoubtedly contain errors. The pricing document may also have been priced to generate early cash or to profit from variations to the contract so can, therefore, be misleading.
- Actual values are based on the external valuation – the valuation undertaken by the clients representative usually – and do not reflect true value as derived from the contractor's internal valuation.
- Earned values are skewed by extra work and variations to the contract. When included in the external valuation, these sums of money falsely indicate financial progress as the actual value figures are no longer comparable with the forecast.
- Forecast costs are calculated by deducting margin (overheads and profit) from value and not on the cost of orders placed with suppliers and subcontractors.
- Actual costs will be based on invoices received and will probably not include allowances for goods and services provided which have not been invoiced (accruals).
- The costs and values are not adjusted to a common reporting date as they are in the CVR process.

17.3.3 Benefits

Despite its limitations, EVA at least provides a rough means of control by comparing forecast and actual value and cost. It also provides a means of determining variances which can be investigated in more detail should there be cause for concern.

A major benefit of EVA, of course, is that it is relatively simple. Once the tender figure has been allocated to the activities on the programme, a spreadsheet can be used to produce graphical displays that are easily understood – especially by contract staff who might not be familiar with the more complex CVR process.

It would be misleading to suppose, however, that positive variances indicate profitability. There are many issues that remain unresolved until the close out of a contract and these are not provided for in the EVA method.

17.4 Cost Value Reconciliation

CVR is the comparison of contract value with contract cost at predetermined intervals during the progress of a project. It provides a means of matching costs and values at a common date where value is determined by adjusting certified contract payments appropriately and costs are established from accounting records. All figures are brought to a common date.

The CVR interval is normally monthly in line with usual contract payment arrangements and is tied in with the company's internal valuation and accounting procedures. The purpose of CVR is to allow management accounts to be prepared on a more meaningful basis. In larger companies, forecast statutory accounts are also prepared based on CVR data.

17.4.1 CVR Terminology

In order to appreciate how CVR works, a few key terms need to be understood.

They are explained in Table 17.1 and should be read in conjunction with Figure 17.3 which illustrates how the reconciliation procedure works.

17.4.2 Reconciliation Date

This is the date agreed by management when the comparison or reconciliation of cost and value is to take place. It is usually the date when the monthly accounting period is closed and is frequently referred to as the **cut-off or CVR date**. This might be the last Friday in each month, the 30th of each month or simply the last day of the month.

> It must be emphasised that the cut-off date is a company-wide date and is not project specific. Using a common date brings the value of each and every contract to the same date, which can then be matched with the costs produced by the accounts department on that date.

This is quite logical as the time and effort needed to produce costs for each contract at different dates would be unthinkable. It would also make it impossible to accumulate the cost and value figures across the entire company or to determine the company-wide cash and profit situation at a given point in time. This would not be a basis for sound accounting.

17.4.3 CVR Process

The process for reconciling cost and value at a cut-off date at the end of each month during a contract is illustrated in Figure 17.3. The timings shown are an example. There may be several different scenarios in practice including different valuation dates and valuations

Table 17.1 CVR terminology.

Term	Definition
Cost	The amount of money needed to supply the labour, plant, materials, subcontractors and preliminaries items required to complete a contract. Cost is the actual cost invoiced and, therefore, excludes margin – i.e. overheads and profit. Costs are established from accounting records to which accruals are added.
Accrual	The value of goods or services received but not invoiced or paid for. Accruals can appear in a CVR or in the annual accounts.
Value	The payment that the contractor receives from the client for work carried out to date. 'Value' is also referred to as 'earned value'. Value = Cost + Margin. Value is determined by adjusting the external valuation and adding work in progress.
CVR	Cost-value reconciliation is the process of matching the costs and values of a contract at a common date (the cut-off date).
Monthly valuation	An assessment of the value of work carried out to date on a contract. This is normally an approximate figure based on the percentage of work done as described in the bills of quantities or activity schedule. It is a financial assessment of the value of construction work carried out to date based on the quantity of work done at the prices stated in the contract bills of quantities or other pricing documents.
Interim payment	The transfer of monies to a main contractor or subcontractor. Interim payments are certified or authorised for payment at the intervals defined in the conditions of contract.
Payment certificate	A formal notice that identifies the amount to be paid according to the conditions of contract. Under the Housing Grants, Construction and Regeneration Act 2006, certain statutory notices are issued as part of the certification process. The certification process includes checking the valuation for correctness.
Work in progress	Work carried out but not yet valued or certified. When work in progress has been valued and certified it becomes a debtor item until paid for. When work in progress has been paid for it becomes a progress payment on account because all interim payments are subject to final review on completion of the contract. Work in progress is shown as a current asset in the balance sheet where it might be referred to as 'stocks and long-term contracts' or 'inventories'.
Materials on site	The value of goods delivered to site for incorporation in the works based on the invoice from the supplier to the contractor.
Cut-off date	The reporting date for all contracts undertaken by a construction company irrespective of when they started or the valuation date. It might be a particular date each month or the end of the last week of a month, for instance. The cut-off date is set by the directors and the financial position of all contracts is reported at this date. This enables the profit/loss situation on each contract to be evaluated at a common date on a like-for-like basis and accumulated to give a total for the company as a whole.
External valuation	The process of establishing the value of work in progress on a contract to certify and pay for the work carried out to date.
Internal valuation	The process of adjusting an external valuation for errors and inconsistencies. It includes the value of work in progress after the external valuation date.
Accumulation	The cumulative value of work carried out to date at a particular point in time such as a monthly valuation date.
Previous payments	Monies already paid to a contractor for work carried out to date. Previous monthly payments added together make the total of previous payments to date.

17.4 Cost Value Reconciliation

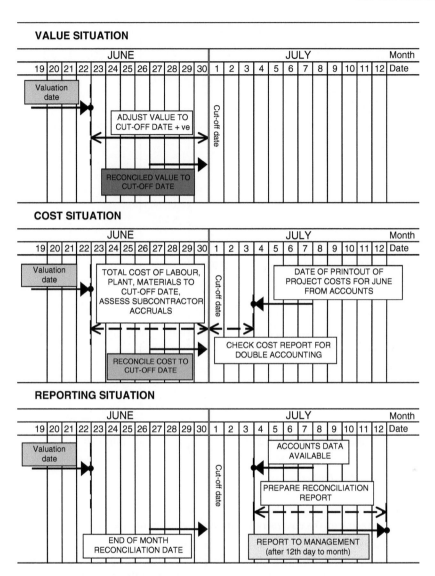

Figure 17.3 Reconciliation procedure.

before or after the cut-off date. The principles, however, remain the same. There are three steps to follow:

1) **Step 1**: Value situation: The external valuation precedes the cut-off date and must therefore be adjusted for work in progress between the valuation date and cut-off date. This is work done at cut-off but not included in the external valuation. The external valuation must also be adjusted for inaccuracies, tactical pricing, over-valuation and for the true (not estimated) quantity of work carried out. This is illustrated in Figure 17.4.

By adjusting the external valuation as described, an internal valuation is generated without which there will be no true reconciliation with costs.

This step reconciles contract value to the cut-off date.

GROSS VALUE		VALUATION DATE May			
Gross Value Certified					176 000
			ADD	OMIT	
ADJUSTMENTS TO RECONCILIATION DATE May					
PLUS or MINUS					
Under valuation +ve					
Over valuation –ve				5 000	
Adjusted value to date of reconciliation +ve			15 000		
Variations issued (not yet included in valuation)			2 000		
Dayworks – ditto.			1 000		
Remeasured work sections			3 000		
Preliminaries adjustments – under valuation +ve or over valution –ve				5 000	
Materials on site adjustments			1 000		
			22 000	7 000	15 000
	RECONCILED VALUE				£191 000

Figure 17.4 Assessment of reconciled value.

2) **Step 2**: Cost situation: Costs are available from the accounts department after the cut-off date. This has no bearing on the cost reconciliation but is influential as to when the report to management is available.

As the external valuation takes place before the cut-off date, this means that the costs of work in progress must be assessed. Figure 17.3 indicates that this consists of the costs of labour, plant and materials produced by the accounts department together with accruals for subcontract work in progress.

An accrual is required because the subcontract work will not have been invoiced post the external valuation date.

This step reconciles contract cost to the cut-off date.

3) **Step 3**: Reporting situation: The CVR report date to management is shown in Figure 17.3. The report will be available after 12 July.

The reason for the delay is because cost information is not available until the 4th and time will be needed to prepare the individual contract CVRs and for the managing quantity surveyor (QS) to bring these together into a company-wide report.

This report will be available some two weeks after the cut-off date, but management will at least be presented with an accurate company-wide situation report as a basis for action.

17.5 Cost Value Reports

Figure 17.5 illustrates a CVR report format that might be used in a large contracting organisation. This form brings together the value and cost sides of the CVR process and provides the opportunity to add accruals and provision for possible future liabilities.

This is the basis of prudent accounting required by law, good accounting practice and sound corporate governance.

17.5.1 Cumulative Value

On most contracts, value is determined by a process of valuation. Commonly the client's quantity surveyor will agree the valuation with the contractor, or the project manager will make an assessment. The payment will then be certified by the contract administrator.

Due to the way that construction tenders are priced, this valuation is not accurate – it is only a way of paying the contractor for work done that is fair and agreeable. In practice, tender prices are manipulated by contractors to generate early cash flow or to benefit from mistakes in documentation and, consequently, valuations based on these figures are not reliable enough to compare with actual cost.

17.5.2 Reconciled Value

Reconciled value is the contract value assessed at the cut-off date. It is based on the value agreed at the monthly valuation date but must be adjusted internally by the contractor to reflect the true situation on site.

This value – for the purposes of comparison with the project costs – must also be reconciled to reflect the time variance between the valuation date and the reconciliation date:

- Work in progress carried out between the valuation date and the cut-off date must be included in value. This involves assessing the value of the work done in the intervening period between the external valuation date and the cut-off date. If the cut-off date is after the external valuation date, the adjustment will be positive.
- When the external valuation takes place after the cut-off date, this also has to be reconciled. Therefore, work done after the cut-off date must be valued by the contractor as an internal value which is then deducted from the reconciled external valuation (i.e. the internal valuation) to give true value at the cut-off date.

The principle is to always value the contract works at the cut-off date, irrespective of whether the external valuation is before or after this date. This ensures that a true value is calculated which can then be compared to costs at the cut-off date.

Figure 17.4 shows a typical format for calculating the reconciled value.

COST–VALUE RECONCILIATION REPORT		
Contract _____ Valuation No. _____		
Contract No. _____ Date of valuation _____		
Contract duration _____ Month _____		
VALUATION ASSESSMENT	**CUMULATIVE**	**THIS MONTH**
Value of certificate to / /		
ADJUSTMENTS :		
Adjustment to valuation date		
Prelimination adjustment		
Overvaluation		
Variations		
ADJUSTED VALUATION TOTAL		
CONTRACT COST ASSESSMENT	**CUMULATIVE**	**THIS MONTH**
Contract costs to / /		
ADJUSTMENTS TO COST (ACCRUALS)		
Plant		
Materials		
Subcontractors		
Inter-site costs		
PROVISIONS		
Subcontractors liabilities		
Future losses		
Maintenance/defects costs		
Cost of delays		
Liquidated damages		
ADJUSTED COST TOTAL		
PROFIT (LOSS) As a value		
Percentage		
Date of reconciliation / /	Prepared by :	

Figure 17.5 CVR report.

17.5.3 Reconciled Cost

Cost is the money spent at the date of the reconciliation. It is the cost in the contractor's cost ledger, adjusted to the cut-off date. The problem here is that the cost cut-off date may not coincide with the submission of invoices and there will consequently be an element of cost that is not accounted for. This is the reason why 'accruals' appear in the CVR calculations.

Accruals are costs which have been incurred but for which invoices have not been received by the accounts department. They represent provisions for future costs which have to be assessed and taken into account in order to report true value.

The cost assessment must include all accruals for materials, hired plant and subcontractors which have not been included in the cost ledger at the date it was closed.

Materials accruals represent the cost of materials delivered to the site but not yet included in the cost ledger. For this purpose, it may be necessary to put a value on goods received records and material delivery notes.

Assessment of the reconciled cost is one of the major areas of error in the reconciliation process, especially in the assessment of the subcontractor accruals.

17.5.4 Cumulative Value and Cumulative Cost

The comparison of cumulative value, reconciled actual value and reconciled cumulative cost can be presented as a spreadsheet – as shown in Figure 17.6 – or graphically, as illustrated in Figure 17.7.

In Figure 17.7, the value variance and time variance have been highlighted.

17.5.5 Provisions

When undertaking contracts, the prudent contractor will recognise potential future risks that might affect the profitability of the contract. There may, for example, be a dispute with a subcontractor regarding payment which has yet to be agreed. If the subcontractor's case is proven, this represents a potential liability for the contractor who will have to settle the agreed amount. This liability must be recognised in the CVR.

In some cases, it becomes clear on a contract that future work to be undertaken will be unprofitable. This might be a question of underpricing in the tender or could be the result of profits on this work being taken in advance to improve early cash flow or to load certain rates for tactical purposes. The CVR process recognises that prudence is required in the preparation of accounts and that such potential losses should be accounted for as soon as they become apparent.

Other potential liabilities include liquidated damages payable to the client if the contractor is in culpable delay or the cost of accelerating delayed work where this is the fault of the contractor.

17.5.6 Profit

Figure 17.6 shows the monthly cost report for a contract, together with a forecast of its cumulative value. The relationship between forecast value, actual value and cost has been presented in graphical form in Figure 17.8 and the percentage profit release situation can be seen in both cumulative and monthly terms in Figure 17.9.

Figure 17.9 shows that the cumulative profit release has been slowly declining each month of the project which, in five months, has fallen from 16% to 5.6%. During this period:

- The monthly profit has declined from 16% to 2.9%.

MONTHLY CVR REPORT

Date	Val. no.	Certified value	Recon. value	Cum. cost	Cumulative		Monthly			
					Profit	%	Value	Cost	Profit	%
30/5	1	27 000	29 000	25 000	4 000	16.0	29 000	25 000	4 000	16.0
30/6	2	64 000	67 000	59 000	8 000	13.5	38 000	34 000	4 000	11.7
28/7	3	110 000	112 000	102 000	10 000	9.8	45 000	43 000	2 000	4.6
26/8	4	170 000	175 000	163 000	12 000	7.4	63 000	61 000	2 000	3.3
25/9	5	270 000	280 000	265 000	15 000	5.6	105 000	102 000	3 000	2.9

Forecast cumulative value
(based on programme)

Val. no.	Forecast cumulative value	Val. no.	Forecast cumulative value
1	30 000	5	310 000
2	70 000	6	400 000
3	130 000	7	480 000
4	200 000	8	560 000

Figure 17.6   CVR report – spreadsheet format.

- During months 3, 4 and 5, the average monthly profit release has been 3.6%.
- At month 5 of the 8-month contract, the cumulative profit release of 5.6% is well below the forecast profit of 10%, as indicated by the horizontal dotted line on the graph in Figure 17.9.
- It is doubtful that the contract will achieve its forecast margin without some drastic action by senior management.

In larger contracting firms, the cost-value reporting procedures will often include a forecast of the cost and profitability of a contract to completion. This will be the profit level that

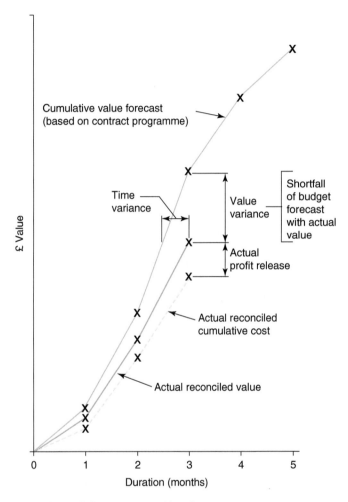

Figure 17.7 CVR report – graphical format.

management considers is achievable on the remaining activities to be completed between month 5 and the end of the contract.

17.5.7 Variance Analysis

Variance analysis – which highlights the difference between actual and expected figures – forms an essential part of the cost-value reporting procedure.

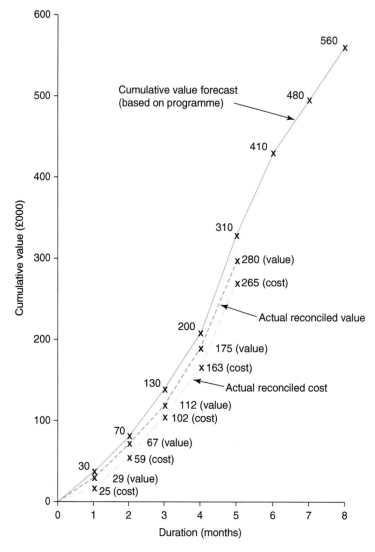

Figure 17.8 Value/time-cost/time relationship.

This is illustrated in Figure 17.7 which contrasts the time variance (programme) and value variance. The time variance reflects the actual value achieved compared to that forecast from the contract programme. The value variance is the shortfall between forecast and actual value determined by the CVR process.

However, where there are differences between forecasts and reality – variances – it is important to look for reasons, particularly where there is a shortfall in the contract value.

Variances should prompt management to review the master programme, for instance, as slow progress, late information, inefficient subcontractors or on-site problems may be some of the many reasons why things are not as expected.

PERCENTAGE PROFIT RELEASE

Date	Valuation No.	CUMULATIVE		MONTHLY	
		PROFIT	%	PROFIT	%
30/5	1	4 000	16	4 000	16
30/6	2	8 000	13.5	4 000	11.7
28/7	3	10 000	9.8	2 000	4.6
26/8	4	12 000	7.4	2 000	3.3
25/9	5	15 000	5.6	3 000	2.9

Profit forecast at tender stage 10%

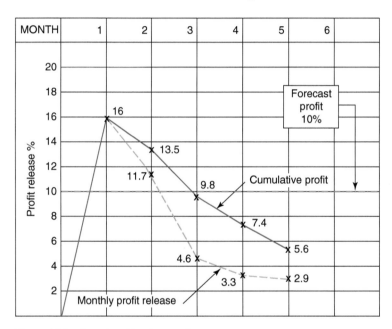

Figure 17.9 Actual profit release v. forecast.

17.5.8 Date of Report to Management

The report date in principle should be as close to the cut-off date as possible but will depend on when the project cost figures are available.

Figure 17.3 illustrates this as being achieved by the end of the 12th day of the month.

References

Barrett, F.R. (1981) *Cost Value Reconciliation*. Chartered Institute of Building.
Financial Reporting Council (FRC) (1975) *SSAP9: Stocks and Long-term Contracts*.

Ross, A. and Williams, P. (2013) *Financial Managemen in Construction Contracting*. Wiley Blackwellt.

Wylie, B. (2020) *Bandit Capitalism – Carillion and the corruption of the British State*. Birlinn.

Note

1 https://researchbriefings.files.parliament.uk/documents/CBP-8206/CBP-8206.pdf and https://commonslibrary.parliament.uk/carillion-collapse-what-went-wrong/.

Chapter 18 Dashboard

Key Message		o The management of resources is the essence of the site manager's job.
Definitions		o **ERP**: Enterprise resource planning software. o **Attendances**: Facilities provided by the main contractor to enable subcontractors to complete their part of the works.
Headings		**Chapter Summary**
18.1	Introduction	o The resources under the control of the site manager are principally labour, materials, plant and subcontractors. o To achieve a successful project outcome, the site manager will need to marshal the resources available to best effect. o Subcontractors need continuity of work, hired plant must not be idle and materials waste must be properly managed. o Larger contractors manage resources within an ERP or construction management software system.
18.2	Sub-contractors	o Subcontractors are the backbone of the construction industry. o It is mainly subcontractors who carry out the site work. o Subcontractors should be pre-qualified. They need to be of good financial standing and be able to supply the expertise that the main contractor requires. o The subcontract programme of works is dictated by the main contractor's master programme. o The performance of subcontractors should be benchmarked for future review of tender lists and frameworks.
18.3	Labour Control	o Some major contractors employ their own workforce to carry out self-delivery of items of work that subcontractors either do not do or do not wish to do. o A resource histogram is a useful visual representation of labour demand which shows how much labour is required and when. o The peaks and troughs in resource demands can be managed by resource smoothing or levelling.
18.4	Materials Control	o Materials expenditure represents a major proportion of contract value. o The purchasing, scheduling, delivery and handling of materials on site must be carefully managed. o Effective materials management relies on good site layout planning and the choice of suitable materials handling equipment. o Delivery and site waste can be reduced by good site layout planning and materials management. o Everyone who produces or handles waste from construction activities has a legal duty of care for its safe keeping, transport and disposal. o Whilst no longer required by law, site waste management plans are good practice.
18.5	Plant Control	o Plant includes small tools and tackle and mechanical and non-mechanical equipment. o The way in which plant is used on site is controlled by legislation. o Plant operators are trained to the Construction Plant Competence Scheme and Construction Skills Certification Scheme standards. o Every lifting operation must be properly planned by a competent person, appropriately supervised, and carried out safely. o Plant must be regularly inspected and maintained according to statute and scaffolding must be checked before use.
Learning Outcomes		o Appreciate the complexity of resource management and the challenges facing site managers. o Understand the interface between subcontractors and self-delivery of certain work activities. o Understand how resource smoothing can improve continuity of work. o Understand the basics of materials handling and waste management. o Be aware of the legislation that applies to waste and plant management.
Learn More		o See also Chapter 13.3 (Pre-construction planning) and Chapter 13.5 (Requirement schedules).

18

Controlling Resources

18.1 Introduction

Most people in the construction industry would acknowledge that the contractor's site manager has one of the toughest jobs in the business.

Armed with a set of contract documents, a programme and a health and safety plan, the site manager requires knowledge, experience and management skills to bring the project to a successful conclusion – on time, on budget and zero accidents – as well as making sure that the client is satisfied with the quality of the finished product.

18.1.1 Targets

The site manager has many issues to deal with. Managing the effects of the weather, chasing designers for information and clarification, managing direct labour and subcontractors, managing site safety and dealing with statutory building control are just a few.

Added to this is that the site manager usually has a contract profit target to achieve. However, whilst being responsible for contract profitability, this is something that site managers often have little or no involvement with nor authority to influence. The mechanics of the financial running of contracts are usually dealt with by a quantity surveyor – visiting, head office based, or site-based on larger contracts – and the site manager may never see the contract prices, profit margins and risk allowances.

Site managers of speculative housing developments normally have sales handover targets to meet rather than profit targets which are a function of sales. However, individual plot start dates and handovers are often decided at head office level, especially for large developers. This means that the site manager has no influence over the order of working on site and must manage available resources to achieve the targets set.

18.1.2 Head Office

The site manager is not alone of course and has the full backing and resources of the company available. This will include subcontract procurement, materials purchasing and plant hire.

Labour, materials and plant are resources that site managers must manage, therefore, but they will often have little or no involvement in choosing them, especially in small and medium-sized enterprises (SMEs). As far as planning the work is concerned, many site managers are simply provided with a programme and have no involvement in its preparation.

> More enlightened contractors will often seek 'buy-in' from their site managers, and site managers for large contracts will have much more 'say' and involvement, especially in the planning process, as they are likely to be senior and highly experienced.

Head office will provide the health and safety audit function – usually through a visiting health and safety manager – unless the contract is large or complex when they will be resident on site. Other head office backup include quantity surveying and accounts.

Site managers are subject to a great deal of scrutiny, with middle and senior management taking a 'hands-off' but very much an 'eyes-on' role in the project. It is a pressurised job!

18.1.3 Resources

The resources under the control of site managers are principally labour – including directly employed labour and labour-only subcontractors – materials, plant and subcontractors. These resources will be procured at head office level and the site manager must simply to get on with the choices made.

To achieve a successful project outcome, the site manager will need to marshal the resources available to best effect and ensure that subcontractors have continuity of work, that labour and hired plant are gainfully employed and that materials waste is properly managed.

18.1.4 Preliminaries

A key element of any project is the site on-costs – preliminaries. This represents the supervision, site security, site accommodation, certain temporary works, site telecoms and so on needed to run the site.

Preliminaries allowances are included in the contractor's tender figure but the 'prelims budget' is often not revealed to the site manager.

Unless involved pre-construction, site managers will face a *fait accompli* regarding the choice of site offices, welfare facilities, fencing and security as well as site management personnel such as foreman, engineers and administrative staff.

These are budget-driven decisions linked to the contractor's tender figure or cost plan, and the site manager may have to make do with what is made available.

18.1.5 Resource Management

For companies able to afford the capital investment and operating and support costs, resources can be managed within an ERP – 4PSConstruct, for example – or construction management software system such as Zutec, Oracle Aconex or Procore.

This can provide site managers, quantity surveyors and procurement managers with access to the company subcontractor database, subcontract quotes and contracts, estimator's pricing notes, materials quotes and orders and financial management data – according to their security status.

Workforce management capability within the software helps contractors to forecast, plan and track labour requirements quickly and efficiently. Workforce planning data show the availability of various trades and classes of labour and where they are currently working. This eliminates the time and effort needed to coordinate traditional labour schedules and resource histograms.

Labour assignment alerts can be sent directly to an operative's mobile phone or inbox, ensuring that the right worker is in the right place at the right time. The software also facilitates data collection regarding labour time and productivity which can then be compared to the labour budget in real time. Using a mobile app, daily timesheets and daywork records can be logged and integrated directly with project financial data.

Collaborative access to such systems can also be provided via a common data environment module as required. With appropriate login status, client team members, project managers and subcontractors can be granted access to documents, budgets, subcontract quotes and project financial data held within the system.

18.2 Subcontractors

Subcontractors – many of whom are micro-firms and SMEs – are the backbone of the construction industry. On many projects it is the subcontractors who carry out the site work, provide specialist services – such as scaffold erection, temporary works design and installation – and largely dictate the project schedule according to their availability.

Contractors procure their subcontractors in two main ways:

- Traditionally, by tendering each contract individually using either a select list of subcontractors or by drawing up an ad-hoc list for specific projects.
- By pre-qualifying subcontractors onto a framework.

The first option is usually chosen by smaller contractors who do not have the resources of larger contractors to have their own framework. A compromise solution is to use a generic framework of pre-qualified subcontractors or trades.

18.2.1 Prequalification

A major prerequisite for contractors is their previous experience with the subcontractor. They need to be sure of the subcontractor's ability to manage its resources and liaise with the main contractor's staff. Good relationships are an essential ingredient to developing a collaborative approach to a project.

Subcontractors also need to be of good financial standing and be able to supply the expertise that the main contractor requires. The subcontractor's current commitments with other contractors should be assessed as this determines their ability to cope with the workload required.

18.2.2 Subcontract Prices

A risky element of subcontracting work is the question of price.

When main contractors obtain subcontract quotes at the tender stage, they must be sure that the subcontractor's prices will still be valid several weeks or months later, when the main contract is signed. If not, risk allowances must be made in the contractor's tender.

A further issue is whether the subcontractor has priced all items in the enquiry and, if not:

- Is the work included in the other rates?
- Has the work been omitted from the quote?

Where a subcontract quote is not fully comprehensive, this will give the main contractor an interface to be managed. A decision will have to be made:

- Whether to engage another subcontractor to carry out this work or
- Whether to price for self-delivery – where the main contractor engages directly employed labour and plant to do the work itself.

Other questions for the main contractor to ponder include:

- Are unit rates consistent throughout the quotation – if not, this could lead to a future dispute.
- Does the quotation constitute a counteroffer, or has the subcontractor accepted the main contractor's terms and conditions?
- Has the subcontractor offered a discount? Discounts are commonly expected in the industry and may be in the order of $2\frac{1}{2}$–5% of the subcontract sum.

18.2.3 Subcontract Attendances

Subcontractors often expect the main contractor to provide certain facilities to enable them to complete their part of the work. These facilities – attendances – may be general or specific in nature but must be detailed in subcontract quotes and incorporated in the eventual subcontract agreement.

General attendances include:

- Provision of site security and storage facilities.
- Safety, health and welfare provisions.
- Water and temporary power supply.
- Unloading, hoisting and distribution of materials.
- Use of standing scaffolding.
- Removal of site waste.

Specific or 'special' attendances include:

- Provision of hardstandings such as piling mats.
- Task lighting.
- Special scaffolding.
- Provision of access equipment or cranes.

18.2.4 Subcontract Programme

The subcontract programme of works is dictated by the main contractor's master programme, but subcontractors are often responsible for planning their own work on major projects.

For instance, a management contractor might be engaged to manage several work packages on a large hospital project where planning of the work packages and work package stage programmes is carried out by the management contractor, but short-term programming of the work is the work package contractor's responsibility. This implies management input from works package contractors which must be allowed for at the tender stage.

Figure 18.1 shows how this might work for a piling and substructure work package where the workpackage contractor takes responsibility for the six-weekly short-term programme. This process could equally well apply to subcontracts on a conventional or design-and-build project.

18.2.5 Subcontract Liaison

Figure 18.2 indicates the meetings that might be held during the construction stage, where subcontractors will discuss problems and information requirements and iron out any difficulties which may be affecting the progress of operations on site. This will include:

- Review progress, site problems, hold-ups and quality of work.
- Review of the programme including interfaces with other subcontractors.
- Action to maintain or accelerate progress.
- Review of the labour, plant and material supply situation.
- Health, safety and welfare issues including any site audit results or recommendations.
- Review of the valuation and payment situation to date including site instructions and variations to the subcontract.

On some projects, a system of short-term planning may be needed to keep subcontract works under review. This will involve preparing weekly or two-weekly programmes which will be discussed at the progress meetings and also work plans for the next short-term period.

Figure 18.3 illustrates a two-week programme for a ceiling installation subcontract. It is important that the subcontractors' site representatives participate in the short-term planning procedures and that a good working relationship is established at site level.

18.2.6 Benchmarking

With future projects in mind, it is a good idea to record the performance of subcontractors. This will provide useful feedback for reviewing tender lists/frameworks to ensure that subcontractors are competent and adequately resourced and that there is an up-to-date database for new projects.

One possible approach to this is to list key performance criteria and give each one a score or rating on a scale of 1–5. The performance criteria might be:

- Price.
- Quality and workmanship.

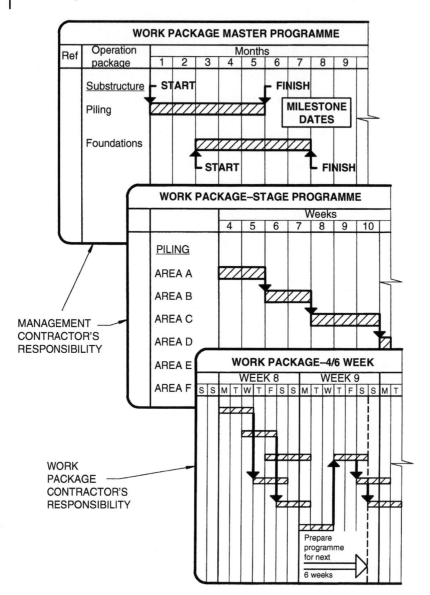

Figure 18.1 Programming stages during construction.

- Health and safety performance.
- Standard of cooperation.
- Time-programme performance.

The maximum score would be 25, but a rating of, say, less than 10 might lead to exclusion from the tender list/framework. Certain criteria, such as health and safety, for instance, could be weighted for importance where appropriate.

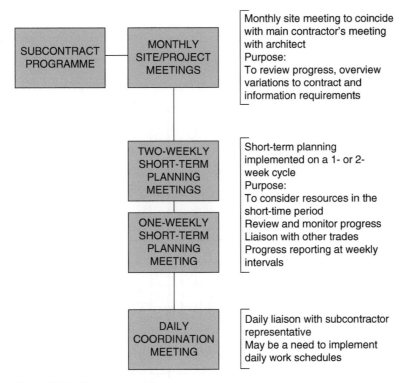

Figure 18.2 Subcontractor liaison meetings.

18.3 Labour Control

Many contractors prefer to have little or no site-based labour at all, and they package up the entire project with a variety of subcontractors.

Other contractors like a combination of directly employed (cards-in) labour and self-employed labour for general building operations such as concrete work, drainage, brickwork, carpentry and joinery, etc., and sublet the remainder of the work. Some contractors, on the other hand, employ the entire workforce directly.

Despite subletting much of their work, some major contractors employ their own workforce to carry out self-delivery of items of work that subcontractors either do not do or are not interested in pricing.

There is nothing typical about construction!

18.3.1 Directly Employed Labour

Despite the predominance of specialist subcontractors in the industry, some contractors employ their own labour force and few subcontractors or none at all. This is often the case with contractors who specialise in particular markets such as pub refits. In this market, the contractor has to be 'in and out fast' and therefore needs to employ all the trades directly to ensure that they will be available when required.

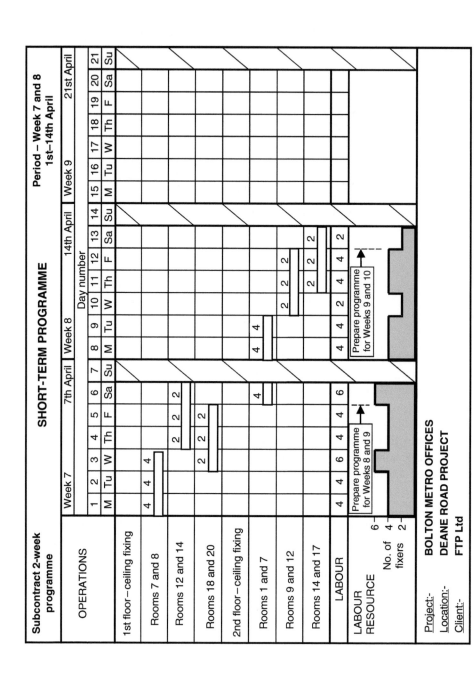

Figure 18.3 Short-term planning.

There are also examples of very large contractors who employ their own workforce – both general building trades and specialist trades – because their business model consists of general construction, plant hire, precast concrete manufacture and specialist trading companies (such as demolition, piling, mechanical and electrical).

On large projects, the contractor is more often concerned with keeping the overall labour expenditure within the estimated allowance. The monitoring of labour expenditure involves recording the actual man-weeks expended each week on the project and matching this with the forecast, as described in Chapter 15.

It is important to relate the analysis to the actual contract progress situation because the reason for an apparent overspend can simply be due to the project being ahead of schedule.

18.3.2 Self-employed Labour

It has always been common practice in construction for self-employed people looking for employment to call at a new site and ask for 'a start'. However, the site manager cannot take people on in this way anymore because of health and safety competence issues and income tax legislation.

One of the problems with self-employed labour is the Construction Industry Scheme which is designed to prevent abuse of the income tax system. The intention is not to prevent self-employment but to prevent the employment of self-employed people on a permanent basis thereby avoiding the PAYE and National Insurance system.

> Self-employed labour therefore must be engaged on a contract-by-contract basis on terms that clearly indicate a contract *for* services rather than a contract *of* service (which implies direct employment under a 'master–servant' arrangement).

The issue is one of 'control'. If operatives organise their own work, provide the necessary tools for the job, make their own decisions as to how the work is to be done, and there is a clear agreement as regards the price to be paid, then this implies self-employment. Operatives are thus classified as 'subcontractors' under tax rules but, if their tax status with HM Revenue and Customs has not been 'verified', they will suffer a 30% reduction in the labour element of any payments.

The widespread use of subcontractors in construction has shifted the emphasis away from the need for contractors to employ sophisticated labour control procedures, as was the case in the days of general contractors who directly employed their own workforce.

18.3.3 Resource Histogram

When planning a project, it cannot be assumed that resources will be available just as required for the contractor's programme, and it may be that the programme will have to be adjusted to suit the resources available.

A resource histogram is a useful visual representation of labour demand which shows how much labour is required and when. The histogram shows where the peak labour demands will be and, from this, the site manager can see whether sufficient labour will be available to meet the demand. It is usual to have a separate histogram for each type of labour or plant.

Resource histograms can be easily prepared using standard project management software such as Asta Powerproject, Project Commander or Microsoft Project. The procedure is as follows:

1. Create the resource categories, e.g. bricklayers, joiners and general operatives.
2. Decide on the number of operatives available in each 'pool' or category.
3. Allocate the labour demand from the method statement to each activity on the master programme.
4. Ask the software to produce a resource graph for each of the labour categories.
5. Look at the histograms in the same view as the bar chart programme in order to see which activities on the programme as causing the peak demands.
6. Inspect the resource histograms to see where resources are 'overloaded', i.e. where the peak demand exceeds the resources available.

> Figure 18.4 illustrates a 45-day project with nine activities A–I.
>
> Activities A, C, D, F, H and I are on the critical path (i.e. there is zero float), and activities B, G and E have 15, 24 and 9 days of float, respectively.
>
> The contractor allocates resources based on the earliest start for the activities as shown in Figure 18.4 and also in the labour histogram in Figure 18.5.
>
> This indicates that there is a peak demand for joiners between days 5 and 15.
>
> The problem for the contractor is that there are only six joiners available to work on this project.

Once the resource allocation procedure has been followed, activities can be inspected for criticality and perhaps rescheduled to overcome over-allocation problems. This might result in a longer construction period, however.

Failing this, the site manager might get round the problem by considering working extra hours each day or working overtime on weekends. To do this, simply ask the software to change the default calendar for individual activities to create a longer working day or change weekends from 'non-working' to 'working' for that activity.

18.3.4 Resource Levelling

An additional problem concerning resources is that of 'peaks and troughs' in resource demands. It might be the case that the contractor is able to resolve resource over-allocation problems for bricklayers and joiners, etc., but this does not necessarily ensure efficient resource usage in itself.

Consequently, the labour configuration may still fluctuate such that six joiners are required on site one day and four the next, none the next day and then six the following week. This is not an efficient use of available resources, and ideally, there should be a gradual increase in resource demand in the early stages of a project which builds to a peak and then gradually slows down towards the end of the job.

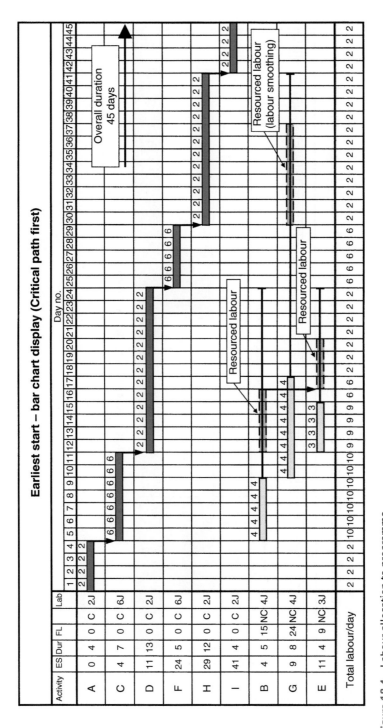

Figure 18.4 Labour allocation to programme.

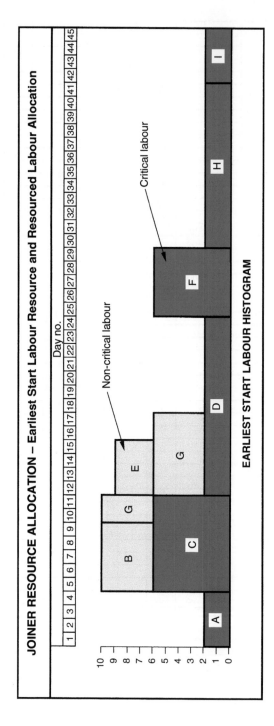

Figure 18.5 Resource histogram – joiners.

Sorting out this type of problem is called Resource Smoothing or Levelling. This is not easy to do and an ideal solution is difficult to achieve in practice. However, by looking at float in the programme and working overtime where possible, the demand may be 'smoothed' to some extent. The choice is whether to level within the slack (float) available or do so irrespective of delay to the project completion time. This would make the project either 'time critical' or 'resource critical'.

> Referring to Figures 18.4 and 18.5, if the project completion date is not to be extended, the only solution to the problem of over-demand for joiners is to level the labour demand within the programme according to the number of joiners available i.e. 6.
>
> This means adjusting the work activities within the available float (but obviously not the critical ones) to reduce the peak labour demand situation – as shown in Figure 18.6 – which also indicates improved continuity of work for the joiners.
>
> The solution is not ideal, however, because there is a fall in demand during days 17–24. This gives the site manager a problem from a labour control point of view because the joiners may find work on another site and decide not to come back.

The software will level the resources, which can be done automatically or manually.

Automatic levelling is best avoided as the computer will make decisions which a human would probably not make. For instance, the computer might decide to 'split' an activity into two parts with a time gap in between so as to level resource demands. This might be OK in theory but is probably impractical in reality as it may not be feasible to split a large concrete pour or a slipform operation for example.

Also, the computer does not optimise its decisions and does not consider and 'weigh-up' all the options as a human would. Manual levelling gives the option to select specific activities to be levelled and allows more control than automatic levelling.

It should be noted that it may be difficult to smooth or level more than one resource because revising the programme again for another resource may reverse the initial smoothing decisions and cause further resource overloading. The resource to smooth will be either the most used, or the most expensive or the least flexible or the least available. The eventual outcome will inevitably be a compromise.

18.4 Materials Control

Materials expenditure represents a major proportion of contract value. Therefore, the control of purchasing, scheduling, delivery and handling of materials on site is an essential part of the control process.

Within small organisations, the responsibility for all aspects of material control lies with the owner/director who is the estimator, buyer, surveyor and contracts manager all rolled into one! In medium-sized companies, responsibility for the purchase of materials may rest with the estimator/buyer or may be part of the surveying function.

Large companies, on the other hand, usually have departments responsible for buying and procurement, estimating, surveying, contracts and administration and therefore a formal approach to the procurement and management of materials is necessary.

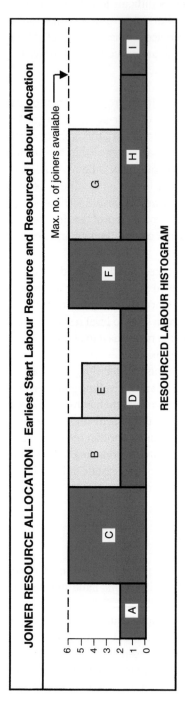

Figure 18.6 Resource levelling – joiners.

Good communications are important, and the buying and contracts departments need to liaise closely in order to ensure that materials arrive on site on time. Contact needs to be established between the contract buyer and the site manager to ensure that the material call-off schedules clearly tie in with the programme of work.

Responsibilities must also be clearly defined. The responsibility for buying materials within the estimated allowances lies with the buyer, and any resulting savings created by efficient buying contribute to the profitability of the contract.

18.4.1 Materials Management

Once materials have been delivered to site, they become the sole responsibility of the construction site manager, who must make arrangements for handling, distribution around the site and fixing them in position. The site manager is also responsible for material loss and accounting for waste.

An essential component of materials management begins with site layout planning. Designated areas must be allowed for storage to reduce waste and damage to both permanent and temporary materials, and these areas should be located away from site traffic routes where materials could be splashed with water and mud. Good practice is illustrated in Figure 18.7.

There is a wide variety of mechanised materials handling equipment available to the site manager. Some enable materials to be offloaded and placed into the exact position required for final fixing – such as rough terrain forklift trucks and telescopic handlers – and others facilitate the direct placing of ready-mixed concrete and screeds using concrete or screed pumps. Elevating materials from ground level may be accomplished with telescopic

Figure 18.7 Materials management.

elevators, tile and block loaders as well as scissor lifts, mast climbers and cranes with various lifting and mobility capabilities.

Effective materials handling requires planning and organisation and, by choosing the right piece of equipment for the prevailing conditions, site efficiency can be increased, and costly waste reduced. There is really no excuse for inefficient materials handling, but on many sites, numerous examples of the mismanagement of materials can be observed, as illustrated in Figure 18.8:

- Excessive waste left under scaffolds, including bricks, blocks, skirting boards, fascia boards, drainage fittings, etc.
- Expensive facings and engineering bricks being bulldozed into the ground and then covered over with topsoil to provide 'instant brick gardens'.
- Materials being stored on uneven ground, adjacent to unprepared access roads, allowing the material to become contaminated with mud and water.
- Pallets of bricks and blocks unloaded directly onto unprepared ground, away from the workplace.
- Damage to materials whilst unbanding the packs.
- Roof trusses being stacked on unprepared areas, allowing them to distort and twist.
- Lack of covering and protection to internal timber floor joists, door frames and finishing joinery items. Structural timbers left unprotected in the rain.
- Excessive thickness of ready-mixed concrete to in situ concrete kerb beds.

Typical examples of poor materials control

Figure 18.8 Mismanagement of materials.

- Out of sequence working, resulting in the excessive waste of stone filling materials, bricks and blocks, etc.
- Commencing foundation work with no provision for adequate access to the works, resulting in chaos for the storage of materials around the work area.

18.4.2 Waste

Historically, waste has always been an 'on-cost' in construction (i.e. a percentage added to the net cost of materials), but modern thinking suggests that waste should be seen as a resource. It is also a legal requirement, under CDM 2015 regulations 18(1) and 13(1) respectively, to keep construction sites in *good order* and to *plan, manage and monitor* work so as not to create risks to health and safety by good housekeeping. By improving on-site practices, the industry can benefit the environment and improve its performance and 'bottom line'.

According to the Department for Environment, Food and Rural Affairs (DEFRA), controlled non-hazardous waste from construction, demolition and excavation amounts to over 66 million tonnes per year – around 62% of all waste generated in the United Kingdom. The impact of this waste is considerable and is estimated to represent:

- Around 4.5% of a construction company's profit.
- 13% of all the solid materials delivered to construction sites.
- One-third of all fly-tipping waste.
- Over £4000 worth of materials in a typical skip (dumpster in the United States).

18.4.3 Planning for Waste

Delivery and site waste can be reduced by considering site layout planning at the pre-contract stage and by allowing the site manager to become involved in the decision-making at this stage of the project. Site managers should be encouraged to manage their materials properly – they can be assisted by establishing in-company courses on materials management. The company may offer some degree of incentive to managers who achieve the minimum waste targets on their projects.

All staff should be made aware of the company materials management policy, and of the allowances built into the estimate – although site managers are frequently excluded from this conversation. Senior managers could be made responsible for the losses occurring on their projects as well as site managers.

The company may also consider producing a simple site guide on its waste management policy, and recommendations could be included on good practices regarding materials handling and control. This could also be issued to subcontractors and to the company's directly employed labour force.

Poor control of material waste can create health and safety problems – nails in discarded timber offcuts, discarded bricks/blocks causing a trip hazard, etc. It is very noticeable that many construction sites have poor standards of site housekeeping, and untidy sites set the wrong tone for good health and safety management. Procedures should be included in the contractor's safety management system to eliminate this problem.

18.4.4 Improving Waste Management

In years gone by construction waste was controlled simply by comparing the actual wastage on site with the estimator's standard waste allowances in the tender. However, it is clear from current thinking and legislation that construction site managers must do more to develop their understanding of the impact of the industry on the environment, and they will need to use their knowledge and skills to find ways to improve recycling and waste reduction.

Examples of poor waste management can easily be found in construction, but there are many instances where the opposite is true. Waste material from tunnelling operations can be removed from site by conveyor and then transported by train to be reused elsewhere. This reduces the need for imported materials and removes large numbers of wagon movements from the road. Many sites now use mobile site-based screening plants which segregate hard materials from re-useable filling material. The hard material can then be crushed on site for reuse or removal for fill on another site.

Site managers must be much more proactive in the way that they plan and manage materials ordering, delivery, handling and storage, and they need to be more environmentally aware and think of ways to manage waste more efficiently. Waste on construction sites is caused by a number of factors which may be conveniently categorised under four headings:

- Design waste – where the design results in wasteful cutting on site.
- Take off/specification waste – where additional materials are ordered 'to be on the safe side' or where materials are over-specified for the job in hand.
- Delivery waste – materials damaged in transit or during offloading.
- Site waste – waste resulting from the production process or where incorrect materials are used.

Figure 18.9 illustrates a proposed strategy for improving site waste management by focusing on three issues – management procedures, procurement and contracts and site practices. By integrating waste management into everyday site management thinking, site waste can be better controlled, and a positive impact made on both the environment and the contractor's profit margin.

18.4.5 Duty of Care

The Construction and Demolition (C&D) 'waste stream' is a priority of the UK Environment Agency due to the amount of waste it generates, its sometimes hazardous nature and the potential for recycling.

'Waste streams' refer to the process by which waste is grouped together to identify where it comes from and how it should be treated and disposed of safely without risk to human health. C&D waste consists of metals, asphalt, tar and tar products, concrete, bricks, tiles, ceramics, gypsum-based products and soil and stones.

C&D waste is normally classed as 'commercial' or 'industrial' and therefore can be designated as a controlled waste subject to waste-related legislation. When waste has properties that make it especially hazardous or difficult to dispose of it is referred to as 'hazardous waste' which requires a pre-consignment note system for recovery or disposal.

Management procedures	Procurement and contracts	Site practices
Notify site personnel and subcontractors of company Site Waste Management Plan (SWMP).	Client and subcontractors to sign the SWMP.	Keep the site tidy to encourage the right attitude.
Observe correct environmental procedures.	Include Site Waste Management Plan in all subcontracts.	Protect materials from the weather.
Allow an adequate budget in the tender for good site layout planning.	Make subcontractors contractually responsible for waste: • Contra-charge subcontractors for waste removal. • Give labour-only subcontractors a waste allowance and contra charge for exceeding it. • Make a charge for collecting and separating waste.	Improve site materials management: • Provide bins/enclosed areas for loose materials. • Provide clean hard standings for lay down areas and materials storage. • Use lockable secure containers for components e.g. windows, sanitary fittings. • Use storage racks for keeping materials off the ground.
Give environmental induction and toolbox talks (including subcontractors).		
Improve site security and materials storage facilities.		
Change work practices to reduce waste.	Partner with suppliers for 'just-in-time' deliveries.	Mechanise materials handling and distribution to reduce damage and waste: • Tele-handlers. • Rough terrain forklift trucks. • Conveyors. • Mast climbers and hoists.
Identify waste streams and recyclable materials.		
Tighten up site procedures for materials deliveries and checking and signing delivery tickets.	Evaluate materials procurement procedures to avoid over-ordering and site waste.	Remove and organise waste regularly.
		Collect and recycle timber.
Adopt simple procedures for reconciling materials orders and usage.		Take more care in disposing of construction waste.
Designate a waste segregation area on the site.	Build materials and packaging return provisions into purchase contracts.	Separate different wastes into separate skips clearly labelled: • Cardboard. • Plastics. • Metals. • Wood. • Plaster/Plasterboard. • General waste, etc.
Conduct regular waste audits.		Reduce number of skip collections: • Make sure skips and waste disposal containers are full. • Use Roll Packer or similar drum compactors for removing air pockets in skips.

Figure 18.9 Site waste management strategy.

Hazardous wastes, such as asbestos, chemicals, oils or contaminated soils, may require a detailed waste transfer consignment note.

Everyone who produces or handles waste from construction activities has a legal duty of care for its safe keeping, transport and disposal. Failure to comply can result in an unlimited fine under Section 34 of the Environmental Protection Act 1990. The duty of care requires the principal contractor, and therefore the site manager, to ensure that:

- The waste is properly cared for whilst on site.
- The waste is passed on to an authorised body (a registered waste carrier or holder of a waste management licence).
- A waste transfer note is made out when the waste is handed over.
- All reasonable steps are taken to prevent unauthorised handling or disposal of the waste.

18.4.6 Site Waste Management Plans

Whilst no longer required by law, site waste management plans (SWMPs) for works involving construction or demolition waste are considered to be good practice, nonetheless.

SWMPs provide a structure for the systematic management and disposal of waste created during construction or demolition work including the design and pre-construction stages. The aim is to encourage contractors and others to predict how much waste will be created on a project and how it will be recycled or disposed of responsibly.

Figure 18.10 illustrates a pro forma approach to preparing a SWMP. There are many other ways of doing this, and all contractors will have their own preferred method, but it should include:

- A description of the construction works proposed including the location of the site and estimated value.
- The contractor's identity or, if there is more than one contractor, details of the principal contractor.
- The person in charge of the project and details of the person who drafted the SWMP.
- A record of decisions made before the SWMP was drafted concerning the nature of the project and its design, construction method or materials employed.
- The types of waste that will be produced and an estimate of the volume of each.
- The waste management action proposed for each different waste type including reusing, recycling, recovery and disposal.

C&D waste should be properly managed in accordance with the waste hierarchy shown in Figure 18.11 which indicates best and worst environmental options for waste.

Figure 18.12 shows a sample waste datasheet which can be employed for estimating and planning waste arisings and for comparing actual waste figures with target. Each time waste is removed from site, the SWMP should be updated with further information, including the type of waste removed, where the waste is being taken to and the identity of the waste management contractor removing the waste.

The principal contractor should also monitor the plan during the project and record any lessons learned from the process. An action plan to address lessons learned is good practice along with a comparison of actual quantities of each waste type compared to the estimated quantities.

SITE WASTE MANAGEMENT PLAN

Contractor		Prepared	
Project		Updated	
Waste coordinator			

Waste management goals

Waste prevention measures

Re-use and salvage items

Communication plan

Contamination prevention measures

Documentation

Project waste estimate			Design/Demolition/Construction phase	
Material	Quantity	%	Waste management contractor	Site handling procedure

Waste summary		
Total C&D waste		
Total recycled		
Total to landfill		

Signed		Date	

Figure 18.10 Site waste management plan.

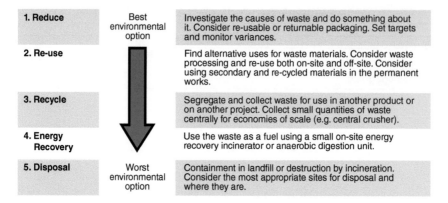

Figure 18.11 Waste hierarchy.

SITE WASTE MANAGEMENT PLAN – DATA SHEET									
Project				Report no.					
Stage				Date					
Address				Responsible person					
Main contractor									
TYPES OF WASTE ARISING									
	Quantity m³								
	Re-use		Recycle		Sent to recycling facility	Sent to WML exempt site	Disposal to landfill	Waste Transfer Notice completed?	
Material	On-site	Off-site	On-site	Off-site				Y	N
Inert									
Non-hazardous									
Hazardous									
Totals	m³								
Target	Tonnes								
Actual	Tonnes								

Figure 18.12 Waste datasheet.

18.5 Plant Control

Contractors rely heavily on all sorts of construction plant and equipment for the efficiency of site operations. This requires careful site layout planning and budgetary control if the plant is to work safely and effectively – and if the budget is not to run out of control. Plant may be categorised as:

- **Small tools and tackle**: hand tools, small power tools, lifting tackle, slings.
- **Mechanical**: excavators, cranes, hoists, mobile elevating work platforms (MEWPs), concrete placing and finishing equipment.

- **Non-mechanical**: scaffolding, formwork, patented formwork systems, falsework (for supporting formwork, e.g. suspended slabs, bridge decks).

The site manager must plan, organise and control the plant required for the project to ensure that:

- The estimator's plant allowances (i.e. the plant budget) are not exceeded.
- Plant is on site and available for use as and when needed.
- Plant is properly inspected and maintained according to statutory requirements.
- Operatives are trained and competent to use the equipment.
- Plant is immediately 'off-hired' when not needed (the dreaded plant budget again!).

18.5.1 Legislation

The way in which plant is used on site is controlled by legislation and in particular by:

- **Provision and Use of Work Equipment Regulations 1998 (PUWER)**: General legislation that applies to all tools, plant and equipment from hammers and chisels to tower cranes. PUWER may be supplemented by other legislation.
- **Lifting Operations and Lifting Equipment Regulations 1998 (LOLER)**: Legislation over and above PUWER that applies specifically to lifting equipment and operations such as cranes, MEWPs, mast climbers and hoists.
- **Construction (Design and Management) Regulations 2015 (CDM2015) Part 4**: Applies over and above to PUWER with respect to the inspection of excavating equipment and cofferdams, etc., the use and loading of vehicles (e.g. wagons), site lighting, task lighting, etc.
- **Work at Height Regulations 2005 (amended 2007)**: Applies specifically to access scaffolding, safety barriers, fall arrest equipment (harnesses, air bags, nets, etc.) and the like.
- **Management of Health and Safety at Work Regulations 1999**: Requires risk assessments and appropriate health and safety arrangements together with a trained and capable workforce, etc.

18.5.2 Organisation

Contractors commonly allocate major items of plant to the master or stage programme which shows when plant is expected to arrive on site and how long it will be there. This helps with entitlement claims when there are delays to the programme.

Plant is often delivered to site outside working hours and this needs to be accommodated in the site access and safety planning arrangements for the job. Other issues that must be anticipated include noise suppression (hydraulic breakers, compressors, etc.) and dust control where abrasive saws and chasing equipment is used.

To ensure that operatives are suitably trained and/or aware of the risks of using site equipment, the contractor should hold regular toolbox talks. They will cover issues such as abrasive wheels, cartridge-operated tools, hand tools, dumpers and lifting gear, etc. Additionally, to be able to work on controlled sites, operatives must hold a Construction Skills Certification Scheme (CSCS) card which requires a suitable qualification of completion of an apprenticeship.

Plant operators are trained to the Construction Plant Competence Scheme and Construction Skills Certification Scheme standard, and this is essential for safe operation of the equipment.

Significant items of plant, such as excavators, MEWPs, cranes, etc., are usually supplied to the contractor under Construction Plant Association (CPA) terms and conditions and crane hire companies also work to BS 7121 – The Safe Use of Cranes.

CPA conditions include requirements for the hirer (i.e. the contractor) to provide suitable access and a stable working area. Some items of plant come with a trained operator according to the hire agreement.

18.5.3 Planning

Under LOLER, every lifting operation where lifting equipment is used must be properly planned by a competent person, appropriately supervised and carried out safely. In cases where two or more items of lifting equipment are used together, such as a tandem crane lift, a written plan should be prepared.

Where crane facilities are required on site the site manager has two choices regarding crane procurement:

1. A CPA Hire.
2. A CPA Contract Lift.

When a crane is simply hired, the contractor must provide a qualified and experienced 'appointed person' to plan and take responsibility for the lifting operation. Method statements and risk assessments (RAMS) must be provided for the crane operator.

When the contractor chooses a contract lift, the crane hire company takes full responsibility for the planning and management of the lift and provides the appointed person/crane supervisor and all necessary fully trained personnel.

In the case of scaffolding – and depending upon the complexity of the scaffold layout – an assembly, use and dismantling plan must be prepared by a competent person in accordance with the Work at Height Regulations.

18.5.4 Inspection and Maintenance

Plant must be regularly inspected and maintained according to statute and scaffolding must be checked before use.

It is common practice to use a tagging system to say that the scaffold is safe and has been properly passed for use. Tagging may also be used for mobile towers, ladders, safety harnesses, hoists, MEWPs, etc., as well as for portable equipment and lifting equipment.

19

Hotel and Commercial Centre

19.1 Project Description

This case study is based on a hotel development in Kingston upon Hull in East Yorkshire, United Kingdom. The £25 million development is in a prime location in the city centre close to the railway station.

The hotel incorporates premium leisure and business facilities, including 165 upscale guestrooms with a bar and signature restaurant, fitness club and an attached 1000 m^2 events and conference centre. This has seating for 750+ delegates and there is also a pillar-free ballroom as well as a top-floor sky bar accommodating 175 people. This features large light-box windows, an external terrace and views over the city and the famous Humber Bridge. The development also includes commercial office space staircases and lifts.

The design of the building façade reflects Hull's industrial heritage and internally its maritime traditions are emphasised.

19.1.1 Project Details

The hotel and ancillary elements of the project are configured to fit the site limits as illustrated in Figure 19.1. This indicates the modular elements of the build which comprise the hotel rooms and corridors. The remaining structures are a mixture of conventional steel frame and load-bearing brick construction.

There are two lift shafts – the first being slip-formed and the second conventional reinforced concrete to first-floor level with the modular units forming the lift shaft on the upper floors.

The number of storeys varies from 1 to 5 as indicated in the cross-sections.

19.1.2 Site Constraints

The project is located on the site of a former nightclub which was demolished in 2009. In common with most demolition work of this nature, the building was demolished to ground level leaving existing ground slabs and foundations in place. The removal of this hard material is invariably part of the ensuing redevelopment of the site and must be accommodated within the development schedule, the cost plan and in the contract pricing document.

Construction Planning, Programming and Control, Fourth Edition. Brian Cooke and Peter Williams.
© 2025 John Wiley & Sons Ltd. Published 2025 by John Wiley & Sons Ltd.

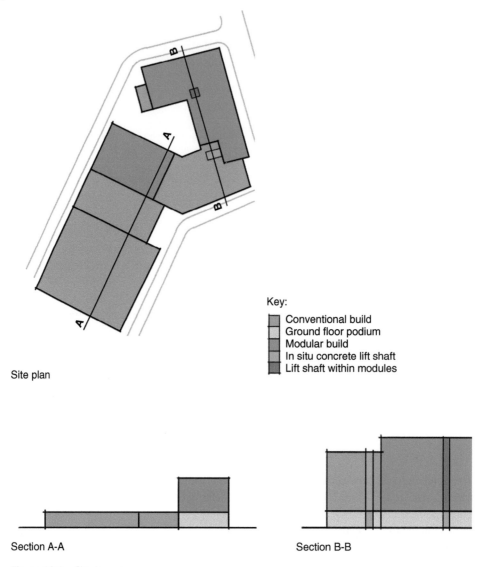

Figure 19.1 Site layout.

The site is very restricted being bounded by a main road to the front, a road to each side and a narrow access road to the rear along with existing buildings including a church, fire and rescue station and low-rise offices. The only access in and out of the site is via the access road to the rear. No access is available on the other sides of the site due to the existing roads.

Working space on site is very limited with little room available for the storage of materials or for site accommodation.

19.1.3 Procurement Strategy

This project is unconventional as it combines a traditional steel framed structure with volumetric construction – modules prefabricated off-site, delivered on low loaders and installed like 'Lego' blocks using a mobile crane.

This method of construction posed some challenges, not least that design of the modular units was undertaken in China and that early engagement of the module supplier was needed to profit from the time advantages of using off-site manufacturing methods. For a hybrid project of this nature, there are several issues to consider:

- Design for the traditional element of the project.
- Design for the modular element.
- Coordination and integration of the two design elements.
- The construction phase and management of the interface between the construction of the traditional works and installation and cladding of the modular elements.

Bearing in mind that a major advantage of modular construction is speed – and getting the completed hotel speedily to the market – the client must procure the design and production of the off-site manufactured elements as soon as the building design has reached an appropriate stage of development.

Additionally, arrangements need to be made for the on-site installation of the modules. Whilst this is relatively straightforward, the installation team needs to be familiar with the product and must be experienced in how modular buildings are put together. If the manufacturer offers a supply and install service, this is probably an advisable option because of the steep learning curve that an untrained team would be faced with. This is an important safety issue for the client to weigh up.

The client's procurement options include:

1. Traditional procurement using single-stage or two-stage tendering for the traditional elements of the project. The modular build could be a separate contract with the client. The main contract would need to incorporate provisions for the contractor to coordinate the modular element into the works as a whole.
2. Appoint a project manager to manage the traditional design concurrently with the modular design and to oversee the construction phase once a main contractor has been appointed.
3. Appoint a development company – with design and construction capability – at an early stage to manage the traditional and modular design processes and then undertake the construction works. Contracts for the traditional and modular design could be with the client and the module installation contract could be placed with the main contractor under a nomination arrangement in the main contract.
4. Appoint a main contractor under an early contractor involvement (ECI) arrangement with separate appointments for a design team to undertake the traditional elements of the design and the modular manufacturer for the prefabrication design.
5. Design and build where the client's requirements include a requirement for modular construction of certain aspects of the design.

With any of these choices – and others, maybe – considerable effort is required to integrate the traditional elements of the project with the modular elements. This requires coordination and design understanding as well as full appreciation of the client's brief and business case.

It is fundamentally important to understand that procurement of the modular elements of the project involves considerable up-front costs as most of the cost of the modules must be met before delivery to site. This risk could rest with the client or could be included in the main contract. However, a contractor would have to be appointed early in the project and would be likely to require advanced payments during the pre-construction period to meet the manufacturer's stage payment requirements.

19.1.4 Project Organisation

Brayford Hotels has an agreement with the Hilton Hotel Group to bring its DoubleTree by Hilton brand to the City. When completed, the hotel will be managed by UK hotel management company Leaf Hospitality under a management agreement with Brayford Hotels.

The construction contract was managed by Manorcrest Developments who also acted as Construction Design and Management Regulations (CDM) principal designer and principal contractor through its development and construction divisions.

The CDM principal designer is responsible for controlling the pre-construction phase of a project to ensure, as far as is reasonably possible, that the project is carried out without risks to health and safety. This requires involvement from the very earliest stages of the design concept through to design for the delivery of construction work.

As illustrated in Figure 19.2, the architectural design work was undertaken by Aros Architects in conjunction with:

- GGP Consult (structural engineering and drainage design).
- China International Marine Containers (CIMC)-MBS (modular design).

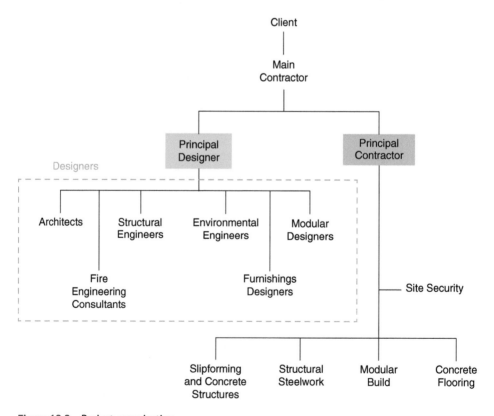

Figure 19.2 Project organisation.

- Compass Energy (environmental, utilities and energy).
- Cahill Design Consultants (fire engineering and acoustic design) and
- Hampton Furnishings.

For the construction phase of the project, the principal contractor's role is to plan, manage, monitor and coordinate the health and safety aspects of the construction phase. The role is normally undertaken by the main contractor who is also responsible for site organisation and for the management of work on site including site clearance, piling and the key work packages identified in Figure 19.2:

- Slipforming and concrete structures (David Ashley Construction).
- In situ concrete floors (Jamie Millar).
- Structural steelwork (P&D Engineering).
- Modular build (CIMC-MBS).
- Site security (White Star Security).

19.2 Design

It is important for designers to understand the interface between modular and conventional construction and how stairwells, lift shafts and utility rooms are designed and built into the modular units. The site shape is also important as curved or angled walls and oddly shaped rooms must be built by conventional methods.

Also, due to the shape of the site for this project, full-height multi-storey traditional construction was required for part of the building to integrate the volumetric units into the overall design.

19.2.1 BIM

The project was fully designed in Revit and the modular design – undertaken by CIMC-MBS – also utilised Building Information Modelling (BIM) technology. 3D models of each module type were produced which helped feed into the architectural, structural and mechanical and electrical (M&E) design models. The front elevation of the hotel is shown in Figure 19.3.

At the design stage, building information modelling and virtual reality software is used to create the internal and external design. Following approval by the client, the designs are then submitted for planning approval. Once the design is finalised, manufacturing of off-site components can commence.

19.2.2 Modular Design

Modular units are constructed as six-sided boxes which stack together in the desired configuration. Consequently, each box sits next to or on top of other boxes with the result that there is a double roof/floor assembly and a double wall assembly to account for when designing the building mass.

Figure 19.3 3D view.

The size of the modules chosen must also be considered in connection with road transport limitations, as there is a long distance between the dock and site.

19.2.3 Developer/Client Role

A key requirement of modular construction is to engage the entire project team as early as possible. The ability to run design and construction concurrently with module manufacture is crucial to giving the time savings and economies promised by off-site manufactured building solutions.

Early commitment to modular construction – and the early appointment of the project team – is essential to avoid redundancies in the design. Choosing modular construction as a solution for an over-budget conventional design will not be cost-effective because of the extent of redesign needed to accommodate modular requirements.

It is also important to engage with lenders for construction projects involving modular because there might be resistance to this construction solution in terms of risk. Also, factory-built construction requires drawdown of funds during the early stages of projects due to the need to fund the manufacturing process. This would not be the case with a traditional design.

Where hotels adopt modular build, early decisions on furnishings, fittings and equipment (FFE) are required as this has to be provided to the manufacturer of the modules so that they can be built fully fitted out. The project programme must incorporate ordering dates and schedules for the FFE, as this could otherwise lead to delays.

19.2.4 Advantages of Modular

Using modular construction in appropriate circumstances provides several significant benefits including bringing the finished project more speedily to market, cost savings and, being a progressive construction method, reduced inefficiencies, less sub-trades on site and improved quality control. Additionally:

- Paying for the majority of the cost of the modular units prior to site delivery reduces materials cost inflation.
- An entire project can be completed between 20% and 40% faster than using conventional methods.
- There is less congestion on site – particularly where space is limited and fewer resources are needed for the build.
- Being largely complete on delivery, the modular build requires fewer finishing trades. This brings its own economies because factory labour rates are normally substantially lower than on-site rates for the same trades.
- Access for installation of external cladding requires less external scaffolding. Scissor lifts are used which takes the scaffolding activity off the programme and saves a considerable amount of time.
- Being faster to build – as many as 15 modules can be installed per day – modular construction means that the project is watertight and secure in a matter of days/weeks instead of months.
- Environmental benefits include less noise and dust on site, fewer site deliveries and significantly reduced waste, as some 70% of the build is off-site.

19.3 CIMS-MBS Off-Site Manufacture

The procedures for the manufacture, delivery and installation of the modular units on the Hull hotel project are those of CIMS-MBS. Other manufacturers may operate differently, and clients on other projects may choose different suppliers and installers for their particular scheme.

For the Hull project, procurement and delivery of the 94 volumetric units needed for the modular part of the hotel build required considerable planning and organisation. There are five stages to this process, as illustrated in Figure 19.4:

- Design period.
- Prototype.
- Procurement.
- Mass manufacture.
- Shipping and customs clearance.

19.3.1 Modular Procurement

The procurement of the modular units relies upon completion of the client's design for the project up to the point where floor plans and elevations are available on which to base the sizing and arrangement of the volumetric units.

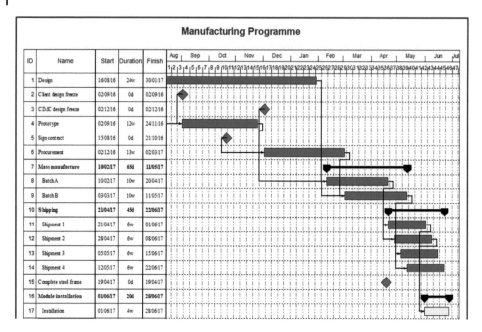

Figure 19.4 Module programme.

At this point, the client's design must be frozen to allow the module manufacturer to commence its design phase, during which a formal contract will be signed by the client and module manufacturer.

19.3.2 Design Stage

Freezing of the client's building design signals the commencement of the modular design phase. This requires a lot of communication and coordination to make sure that the client's requirements are fully assimilated into the modular design.

Hotel chains invariably have brand standards to ensure that customers receive the quality of accommodation, convenience and service associated with that particular brand. To ensure consistency across the brand, there is usually a standards manual which sets down detailed requirements to make sure that each hotel built – irrespective of location – is built to the same standard. The interior of a completed module is shown in Figure 19.5.

Standards manuals include such matters as the dimensions and volume of the accommodation, door styles and sizes, quality and positioning of bathroom appliances, type of shower and wall finishes in both wet and non-wet areas, floor, wall and ceiling finishes in other rooms, mechanical, electrical and air-conditioning requirements and so on.

These brand standards must be incorporated into the module design to ensure that the ensuing client review and approval of the modular design goes smoothly. Upon approval, the modular design will be frozen prior to the production of shop drawings and manufacture of a prototype module.

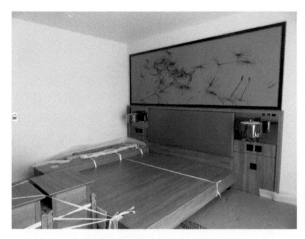

Completed module as delivered to site complete with bedroom fixtures and fittings.
Soft furnishings is a separate work package post module installation.

Completed bathroom within the modular unit.

Figure 19.5 Completed module.

As shown in Figure 19.4, the client's design must be frozen before commencing the design phase of the modular prefabrication. At the end of CIMC's design period, there must be a client review and approval of the design following which the CIMC design will be frozen.

This enables CIMC to produce shop drawings for the next stage of the process – construction of a prototype.

19.3.3 Prototype

The prototype stage includes a period of procurement for steel sections to fabricate the modular structure along with fit out materials for constructing the floor, walls, ceilings and finishings.

During this period, the client also ships FFE items from the United Kingdom to China in compliance with a 'drop-dead' date agreed by the parties so that the finished prototype mirrors exactly what the client is expecting when the units are delivered to site.

Following construction of the prototype unit, the client must sign off the work following any last-minute changes. At this point, the design and materials used for the mass production of the units are frozen.

19.3.4 Mass Manufacture

Mass manufacture is preceded by a procurement stage for steelwork and fit-out materials, following which production of the first of four batches of modules is commenced. This process ends with final inspections and sign-off.

Shipping of the units takes six weeks, including customs clearance, following which the modules are transported to site in construction order on low-loaders. Transportation

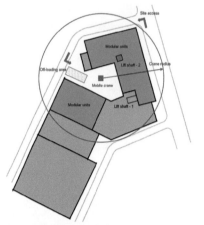
Site layout showing delivery/off-loading area and position of mobile crane.

Lifting modular unit from low loader by mobile crane

Positioning modular unit showing crane boom and lifting beam

Work at height – access by cherry picker with operative wearing harness and lanyard

Figure 19.6 Installing modular units.

of the modules from the dock should ideally be between 250 and 400 miles, after which transportation costs increase. The off-loading and positioning of modules on site are shown in Figure 19.6.

19.3.5 Payment

Contract payments for the modules are made on a stage or milestone basis, with some 90% of payments being made prior to delivery to site. This reduces the risk of bad debt.

Contracts are entered into where security of payments is available for the full contract sum in the form of payment guarantees, secured monies deposited in escrow accounts or confirmed letters of credit – where an advisory bank agrees to guarantee payment should the issuing bank default.

Preliminary work on off-site modular prefabrication is funded under letters of intent with mass production following once an agreed construction contract has been formulated.

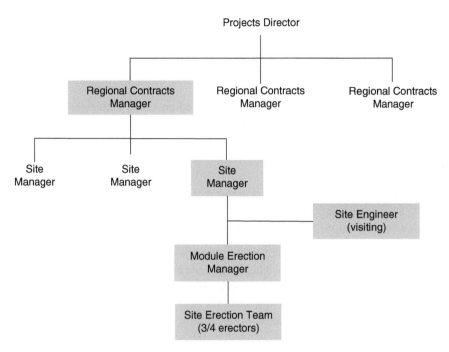

Figure 19.7 CIMC site organisation structure.

Contract retentions are released 50% on practical completion and the remaining 50% 12 months later in common with usual practise in the construction industry.

19.3.6 Installation

Modular installation contracts are controlled and managed by allocated project teams. Senior management undertakes monthly contract reviews of both operational and financial areas of the business. Financial reviews include scrutiny of actual costs to date and forecast cost to complete compared to budgets. Associated movements in each month are reviewed and interrogated.

Figure 19.7 shows the CIMC modular erection team for the project which CIMC-MBS has developed to manage all its contracts. The erection teams, comprising site managers and three or four experienced module erectors, move from project to project.

Health and safety management on site is the responsibility of site managers but health and safety audits of CIMC operations are carried out by third party consultants on a project-by-project basis.

19.4 Construction

Whilst modular components are being assembled in a controlled off-site factory environment, site preparation and groundwork can be carried out. This is then followed by construction of the project frame and structural elements in readiness for module installation.

Once on site, the modules are offloaded by crane and stacked, bolted together and sealed for weatherproofing. The modules are connected together, utilities are hooked up, external cladding and roofing are completed, and site works are finished. Internal works can include painting, joinery items, cabinet worktops and the installation of appliances that were not fitted at the factory.

During the build, the total number of people expected on site is 40 and the number of contractors is 12. The planned project duration is 54 weeks.

19.4.1 Construction Programme

The construction programme shown in Figure 19.8 indicates the sequence of activities prior to and during module installation. It can be seen that a considerable amount of conventional construction work is required prior to module installation, including:

- **Site clearance and groundworks**: Before piling work can begin, foundations and ground slabs to previously demolished buildings must be removed. Foundation work for the new build can then commence which will also incorporate the two lift shaft bases to Block A and the conference centre building/Block B.
- **In situ concrete ground floor slab**: This provides the base for the erection of the slip-forms to the five-storey lift shaft core which must be completed before the steel frame is commenced. The second lift shaft is constructed with conventional formwork as this is only one storey high. The remainder of this lift shaft core is formed by the modular units. Figure 19.9 illustrates the sequence of construction, and Figure 19.10 shows the two lift shafts and the modular units in place.

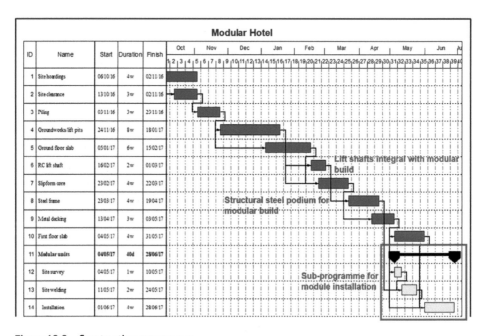

Figure 19.8 Construction programme.

19.4 Construction | 611

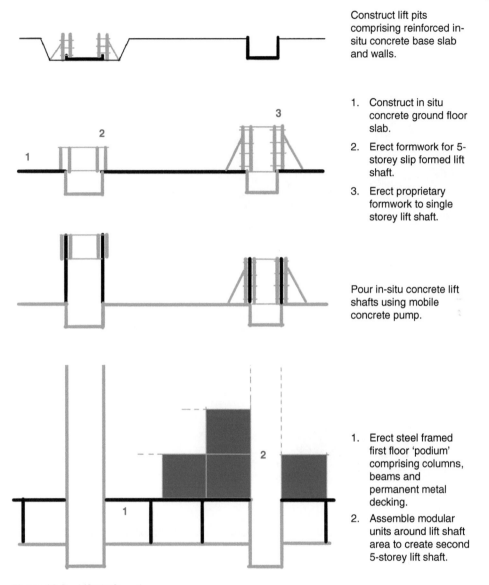

Construct lift pits comprising reinforced in-situ concrete base slab and walls.

1. Construct in situ concrete ground floor slab.
2. Erect formwork for 5-storey slip formed lift shaft.
3. Erect proprietary formwork to single storey lift shaft.

Pour in-situ concrete lift shafts using mobile concrete pump.

1. Erect steel framed first floor 'podium' comprising columns, beams and permanent metal decking.
2. Assemble modular units around lift shaft area to create second 5-storey lift shaft.

Figure 19.9 Lift shafts – 1.

- **Structural steelwork**: Following completion of the lift shaft cores, structural steelwork can commence comprising the five-storey steel framed conference/commercial centre and staircase, the rear five-storey steel framed access staircase, and the single-storey steel framed podium. Precast concrete stairs and landings are incorporated into the steel frame staircases. Refer to Figure 19.11.
- **Upper floors**: In the modular section, floors are integral to the volumetric units. In the steel framed section, upper floors are power floated in situ concrete on steel metal decking.

Slipformed staircase core

Traditional poured concrete staircase core

Lift shaft formed by modular units

Figure 19.10   Lift shafts – 2.

- **Pre-module installation**: Prior to module delivery, some preparatory works are needed ahead of the installation stage:
  - For safety reasons, scaffold edge protection must be erected around floor perimeters and lift shafts prior to work commencing on the reinforced concrete floor slab and module installation. See Figure 19.12.
  - Metal decking (acting as permanent formwork) and shear studs must be installed in all podium areas.
  - The first-floor reinforced concrete slab must be poured and allowed time to cure.
  - Bearing plates must be welded to the steel frame to locate the first-floor modules, as shown in Figure 19.12.
- **Installation of the modular units**: This follows erection of the steel frame to the conference centre, rear access staircase to Block A and the first-floor podiums, and construction

External structural steelwork staircase tower.
Precast concrete stairs and landings.

First floor structural steel podium receiving deck for modular units.
Rear access to site and edge protection to slab shown

Figure 19.11 Structural steelwork.

and curing of the first-floor in situ concrete slabs. See Figure 19.6. Further safety provisions are shown in Figure 19.12.

- **Miscellaneous**: External facing brickwork to ground floor level.

  A variety of cladding to upper floors including vertical aluminium-faced panels and rendering. See Figure 19.13.

  Drainage, external works and landscaping.

19.4.2 Module Installation

Figure 19.6 shows the module installation sequence and also indicates the position of the 100 Tonne mobile crane located at the rear of the conference centre and between the two hotel buildings. This is also the delivery station for the module delivery vehicles.

Edge protection to first floor slab for module erection

Edge protection for teams working on module installation at 4th and 5th floor levels

Figure 19.12 Safety provisions.

Up to 12 modules per day are positioned during the eight-day erection period, and each floor plan contains 15 modules per floor on Block A and six modules on Block B.

The modules vary in size from approximately 12 m to 6 m in length, depending on their location on the floor plan. A total of 94 modules are needed to the complete five floors of the building.

19.4.3 Modular Interface

It is important that site managers understand what portion of the project will be manufactured off-site and its scope, specification and extent of FFE. Work packages for subcontractors also need to take into account that trades such as electrical, tiling and plumbing are fully installed in the modules and that plumbers only have to make horizontal and vertical connections in corridor risers.

Scissor lift access for cladding. Avoids extensive scaffolding.

Cladding rails incorporated into module construction

Figure 19.13 Cladding.

Whilst the modular installer will normally organise crane hire and lifting arrangements, main contractors normally allow for attendances such as provision of crane hard standings, scaffolding for edge protection, storage and site security.

Site safety and induction procedures are also the province of the principal contractor albeit that the modular installer will have their own regular site safety audits.

19.4.4 Defects

As far as defects are concerned, the contractual arrangements should reflect responsibility for their rectification and liability for any costs. Snagging of the modular units may reveal defects missed at the factory or those resulting from transit or site damage.

Such work will require scoping for time and resources, and a budget or contingency will need to be included in the main contract. Volumetric manufacturers do not carry out site repairs as a rule.

19.5 Legal Matters

Sometime after completion of the Hull DoubleTree by Hilton Hotel project, issues came to light concerning the modular supplier/installer CIMC-MBS. These issues concern their payment mechanism for the supply of modules and the insolvency of the company.

19.5.1 Payments

The payment terms under a standard construction contract usually require an interim valuation of the work carried out to date which is then checked and certified by the contract administrator and paid by the client.

Where off-site fabrication of construction elements is concerned, however, this method is inappropriate. Regular visits to the factory would be required, and a detailed breakdown of the agreed price for supply and delivery of the product would be needed for conventional payments to be made. This would be impractical.

Consequently, different payment terms are usual in these circumstances. These payments are called milestone payments and are subject to agreement between the parties. Such an agreement would require an amendment to any standard form of contract used, such as the JCT SBC.

19.5.2 Milestone Payments

In the case of *Bennett (Construction) Ltd v CIMC-MBS Ltd (2019)*, a dispute arose regarding the system of milestone payments employed to pay the manufacturer for the design, manufacture, delivery and installation of modular units for a hotel project in Woolwich, London.

The Technology and Construction Court (TCC) examined the relationship between the agreed payment terms and the provisions of the Housing Grants, Construction and Regeneration Act 1996 (as amended) – the Construction Act. The agreed payment terms were:

- **Milestone 1**: 20% deposit payable on execution of the contract.
- **Milestone 2**: 30% on the sign-off of a prototype at the factory by the client.
- **Milestone 3**: 30% on sign-off of all snagging items by the client.
- **Milestone 4**: 10% on sign-off of units delivered to Southampton dock.
- **Milestone 5**: 10% on completion of installation and snagging.

However, should the work in progress not be signed off at each stage – perhaps due to defects in the work – this would mean that the manufacturer would receive no payments at all for work carried out to date.

The TCC decided that payments relying upon the sign-off of a particular stage of the works did not comply with the Construction Act. It also decided that use of the term 'sign-off' was a generic reference to the completion of a stage of construction but was not sufficiently precise as a payment certificate. In usual practice, this is required as a condition precedent of payment becoming due. The TCC also decided that, to be acceptable, a sign-off arrangement needs to be clear, with specific payment criteria and timescales and not a discretionary date based upon a unilateral opinion.

On appeal, however, the Court of Appeal overturned the decision of the TCC and decided that having no express date for payment was irrelevant because the milestones related to the completion of the relevant stage which would then trigger payment. The Court of Appeal also decided that sign-off of each stage would be an objective assessment and that a certificate would not be required as a condition precedent to the amount becoming due.

The case illustrates that the courts will seek to preserve the intentions of the parties to a contract, including the agreed payment mechanism.

19.5.3 Insolvency

China International Marine Containers (Group) Ltd was founded in Shenzhen, People's Republic of China, in 1992. It diversified into modular building systems and opened regional offices in the United States, Australia and the United Kingdom. Its head office is in southern China.

In the United Kingdom, CIMC Modular Building Systems (CIMC-MBS) – a business division of the CIMC Group – carried out several hotel projects for major brands in major cities and at various airports. Several of its contracts were repeat orders from existing clients.

The company also developed university student accommodation in the United Kingdom including the £75 million student village project for the University of Newcastle, where 800 modules were installed, creating 1243 student rooms.

Some two and a half years after completing the Hull hotel project, CIMC-MBS entered voluntary liquidation. A statement of affairs was recorded at Companies House indicating that the company owed over £14 million, including over £450 000 to Manorcrest Construction Ltd – the main contractor for the Hull hotel project.

Company sources were quoted in the press as suggesting that the reasons for the insolvency were a downturn in the hospitality sector, leading to reduced orders, bad debts from clients who had not paid for completed schemes and the global pandemic.

Scrutiny of the last published accounts supports this view as:

- Annual turnover had fallen from £43.9 million to £4.2 million in 12 months.
- Profits for the same period had fallen from £54 000 to a loss of £2.4 million.
- Amounts recoverable on contracts had increased from £0.5 million to £2.2 million albeit that trade debtors had fallen from £533 086 to £42 494.

The effect of the pandemic is a matter for speculation, but it is likely that future orders were impacted by the resulting global downturn and that the dramatically reduced turnover and profits figures reported in the published accounts were a prelude to this.

Prior to its insolvency, CIMC diversified into modular build rent-to-buy housing schemes in the United Kingdom and currently has a major presence in the Hong Kong housing market.

To a degree, the CIMC situation is symptomatic of the modular market and perhaps other modern methods of construction. The sector is in its infancy and trying to make headway against a reactionary industry and difficult economic times. Two of the UK's market leaders in volumetric production have, however, grown their businesses and increased turnover significantly, according to the latest published accounts.

Reference

Bennett (Construction) Ltd v CIMC MBS Ltd (formerly CIMC Systems Ltd) (2019) EWCA Civ 1515.

20

Motorway Bridge Replacement

20.1 Project Description

This case study concerns the complex and technically challenging design and construction of a bridge to carry the A533 in Cheshire across a motorway on a restricted site and in live traffic conditions.

As illustrated in Figure 20.1, the new bridge is being constructed adjacent to the existing A533 bridge which is to be demolished towards the end of the contract.

20.1.1 Project Background

This project was originally part of a larger scheme to upgrade an adjacent roundabout and construct a new east/west-bound access to the M56 motorway junction between existing junctions 11 and 12 in Cheshire.

The larger scheme was put to public consultation in 2014 but was later rejected on its business case.

20.1.2 Project Location

The A533 is a primary route between Widnes and Middlewich in Cheshire. It crosses the River Mersey from Runcorn to Widnes via the Mersey Gateway Bridge and the Silver Jubilee Bridge,[1] carrying traffic west to Liverpool and north to the M62 and M57.

The A533 Expressway south of Runcorn links to the M56 motorway at junctions 11 and 12, enabling traffic to travel west to Chester and North Wales and east to the M6, to Manchester and beyond.

20.1.3 The Existing Bridge

The A533 crosses the M56 motorway near Preston Brook on a 70-m twin box-section single-span skewed two-lane bridge that runs adjacent to a railway bridge carrying the West Coast Main Line from the south to Liverpool Lime Street.

The single-span structure was built in 1971 but has reached the end of its life cycle and needs to be replaced. The bridge is supported on abutments on either side of the M56 but,

Construction Planning, Programming and Control, Fourth Edition. Brian Cooke and Peter Williams.
© 2025 John Wiley & Sons Ltd. Published 2025 by John Wiley & Sons Ltd.

Key:
   1a: M56 Motorway westbound
   1b: M56 Motorway eastbound
   2. West coast mainline
   3. A533
   4. Site of new A533 bridge

Figure 20.1 Site plan. Photograph courtesy of Amey/SRM JV.

in recent years, has required heavy-duty mid-span support as a precautionary measure.

20.1.4 Project Procurement

National Highways (formerly Highways England and earlier the Highways Agency) is client for the project. It is a government-owned, arms-length company established under the Infrastructure Act 2015.

As an incorporated (limited) company, National Highways is governed by a board of directors and is answerable to the Secretary of State of the Department for Transport and, ultimately, Parliament.

The company is responsible for maintaining and renewing England's motorway and major A-road network and awards contracts to supply chain partners who are approved for the National Highways procurement framework.

The A533 bridge replacement contract was awarded under National Highway's Regional Delivery Programme Framework to the AmeySRM JV. The form of contract used was a bespoke version of the NEC4 form of contract, Option C: Target price – as amended by National Highways. The contract includes for detailed design and construction and for value engineered design and construction solutions with a shared gain/pain arrangement with the client.

AmeySRM JV is a joint venture between Amey plc and Sir Robert McAlpine Ltd. Amey is an infrastructure support service provider, and Sir Robert McAlpine Ltd is a construction and civil engineering company. The two companies have a history of working closely together on other projects and formed their joint venture in 2018. The JV will operate as a fully integrated design and construction team, addressing mutual risks and providing efficient and effective ways of working.[2]

Under the contract, Amey and Sir Robert McAlpine are jointly responsible for the design and construction of the works. National Highways is effectively the client for the project but works collaboratively with the JV, and all three partners have a presence on site.

The approximate cost of the scheme is £27 million, with a construction period of 14 months. Prior to starting work on site, several years of planning, consultation and design work have been necessary to make the project a reality.

20.2 Project Scope

The existing bridge is to be replaced with a new 67 m two-span structure sitting on new abutments and on a central pier sitting in between the two 3-lane carriageways. The new bridge will be constructed adjacent to the existing bridge and will provide improved cycle and pedestrian facilities as well as a single two-lane carriageway for traffic. It has a design life of 120 years.

As indicated in Figure 20.2, there is limited working space on the site of the new bridge, and although National Highways has made land available to the contractor for a site compound, a spoil heap and access to the bridge abutment sites, these locations are separated from the construction site by the A533. This road is to remain 'live' until the new bridge is commissioned. Added to this, the new north and south abutments are separated by the east and west-bound carriageways of the M56, and the central pier is to be constructed in the central reservation between the two live carriageways.

The scope of works is considered below in four main sections:

1. Enabling works.
2. Permanent works.
3. Off-site works.
4. Demolition.

20.2.1 Enabling Works

Working in live traffic conditions, and with limited working space available, it is crucial to establish appropriate traffic management arrangements to facilitate access to the bridge site both from the M56 and from the A533.

On the M56, three lanes will be maintained, but the works will result in narrower running lanes, reduced speed limits for traffic over a 2.4 km stretch of the motorway and loss of the hard shoulder on both sides of the carriageway due to the existing bridge deck propping and protective measures. As for the A533, temporary traffic lights will be in operation to enable

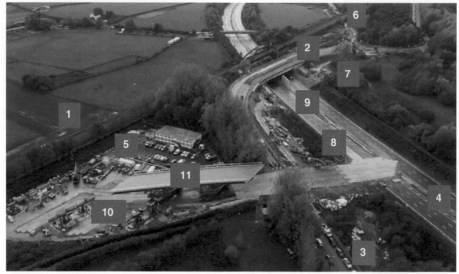

KEY:
1. West Coast Main Line
2. Existing A533 Bridge
3. A533
4. M56 Motorway
5. Site compound
6. Temporary spoil heap
7. North abutment access
8. South abutment access and access to M56
9. New bridge abutments and central pier
10. Bridge deck construction area
11. Bridge deck being moved to permanent position

KEY:
1. New bridge deck in position and open to traffic
2. Old bridge awaiting demolition

Figure 20.2 Site layout. Photographs courtesy of Amey/SRM JV.

construction traffic to cross from the site compound across the A533 to the new bridge south abutment site.

For projects of this nature, additional land is required for lay down areas, site access roads, site establishment and for on-site disposal of excavated material for later re-use. This is often the contractor's problem but, in this case, the land was provided to the contractor by National Highways as part of the compulsory purchase-general vesting declarations process.

The availability of land for site offices and parking, etc., is central to the chosen construction method as this is where the new bridge deck will be constructed offline. The land was cleared and prepared for use as part of the project enabling works and will be reinstated upon completion of the works.

20.2.2 Permanent Works

The permanent works comprise:

- Contiguous bored piling and capping beam to extend the east wing wall of the existing bridge in order to support the existing road and facilitate excavation of the abutment.
- Bored piling to the north and south abutments and central pier foundations.
- Sheet piling around the foundations to act as permanent formwork for the reinforced concrete slab and to act as earthwork support to the motorway cutting.
- Associated excavation and disposal to temporary spoil heaps adjacent to the site.
- Reinforced concrete abutment walls and central pier.
- New composite steel and concrete bridge deck.
- Associated roadworks, parapets, barriers, street lighting, road markings and street furniture, fencing and landscaping.

Figure 20.3 shows the scope of the permanent works and illustrates how little working space is available to the contractor.

20.2.3 Off-site Works

A crucial aspect of the contractor's programme is the concurrent construction of the bridge foundations and the bridge deck which is being built adjacent to the site.

The offline bridge deck construction idea will take a considerable amount of time off the programme and will result in much less traffic disruption and fewer lane closures and possessions. The deck will be installed in a single weekend possession using self-propelled modular transporters (SPMTs).

20.2.4 Demolition

Once the new bridge is commissioned, the existing bridge must be demolished. This is a complex operation as the old bridge will be sandwiched, in close proximity, between the new bridge to the east and the West Coast Mainline railway bridge to the west and there is also a live 3-lane motorway below.

Key:
1. Existing bridge
2. Temporary spoil heap
3. Site compound
4. Central pier
5. South abutment
6. North abutment

Key:
1. Site access to south abutment
2. Site access to central pier
3. Contiguous piled support to existing bridge
4. Existing bridge

Figure 20.3 Site organisation. Photographs courtesy of Amey/SRM JV.

The demolition works will require a full weekend closure of the M56 and A533.

20.3 Project Organisation

The A533 project is organised on the basis of:

- Scope options and preliminary designs prepared on behalf of the client.

- A design and construct contract to develop the detailed design and construct the permanent works.

National Highways initiated the scope design of the scheme in partnership with design consultancy Jacobs and Arcadis who were engaged under an NEC4 Professional Services Contract (PSC), as illustrated in Figure 20.4.

Following contract award, joint venture partner Amey commenced detailed design work. This was done in collaboration with the scope designers, a technical assurance partner, National Highways and JV partner Sir Robert McAlpine. This relationship is illustrated in Figure 20.4.

The collaborative nature of the project is also illustrated in Figure 20.4 which shows how National Highways interacts with the joint venture partners both at a senior board level and with the design and construction teams. This collaboration is emphasised by the use of a Common Data Environment (CDE) for document and data management for the project.

Also evident in Figure 20.4 is the site management team led by the project manager. The importance of temporary works, and particularly the offline bridge deck

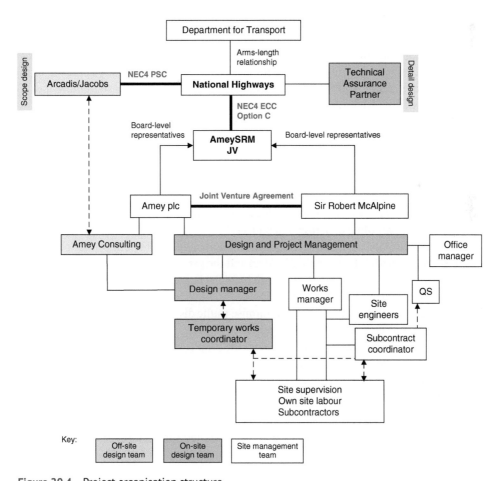

Figure 20.4 Project organisation structure.

construction, is emphasised by the appointment of a senior foreman specifically for this aspect of the project who is answerable to the works manager and, ultimately, the project manager.

20.4 Programme

For any public sector project, a vital phase of the programme is consultation with the public, with key local stakeholders and affected landowners and for making a convincing business case. The business case establishes the costs of the project, and the disruption caused by the works, balanced against the benefits of the completed scheme.

The A533 bridge replacement project programme is considered in two basic parts:

- Pre-construction phase.
- Construction and commissioning phase.

20.4.1 Pre-construction Phase

As project client, answerable to the Secretary of State (SoS) for Transport, National Highways is responsible for ensuring that the correct statutory processes are enacted to pave the way for design and eventual construction of the bridge. Parliamentary approval for the scheme is not required, nor is an Act of Parliament, but the SoS for Transport must approve an Order under the Highways Act 1980 authorising the works.

Additionally, National Highways must issue statutory notices and general vesting declarations for the compulsory acquisition of land and affected local highways.

A project of this nature is planned years ahead under the control of the Network Planner for National Highways. This work includes:

- Structural analysis of the existing bridge and the provision of temporary support.
- The outline development of possible design alternatives to be costed by National Highways against prevailing departmental budgets.
- The preparation of public consultation documents and conducting private and public consultations with interested stakeholders and the general public.
- Modelling of worst-case scenarios to anticipate the impact of the project on road users.
- Minimising road closures as much as possible.
- Planning appropriate mitigations to minimise traffic disruption:
  - Notifications targeted at local road users.
  - Motorway gantry messages weeks before the project starts on site.
  - Variable message signs on linking roads up to 100 mi away.
  - Trailer-mounted message boards.
  - Time over-distance advice to road users to help them decide on alternative routes.

For the A533 project, the key to a successful outcome is collaborative planning involving local stakeholders, National Highways and the appointed contractor.

Once the project business case has been made out, and public consultation and land acquisition successfully completed, the outline design can be developed and brought to a suitable stage to invite tenders.

20.4.2 Tender Period and Contract Award

In common with similar public works projects, the A533 scheme necessitated an outline design, developed by National Highways, to set a budget so that the required funding could be secured prior to engaging with contractors.

Following this initial stage, the project entered a development phase where a detailed design was agreed along with a budget to include design and the surveys necessary to inform the detailed design.

The scheme was awarded to the Amey/SRM Joint Venture based on a total of prices exercise through the framework agreement under the regional development partnership.

20.5 Detailed Design Development

The detailed design for the A533 bridge replacement is in two parts:

- Design of the permanent works.
- Temporary works design – contractor-led and delivered:
  - Earthwork support.
  - Bridge deck assembly.

20.5.1 Design of the Permanent Works

The outline design for the scheme was developed by National Highways in partnership with leading design and engineering consultancies Jacobs and Arcadis. This design was brought to the point – equivalent to Royal Institute of British Architects (RIBA) Stage 3 – where tenders could be invited from interested contractors on a design and construct basis.

The size, complexity and engineering challenges posed by the project made the case for long-standing joint venture partners Amey and Sir Robert McAlpine to combine their considerable resources and in-house skills and to develop the works. They were appointed in June 2020.

It would be inaccurate to say that Sir Robert McAlpine is responsible for the construction works and Amey for undertaking the detailed design work. The nature of the joint venture, and its relationship with National Highways, means that the permanent works design is a collaborative process involving both the client and the JV partners and the construction works are actively managed by both Sir Robert McAlpine and Amey.

The relationship between the project partners is illustrated in Figure 20.4. It can be seen that the design has been partially developed off-site but that further detailed design development and modifications have been made in collaboration with the other partners by designers on site.

It has been essential for the developing design to take into account proposed construction methods and the order and sequence of work and to do so in collaborative discussion with the JV partners and National Highways on site. In this regard, the design has benefitted from the use of Stage 2 BIM which has enabled team members to work collaboratively and agree on the implementation of changes to the design. The BIM software being used is Bentley Systems OpenRoads software suite.

Bearing in mind that a budget has been agreed as a basis for the contract, alternative designs for the various elements of the bridge must be costed and the cost plan targets maintained. Where design elements are over budget then savings are needed in other elements in order to maintain the integrity of the overall budget but the budget can only increase with changes to high-level requirements.

Combined with these considerations is the incentive to develop novel design solutions that enable cost savings to be made within the gain/pain arrangement with the client. Even though gains are shared, they nevertheless add to the profitability of the contract for the JV partners. This, consequently, incentivises the JV to come up with cost effective design and construction solutions that save money either on direct costs or by making savings in the programme and thereby the time-related elements of the target price.

20.5.2 Temporary Works Design

Temporary works are invariably an important part of civil engineering works and the A533 bridge project is no exception. There are two key elements to the temporary works requirements for the project:

- Bridge abutments and central pier.
- Bridge deck.

20.5.2.1 Bridge Abutments and Central Pier

The north and south abutments and central pier of the bridge are to be constructed in cofferdam, the temporary works design and installation for which has been undertaken by MGF, a specialist supply chain partner engaged by the JV earthworks contractor. This necessitated close collaboration between National Highways, the contractor AmeySRM JV and the earthworks contractor P. P. O'Connor.

Figure 20.5 shows the cofferdam arrangement for the south abutment and illustrates the stages involved:

1. Install sheet piling.
2. Carry out stage 1 excavation.
3. Install steel walings and braces.
4. Carry out stage 2 excavation to expose piles.
5. Excavate around piles.
6. Pile cropping.
7. Lay blinding concrete.

Due to the pandemic, virtual meetings were arranged at an early stage with members of the design and site teams, together with the temporary works contractor MGF, who

Cofferdam prior to excavation

Cofferdam excavated to expose piles

Piles cropped and set aside for disposal

Figure 20.5 Cofferdam to south abutment. Photographs courtesy of MGF Ltd.

employed 3D imagery to model the interaction between the temporary works and existing structures.

20.5.2.2 Bridge Deck

In order to remove the bridge deck construction off the critical path, the decision was made to fabricate this element of the structure off-site and to later move it into position during a full weekend road closure.

The two-span deck is of composite construction consisting of steel beans and a reinforced concrete slab. The structural steelwork is 'weathering steel' which eliminates the need for painting and develops a stable rust-like appearance once exposed to the weather for some time.

Approximately 293t of structural steelwork has been delivered to site on low-loaders by Briton Fabricators of Nottingham. Britons erected the steelwork on temporary trestles constructed on specially prepared land close to the site. There are three trestles at both ends and mid-span of the deck. The reinforced concrete trestle bases have been constructed in situ on site as part of the contractor's temporary works operations on top of which are steel trestles which elevate the deck to the correct level for the SPMTs to sit under.

Figure 20.6 shows the steelwork being delivered to site and assembled on trestles using a 500 Tonne Liebherr mobile crane with genieboom and cherry picker access for the final bolting and assembly.

Prior to delivery, the steel structure was fully trial-erected at the fabrication yard in order to ensure that there would be no on-site assembly problems.

The next stage is to construct the deck slab before the bridge deck can be moved into its final position.

20.5.3   4D BIM

The ability of 4D BIM software to model the construction process at the design stage is a crucial element of the detailed design work carried out by the JV.

Steelwork arriving on site

Offloading steelwork

Placing steelwork on trestles

Bolting together

Steelwork on trestles

Figure 20.6   Structural steelwork to bridge deck. Photographs courtesy of Amey/SRM JV.

Using National Highway's Bentley OpenRoads outline design model, Amey was able to develop the detailed design in conjunction with the SRM construction team thereby permitting a two-way dialogue to ensure that the most efficient construction method was integrated into the eventual design.

Both the design model and the Primavera P6 construction schedule developed by SRM can be imported into Bentley Synchro Pro to facilitate 4D simulations of the construction process.

20.6 Construction Planning

Following preparatory work, the bridge replacement scheme is expected to take 14 months to complete.

A major feature of the contractor's programme is the offline construction of the bridge deck which will be moved into place once the bridge abutments, pier and bridge bearings have been completed. This will not only benefit road users but also take a considerable amount of time out of the works programme and mean that completion of the scheme will take much less time than using conventional construction methods.

Planning the construction stages of the project have been undertaken using Oracle Primavera P6 software. An indication of the work order and sequence is illustrated in Figure 20.7 which also shows key milestone dates. Figure 20.7 further illustrates how the requirements of Engineering and Construction Contract (ECC) Clause 31.2 might be met which requires the construction method to be shown on the accepted programme.

Key activities on the construction programme are:

- Major milestones.
- Motorway possessions and road closures.
- Bridge abutments and central pier:
  - Piling.
  - Temporary works.
  - Formwork and concrete.
- Bridge deck:
  - Offline fabrication.
  - Transportation.
  - Installation.
- Demolition.

20.6.1 Major Milestones

When developing a programme for any project, milestone dates are fundamental to a successful outcome. For a project such as the A533 bridge replacement this is doubly important because failure to identify and hit key dates could cause inconvenience to road users and major inconvenience for National Highways' early warning and road diversion mitigation measures.

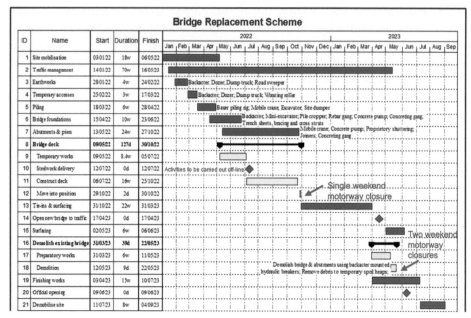

NB:
This is a simplified interpretation of the project programme.
It is not the contractor's programme which is confidential and much more complex.

Figure 20.7   Clause 30 programme.

For the A533 bridge, the key milestones are:

- Commencement following development of the detailed design.
- Installation of traffic management measures.
- Partial road closures for concrete pours to bridge foundations and walls.
- Delivery of bridge deck steelwork.
- Road closure for bridge deck installation.
- Commissioning of the new bridge.
- Road closures for demolition of the old bridge.
- Completion and handover.

Some of these key milestone dates can be seen in Figure 20.7.

20.6.2   Motorway Possessions and Road Closures

In order to carry out certain construction operations, road closures are unavoidable on a project of this nature. Whilst this requirement has been kept to a minimum, the following operations necessitate either a full possession or partial closure:

- Installation of traffic management measures (partial night-time closure).
- Temporary works and concrete pumping to bridge abutments and central pier (partial night-time closures).
- Bridge deck installation (full weekend closure).
- Demolition of old bridge (full weekend closure).

The contractor's working methods, and order and sequence of working, have avoided weeks of lane restriction/closures.

20.6.3 Bridge Abutments and Central Pier

The north and south bridge abutments are located on the slope of a cutting on either side of the motorway close to moving traffic. The central pier is at road level. The bridge foundations are piled with some 50 No 1200 mm diameter × 15 m long bored piles. At each abutment, the piling rigs were located on a stoned platform, as illustrated in Figure 20.8 whilst piles for the central pier were installed from a piling mat within the central reservation.

It can be seen in Figure 20.8 that a mobile crane is working alongside the piling rig. This is needed to move piling equipment, casings, etc., to install rebar and to facilitate concreting to the piles using tremie pipes. An excavator is also in attendance, removing pile arisings and keeping the site clear.

Figure 20.5 shows the piles cropped to their required cut-off level and a 1.5Te mini excavator can be seen preparing the formation between the piles ready for blinding and the bridge abutment concrete pour.

Prior to the main piling work, a contiguous piled retaining wall, with reinforced concrete capping beam, was constructed on the south side of the bridge to stabilise the A533 behind and above the abutment works.

At the bottom of the cutting on each side of the motorway, the ground is retained with high-strength concrete 'Lego' blocks weighing 2.4Te each stacked to form a short retaining wall. Nearer to the carriageway, a further row of 2.5Te Temporary Vertical Concrete Barriers (TVCB) barriers with integrated fencing and debris netting protect the road from any falling

Rotary bored piling to south abutment contiguous retaining wall.

Figure 20.8 Bored piling. Photograph courtesy of Amey/SRM JV.

materials. Precast concrete 'Jersey' barriers protect either side of the working area in the central reservation.

Following the bored piling operation, both abutment sites were excavated to form a bench to facilitate access for construction plant. Each foundation is constructed within an interlocking sheet piled cofferdam with proprietary steel walings and braces. This arrangement facilitates safe excavation of the bridge foundations to expose the tops of the bored piles. The piles were then cropped down to the required cut-off level using the Coredek 'Corbreak' passive cut-off method and the formation blinded with concrete.

20.6.4 Bridge Deck

Construction of the bridge deck is arguably the most complex element of the project and carries the most risk. However, the downside risk must be weighed against the use of traditional construction methods which would require significant temporary works, a much greater working area around the site footprint and would also create much greater disruption to road users. Additionally, the overall construction period would be much longer.

Constructing the deck offline has significant advantages, therefore, but moving the structure into position is a tricky operation. Whilst the use of SPMTs is not a new idea, every site is different and so are the risks. Not least of the contractor's problems is the weather which could increase the risk factor if wet or windy conditions prevail. In view of the criticality of the deck installation milestone, bad weather is unlikely to delay the operation but it would make the task more difficult. However, a contingency possession is nevertheless built into the contractor's programme should it be necessary to delay the work.

The deck weighs in the order of 1100 tonnes, and the SPMTs must negotiate a way onto the motorway carriageway, then straddle the central reservation and finally move the deck into position and on to its bearings. To facilitate this part of the move, the motorway carriageway will be covered with heavy-duty proprietary load-spreading protective matting which must be put in place in readiness for the move. Key to the move is good quality off-site production both at the fabrication yard and on site to ensure that everything 'fits' first time.

The deck construction and installation sequence is as follows:

- Plan/operate a safe system of work.
- Prepare and surface the deck construction area to take high axle loadings.
- Construct concrete foundations and assemble temporary steel trestles.
- Bring mobile crane to site.
- Transport steel beams to site and crane offload onto trestles.
- Bolt steelwork together – some 293 tonnes in all.
- Erect falsework for construction of the reinforced concrete deck.
- Fix rebar, edge formwork and joints.
- Concrete the deck.
- Assemble SPMTs under the deck.
- Carry out trial lift/movement prior to closure of road.
- Lift deck using SPMT hydraulic jacks and move into final position.

Figure 20.9 illustrates the bridge deck installation process from initial mobilisation of the SPMTs in the site compound. It also illustrates the temporary works required to give access to the motorway carriageway. This included 1000 tonnes of imported stone for the A533

Bridge deck ready to be moved.
SPMTs in foreground ready to be positioned underneath the bridge deck.
The bridge deck will be hydraulically lifted off its temporary supports when ready to move.

Bridge deck moving from site compound towards the M56. A stoned road crossing has been layed across the A533 and temporary matting laid. Temporary matting is also laid on the motorway carriageway.

Bridge deck nearly onto motorway carriageway in background. The prepared bridge abutments and pier are in the foreground. The M56 and A533 are closed to normal traffic.

Bridge deck approaches bearings. Temporary site lighting in place for night working.

Figure 20.9 Moving the bridge deck. Photographs courtesy of Amey/SRM JV.

road crossing and 180 2.4 m × 1.2 m heavy duty, high visibility high-density polyethylene mats to provide smooth access for the SPMTs onto the carriageway.

The deck must cross the A533, down a ramp onto the M56 main carriageway and, straddling the central reservation, move along the carriageway to its final position on the abutment bearings. It can be seen that part of this complex operation was undertaken at night under floodlights.

20.6.5 Demolition

The existing A533 bridge consists of a wide single span deck sitting on twin post-tensioned box-section beams supported by concrete abutments on either side of the motorway. The deck forms a continuous span over both carriageways. Temporary steel propping has been installed mid-span to support the weak deck prior to construction of the new bridge.

Demolition of the bridge is planned in two phases:

- Removal of the concrete deck, a considerable quantity of post-tensioning cables and parapets.
- Demolition of the bridge abutments.

This work will require two 56-hour weekend possessions of the M56 between junctions 11 and 12 and the affected part of the A533. Following substantial completion of the works to the new bridge, and based on vibration calculations, it has been decided that possession of the West Coast Main Line rail bridge will not be necessary despite its proximity to the demolition site.

Prior to demolishing the bridge deck, the new road alignment must be completed as far as possible to facilitate diversion of traffic onto the new bridge at the end of the possession period. Having already diverted utilities into the new bridge deck, the demolition team is able to concentrate on the demolition process which must be carefully managed to avoid:

- Damage to the new bridge.
- Damage to the West Coast Main Line bridge, catenary and track.
- Uncontrolled collapse of structural elements.

The demolition work will be carried out by the JV Tier one demolition subcontractor who has the expertise, experience and equipment to carry out such work. The work is priced using NEC4 Option A: Lump sum as the subcontractor is in possession of all the as-built drawings.

Health and safety is the prime consideration, of course, and it is also essential to avoid any damage to the motorway or to the rail bridge which would prevent reopening at the end of the possession periods.

Well in advance of the demolition work, diversion planning in conjunction with National Highways and devising a safe system of work is essential. The safe system of work will be approved by Network Rail and will be based on:

- A detailed risk assessment.
- A work/safety method statement.
- Task talks with the operatives.

20.6 Construction Planning

The two separate 56-hour weekend possessions are not much time to demolish the bridge and to reinstate the affected area of the motorway, especially as night-time working will be necessary. Consequently, detailed plans must be put in place to mobilise:

- Labour and supervision working 12-hour shifts.
- Heavy-duty timbers to construct a protective 'crash deck' on the motorway.
- Shipping containers to be placed under the bridge deck to guard against uncontrolled collapse of the structure.
- Plant and equipment:
  - Several large crawler excavators variously equipped with hydraulic breakers, pulverisers, nibblers, grabs and rock buckets.
  - Articulated loading shovels.
  - Steel-bodied road waggons.
  - MEWP (Mobile elevating working platform).
  - Telescopic materials handler (Telehandler).
  - Bobcat skid-steers to assist clean up.
  - Road sweepers to clean the carriageway.
- Water canons for dust suppression.
- Site lighting towers.
- Catering arrangements for 24-hour working.

Additionally, the contractor must arrange for backup equipment, and plant mechanics will be needed on site to deal any mechanical breakdown.

Figure 20.10 illustrates the position of the three bridges with the old bridge sandwiched in the middle and also identifies the access point on the A533 down to the M56. The two crucial limiting factors concerning the proposed method of demolition are:

- The proximity of the newly constructed bridge which is 5 m to the east.
- The West Coast Main Line bridge which is 20 m to the west.

Despite the proximity of these two structures, demolition of the bridge deck and parapets, together with the two single-span box-section supporting structures, will be carried out from the M56 below as would be normal practice.

Working from both sides of the bridge, the road surfacing, kerbs and parapets would be removed first. This would be followed by hydraulic breaking of the concrete elements of the box-structure in order to expose the post-tensioning cables. Once the deck box beams are demolished, the mid-span temporary steel supports to the bridge can be removed.

The existing reinforced concrete bridge abutments can be demolished from below with tracked excavators equipped with hydraulic breakers working off the protective timber 'crash deck' on the M56. Debris will be removed by loading shovels onto articulated dump trucks for temporary disposal on site.

Figures 20.10 and 20.11 illustrate the demolition work in progress. The sequence of work is:

- Partially remove traffic management measures near to the bridge.
- Install protective membrane and heavy-duty timber 'crash deck' protection to M56 carriageway.

638 | 20 Motorway Bridge Replacement

Bridge deck demolition in progress. Illustrates proximity of new bridge deck and west coast mainline.

Demolition of south abutment in progress. Demolition waste is stockpiled away from working area and removed to site compound with articulated dump trucks (ADTs). Demolition waste is later crushed on site.

A variety of hydraulic excavator attachments are used for the demolition work including nibblers, pulverisers, shears, breakers and grabs.
Standby plant is available in case of breakdowns.

Figure 20.10 Demolition 1. Photographs courtesy of Amey/SRM JV.

Shipping containers installed beneath existing bridge deck in the event of premature collapse.

Water tankers positioned on new bridge deck to service water canons. Green debris netting protection in place.

Night-time working. Lighting towers in place to illuminate demolition zone.

Figure 20.11 Demolition 2. Photographs courtesy of Amey/SRM JV.

- Install shipping containers under the bridge deck.
- Excavate to expose north and south abutment wing walls.
- Remove street lighting columns.
- Remove steel and concrete bridge parapets.
- Break out tarmac and concrete deck carriageway.

- Remove temporary supports to bridge.
- Demolish abutments (there is no central pier on the old bridge).
- Remove debris to temporary spoil heaps.
- Remove motorway 'crash deck'.
- Install new Armco barriers as required.
- Reinstate affected areas and re-install traffic management measures.
- Re-open M56 and A533 (on new bridge).

The final few weeks of the project will be concerned with completing roadworks, finishing works, fencing and landscaping, de-mobilisation of the site compound and reinstatement of land used for access and spoil heaps, etc.

Notes

1 Formerly the Runcorn-Widnes Bridge.
2 https://newsroom.ferrovial.com/en/news/amey-and-sir-robert-mcalpine-join-up-to-deliver-improvements-across-the-northern-road-network/.

21

High Speed 2

21.1 Project Description

High Speed 2 is a mega-rail project being undertaken by the UK government which was planned to run from London to Birmingham and beyond to destinations in the north-Midlands and to the north of England. The new railway was also intended to link up to existing infrastructure for connections to Scotland.

The scope of HS2 has changed several times over the years, but most markedly in October 2023, when the government of the day decided to curtail the project north of Birmingham due to concerns over cost. The cost of Phases 1 and 2 of the scheme has spiralled from around £30 billion originally to well over £100 billion recently.

The original project would have delivered over 400 km (260 mi) of new high-speed twin track[1] but, as it stands, HS2 currently comprises 225 km (140 mi) of dedicated twin track, four new stations and two rail depots. It will initially run from Old Oak Common near London to a new station at Curzon Street, Birmingham. Services are planned to start between 2029 and 2033. Work on a new Euston station has commenced but Stage A is not expected to complete until 2036, when Stage B1 is planned to start, whereas Stage B2 has no planned start or completion date as yet.

21.1.1 Business Case

Justification for HS2 – albeit politically, economically and environmentally controversial – was set out in the Department for Transport (DfT)'s Full Business Case published in 2020 (DfT 2020). Essentially, the purpose of the railway was to reduce journey times between major UK cities, to add capacity to the existing rail network – principally the West Coast Mainline and the East Coast Mainline – and to create investment, employment and *form the spine of the UK's future transport network*.

Establishing greater connectivity in the north–south rail network was expected to generate economic benefits and shift reliance on road and air traffic to the railway system. Building a new passenger-dedicated railway would also have released capacity to increase freight traffic on existing routes (Network Rail 2009).

The government's business case for Phase 1 of HS2, which set out the economic, financial, commercial and management justifications for the project, concluded that the potential benefits of the high-speed line outweigh its costs.

However, proposed changes to the route announced in 2021 met with widespread disapproval, especially as the independent Oakervee Review supported the original route. No doubt cost was a significant factor in the decision, but the majority of informed opinion felt that the revisions to Phase 2 compromised the 'levelling-up' of the north of the country. The government's decision to abandon Phase 2 in 2023 ended this debate with the intention that money will be directed to improving existing local transport connectivity in the north of England in a variety of supposedly more cost-effective ways.

21.1.2 Project Delivery

In order to facilitate a project such as HS2, Acts of Parliament are needed which set out the scope of the scheme and provide the statutory power to acquire land and carry out the necessary works. The HS2 rail project was sponsored through the Parliamentary process by the Secretary of State for Transport (SoST) who appointed a body to be responsible for delivering the proposed scheme – HS2 Ltd.

The SoS for Transport and HS2 Ltd entered into a Development Agreement with the DfT to deliver the railway – 'the Programme' – in December 2014, with amendments in 2017 and 2018.

21.2 Route

HS2 was originally planned to be completed in two phases – Phase 1 between London and Birmingham and Phase 2 consisting of two branches, one from Birmingham to Manchester (Phase 2a) and another to Leeds (Phase 2b).

This proposal was changed in 2021 when the Government published its Integrated Rail Plan[2] which outlined the government's intentions for delivering major rail investment in the North and Midlands. Subsequently, the scope of HS2 has changed such that only Phase 1 of the original 'Y-network' will be completed as illustrated in Figure 21.1.

21.2.1 Oakervee Review

Following a period of debate and uncertainty surrounding the future of HS2, Sir Douglas Oakervee CBE FREng was asked in 2019 to conduct a review of the project by then Prime Minister Boris Johnson. The purpose of the review was to better inform the government as to how best to proceed with HS2 or whether to cancel it.

Sir Douglas is a civil engineer and former chairman of HS2 and Crossrail, and his review, which was published in 2020, considered that the project should proceed subject to some reservations.

The Oakervee Review (2020) concluded that considerable design and enabling works had already been carried out and that cancelling the project would incur additional direct costs of between £2.5 billion and £3.6 billion on top of the £9 billion already spent. Savings of between £2 billion and £3 billion of recoverable costs would be expected making the overall

21.2 Route

High Speed 2 – Original route

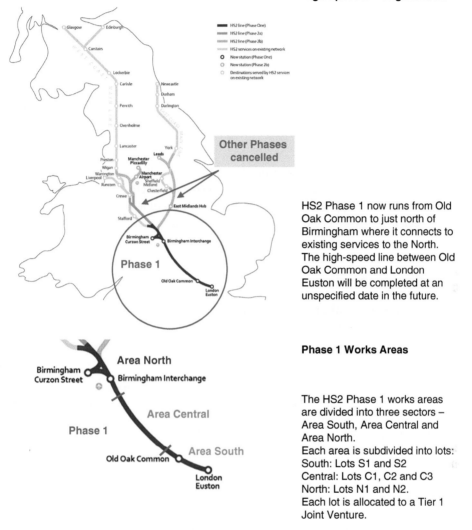

Other Phases cancelled

HS2 Phase 1 now runs from Old Oak Common to just north of Birmingham where it connects to existing services to the North. The high-speed line between Old Oak Common and London Euston will be completed at an unspecified date in the future.

Phase 1 Works Areas

The HS2 Phase 1 works areas are divided into three sectors – Area South, Area Central and Area North.
Each area is subdivided into lots:
South: Lots S1 and S2
Central: Lots C1, C2 and C3
North: Lots N1 and N2.
Each lot is allocated to a Tier 1 Joint Venture.

Figure 21.1 HS2 original route and works areas. Source: Department for Transport (2017).

cost to the Exchequer close to £10 billion. In addition to the wasted expenditure, it was also felt that significant damage could be caused to the construction industry supply chain and to fragile industry confidence as Oakervee's deliberations followed on closely behind the collapse of Carillion in 2018.

The review also confirmed that the rationale for HS2 was still valid but that certain issues should be addressed including:

- Reduce costs by removing over specification in the project and making changes to alignment design.
- Altering the project procurement and contracting strategy adopted by HS2 Ltd.

- Reduction of the speed and frequency of trains and
- Construction of Phase 2a in parallel with Phase 1.

Oakervee noted that the key decision to be made by the government was whether to issue the Notice to Proceed (NtP) which was a necessary step in order to finalise contracts for major construction works to be started on Phase 1 of HS2. This would effectively be a 'go/no-go' decision for the entire project as it was considered that it made no sense to construct Phase 1 unless the phases planned for the north of the route were also to proceed.

21.2.2 Integrated Rail Plan

The 'rethink' prompted by Oakervee resulted in publication of the Government's Integrated Rail Plan in November 2021, which significantly altered the original HS2 programme.

Despite Oakervee's support for constructing the full eastern leg of the 'Y', this proposal has now been curtailed and HS2 will terminate at Birmingham.

21.3 Project Scope

Even at its reduced scope, the projected cost of HS2 Phase One of £45–54 billion (US$57–68) places it as one of the largest construction projects in the world:

Project	Description	Estimated cost Billion US Dollars	Estimated cost Billion £ Sterling
California High-Speed Rail	High-speed rail project California, USA	98	80
Dubailand	Disney world-type Entertainment city, Dubai	76	62
Hinckley Point C	Nuclear power station Somerset, England	43	31–34
Plant Vogtle – Units 3 and 4	Nuclear plant Georgia, USA	30	25
Samsung factory	Semi-conductor plant Arizona, USA	17	14

At the time of writing, construction work on Phase 1 is around 40% complete with over 350 individual construction sites underway.

21.3.1 Phase 1

The route for Phase 1 is between London Euston (eventually) and Birmingham Curzon Street with interchanges at Old Oak Common near London and Birmingham Interchange. This route contains several complex structures including some 500 bridging structures and 72 km (45 mi) of tunnel. Of the 50 viaducts on this route, the Colne Valley Viaduct is the longest at 3.4 km (2.1 mi) and the longest tunnel is the 16 km (10 mi) Chiltern Tunnel which is up to 90 m (295 ft) deep.

Near Birmingham a delta junction will be constructed over a length of 9.5 km (5.9 mi) consisting of seven bridges and viaducts crossing three rail lines, eight roads and five rivers and canals.

21.3.2 HS2–HS1 Link

Plans for a rail link connecting HS2 and the Channel Tunnel Rail Link (HS1) north of London have been shelved to make cost savings in the order of £700 million.

Instead, the 640 m (0.4 mi) from London Euston (HS2 terminal) and London St Pancras (HS1 terminal) will be connected by improved pedestrian links and automated people movers.

21.4 Legal Framework

The role of legislation in a project such as HS2 cannot be underestimated. The various Acts of Parliament needed to make the project a reality set out a detailed specification of the works to be undertaken and delegate the necessary authority not only to construct and maintain the railway but also to acquire land, conduct surveys and ground investigations, divert existing services and carry out the necessary enabling works prior to construction proper.

Despite their great potential for positive outcomes for society generally, mega-projects such as HS2 divide public opinion as they have an enormous impact on large parts of the countryside, on towns and cities, on local communities and on individuals and businesses whose lives and livelihoods can be changed irrevocably and, possibly, detrimentally. It would be impossible, therefore, to undertake such large and complex projects without political consensus.

The role of Members of Parliament is to represent the electorate, and it is essential that projects such as the Channel Tunnel Rail Link (High Speed 1), Crossrail in London, and, indeed, HS2, are undertaken according to the powers granted by Parliament and in the best interests of society.

New laws to be passed by Parliament give permission for construction to be started and also give the government the requisite statutory powers to:

- Operate and maintain the project and its associated works.
- Carry out necessary protective works to buildings and third-party infrastructure.
- Permanently or temporarily divert highways, waterways and the like.
- Make modifications to infrastructure belonging to utility companies and the like.
- Carry out work on listed buildings.
- Demolish buildings in conservation areas.
- Compulsorily acquire land and property.
- Alter rights of way and so on.

21.4.1 Parliamentary Bills

Proposals for new laws (Acts of Parliament), are presented to Parliament in the form of draft legislation called Parliamentary Bills. The HS2 Bills set out the government's plans for the project together with the powers being sought from Parliament to build and run the new high-speed railway.

For projects such as HS2, a great deal of preparatory work is required to produce the Bills. Supporting documents required include details of the route, with plans and sections showing the land required for the railway, environmental statements, including a draft code of construction practice and an estimate of cost.

There are various types of bills but proposals for HS2 took the form of two hybrid Bills which were presented to Parliament in 2013 (Phase 1) and 2017 (Phase 2). Although relatively rare, hybrid bills are often used to secure powers to construct and operate major infrastructure projects of national importance:

- Public bills propose laws which apply to everyone within their jurisdiction in equal measure.
- Private bills seek to change the law in a way that affects some people differently to others and may impact specific individuals or groups.
- Hybrid bills address both public and private matters and, whilst they may propose works of national importance, their impact is limited to specific parts of the country.

The Channel Tunnel Bills in the 1970s and 1980s, the Rail Link Act 1996 (High Speed 1) and the Crossrail Bill (2008) are examples of hybrid bills because the ensuing legislation impacted the south–east of England and London respectively but not did not directly affect other parts of the country.

The time required for hybrid bills to pass into legislation is longer as they combine the features of both public and private bills.

21.4.2 Acts of Parliament

Once passed into law, the hybrid bills for HS2 not only grant the Government planning permission to build the HS2 network but also grant the statutory powers needed to make the project a reality. This provides the government with sweeping powers to implement the project but the Acts of Parliament are very specific about the scope and detail of what is allowed to be built. In view of the opposition that projects such as HS2 can engender, this is an important measure of restraint on what the government can do.

In January 2009, HS2 Ltd was set up with the remit to develop designs for a new high-speed railway connecting London with the Midlands, the North of England and to Scotland. However, making the case for such a project is a lengthy process requiring political discussion, public consultation and impact assessments regarding aesthetics, noise and other environmental issues and it was not until January 2012, therefore, that the SoST announced that HS2 would go ahead.

The High-Speed Rail (London-West Midlands) Act 2017 was passed by both houses of Parliament and received Royal Assent in February 2017. This authorised construction of HS2 Phase 1. Legislation for other phases – now abandoned – remain on the statute book for the present time.

Under Section 1(a) of the High-Speed Rail (London-West Midlands) Act 2017, Phase 1 of High Speed 2 is defined as:

> *a railway between Euston in London and a junction with the West Coast Main Line at Handsacre in Staffordshire, with a spur from Water Orton in Warwickshire to Curzon Street in Birmingham.*

The statutory provision to make this happen is devolved under Section 45(1) of the Act to the SoST who may appoint a 'nominated undertaker' to *construct and maintain the works specified in Schedule 1*. The nominated undertaker is HS2 Ltd and Schedule 1 of the Act provides a very detailed list of the works required for the construction of Phase 1 of High Speed 2.

21.4.3 Planning Permission

Permission under Town and Country Planning legislation is not required for HS2 Phase 1 because 'deemed planning permission' is granted under the High-Speed Rail Act of 2017 (Section 20). This might be regarded as deemed 'outline' approval because:

- Permission is granted for the development authorised by the legislation which is specified in the detailed provisions of various Schedules to the Acts and
- Schedule 17 of the 2017 Act imposes conditions on the deemed planning permission such that local planning authorities are able to influence the proposed work concerning what might be called 'reserved matters'.

The logic behind the legislation becomes clearer by referring to Schedule 1 – Scheduled Works – of the 2017 Act. This refers to 'tunnels', 'viaducts', 'bridges', 'sewers', etc. which illustrates the level of detail contained within the Statute which is extensive but not specific. When the law was passed through Parliament it is probable that the design of HS2 was at RIBA Plan of Work 2020 Stage 2 – Concept Design – i.e. in its relatively early stages.

Clearly then, although the principle of the railway line has been agreed by Parliament, together with its attendant stations, tunnels, bridges and other structures, it is logical that HS2 Ltd should have to apply to local planning authorities for the approval of certain details associated with finalising the design and with regard to its proposals for constructing and delivering the railway.

21.4.4 The Secretary of State for Transport (SoST)

The Department of Transport is the government department responsible for HS2. As its Minister of State, the SoST, who promoted the HS2 Bill through Parliament, is responsible for keeping Parliament informed about progress with the railway.

Under the High-Speed Rail Act of 2017, the SoST may appoint a nominated undertaker to assume responsibility for the performance of such parts of the legislation as the SoST shall decide. This means, for instance, that the nominated undertaker shall construct and maintain the works specified in Schedule 1 of the 2017 Act (i.e. Phase 1), and any works consequent on and incidental thereto, in accordance with Section 1(1)(a) and (b) of the Act. The appointed nominated undertaker is HS2 Ltd.

As sole shareholder of HS2 Ltd, the SoST is responsible for ensuring that the company is managed in the interests of the public and the taxpayer and for holding the company to account regarding its performance and governance. In carrying out these and other duties regarding the HS2 programme, the SoST is supported and advised by Department of Transport officials.

Due to the singular nature of the HS2 programme, and the fact that HS2 Ltd is not a contractor in commercial sense, the relationship between the SoST and HS2 Ltd is both special and crucial. They are bound by a legal Development Agreement to deliver the project as agreed but, perhaps more importantly, by a non-binding Framework Agreement – agreed between the DfT and HS2 Ltd – which sets out what the parties can expect from each other based on trust, mutual respect, communication and evidence-based assurance.

21.4.5 HS2 Ltd

HS2 Ltd is a government-owned company, incorporated in January 2009 and also has the status of an executive non-departmental public body. This means that it can undertake commercial activities but is not part of the DfT or any other government department and operates more-or-less at arm's length from ministers of the Crown.

HS2 Ltd is sponsored by the DfT which manages relationships between the Department and the company according to a Code of Good Practice thereby ensuring *purpose, assurance, value and engagement*. This partnership is defined in a Framework Document (Department for Transport 2018) – which is regularly reviewed and updated. The governance arrangements for HS2 Ltd comply with Managing Public Money (HM Treasury 2022), thus ensuring high standards of probity and harmonious working with Parliament.

The company also works under the terms of a Development Agreement entered into with the SoST which creates the legal relationship between the DfT and HS2 Ltd. This is a contractual agreement but not a construction contract in the generally accepted sense – i.e. there is no offer, acceptance or consideration[3] as there would be in a 'normal' engineering/construction contract. This agreement sets out the rights and obligations of the parties.

HS2 Ltd is distinguished from a contracting company because:

- It is responsible for project delivery but is not a commercial enterprise and it is non-profit making and subject to public audit.
- It is a non-departmental public body that operates at arms-length from the DfT, but it is not fully autonomous.
- Funding for the HS2 project is provided by the government through the SoST whose department (the DfT) sponsors the project.
- It is nonetheless an incorporated company, with its own legal identity, and can therefore enter into commercial contracts for the design and construction of the railway with other companies.

HS2 Ltd has one shareholder – the SoS for Transport – but has no share capital as it is funded directly by the DfT. Accounts for the year ending March 2023 state that the company had assets of £21.36 billion, £128.6 million of cash at bank and 1941 whole-time equivalent employees. It is classed as a large company.

The liabilities of most incorporated companies are limited by shares such that each owner of the business is liable for the debts of the company up to the value of the shares held. HS2 Ltd, however, is limited by guarantee meaning that the debts/liabilities of the company will be funded by the public exchequer.

As an incorporated company, HS2 Ltd is a legal entity capable of entering into commercial contracts with its delivery partners such as the Joint Ventures who are Main Works Civils Contractors (MWCC) for Phase 1.

21.4.6 Formal Agreements

The relationship between the SoS for Transport (who is represented by the DfT) and HS2 Ltd is determined by two principal agreements:

1. **Framework document**: This is effectively a partnership agreement which conveys no legal powers or responsibilities. It sets out the requirements and expectations of HS2 Ltd as an arms-length public body and includes issues such as appointments to its board of directors, risk management protocols and expectations for financial management and the control of expenditure for the project.
2. **Development agreement**: This is a contractual agreement, without consideration, which determines HS2 Ltd's role in the design, construction and operation of the railway and the role of the SoST as sponsor and client.

21.4.7 Development Agreement

Most construction projects – even very large ones – are delivered by a contractor to a client on the basis of a price or estimate of final cost, an agreed delivery date or programme of work and a formal contract (either a bespoke or standard contract) which sets out the terms and conditions of the agreement and the rights and obligations of the parties.

Arrangements for the delivery of HS2 are different, however, as it is one of the largest infrastructure projects ever undertaken in the UK.

The size and complexity of the project necessitates a bespoke delivery organisation which is why HS2 Ltd was incorporated. Its role is that of a facilitator with the expertise to arrange the planning and design of the new high-speed railway and to integrate the hundreds of component parts that make up the project such as tunnels, bridges, viaducts, and stations.

The Development Agreement defines two distinct programmes for which HS2 Ltd is responsible: the Core Programme and the Wider Programme.

21.4.8 The Core Programme

The Core Programme is defined in the Development Agreement Recital A as:

- The High Speed 2 project.
- The design, procurement, construction, commissioning, operation and maintenance of a new Railway, and includes:
- All the work and functions to be carried out by HS2 Ltd in accordance with the Agreement.

The Development Agreement also states that the Core Programme will deliver the benefits set out in the DfT's Business Case for the project which comprises the strategic case, financial case, management case, commercial case and economic case.

21.4.9 The Wider Programme

The SoST has initiated further objectives for the project defined in Recital B of the Development Agreement as the Wider Programme.

This is to ensure that HS2 Ltd helps to deliver benefits beyond the Core Programme in respect of the government's objectives for growth and regeneration resulting from the HS2 programme.

21.5 Funding and Governance

Funding for HS2 comes from 'grant-in aid' directly from the Government. This money is paid through the 'sponsor department' – in this case, the DfT – to support the costs of HS2 Ltd.

Whilst expenditure is subject to Parliamentary controls, HS2 Ltd has a high degree of autonomy as it operates at 'arms-length' from the DfT which has little involvement in the detailed day-to-day control of operations. A high level of trust is the basis of the partnership between the two bodies but a complex system of governance has also been established to ensure delivery of the completed railway and make sure that the expected project outcomes are realised.

21.5.1 Spending Oversight

The budget for HS2 is set by the government, consistent with its overall spending plans, based on estimates provided to it by the DfT and HS2 Ltd. Spending, and progress, is scrutinised on a regular basis via six-monthly reports to Parliament and to the House of Commons Public Accounts Committees where representatives of the DfT, HS2 Ltd and the High Speed Rail Group are questioned by Members of Parliament.

Further oversight is exercised by the National Audit Office (NAO) which scrutinises public spending for Parliament independently from government and the civil service and helps Parliament hold government to account.

HS2 Ltd is, therefore, routinely held to account for its stewardship of HS2.

21.5.2 Programme Governance

A project of the scale and importance of HS2 requires a complex and robust system of controls to ensure that the desired project outcomes are realised and that the spending of enormous sums of public money is closely monitored, even if costs cannot be entirely controlled.

Figure 21.2 illustrates the HS2 governance structure, although it must be remembered that arrangements may change from time to time to suit the needs of the programme and

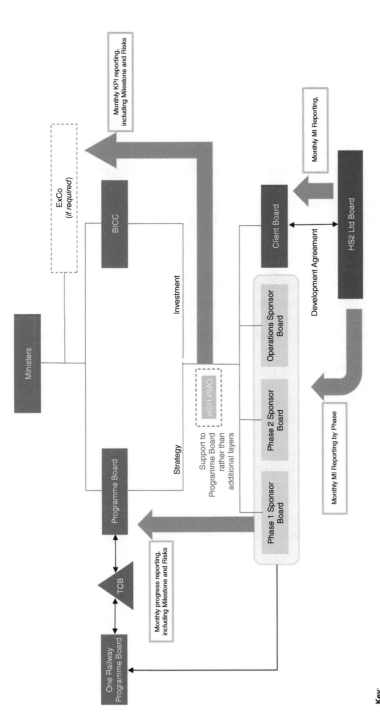

Key:
Programme Board – for strategic decisions
BICC (now Investment, Portfolio and Delivery Committee) – for investment decisions
One Railway Programme Board
  • Part of DfT
  • Coordinates the existing railway with HS2
TCB – Tripartite Cooperation Board
  • Made up of DfT, Network Rail, HS2 Ltd
  • Provides strategic advice to Programme SRO and Programme Board
HSR PMO – High-Speed Rail Programme Management Office – supports the Senior Responsible Owner (SRO)
Client Board – the primary channel of communication between the SoS and HS2 Ltd
HS2 Ltd Board – responsible for ensuring that HS2 Ltd fulfils its obligations under the Development Agreement

Source: Department for Transport (2017)/Crown copyright/OGL v3.0.

Figure 21.2 HS2 governance.

in the light of political decisions. For instance, the DfT Board Investment Commercial Committee (BICC) has been renamed the Investment, Portfolio and Delivery Committee (IPDC) and Phase 2 of the project has now been scrapped.

The DfT is sponsor of HS2 and is responsible for funding and ensuring that the expectations from the completed railway are achieved. Within the DfT, the HS2 programme is led by the Senior Responsible Owner (SRO) – the DfT Director General for High-Speed Rail – who chairs the High-Speed Rail Programme Board. This board makes strategic decisions concerning the project, and the BICC makes investment decisions. For major decisions, approval is sought from HM Treasury.

The DfT has a Project Representative within HS2 Ltd who provides the Department with independent assurance that both the project and programme are deliverable, affordable and represent value for money.

An Integrated Programme Management Office (IPMO) has been established to coordinate the management of the programme between HS2 Ltd, DfT and Network Rail.

21.5.3 Programme Governance at HS2 Ltd

HS2 Ltd is the nominated undertaker responsible for programme delivery and acts in an advisory role to the DfT by providing advice as part of the development of the high-speed rail network. As such, the company advises the Secretary of State on matters relating to HS2 Ltd, and on the development and delivery of the high-speed rail network as set out in the Development Agreement.

Although the relationship between HS2 Ltd and the DfT is contractual in terms of the Development Agreement, it must be remembered that HS2 Ltd is not fully autonomous in the sense of an independent limited liability company. It is an executive non-departmental public body which operates at arms-length from the DfT.

HS2 Ltd is managed by the HS2 Ltd Board which comprises an executive board of directors and a number of non-executive directors with extensive senior-level experience of very large infrastructure project delivery. The Board has corporate responsibility for ensuring that the company fulfils the remit, aims and objectives set by the DfT and for ensuring the organisation is fit for purpose.

The extensive experience of both executive and non-executive members of the Board provides both HS2 Ltd and the DfT with access to the strategic guidance needed to ensure effective delivery of the HS2 project.

21.6 Roles and Cooperation

HS2 is not a commercial project – it is a public sector investment in rail infrastructure paid for with public money. As such, government ministers are responsible for seeing that the stated project outcomes are delivered and, in simple terms, HS2 Ltd exists to ensure that the project is built in accordance with statutory requirements and with the Development Agreement signed with the SoST.

It is clear from Figure 21.2 that the main hub of control for the HS2 programme focuses around the interaction between the Board of HS2 Ltd and the DfT Client Board. This relationship centres around the Development Agreement, which defines the contractual arrangements between the SoST and HS2 Ltd, and the reporting protocols between HS2 Ltd, the Client Board and the Sponsor Boards.

As with any contract, the Development Agreement states the terms and conditions agreed by the parties and includes *inter alia* undertakings to:

- Act in good faith and cooperate with all interested third parties and in a reasonable and transparent manner (Clause 7.1).
- Act in good faith and promptly give early written notice to the other party if it becomes apparent that there is any default, delay or breach by either party which could adversely impact the project (Clause 7.2).

The Development Agreement defines the roles and responsibilities of the parties to the agreement and also identifies other bodies appointed by the SoST to undertake roles defined in the Development Agreement. Defined roles under the Development Agreement are, therefore:

- The signatories to the Agreement:
  - The SoST and
  - HS2 Ltd.
- The Client Board.
- The Programme Representative.

The SoST, as represented by the DfT is, of course, responsible for the overall management of the HS2 Programme and HS2 Ltd is responsible for its delivery and for acting in an advisory role.

In addition to these defined roles, sandwiched between the SoST and HS2 Ltd is a raft of boards, panels and advisors all of whom play a role in the monitoring, control and governance of the project in order to protect the public interest and make sure that the project is successful, bearing in mind its scale, cost, timeline and complexity. This includes:

- The Senior Responsible Owner (SRO).
- The Programme Board.
- DfT BICC.
- The HS2 Ltd Board.
- The Sponsor or Project Boards.

Further to this governance infrastructure are two individuals who play important roles in the financial governance of the programme:

- The Principal Accounting Officer.
- The HS2 Ltd Accounting Officer.

21.6.1 Secretary of State for Transport

The responsibilities of the SoST are:

- To act as funder and sponsor of the Core Programme.
- To set the Sponsor's Requirements and approve changes to the Sponsor's Requirements as appropriate.

- To be accountable for the delivery of the benefits of the Core Programme.
- To keep the Business Case under review and up to date.

21.6.2 HS2 Ltd

As a government company and non-departmental public body, HS2 Ltd is managed by a board of executive and non-executive directors appointed by the SoST. The board is responsible for ensuring that HS2 Ltd fulfils its obligations to the DfT under the Framework and Development Agreements. The company is advised at corporate level by the HS2 Ltd Executive Committee and, at programme level, by various project boards (the original phases 1, 2, etc. as shown on the diagram). The project boards focus on day-to-day management of specific areas of the programme.

The main decision making, executive and managerial bodies at HS2 Ltd are the executive management team and the board. In this regard, HS2 Ltd has extensive responsibilities under the Development Agreement including:

- Delivery of the Railway.
- The execution and completion of the Works.
- Acting as a Proxy Operator of the railway until a Train Operating Company (TOC) is franchised and to subsequently manage the infrastructure of the railway.

21.6.3 The Client Board

It can be seen in Figure 21.2 that the Client Board and the Board of HS2 Ltd are directly linked by the Development Agreement, i.e. the contractual agreement for the delivery of HS2. Consequently, the Client Board is an intermediary and primary channel of communication between the SoST and the management of HS2 Ltd who are the signatories to the Agreement.

In a sense this is a similar role to a contract administrator in a construction contract whose duties include overseeing the performance of the contractor, ensuring that contractual obligations are fulfilled and reporting on progress to the client.

The Client Board acts in a mainly monitoring and overseeing role but has no executive powers under the Development Agreement in contrast to those of an architect or project manager. In all probability, however, having the ear of the Cabinet Minister responsible for HS2 must bring with it considerable status and authority in its dealing with the HS2 Board.

21.6.4 Programme Representative

The programme representative is appointed by the SoST to act as his/her agent and:

- To monitor the management, development and delivery of both the Core Programme and the Wider Programme.

- To review and challenge the proposals and performance of HS2 Ltd in respect of costs, time, quality and organisation.
- To work collaboratively with the SoST and HS2 Ltd to provide independent and objective assurance and advice.
- To monitor the management, development and delivery of both Programmes and give advice and assurances to the SoST. It is a collaborative role but able to review and challenge the proposals and performance of HS2 Ltd regarding costs, time, quality and organisation.

This is an overseeing role aimed at providing the SoST with assurance and advice that the development and delivery of HS2 programme is meeting requirements, is affordable and is providing good value for money. The programme representative works collaboratively with the SoST and HS2 Ltd.

21.6.5 Senior Responsible Owner

The DfT's Director General for the High-Speed Rail Group is SRO of the HS2 Programme whose responsibilities include:

- Ensuring delivery of the HS2 programme.
- Securing the necessary resources for this to happen.
- Being accountable to various Parliamentary Select Committees.
- Ensuring the effective governance of the programme and making sure that assurance and programme management arrangements are effective throughout the life of the programme.
- Reporting to the HS2 Programme Board and to various project boards and working groups in order to ensure that risks, issues and changes are effectively managed.

21.6.6 Programme Board

The Programme Board is not defined in the Development Agreement but is nonetheless a key high-level board in the governance of HS2 (DfT 2016).

It is instrumental in facilitating the strategic management of the Programme and in retaining oversight of the Wider Programme as well. The HS2 Programme reports to the Programme Board for strategic decisions.

The Board supports the HS2 Senior Responsible Owner – the Director General for the High-Speed Rail Group – who leads the HS2 Programme. The SRO chairs the Programme Board.

21.6.7 DfT Board Investment Commercial Committee

The BICC – which has been renamed as the IPDC[4] is the second high-level board to which the HS2 programme reports, in this case, regarding investment decisions.

For major investment decisions, HM Treasury approval is sought.

21.6.8 The Sponsor or Project Boards

The Phase 1, Phase 2 (now scrapped) and Operations Sponsor Boards operate at programme level and are responsible for managing day-to-day activities of the HS2 project. This includes work on property acquisitions and preparing for major construction activity.

The interaction between the Project Boards, the HS2 Ltd Board and the High Speed Rail (HSR) Programme Board can be seen in Figure 21.2.

21.6.9 Principal Accounting Officer

The Principal Accounting Officer – a Permanent Secretary to the DfT – is accountable to Parliament regarding the issuing of capital or other resources to HS2 Ltd, for advising the responsible minister and for monitoring and reporting on HS2 Ltd activity and spending.

21.6.10 HS2 Ltd Accounting Officer

Designated by the Principal Accounting Officer, the HS2 Ltd Accounting Officer is the Chief Executive Officer of HS2 Ltd who is personally responsible for safeguarding the public funds in his/her charge.

21.7 Order of Cost

History proves that it is notoriously difficult to predict the cost of mega-projects, and it is no surprise that, in March 2019, HS2 Ltd formally advised the DfT that it would be unable to deliver the HS2 Phase One programme on time or within available funding.[5]

21.7.1 National Audit Office Reports

The NAO has made it clear in its four reports to date that the government, HS2 Ltd and the DfT have significantly underestimated the risks associated with such an enormous and ambitious project and that estimates of cost and delivery dates had been over-optimistic.

The fourth NAO report (National Audit Office 2020), published in January 2020, also examined whether the DfT and HS2 Ltd had safeguarded value for money in their stewardship of the project so far and what the risks to value for money would be as the programme proceeds.

The NAO concluded that there were flaws in the way that HS2 Ltd was managing its cost model and that it had underestimated the risks and complexity of the programme which confirmed that Phase 1 was likely to exceed available funding and be delivered late.

21.7.2 Parliamentary Reports

A report to a Commons Select Committee in 2021 (HC 329 2021) confirmed that the cost of the construction element of Phase 1 is now on a firmer footing but that the scale of work remaining on the project means that future costs are uncertain.

Despite the best efforts of HS2 Ltd to control the budget, empirical evidence would suggest that an out turn cost for HS2 of £100+ billion is likely. This was confirmed in 2019 by the then Prime Minister Boris Johnson[6] but this pronouncement was made prior to the Oakervee Review, the Integrated Rail Plan and prior to the resulting changes to the scope of the project announced in 2023.

Further concern about the spiralling costs and delays to progress were expressed by the House of Commons Public Accounts Committee in May 2022.[7]

21.7.3 Funding Envelopes

The most reliable method of estimating the cost of construction work is the 'bottom-up' method which relies on contractor pricing. In the early stages of a project, this is not possible, however, due to the lack of a completed design upon which to base quantities and prices. As a consequence, top-down methods of estimating are employed, which are less reliable, until the design is sufficiently advanced.

This is long-established industry best practice for order of cost estimating which begins with a 'high-level' estimate usually expressed as a cost range or 'cost bracket'.

HS2 is no different and project delivery best practice aims to set an initial cost range for proposed works which is then refined and narrowed down over time into a target cost. For HS2 the estimated cost range is referred to as a 'baseline'. Developing baseline costs for HS2 is an iterative process which is subject to governance and independent assurance controls.

21.7.4 Oakervee Review

The Oakervee Review was critical of HS2 Ltd's cost estimates but accepted that, at the date of publication of the Report in early 2020, current cost estimates for Phase 1 were more reliable than previous baselines as they were based on contractor pricing. However, estimates of cost and time for Phase 2 were less reliable as this part of the project was less mature than Phase One. This is now academic, of course.

At the time of the Oakervee Review, separate benchmarks had been prepared by an external consultant and by the DfT. Evidence provided to Oakervee by Network Rail suggested that the external consultant's cost estimates for major construction works were too low and those for rail systems were too high. The alternative approach by the DfT relied on outturn data from previous projects (multi-project benchmarking), rebased to bring it up to date, for works such as tunnelling and earthworks.

Other benchmarks, developed by HS2 Ltd, were based on comparisons with the costs of high-speed lines in other countries. Oakervee concluded that, whilst baselines were a useful tool, they had not been applied on HS2 in a consistent manner and that HS2 Ltd and the DfT could do more to ensure uniformity on the HS2 project.

21.7.5 Forecasting Costs

It is easy to criticise the budgeting and estimated time-lines for HS2 but mega-projects are always beset by problems with forecasting and there are many contributory factors that make the control of costs and time so difficult including:

- The project timescale, which is measured in decades.
- Competition for resources with the wider construction industry.
- At the peak of construction activity hundreds of live projects will be underway, many of which will be very large projects in their own right.
- World-wide political and economic uncertainty.
- The time required for parliamentary approval, public consultation, statutory planning approval and decision making.
- The cumulative effect of risk and uncertainty attached to aggregation of multiple major projects into one mega-project.
- Inflation, interest rates and the cost of borrowing huge sums of money over a protracted period.
- Supply chain uncertainty attached to availability of resources in the world-wide marketplace especially following a world pandemic and the war in Ukraine.
- The availability of labour and skilled trades.
- The fragile and risk-averse domestic construction industry in the wake of the collapse of Carillion and others.

21.7.6 Contingency

As with most projects, budgeting for HS2 includes a contingency figure to cover risks and uncertainties. This contingency is in two parts:

- **Delegated contingency**: A sum of money included in the target cost which HS2 Ltd has discretion to spend as required or add to with cost savings achieved through efficiency gains, savings in land and property acquisition and procurement savings.
- **Retained contingency**: A sum of money retained by the government that is additional to the delegated contingency but is not included in the target cost.

The provision of a delegated contingency is clearly a sensible arrangement, as delivery of the project within target cost would not be realistic without providing HS2 Ltd with some breathing space to pay for unforeseen eventualities.

The additional contingency in the form of government-retained money provides the project with latitude to cover major risk events or scope changes that would be outside the remit of HS2 Ltd.

The DfT provides Parliament with six-monthly reports (DfT 2022) concerning costs, programme and other matters such as community impact, environment and project governance.

21.7.7 Schedule

Parliamentary approval to begin work on Phase 1 of the project was given in February 2017 which initiated a raft of scheme design and preparatory work. This 'enabling work' included ground investigations, conservation work, archaeological investigations, utilities diversions and road improvements.

Following a period of detailed design work, approval to commence full detailed design and construction work was given by the DfT to HS2 Ltd which issued a NtP to the four MWCC on 4 September 2020.

21.7.8 Oakervee Review

The Oakervee Review in 2020 expressed concern that uncertainty remained over the schedule for the HS2 programme and that schedule ranges being considered by HS2 Ltd and the DfT are wider than those announced by the Chairman of HS2 Ltd[8] in August 2019. This was confirmed in the 2021 NAO Report which stated that:

- Partial Phase 1 passenger services between Old Oak Common and Birmingham Curzon Street should commence between 2029 and 2033.
- Full Phase 1 services from London Euston were forecast to start between 2031 and 2036.
- This is five years later than originally planned.

It is clear from the Oakervee Review that slippage in the schedule, and indeed the cost of the HS2 programme, are a function of both a flawed contracting model and over specification, and 'gold-plating' of HS2's design. Oakervee also suggested that more efficient designs and less ambitious design standards and specifications, resulting in less engineering work, could be adopted *without major risk* and but only within the limitations of the Phase 1 Act.

21.7.9 Phase 1 Programme

The programme for Phase 1 is approximately eight years from commencement of site clearance to completion of railway installation (HS2 2017). The key work stages are:

- **Environmental mitigation**: species translocation, ground investigation and archaeological surveys.
- **Site clearance**: site possession, set up site compounds, site clearance, enabling works.
- **Earthworks**: cuttings, embankments, tunnelling.
- **Civil engineering**: track bed preparation, bridges (over or under the railway), tunnels, viaducts, retaining walls and stations.
- **Railway installation**: install the railway systems, including ballast, slab, tracks, signalling and the power supply. This stage will also include the final finishes to station buildings.
- **System testing and commissioning**: four years for the railway to be fully tested to ensure it can operate safely and reliably.

Within the eight-year timescale, active construction at each location along much of the route will be around two years where there are no major structures prior to railway installation. Where there are major structures and stations along the route, construction will take much longer.

The tunnels into London are expected to take up to five years to complete. After the tunnels have been bored, there will be a period of tunnel fit-out including installation of track, overhead lines and electro-mechanical services (such as ventilation fans and communication equipment).

21.8 Project Design

The design of HS2 is a long and complex process involving years of preliminary design work leading up to the submission of hybrid bills to Parliament and thence to scheme designs (RIBA Stage 3) for Schedule 17 applications and approvals. Once approved, designs can then be tendered in order to appoint design and build contractors who complete the detailed design.

Throughout this process HS2 Ltd's large team of directly employed architectural, engineering and landscape designers play an important role by working with local authorities, private developers, design consultants – such as station and interchange scheme designers – and contractors. This involvement helps to ensure that the design and specification of stations, viaducts, bridges, tunnels, etc., meet statutory and Schedule 17 requirements and remain consistent with HS2's Design Vision.

21.8.1 Design Vision

Published in 2015 and developed by HS2 Ltd in collaboration with industry leaders and infrastructure specialists, the HS2 Design Vision underpins all facets of the design of HS2 and firmly establishes *the role that design can play in making High Speed Two a catalyst for growth across Britain.*

At the core of the Design Vision are three principles each aiming to drive design excellence:

- **People**: design for people to enjoy and benefit from.
- **Place**: design to create a sense of place and space that support quality of life.
- **Time**: design to stand the test of time.

Each of the three core principles has three subsets which, as a complete vision, *aim to enhance the lives of future generations of people in Britain by designing a transformational rail system that is admired around the world.*

The Design Vision underpins all HS2 design standards, guidance and contractual and technical requirements. All design partners and contractors have to assure that this is taken into account in undertaking their design work which is monitored by HS2's in-house designers who act as Design Vision guardians.

21.8.2 Design Panel

The design of HS2 has been closely monitored by a Design Panel since its inception in 2015. The panel is funded by HS2 but is fully independent whilst working closely with HS2. It has some 40 members from an architectural, engineering, landscape and public background. The panel is also involved in relation to design proposals submitted by design consultants prior to the presentation of hybrid bills.

The HS2 Design Panel accords with the government's National Infrastructure Strategy (HM Treasury 2020) which expects *all infrastructure projects to have a board-level design champion in place by the end of 2021 at either the project, programme or organisational level, supported where appropriate by design panels.*

Design champions are also to be found in HS2's in-house design teams such as lead architects and landscape and urban design managers. A Design Panel Handbook explains how the HS2 Independent Design Panel works (HS2 Ltd 2016).

21.8.3 Design Review

The Design Review Panel is mandated by the DfT to provide independent expert advice and act as an non-executive critical friend to HS2's designers.

The rigorous design review process involving the HS2 Design Review Panel advises and holds to account HS2 leaders, project teams and partners and helps them to make the right design choices consistent with the project Design Vision (HS2 Ltd 2017).

Despite having no executive powers, the Design Review Panel has influenced the design of structures, buildings and landscapes along the line of the route and has also guided wider aspects of the project such as procurement strategies, operational design, urban design and integration and sustainability.

21.8.4 Design Strategy

As a government sponsored programme, naturally HS2 has Building Information Modelling (BIM) and digital construction at its core (Infrastructure and Projects Authority 2016). However, the scale and complexity of HS2 is such that industry-leading digital technology has been developed to build the project twice – once virtually and once in reality. The HS2 design strategy consists of four pillars:

- Building Information Modelling – BIM Level 2.
- Asset Information Management System.
- Geographic Information System (GIS).
- Visualisation Hub.

The four pillars make up the HS2 digital engineering model which is designed to capture huge amounts of data which can be stored, retrieved, reused and shared with all designers and contractors involved in the programme. This enabled designers to create a virtual reality version of HS2 – a digital model of physical reality. It also means that real-time data can be fed into the model to inform construction phase decision making.

Figure 21.3 shows examples of the life-like animations possible with the model which uses photo realism, virtual and augmented reality and sophisticated visual and aural

High-speed train emerging from a 'green' tunnel.

Colne Valley viaduct – designed to resemble stones skipping over water.

Old Oak Common station

Figure 21.3 Design animations. Photographs courtesy of HS2 Ltd.

representations to create a digital version of the railway to aid the decision making, planning and consultation process.

The model reflects the hundreds of individual 'assets' that make up HS2 – track, tunnels, bridges, stations, interchanges, depots, etc. – and the large quantities of data that each individual 'asset' generates during its design and construction. The idea is to create a 'one-truth' model to be used and shared throughout the many hundreds of designers and contractors involved in the programme.

21.8.5 Design Stages

The HS2 project was effectively initiated in January 2009 when HS2 Ltd began operating and when it commissioned Arup[9] to conduct route option studies for Phase 1 from London to the West Midlands. At the end of 2009 HS2 Ltd submitted a confidential report to ministers to enable a decision to be made on the future of the high-speed railway and provide a basis for possible future public consultation.

Further route option studies followed, and Phase 1 preliminary engineering designs were prepared in readiness for submission of the first hybrid bill in November 2013.

Arup later engaged in the crucial task of modelling exactly what HS2 would look like, sound like and how it would operate before it became a physical reality. HS2 Ltd has defined the following design stages for the purposes of managing the programme of design work for each phase of the project:

Stage		
1	Parliamentary design	Route option and engineering design sufficient to enable the Hybrid Bill to be submitted to Parliament and to achieve Royal Assent.
2	Specification design	For design and construction contracts.
3	Employer requirements design	Scope for construction contracts.
4	Scheme design	RIBA Stage 3 (Outline design) to obtain Schedule 17 and other approvals and permissions and sufficient to form the basis of bids for design and build contracts.
5	Detailed design	Detailed design for manufacture and construction following the appointment of design-build contractors.

Source: HS2 Ltd.

21.8.6 Design Critique

The Oakervee Review (2020) addressed itself, *inter alia, as* to why the costs of HS2 had *escalated over time* and how its design and specification had *influenced the pricing….. received from contractors*.

The Review concluded that… *there has been considerable over-specificaton and 'gold plating' in HS2 contracts with much of the design seemingly done on a worst-case, risk-averse scenario*. The Review considered this to have *been a key driver behind the inflated prices on Phase One*.

Oakervee also remarked that:

- *Stations are a core part of the HS2 project* as they act as a gateway for passengers into city centres.
- *More needs to be done to drive further value* from them and that the assumption that stations should be built entirely at public expense should be challenged.
- Local authority-private sector partnerships would maximise economic value from stations and help mitigate cost and schedule overruns.
- The integration of stations with existing transport networks and the local areas they serve should be at the forefront of station design.

21.8.7 Station Design

Since its inception in 2015, the HS2 Design Review Panel has been active in supporting HS2's in-house design teams and external design consultants, in collaboration with local planning authorities and other stakeholders, to move station designs to the RIBA Stage 3 – *Spacial Coordination* (Scheme) stage and thence to Schedule 17 approval.

External design consultants were awarded contracts for the four HS2 Phase 1 stations in early 2018 following a competition undertaken the preceding year:

- **Birmingham Curzon Street**: WSP UK (working with Grimshaw Architects).
- **Birmingham Interchange**: Ove Arup & Partners International (working with Arup Associates and Wilkinson Eyre Architects).

- **Old Oak Common**: WSP UK (working with Wilkinson Eyre Architects).
- **London Euston**: Ove Arup & Partners International (working with Grimshaw Architects).

The design process for the stations is complex. They are large £multi-million projects in their own right and have to be integrated with the design of the new high-speed railway line which they will serve. Additional complexities include designing around existing road and rail infrastructure and integrating local communities.

For the Birmingham stations, for instance, design ambition documents were produced. For Curzon Street, virtual reality was used to convey the scale and arrangement of the developing design and an integrated 3D model was produced integrating the architects' station design model with that of the main works civils designers.

21.9 Procurement

HS2 Ltd employs around 1900 whole-time equivalent staff – mainly from its headquarters in Birmingham. However, to design and build HS2, more than 400 000 jobs are now supported by the project including private-sector consultants, architects, engineers, surveyors, contractors and subcontractors.

HS2 Ltd in-house staff play an important management and supervisory role, working with consultants and contractors, contributing to design reviews, liaising with local authorities and other stakeholders and championing HS2's Design Vision, values and culture.

As might be imagined, a mega-project such as HS2 needs to procure a vast array of products and services to complete a working railway, as illustrated in Table 21.1. For each procurement category, HS2 Ltd is placing small numbers of high-value direct contracts with Tier 1 suppliers who will source lower-tier contracts directly. This procurement hierarchy is illustrated in Figure 21.4.

Table 21.1 Procurement

Procurement category	Type of work
Design and professional services	• Architectural, civil and structural engineering, landscaping, geotechnical, land survey, etc.
Civil engineering	• **Enabling works** – demolition, archaeology, ecology • **Main civils works** – tunnels, cuttings and embankments, drainage, bridges, viaducts, etc.
Stations	• **Phase One** – Euston, Old Oak Common, Birmingham Interchange and Birmingham Curzon Street.
Railway systems	• Track, overhead catenary, telecoms, traction power, signalling, mechanical and electrical systems.
Rolling stock	• Trains.
Corporate	• Commercial services, assurance, regulatory and business support services

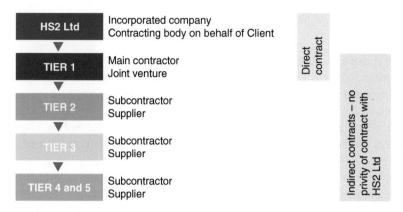

Figure 21.4 Phase 1 framework. Source: Adapted from HS2 Ltd.

21.9.1 Design Procurement

For the hybrid bill stages of HS2, a Development Partner is appointed to advise HS2 Ltd and to manage and supervise HS2 consultants. Other professional service companies (PSCs) are then appointed to undertake the design, engineering, environmental and land referencing work necessary for the hybrid bill submission (Blight 2018). Appointments are made, largely, via a competitive process which results in a formal contract. This team works in conjunction with HS2 Ltd staff which ensures that HS2 Ltd's values and collaborative working culture are embedded in the process.

Following Royal Assent, designs are then progressed to the scheme design stage when tenders can be invited for the detailed design and build stages.

For the design of the station developments at Euston, Old Oak Common and two in Birmingham – which will be central to HS2 – invitations to tender (ITTs) for the station design services contracts were issued. Invitations were then released to shortlisted bidders to participate in dialogue (ITPD) for the appointment of a Euston Master Development Partner.

21.9.2 Construction Procurement

The placing of contracts for the civil engineering and construction aspects of HS2 begins with invitations to tender (ITT). This precedes the shortlisting of potential bidders – commonly a shortlist of three – following which invitations to participate in dialogue (ITPD) are issued. Other categories of HS2, such as the placing of contracts for rolling stock, may be procured differently.

For construction works contracts, the procurement strategy is a two-stage design and build method with the ensuing contract based on the New Engineering and Construction Contract (ECC) Option C: Target price with activity schedule:

- Stage 1 of the procurement process focuses on:
    - Developing the design from scheme stage.
    - Maturing the schedule.
    - Considering cost and risk issues.
    - Agreeing a target price.

- Stage 2 fixes the Stage 2 target prices and proceeds to the detailed design and build stages of the contract.

HS2 Phase 1 operates a procurement Framework providing over 400 000 supply chain contract opportunities to Tier 1, Tier 2 and other suppliers. Tier 1 JV members of the HS2 supply chain are engaged directly by HS2 Ltd in a competitive process with the eventual winners contracted on the basis of New Engineering Contract (NEC) ECC Option C: Target price with activity schedule. There is no privity of contract between HS2 Ltd and the Tier 2 and lower contracts which are placed directly by the Joint Venture partners, as illustrated in Figure 21.4.

All Tier 1 suppliers are mandated to use CompeteFor (www.competefor.com/hs2) to advertise for tendering opportunities and to cascade this requirement through their own supply chains. There is an HS2 e-procurement portal (https://hs2.bravosolution.co.uk) powered by Jaggaer which enables interested parties to respond to tender opportunities as they arise.

Underpinning the HS2 programme are the four core values of safety, respect, leadership and integrity which operate within a culture founded on sustainability, collaboration and innovation. Supply chain partners are expected to share HS2's values and contribute to building the culture in order to jointly deliver the programme of work.

21.9.3 Labour and Plant Procurement

A project such as HS2 places enormous resource demands on the construction industry for skilled labour, management and supervisory staff and training and for plant and vehicles including leading-edge equipment. Local subcontractors and plant hire firms are used as a matter of HS2 policy, but these resources are in finite supply.

Consequently, a collaborative group of companies have formed the Connect Alliance[10] which, along with agencies and plant hire firms from all round the country, help to fill the gaps.

The Balfour Beatty Vinci JV operates a collaborative labour desk with support firms such as Fortel, McGinley Support Services and VGC Group Ltd which also delivers price work projects to the railway and construction industry.

21.9.4 Forms of Contract

The forms of contract used for the design and construction elements of HS2 come from the New Engineering Contract family of documents, principally:

- EEC Professional Services Contract.
- EEC Option E: Cost reimbursable contract.
- EEC Option C: Target cost contract with activity schedule.

The contracts for the pre-hybrid Bill design were let under an ECC NEC Option E (Cost reimbursable) professional services framework under which service delivery contracts were placed for design, engineering, environmental and land referencing work (Blight 2018).

For Phase 1 procurement, there is a Main Works Civils Contracts (MWCC) framework agreement for joint venture contracts, based on Option C, which includes design and build and gain/pain share provisions.

21.9.5 Payments

In accordance with UK government policy, HS2 Ltd operates a project bank account arrangement meaning that contractual payments throughout the project supply chain are assured and regular and less prone to insolvency in the higher tiers of the framework. The ECC Option Y(UK) 1 – Project Bank Account applies.

No retention monies are held by HS2 Ltd and, consequently, contractual payments are made in full. ECC Option X16 is therefore not used in HS2 contracts.

21.9.6 Procurement Critique

The procurement arrangements for HS2 were criticised by Oakervee who suggested that inflated prices on the Phase 1 MWCC were a result of HS2 Ltd's procurement and contracting approach.

In particular, the target cost model with gain/pain elements – designed to incentivise the contracts – was thought to place all the risk on the contractors with consequent impact on pricing levels for the initial draft target prices received.

Oakervee advised that HS2 should endeavour to reach agreement with the MWCC such that acceptable Stage 2 target prices – with a reasonable level of value engineering – might be fixed before the NtP with Stage 2 is issued. Failing this, Oakervee suggested that these contracts might be re-procured at the expense of time delays but with the benefit of cost savings.

Oakervee also criticised the contracting model used to engage design consultants on the Phase 1 MWCC. It was felt that the designers should be working directly for HS2 Ltd as opposed to the contractors and that this has led to risk-averse designs that are not cost-effective.

21.10 HS2 Works Areas

The high-speed route between London and Birmingham is over 200 km long and, therefore, for logistical, procurement and construction reasons, it makes sense to subdivide it into manageable 'chunks'. HS2 Phase 1 is therefore divided into three sectors – Area South, Area Central and Area North, as illustrated in Figure 21.1.

21.10.1 Enabling Works

Prior to issuing the Notice to Proceed (NtP), HS2 Ltd placed enabling works contracts to cover the entirety of Phase 1 of the project. Almost £1 billion worth of contracts were placed with three joint ventures:

- **Area South**: CS JV (Costain, Skanska).
- **Area Central**: Fusion JV (Morgan Sindall, BAM Nuttall, Ferrovial).
- **Area North**: LM JV (Laing O'Rourke, J Murphy & Sons).

The works included:

- Establishing site compounds.
- Archaeology, ecology surveys and site clearance.
- Utility diversions, drainage work, demolitions.
- Highway realignment.

Examples of the work undertaken include:

- Bulk earthworks in connection with archaeological investigations.
- Demolition and site clearance at station sites such as Euston and Birmingham Curzon Street.
- Demolition of over bridges during rail possessions.
- Roundabout relocation.
- M42 and A446 bridge replacements.
- Creation of new accesses to stations.

21.10.2 Main Civils Works

Following on from the enabling works, the work in each Phase 1 area is referred to as Main Civils Works. This work is divided into 'lots'. Thus, Lots C1 and C2 represent two lots or packages of work in the central area of Phase 1.

Stations, interchanges, depots, etc., are a separate entity to the civils works and are tendered separately.

The Civils Works lots are as follows:

- **Area South:**
  - **S1**: Euston Tunnels and Approaches.
  - **S2**: Northolt Tunnels.
- **Area Central:**
  - **C1**: Chiltern Tunnels and Colne Valley Viaduct.
  - **C2**: North Portal Chiltern Tunnels to Brackley.
  - **C3**: Brackley to South Portal of Long Itchington Wood Green Tunnel.
- **Area North:**
  - **N1**: Long Itchington Wood Green Tunnel to Delta Junction and Birmingham Spur.
  - **N2**: Delta Junction to West Coast Main Line (WCML) Tie-In.

Four Tier 1 joint ventures have been appointed to undertake the main civils works in these areas with the central area split between two separate JVs. A considerable amount of work is now underway in all areas on over 350 individual sites. The joint venture companies directly contracted by HS2 Ltd are:

- **Area South**:
  - **S1**: SCS JV – Skanska, Costain, STRABAG.
  - **S2**: SCS JV – Skanska, Costain, STRABAG.
- **Area Central**:
  - **C1**: Align JV – Bouygues, VolkerFitzpatrick, Sir Robert McAlpine.
  - **C2**: EKFB JV – Eiffage, Kier[11], Ferrovial, BAM.
  - **C3**: EKFB JV – Eiffage, Kier[11], Ferrovial, BAM.

- **Area North**:
  - **N1**: BBV JV – Balfour Beatty, VINCI.
  - **N2**: BBV JV – Balfour Beatty, VINCI.

21.11 Construction Management

The huge cost of HS2 and its long timescale are both witness to the size and complexity of the project which draws upon the combined resources of some of the largest contractors in the UK and Europe. The HS2 programme comprises hundreds of individual projects – some very large and others relatively small – which combine to make the completed railway.

21.11.1 Area Organisation

Area North typically illustrates how the main civils works for HS2 are organised, procured and constructed.

HS2 Area North is a 90 km stretch of high-speed track and comprises a considerable number of structures along with some 40 million m^3 of bulk earthworks. The Area is made up of two main work packages – lots N1 and N2 – both of which were awarded by HS2 Ltd to the same joint venture contractor – BBV (Balfour Beatty/Vinci Joint Venture). Area North is subdivided into eight sub-lots which are allocated to three distinct delivery functions:

- Major structures:
  - Delta junction and spur to Birmingham Curzon Street station.
- Tunnels:
  - Twin-bore tunnels at Long Itchington Wood and Bromford.
- Mainline:
  - Major earthworks, new highways and structures.

Each sub-lot constitutes a major contract in the order of £350–500 million.

21.11.2 Area North Joint Ventures

The MWCC for HS2 have been awarded to four joint ventures. The JVs are Tier 1 partners in the HS2 framework and are contracted to HS2 Ltd under the NEC Option C (Target contract with activity schedule) form of contract.

The JVs are responsible for developing the RIBA Stage 3 (Outline) design and for delivering the completed works within the target price, allowing for the value engineering pain/gain arrangement built into the contract.

In Area North, for instance, there are two principal joint venture partners responsible for delivering lots N1 and N2:

- **Main Works Civils (50:50)**:
  - **Balfour Beatty**: the largest contractor in the United Kingdom.
  - **Vinci Group**: the largest contractor in Europe comprising Vinci Construction Grands Projets, Vinci Construction UK (Taylor Woodrow) and Vinci Construction Terrassement.

- **Design (50:50)**:
  - Mott MacDonald.
  - Systra.

Despite having contractual liability for delivery of lots N1 and N2, the joint venture partners work as part of an integrated project team with HS2 Ltd, in which HS2 Ltd staff work in a close collaborative unit directly with the JV team. This is an unusual arrangement where client staff work closely with JV staff but illustrates the cooperative project culture set by HS2 Ltd.

Some Tier 2 suppliers are also joint ventures such as the SB3JV between Bachy Soletanche and Balfour Beatty Ground Engineering. They are supplying geotechnical, foundation engineering, piling and ground improvement services to the Tier 1 JV BBV.

21.11.3 Administrative Support

The line management functions of the Tier 1 JVs are assisted by a central 'core' of administrative support which carries out important functions in support of construction activities, including:

- Commercial.
- Project controls.
- Health and safety.
- Environmental and
- Logistics.

21.11.4 Construction Planning

Conceptually a railway is a simple means of transport with trains travelling on steel rails laid at relatively flat gradients between stations located at various places along the route. In the case of High-Speed 2, the normal limiting gradient is 2.5%, which occurs at various locations along the route.

There are, however, some instances of steeper gradients along the route where there are physical constraints that do not allow gradients of 2.5% or less to be achieved. This is the case between Euston Station and Euston Tunnels (3.4%), in the Euston Tunnels before Old Oak Common (3.5%), at the east end of the Bromford Tunnel near Birmingham (2.9%) and at the approach to Birmingham Curzon Street (3.03%).[12]

Gradient is not the only factor that influences the choice of route and its design, however, and topography imposes requirements for tunnels, cuttings and embankments, retaining walls, viaducts, bridges, culverts and the like along the route. Such civil engineering works punctuate construction of the permanent way (i.e. the railway line) and influence the order and sequence of the work. It might appear, consequently, that construction of the railway is a somewhat haphazard process but there is a certain logic to what is being constructed and when.

Other factors that influence the works programme include creating access to isolated rural locations where a tunnel or viaduct is needed, for instance and design changes that might require statutory approval.

An example of this is the Water Orton location near Birmingham where additional consents have had to be sought to change the wording of the High-Speed Rail (London-West Midlands) Act 2017 because of a change in design (HS2 Ltd 2022). Schedule 1 of the Act states that Work No 3/157 will be 'partly on viaduct' but this has now been re-designed to be in tunnel – the Bromford Tunnel.

21.12 Construction Methods

The HS2 programme is made up of hundreds of individual projects of different sizes, each with its own complexities and challenges. A wide variety of construction methods are being adopted for the construction of HS2 including a number of innovative methods as described below.

21.12.1 Long Itchington Wood Tunnel

Along with the neighbouring Ufton Wood, Long Itchington Wood is an ancient woodland in Warwickshire dating from before 1600 AD. The woodlands are a Site of Special Scientific Interest (SSSI) with complex ecosystems established over hundreds of years.

To preserve this natural heritage, Long Itchington Wood is one of several tunnel drives along the HS2 route. The tunnel is a twin bore tunnel 1 mi (1.6 km) long, driven in two stages by a single tunnel boring machine (TBM). The 2000 tonne TBM – named Dorothy – is 125 m long with a 10 m diameter cutter head weighing 160 Te. The first bore was completed in July 2022 and the second in March 2023.

Dorothy's eight-month first bore started at the tunnel's North Portal. After breaking through the wall of the reception box at the South Portal, Dorothy was relaunched for the second bore (also from the north portal) which took a remarkable four months to complete.

Figure 21.5 illustrates the extensive preparatory and temporary works required for the tunnel bore which includes assembling the TBM 'train', erecting the reaction frame and establishing the waste material recycling plant.

Propulsion of the TBM relies initially on a steel reaction frame as illustrated in Figure 21.5 which also provides support for other parts of the TBM train, including the waste removal pipes and conveyors. The reaction frame is anchored to the ground and is designed to resist the thrust and torque forces developed as the TBM moves forward. When sufficient rings have been installed to resist these forces, the TBM's hydraulic rams act against the concrete rings instead.

The TBM runs continuously with teams working shifts. The hydraulic rams push the TBM forward excavating the ground at the tunnel face. The cutterhead then stops to allow an annulus of precast concrete rings to be placed in position and bolted together. The cutterhead then moves forward again.

The choice of TBM depends upon the geology of the tunnel drive. Earth pressure balance (EPB) TBMs, for instance, are more suited for cohesive soils such as clay, whilst slurry TBMs are used for cohesionless soils such as sand and the mudstone found at Long Itchington Wood. Consequently, the earth face in this tunnel needed to be conditioned with bentonite

Tunnel North Portal showing:
- Excavated cutting.
- Diaphragm wall.
- Concrete base slab for the two tunnel drives.

TBM Dorothy in place for the first tunnel drive south.
Initial delivery of precast concrete segments in lay down area.
Slurry treatment plant in foreground.

TBM Dorothy being prepared for first tunnel drive.
Scaffolding in place as temporary works for fixing the start-up shield.
Behind Dorothy is the steel reaction frame and part of the following 'train'.

Figure 21.5 Long Itchington Wood. Photographs courtesy of HS2 Ltd.

Tunnel North Portal and TBM

Key:
Background right – Tunnel North Portal
Foreground – Slurry treatment plant
Mid-left – Overhead conveyor

Conveyor at Great Union Canal crossing

Disposal zone showing end of conveyor and earthmoving plant – loading shovel and articulated dump truck

Figure 21.6 Spoil disposal Long Itchington Wood tunnel. Photographs courtesy of HS2 Ltd.

and water for easier removal. This suspension is injected at the tunnel face from the logistics section to the rear of the front shield and the excavated material is removed by a slurry circulation system. The slurry treatment plant and waste removal conveyors are shown in Figure 21.6.

The annular gap (overbreak) between the outside of the ring segments and the earth face is filled with grout which is pumped through the tail skin of the TBM shield or through holes in the tunnel lining segments. This provides a bed for the tunnel lining segments and stabilises them in position.

Remobilising the second drive at the north portal necessitated moving the cutter head and other very heavy pieces of the TBM – such as the front and middle shields and tail skin – 2.5 mi by road. This was accomplished by specialist teams, cranes and low-loaders working at night. Some parts of the TBM were brought back through the completed tunnel and reassembled at the north portal.

There are three phases to building a tunnel:

1. **Preparatory phase**:
   a. Prior to bulk excavation down to the invert of the tunnel, diaphragm walls (D-walls) are constructed to form the north and south portals. Glass Fibre Reinforced Polymer (GFRP) reinforcement in the D-walls makes TBM launch and breakthrough easier.
   b. Following bulk muck shifting, the D-walls are stitch-drilled where the TBM will start/breakthrough and a reinforced concrete base slab is constructed at the launch site to provide an invert for the TBM. A steel reaction frame is built to facilitate initial forward motion and a startup shield is installed in the portal diaphragm wall where excavation of the tunnel commences.
   c. Access roads, a batching plant and slurry treatment plant are also part of the tunnelling enabling works.

2. **Tunnelling phase**:
   a. A rotating cutter head removes the excavated material at the tunnel face – a total of around 500 000 Te of mudstone. This works thanks to hydraulic rams acting on the tunnel lining behind the cutter head which push it against the earth face thereby facilitating excavation of the tunnel.
   b. Behind this a protective tube or 'shield'.
   c. For initial propulsion, the TBM relies on the steel reaction frame and then the rams act against the segmental rings once sufficient complete sections have been built.
   d. The excavated mudstone/bentonite slurry is removed to the rear of the 'train' in slurry pipes and on to the on-site slurry treatment plant. Here, the bentonite slurry is filtered, the solids are removed and fresh bentonite is added to the slurry which is then pumped back to the cutter head.
   e. Spoil is then removed along a 254 m long conveyor to be used as fill for embankments along the route of the railway.

3. **Ring building phase**:
   a. Protected by the shield skin, the segmental rings are set in place by the tunnel segment erector. The rings are moved forward from the rear of the train on wheeled transporters, offloaded and stacked ready to be assembled into a ring.
   b. A complete ring comprises eight 2 m-wide segments, each weighing up to 8 tonnes.
   c. When a complete segmental ring is assembled and jointed together, the cutter head is then able to push off and move forward to excavate the next section of tunnel.

Behind the cutter head is the long logistics section of the TBM which contains the central control unit from which the cutter head direction, elevation, earth pressure balance and other tunnelling parameters are controlled.

On completion of the Long Itchington Wood tunnel, the shield was dismantled from the TBM. Dorothy was then transported to another tunnel project near Birmingham where she was adapted and fitted with a new shield for a smaller diameter tunnel. The rest of the TBM train was pulled back to the start of the tunnel, where it was disassembled, ready for transportation.

21.12.2 Colne Valley Viaduct

When completed, the Colne Valley Viaduct northwest of London will be the longest railway bridge in the United Kingdom. The viaduct is more than 3.4 km (2.1 mi) long, part of which crosses a series of lakes and waterways.

The project is being led by main works contractor Align JV which comprise Bouygues Travaux Publics, Sir Robert McAlpine and VolkerFitzpatrick.

The viaduct comprises a total of 56 piers, 11 of which are constructed in water within steel sheet piled cofferdams. Construction of the conventional piers is illustrated in Figure 21.7 which also shows the temporary jetties across the lakes and construction of the V-pier foundations in cofferdam. This is undertaken from four temporary jetties which provide a working platform for piling and other plant. The jetties also provide a means of transporting away excavated material from within the cofferdams rather than using local roads.

Within the lake, the pier foundations consist of six bored piles between 1.5 and 1.8 m in diameter over 60 m deep, installed under bentonite slurry. The piling rig sits on the temporary jetty with jetty panels removed for access. The pier foundations are completed with an in-situ reinforced concrete pile cap.

Each of the 11 V-piers is constructed in situ using special forms, as illustrated in Figure 21.8. The V-shaped piers require approximately 520 m^3 of concrete poured over a nine-hour period using concrete pumps. On top of each V-pier are two separately poured in-situ concrete segment-on-pier structures which resemble the shape of the precast concrete deck units that complete the viaduct. The viaduct over the lakes is designed to resemble stones skipping over water, as illustrated in Figure 21.3.

Following on behind construction of the piers, a 160 m long 700 Te launching girder – Dominique – lifts the concrete deck segments into position to form the arches of the viaduct, as illustrated in Figure 21.9.

The launch girder sits on top of two completed piers and installs deck units simultaneously as two half-arches using the cantilever principle to balance the structure. One half-arch is installed back towards the previously installed half-arch, and the other half-arch is installed towards the next pier. Once each arch section is complete, the machine moves itself forward to the next pier to begin the following arch. A total of 1000 post-tensioned deck segments weighing up to 140 tonnes each will complete the viaduct.

Each deck unit has a standard width but the shape beneath is slightly different to accommodate the gently curved design of the viaduct.

The modular deck units are made on site at a temporary purpose-built factory nearby using the 'match-casting' technique. This is a process where each segment is poured against the previous one, ensuring a perfect fit when the deck is assembled on the piers and was chosen for quality, safety and efficiency reasons. A peak output of 12 segments can be cast each week.

676 | 21 High Speed 2

Placing concrete by pump in a conventional viaduct pier.
Operatives are working from a safe working platform with staircase access from ground level.
The working area is well fenced with plant and vehicles segregated from pedestrain areas.

The temporary jetty across the lakes. This comprises a road system and working platform extended at each V-pier location.
Each V-pier has its own dedicated luffing-jib tower crane.

Steel sheet piled cofferdam at a V-pier location.
Rebar is in place ready for concreting the pile cap foundation.
A mobile crane is used for moving materials.
The jetty and all working areas are fenced off due to the surrounding water hazard.
Scaffolded staircase access is provided into the cofferdam for fixing rebar and concreting.

Figure 21.7   Colne Valley Viaduct – 1. Photographs courtesy of HS2 Ltd.

21.12 Construction Methods | **677**

Falsework and temporary formwork in place for construction of a V-pier.
Fixing of rebar is under way.
High standards of access provision, edge protection and safe working platforms are evident.
The tower crane base is integral with the foundation pile cap.

The V-pier is complete and falsework is in place for constructing the in-situ concrete segment-on-pier deck (SOP) units.
The two SOPs have been concreted and formwork is being stripped for the next use.
The precast concrete deck units connect either side of the SOPs.

The deck launching girder is in position ready to place the deck units to form the two half-spans away from the central pier.
The rear half-span at the previous pier is in place. The completed deck provides access for delivery of the deck units from the precasting yard.

Figure 21.8 Colne Valley Viaduct – 2. Photographs courtesy of HS2 Ltd.

Precast concrete deck units under construction in nearby purpose-built factory.
Modular methods of construction (MMC) are being used employing the 'match-casting' technique for precision.

This animation shows how the deck units are installed.
The launch girder sits over two completed piers for stability.
Inside the main girder is a second retractable girder which places the units simultaneously either side of the forward pier.
This completes the rear half-arch and creates a new half arch for the next span.

Precast concrete deck units being placed by the lauching girder.
Service ducts and holes for the stressing tendons can be seen.
Each deck span is the same width but the cross section below varies according to the shape of the viaduct.

Figure 21.9 Colne Valley Viaduct – 3. Photographs courtesy of HS2 Ltd.

21.12.3 Old Oak Common Station

Oak Common is the largest railway station ever built in the United Kingdom. It is HS2's main transport hub connecting the Midlands and North, South Wales and the Southwest of England. Every train to and from London will stop at Old Oak Common which will provide connections to London and Heathrow airport via the Elizabeth line (Crossrail).

Balfour Beatty VINCI SYSTRA joint venture was awarded the HS2 Old Oak Common station construction management contract for circa £1 billion.

The station is being constructed on a brownfield site that requires remediation before work commences. All excavated material from the remediation and station construction is removed by articulated dump trucks to a nearby 3 km long conveyor which takes the waste to the logistics hub at Willesden. From here, the material is carried by rail to three locations around the country for use in land reclamation. The scale of the construction work and the eventual station is illustrated in Figure 21.10.

Old Oak Common station is being constructed in three main phases:

1. 6 new high-speed platforms below-ground serving HS2 trains.
2. 8 new platforms at surface level serving the Great Western Main Line (GWML) and Heathrow Express services.
3. The station building which will be capable of handling up to 250 000 passengers per day.

HS2 trains will arrive at Old Oak Common from London in tunnel – eventually – and, from the north, in cutting. The high-speed platforms will, therefore, be at a considerable depth below-ground level. For this reason, the platforms are being constructed within a station 'box', the southern end of which will form the portal for the eventual tunnel drive to central London at the proposed new Euston station.

The station box is an underground structure 850 m long × 70 m wide and over 15 m deep to track level. This is a huge structure, considerably bigger than previous boxes constructed at Stratford station for HS1 and at Canary Wharf on the Elizabeth line. It is so big that the contractors have divided the work into three sections, the east box and the west and central boxes. Construction of the station box is illustrated in Figure 21.11.

The perimeter of the box is constructed with 30 m deep diaphragm walls (D-walls) within which large diameter bored piles and steel plunge columns support the reinforced concrete structure above track level.

The D-walls comprise 275 panels, constructed in alternate bays approximately 7 m long, with earthwork support provided by a bentonite slurry which is pumped out, filtered and reused. The rebar cages are lowered in by mobile crane and concrete is placed by tremie pipe.

The internal station structure is being built using the top-down construction method. This requires 161 bored piles of differing diameters to be inserted 57 m into the ground within the station box. The pile bores commence at ground level with some 15 m of empty bore before a rebar cage is lowered into place. The piles are then concreted, and steel plunge columns are inserted which provide superstructure support and ground heave resistance as illustrated in Figure 21.11. The piling work is completed in sections at a rate of one completed pile per day using a Bauer BG39 rotary drilling rig.

At surface level, huge in-situ concrete beams are constructed across the box which provide lateral support for the diaphragm walls and tie in to the concrete ground slab. There are large permanent apertures in the slab which allow plant to be craned in to work at the lower levels of the box and will also allow natural light into the completed platforms below. The lateral beams are heavily reinforced and are built on soffite formwork on falsework with proprietary wall formwork forming the sides.

The high-speed station box is under construction between existing surface lines. Working next to a live railway poses a major safety hazard. The new station will cover both the HS2 stations and those of the conventional lines, illustrating the enormity of this project.

Part of the station box is illustrated which is being constructed in three sections – east, west and central.

Figure 21.10   Old Oak Common – 1. Photographs courtesy of HS2 Ltd.

As excavation proceeds beneath the ground slab – which now acts as a roof – temporary slabs are constructed at intermediate levels below. The slabs provide a working platform for the construction of reinforced concrete transverse beams. The temporary slabs are broken out as the work proceeds.

21.12 Construction Methods | 681

Excavating diaphragm wall to station box.
Bentonite slurry tanks, pumps, filters and pipework are shown.
Excavated material is loaded into ADTs for delivery to the logistics hub at Willesdon Euroterminal.

Excavation of the station box down to platform level.
Plunge columns can be seen emerging from the bored piles.
Tracked excavators and loading shovels move the excavated material to where the long-reach excavator is stationed.
The attached grab bucket loads the waiting ADTs.

Formwork erection and rebar fixing in progress at an intermediate level of the station box.
The concrete slab is temporary works to provide a working platform for casting the transverse beams. A steel plunge column can be seen.

Figure 21.11   Old Oak Common – 2. Photographs courtesy of HS2 Ltd.

Excavation – over 700 000 m³ of London clay – is carried out by tracked excavators and loaders lowered in by crane and the excavated material is removed from the box by long-reach excavators fitted with clamshell buckets. The material is then removed by conveyor to the logistics hub at Willesden Euro terminal rail head.

21.13 Innovation

With a project of the scale and complexity of HS2, innovative solutions to problems are needed. HS2 is being built in the centre of two of the UK's biggest and busiest cities; it is crossing a swathe of highly sensitive rural landscape and villages, it is generating huge quantities of excavated and demolition materials, and it demands very large quantities of construction materials and resources. Central to these problems is the UK government's decarbonising commitment to net-zero emissions by 2050.

Among the initiatives undertaken by HS2 are:

- Creation of a logistics hub.
- The use of conveyors for earthworks disposal.
- Reduction of Heavy Goods Vehicle (HGV) impact on roads.
- Diesel-free construction sites.
- Construction of 'green' tunnels.

21.13.1 Logistics Hub

The Skanska Costain STRABAG JV (SCS JV) is contractor for the 26.4 km (16.5 mi) HS2 route from Euston to West Ruislip. This stretch includes three 8.8 m internal diameter twin-bore tunnels lined with 400 mm thick precast concrete segments at a total length of 21 km (12.9 mi) and comprises 22 operational sites. The SCS JV is also responsible for interfacing with Euston and Old Oak Common stations and with their respective contractors.

During several years of tunnelling activity, along with construction of ventilator shafts and head-houses, spoil will be removed to a 30-acre logistics hub strategically located just north of the Old Oak Common station site. The spoil was initially transported by road, but a 1.7 mi-long network of conveyors, serving three major sites – including Old Oak Common station – now removes material directly to the logistics hub at Willesden for removal by rail. The conveyor is capable of removing up to 10 000 tonnes of material per day.

Hub operations are coordinated from a refurbished building at the Willesden site using bespoke 3Squared BulkSmart software. This facilitates the planning and monitoring of train movements in the yard and tracks the progress of trains to and from the site.

A peak of six locomotives and more than 200 purpose-built wagons will eventually be used to remove over 5.6 million tonnes of spoil from HS2's London sites via the hub at Willesden. The trains will remove spoil for reuse on a housing development at Barrington in Cambridgeshire, to Cliffe in Kent and to Rugby.

In addition to the removal of spoil, a further train will deliver construction materials including precast concrete tunnel lining segments which are being manufactured by Spanish company Grupo Pacadar at its factory on the Isle of Grain in Kent.

21.13.2 Conveyors

The tunnelling operations at Long Itchington Wood will generate some 500 000 tonnes of mudstone and soil with an additional 250 000 tonnes of material coming from excavations for the railway cutting. This is a large quantity of spoil which is estimated to be the equivalent of around 30 000 HGV journeys. To reduce the impact on local roads, and to reduce carbon emissions, the excavated material is to be processed at an on-site treatment plant and then removed by two conveyors to a disposal zone.

The treatment plant and conveyors are located at the North Portal end of the tunnel – which is why the direction of the two bores is north to south.

Spoil generated at the TBM cutter head is removed as a slurry, and this is pumped to the treatment plant. Once at the treatment plant, the slurry passes centrifuges and shaker beds to separate the stones, sand and fines and most of the liquid. The remaining material is then transported by the first conveyor to a temporary disposal zone. This material is then moved by articulated dump trucks to a second, longer, conveyor. This 254 m long conveyor can move large quantities of spoil and is fully enclosed to prevent spillage as it passes over the Grand Union Canal from the treatment plant to the second disposal zone for removal off site.

The process can be seen illustrated in Figure 21.6 which shows the location of the tunnel portal and its cutting, the slurry treatment plant and the conveyor.

Once at the disposal zone, the processed material is transported to other locations along the HS2 route for reuse, creating noise bunds and for landscaping and filling for building embankments.

Following completion of the work at Long Itchington Wood in early 2023, the conveyor was dismantled and re-erected at Water Orton to undergo a further two years of service as part of a 1200 m long conveyor which will remove hundreds more lorries from the roads every day.

21.13.3 Access Roads

Access roads, and haul roads for highway and pipeline projects, are common temporary works in construction. They are often constructed with excavated material topped with a layer of stone hardcore or sub-base. On highway projects, the capping layer – part of the road foundation – is frequently used as a haul road.

In the central area of HS2, some 30 million cubic metres of material is to be excavated from cuttings for building embankments and for landscaping. This large-scale earth-moving operation will obviously take place over a considerable distance and so a series of local haul roads have been constructed with compacted stone for use by heavy earth-moving equipment and articulated dump trucks.

For other road-licensed site traffic – vans, materials deliveries, concrete trucks, etc. – HS2 Ltd's partner EKFB JV has constructed an 80 km (50 mi) long temporary access road linking HS2 construction sites across Buckinghamshire, Oxfordshire, Northamptonshire and South Warwickshire. This is a metalled (tarmacadam) road, which closely follows the route of the new high-speed railway. It is designed to carry up to 500 vehicles per day and will be used for moving materials, equipment and workforce to various sites along the predominantly rural route quickly and efficiently. The road will also relieve the congestion on local roads.

Upon completion of the construction work, the road will be removed, although some sections may be repurposed to provide local transport links. The land will be reinstated, landscaped and planted with new grassland and woodland.

21.13.4 Diesel-Free Site

In May 2022, HS2 Ltd announced its first completely diesel-free site. This is a tunnel vent shaft site situated at Canterbury Road, South Kilburn, in London.

The site features a range of diesel-free technologies and greener equipment which have been introduced by HS2's civil contractor Skanska Costain STRABAG joint venture (SCS JV). These innovations include:

- One of the UK's first 160 tonne emissions-free fully electric crawler cranes.
- The use of biofuels (Hydrogenated Vegetable Oil) to power plant and machinery on site.
- An electric compressor and
- Site access to mains power on a 100% renewable energy tariff.

HS2 Ltd aspires to reduce the carbon footprint of its programme with the stated aim that HS2 will be net-zero carbon from 2035 and that all its construction sites will be diesel-free by 2029.

Other carbon-reduction initiatives are planned including the use of retrofit technologies, biofuels, hydrogen and solar power.

21.13.5 Green Tunnels

Figure 21.12 illustrates the construction process for one of the five 'green' tunnels being constructed on HS2 Phase 1 – this one is at Chipping Warden. This is the first time that this form of construction has been used in the UK although it is common in France. Specialist engineers Matière – a French company specialising in off-site manufactured precast concrete and modular construction – is carrying out the tunnel installation on site for EKFB JV.

The tunnels are designed such that there is a specially shaped 'porous portal' at either end which will reduce the noise of trains entering and exiting the tunnel. The precast concrete units are being supplied by British manufacturer Stanton who have constructed new production sheds, together with casting and storage areas, at their factory in Derbyshire to accommodate the additional work required for this project.

The Chipping Warden tunnel is 2.5 km (1.5 mi) long with 1040 completed segments comprising one central pier, two side walls and two roof slabs – a total of 5020 precast units in all. Each unit is steel reinforced with the largest weighing up to 43 tonnes. It is being built in sections – i.e. not a continuous tunnel – with construction expected to be fully complete in 2024.

As with many of the individual contracts that make up the HS2 programme, enabling works have been necessary at Chipping Warden where the first part of a relief road has been constructed in order to take HS2 vehicles – and other local traffic – away from the centre of the village. This will later be extended, to take the A361 over the top of the green tunnel.

21.13 Innovation

Off-site pre-casting of tunnel units

Earthworks & preparation of cutting

Lifting of wall unit onto concrete slab

Completion of single arch

Erection sequence of twin tunnels

Fixing rebar for track bed

Figure 21.12 Green tunnel. Photographs courtesy of HS2 Ltd.

The green tunnels are constructed using the 'cut and cover' technique. The method requires a cutting to be excavated with sloping/battered sides and the excavated spoil is set aside for reuse. Once the structure has been constructed the tunnel is backfilled with excavated material and the surface is reinstated to original levels and planted with indigenous trees/shrubs.

The 'green' elements of the method are:

- No spoil is removed from site.
- The tunnel components are factory made.
- Components are delivered to site 'just-in-time'.
- The twin arch 'M' shape is more efficient than a standard box structure thus reducing the amount of concrete required and the consequent carbon impact.
- Site erection of the tunnels is simplified to just five precast components.
- Plant and labour requirements are significantly less than for traditional methods.

- CO_2 emissions are considerably less.
- The natural environment is treated sympathetically and restored.

The green tunnel sequence of operations is:

1. Excavation of cutting and on-site disposal to temporary spoil heaps.
2. Preparation of sub-base and laying waterproof membrane.
3. Lay over-site concrete platform.
4. Lift wall units with integrated foundation pad into place.
5. Install curved roof sections.
6. Fix rebar for track bed.
7. Concrete invert slab (rail track bed) using site batched concrete.

21.14 Epilogue

There are many salutary to be learned from the HS2 project.

However, setting aside political controversy, the blight that the project has brought to some peoples' lives, and questions over suffocating bureaucracy, bloated and inefficient management and gold-plated designs, there is no doubting the many outstanding technical achievements and high standards of site safety and public engagement that characterise HS2.

This is testimony to the work of thousands of skilled and dedicated people and to an endlessly resourceful and inspirational construction industry.

The many people and organisations involved in HS2 may well be encouraged by the possibility that the 2024 change of government could result in Phase 2 being revived. Watch this space!!

References

I.C. Blight (2018) *HS2 Railway UK – Route development to Hybrid Bill, A Collaborative Approach*. ICE Publishing.

Department for Transport (2018) *Framework Document – Moving Britain Ahead*.

DfT (November 2016) *High Speed Rail Phase 2b, Strategic Outline Business Case – Management Case*.

DfT (April 2020) *Full Business Case High Speed 2 Phase One*.

DfT (March 2022) *HS2 Ltd and Andrew Stephenson MP, HS2 6-monthly report to Parliament*.

HC 329 (2021) *House of Commons Committee of Public Accounts*.

HM Treasury (November 2020) *National Infrastructure Strategy*.

HM Treasury (March 2022) *Managing Public Money*.

HS2 (2017) *HS2 High Speed Two Phase One Information Paper D6: Hs2 Phase One Construction timetable Version 1.3*. Last updated 23rd February 2017.

HS2 (August 2019) *HS2 Chairman's Stocktake*.

HS2 Ltd (2016) *HS2 Independent Design: Case Studies*. https://www.gov.uk/government/publications/hs2-design-case-studies.

HS2 Ltd (April 2017) *HS2 Design Vision*.
HS2 Ltd (20th January 2022) *Application for an Order under the Transport and Works Act 1992*.
Infrastructure and Projects Authority (2016) Government Construction Strategy 2016–20, March 2016.
National Audit Office (24 January 2020) *High Speed Two – A Progress Update, The Comptroller and Auditor General*
Network Rail (2009) *Meeting the Capacity Challenge: The Case for New Lines*. Network Rail
Oakervee, D.E. (February 2020) *Oakervee Review*. DfT.

Notes

1. https://www.hs2.org.uk/what-is-hs2/.
2. https://assets.publishing.service.gov.uk/government/uploads/system/uploads/attachment_data/file/1062157/integrated-rail-plan-for-the-north-and-midlands-web-version.pdf.
3. The elements normally required in a legal contract.
4. https://www.gov.uk/government/publications/government-major-projects-portfolio-accounting-officer-assessments/high-speed-2-hs2-phase-2a-accounting-officer-assessment-june-2017.
5. NAO.
6. https://www.birminghammail.co.uk/news/midlands-news/boris-johnson-hs2-cost-more-16653605.
7. https://committees.parliament.uk/oralevidence/10231/default/.
8. HS2 Chairman's Stocktake, August 2019.
9. https://www.arup.com.
10. Connect alliance is led by A-Plant, L Lynch, P Flannery Plant Hire and Fortel and also includes Ainscough Crane Hire, Selwood Pumps, MGF Excavation Safety Solutions, Morson Group, VGC and Citrus Training.
11. Kier replaced Carillion on Lots C2 and C3 following its insolvency in 2018.
12. UK Parliament, High Speed 2 Line, Question for Department for Transport, UIN HL2393, 5 October 2015.

Index

a

A533 bridge replacement project 619–640
  background 619
  bridge abutments and central pier 633–634
  bridge deck 634–636
  construction programme 631–640
  demolition 623, 624, 636–640
  detailed design development 627–631
  enabling works 621, 623
  existing bridge 619–620
  4D BIM 630, 631
  key milestones 631–632
  location 619
  off‑site works 623
  organisation 624–626
  permanent works 623, 627–628
  procurement 620–621
  programme 626–627
  road closures 632–633
  scope 621–624
  site layout 622
  site plan 620
  temporary works 628–630
acceleration
  constructive 539
  expedite 539
  project 539–549
  pure 539
access date 422–423
access roads 683–684
accidents 58, 76, 374
  book 417
  cost of 417
  frequency rate 411
  incidence rate 411
  RIDDOR 416–417
  working at height 378
accruals 566, 567

activity
  duration 425–427
  float 529, 530
  key 424–425
  ranking 542
  schedule 175, 222–223, 254, 255
additive manufacturing 161
ad hoc approved lists 176
admeasurement contracts 203, 204
agreement
  contractual 192–196
  framework 176–178
  HS2 project 649
alliances 180–181
APM Competence Framework 117
approvals 267
Approved Codes of Practice (ACoPs) 390
approved lists 175–176
archive drawings 327
arrow diagrams
  comparison of 296–302
  example 291–293
  precedence and 282, 291, 294
  principles 290–291
as‑built programme 435, 453
as‑planned programme 530–532
Association for Project Management (APM) 94
association for project management body of
  knowledge (APMBOK) 94
Asta Powerproject 283, 284, 427
attendances 135, 578
audits 411–412
augmented reality (AR) 129–130
Australian construction contracts 200
Automatic resource levelling 587

b

bar chart 427, 520–523
  colour coding 520, 521

Construction Planning, Programming and Control, Fourth Edition. Brian Cooke and Peter Williams.
© 2025 John Wiley & Sons Ltd. Published 2025 by John Wiley & Sons Ltd.

bar chart (*contd.*)
   delay analysis   535, 536
   earned value   522–523
   programme   467–470
   progress tracking   520, 522
baseline programme   437–440, 531
basements   333
Bennett (Construction) Ltd v CIMC-MBS Ltd (2019)   616
Bentley Synchro   125
bid
   documentation   253–257
   management   255
   preparation   251–252
   submission   253
Bidcon   225–226
bidding   213, 248–257
bid manager   70, 71
bill of quantities (BQs)   69, 136, 175, 359, 425, 426
Board Investment Commercial Committee (BICC)   655
bottom-up construction   336–338
bottom-up estimating   217, 218, 241–248
box culvert   330–331
Bragg Report   323
British Airports Authority (BAA)   63, 104
British Standards Institute (BSI)   126
Bryden Wood   155
BS5975:2019   323, 324
budget   490
   basis of   492
   estimate   229
   labour   493–495
   plant   493, 496, 497
   preliminaries   493, 495, 497–499
   presentation   491–492
   reason for   490–491
   types of   491
budgetary control   489
   reporting   490
   software   492
builders' merchant   50
builders' quantities   242–243, 254
Building a Safer Future (BSF)   106
Building Cost Information Service (BCIS)   231, 235, 269
building information modelling (BIM)   4, 20, 112, 123
   A533 bridge replacement project   630, 631
   Asta Powerproject   124, 125
   benefits   124–125
   definition   124
   downsides   128
   4D   423
   Hull hotel project   603
   Level 3   127–128
   levels   126–127
   managers   128–129
   measure   222
   models   223, 519
   project planning   423, 424
   standards   125–126
Build UK   52–53
business units   100–102

c

calendars   428
capital expenditure   491
cash flow   458
   budget   491
   client   459, 460
   contractor   459, 461
   credit   479
   improvement   476–477
   late payment   480
   money in and money out   477, 478
   payment delay   479
   retention   477
   risk   59
cash flow forecasting   459–461, 470–475
   credit terms   471–472
   earned value analysis   461–470
   illustration   471
   payment terms   473–474
   preparation   470
   retention   472
CATO   226
CDM Regulations   401
   application   402
   client duties   403–404
   construction phase plan   402, 406–408
   designer duties   404
   duty holders   403
   health and safety file   408
   notification   402, 403
   principal contractor duties   404
   principles of   405
   site rules   408
   structure   402, 403
   summary   402
   welfare facilities   405
Certificate of Training Achievement (CTA)   409
change(s)   501
   Accelerating Change (Egan Report)   24, 26
   acceptance   112
   control   516–517

Index

cultural 28
design 9
management 94, 193, 438
method statements 360–361
in procurement methods 242
safe system of work 399
charter 180
Chartered Institute of Building (CIOB) 4, 6, 81–83, 95, 266
  survey 9, 12, 15
  Time and Cost Management Contract 198–199
CIMC Modular Building Systems (CIMC-MBS) 605, 609, 617
  CIMC modular erection team 609
  design stage 606–607
  installation 609
  mass manufacture 607–608
  modular procurement 605–606
  payments 608, 609
  prototype 607
civil engineering 159–161
claims 16
client-led design and build 184, 186
clients 44–46
  cash flow 459, 460
  CDM Regulations 403–404
  definition of 45
  funding 46
  influence of 51–52
  method statements 348–349
  private sector 44, 45
  professional advisers 45
  project manager 106–107
  public sector 45
Clients' Charter 52
Cloud based reporting 519–520
collaboration 109–114
Colne Valley Viaduct 675–678
commercial risk 71
commitment 25
common data environment (CDE) 110–111, 501
comparative cost planning 239–240
completion 78, 79, 422–423
  obligations 209
  time for 512
complexity 10–11, 103
Complex Projects Contract 2013 (CPC2013) 16
component-based systems 146, 155
computer-aided design (CAD) 124
concrete monitoring 164–165
concurrent delay 526, 528
Construction Act 195, 473, 616
Construction Clients' Group 52

Construction (Design and Management) Regulations 43–44, 76, 601
construction firms 46–51
  large firms 47
  micro firms 47
  SMEs 47
  in UK 46
  very large contractors 48
construction industry
  classification 40–41
  clients 44–46
  culture 36–40
  customs and practices 36–37
  leadership 51–53
  output 46
  reports 13–14
  reputation 4
  sectors 42
  types of work 44
Construction Industry Board 21, 24
Construction Industry Council (CIC) 22, 45, 408
Construction Industry Scheme 583
Construction Industry Training Board (CITB) 30
construction inflation 238, 239
Construction Leadership Council (CLC) 30, 38, 39, 51
construction management 133–139, 189, 191
construction management software 112–114, 492
Construction Manager at Risk (CMAR) 202
construction manager (CM) 202
construction market 215
construction method statements 354–355
construction period 269
construction phase plan 402, 406–408
construction planning 274–276
construction plant and equipment 319, 596–598
  budget 493, 496, 497
  categories of 596–597
  control 596–598
  CPA terms and conditions 598
  inspection 598
  legislation 597
  maintenance 598
  organisation 597–598
  planning 598
  schedules 441–442
Construction Plant Association (CPA) 598
Construction Playbook 121, 172
Construction Products Association 50
construction projects 4–5

construction risk   71
construction schedule   166
construction site   94
   induction   409
   layout planning   435
   possession/access   422–423
   records   499–502
   rules   408
   waste   592, 593
Construction Skills Certification Scheme (CSCS)   53, 409
Construction Supply Chain Payment Charter   38–39
Construction Task Force   21
consultants, method statements   349
continuity of work   319–321
continuous improvement   24
contract(s)   14
   admeasurement   183
   awarding   171
   cash flow improvement   476, 477
   certainty   192–193
   change management   193
   CIOB Time and Cost Management   198–199
   collaborative   181
   completion   78, 423
   cost reimbursement   204–205
   CPC2013   16
   ECC   87, 181, 197
   entry into   171
   FIDIC   197–198
   forms of   194, 202
   framework   176
   HGCR Act   195–196
   ICC   195, 198
   IChemE   199
   JCT   195, 196, 527
   LADs in   209
   law of   192–193
   lump sum   183, 203, 204
   management   171
   measure and value   203, 204
   NEC   195–197, 511
   Network Rail   199
   overseas   200–202
   planning   447–453
   provisions   526–527
   risks   59, 71
   risk management   193
   subcontracts   194–195
   target cost   205–206
   time management   206–209
   Transport for London   199
   types of   202–206

contracting
   authority   178
   effective   172
   general   182–183
   management   189, 190
   prime   188
contractor
   cash flow   459, 461
   CDM Regulations   405
   design and build   186–187
   master programme   275–276
   preliminaries   249, 250
   principal   402, 404
   programme   512–514
   project manager   107–108
   risk assessment   386
   risk events   537
   selection   174
   view of planning   263–265
contractual agreements   192–196
control
   labour   581–587
   materials   587, 589–596
   measures   382–384, 393
   plant   596–598
ConX   224
cooling jackets   164
core walls   333–334
corporate governance   555
corporate manslaughter   414–415
Corporate Manslaughter and Corporate Homicide Act 2007   414–415
cost
   of accidents   417
   accruals   566, 567
   actual   560
   checking   240, 241
   cumulative   567
   direct   542–544
   estimates   268–269
   forecast   560, 657–658
   indirect   542–544
   method-related   74
   planning   239–240
   reconciled   566–567
   reimbursement contract   204–205
   slope   542
   time-related   74
   total project   542–544
   and values   560
cost estimation   122–123
cost information   234–239
   historic data   234–235
   indices   236–237

inflation   237–239
price books   235
cost reconciliation   241
cost value reconciliation (CVR)   554, 556, 561–565
  in large firms   558
  process   561–565
  reconciliation date   561
  reports   565–571
  terminology   561, 562
cost value report   566
  cumulative cost   567
  cumulative value   565, 567
  date of report to management   571
  profit   567–569
  provisions   567
  reconciled cost   566–567
  reconciled value   565
  variance analysis   569–571
cranes   356, 598
credit   479
  risk   59
  terms   38, 471–472
critical path method (CPM)   281–284
cross-laminated timber (CLT)   154
Crossrail hierarchy of risks   62
Crown Prosecution Service (CPS)   413
Cubit   224–226
culpable delay   525
culture   36–40
cumulative percentage value   463–467
cumulative value   565, 567
cut-off date   561, 565–567
cycle times   344

d

date
  cut-off   561, 565–567
  reconciliation   561
daywork sheets   502
death   414. *see also* accidents
decision making   105
deeds   193
defects   78–80
degree of control   46
delay   8–10, 75–76, 423, 502, 516. *see also* Late completion
  definition   524, 525
  payment   479
  reasons for   523–524
  types of   525–526
delay analysis   528, 532–539
  as-planned *vs.* as-built   534

'but-for' methods   534
Delay and Disruption protocol   533
  in practice   534–539
  theoretical methods   534
  update methodologies   534
Delay and Disruption protocol   533
delay event   516, 525
delegated contingency   658
deliverables   93
delivery
  of project   93
  waste   592
demolition   317–318, 326–327, 623, 624
Department for Transport (DfT)   648, 652
design
  development   270
  Fletchergate development   272–273
  HS2 project   660–664
  Hull hotel project   603–605
  management   270–271
  planning   269–271
  procurement   268
  team   269–270
  waste   592
design and build (D&B) procurement   183–188
  advantages and disadvantages   185
  client-led   184, 186
  contractor-led   186–187
  novated   186
designers, CDM Regulations   404
Design for Manufacturing and Assembly (DfMA)   4, 143
  HMP Five Wells Prison   31
  MMC   144–145
  overlay   144
  principles of   144
design liability   68
design management   114–115
design managers   115–116
design procurement   173–174
design risk   68–69, 71
developers   60
development
  appraisal   272
  Fletchergate   271–274
digital construction
  augmented reality   129–130
  BIM (*see* building information modelling (BIM))
  digital twin   129
  virtual reality   130
digital technologies   161–165
  concrete monitoring   164–165

digital technologies *(contd.)*
   drones   164
   3D printing   161–163
   wearable technology   164
digital twin   129
directly employed labour   581, 583
disputes   16, 38
disruption   502
   definition   524, 525
   reasons for   523–524
documentation
   bid   253–255, 256–257
   generic documents   352
   hand-down   273–274
document risk   71
drag boxes   327–329
drones   164
due diligence   273
duration of activities   425–427
duty to complete   511–512
dynamic time modelling   84–86

e

E39 highway   3
early contractor involvement (ECI)   175
early cost advice   217
early warning registers   87
early warning systems   514, 515
earned value   462–463
   recording progress   522–523
earned value analysis (EVA)   461–470, 558–561
   bar chart programme   467–470
   benefits   560–561
   cumulative percentage value   463–467
   limitations   560
   process   559–560
   quarter–third rule   463–465
   S-curve   461
Egan, John   8, 22, 63
Egan report   13, 22–25
   Accelerating Change   24
   Rethinking Construction   22–23
Egan targets   26
elemental method   230–231
elemental trend analysis. *see* line-of-balance diagrams
emergent complexity   103
employer delay   525
employer risk events   539
enforcement notices   412–413, 416
Engineering and Construction Contract (ECC)   87, 181, 197

Engineer Procure and Construct (EPC) contract   192
Engineers Joint Contract Documents Committee (EJCDC)   201–202
enterprise resource planning (ERP)   112, 252, 253, 501
Eque2   252
erection sequence   166
estimate   213
   budget   229
   order of cost   230
   time and cost   268–269
estimating   213
   bottom-up   217, 218, 241–248
   methods   216–217
   operational   244, 246–248
   software   223–226
   and tendering   214–215
   top-down   217, 218, 227–234
   unit rate   243–245
estimator   248–249
EU Directive   401
excusable delay   525
extended completion   423
extensions of time   526–530

f

FAC-1   181
façade retention   327
factory-in-a-box technology   30
falls   378
Farmer, Mark   29
Farmer Review   29–31
fast-track construction   138–139
feasibility   66, 67
feel for the project   421–422
FIDIC contracts   197–198
financing   267–268
finishes   431, 434
fire risk   76–77
Fletchergate development   271–274
   design programme   272–273
   development appraisal   272
   due diligence   273
   hand-down document   273–274
   pre-construction programme   274
   project master schedule   271–272
float   529, 530
floor area method   228–229
forecast(ing)
   cash flow   459–461, 470–475
   costs   657–658
   cumulative value   467–470, 475

payment terms   474, 475
revenues and profits   553–554
working capital   480, 481
formal meetings   504
forms of contract   194, 202
4PSConstruct   112, 252
four Cs of health and safety   392–393
4D BIM   423
framework agreement   176–178
free float   529, 530
frequency rate   411
functional unit method   229–230
funding   46, 67–68, 650

g

Gantt charts   280–281
general contracting   182–183
generic documents   352
generic hazards   378, 379, 394, 396
global positioning system (GPS)   164
green tunnels   684–686
guidance
   hazard   380–381
   health and safety   392
   temporary works   323
Guide to Rail Investment Process (GRIP)   120–121
Guide to sustainable procurement in construction (CIRIA)   52

h

hard hats   164
hazard-based approach   394–396
hazards   58, 64, 377
   common   379–380
   definition   378
   generic   378, 379, 394, 396
   guidance   380–381
   identification   380
   planning for safety   395
   and risks   377
   specific   378, 379
   types   378–379
head office   575–576
health and safety   76, 319. *see also* safety
   accidents   416–417
   audit   411–412
   CDM Regulations   401–408
   definitions   375–376
   EU Directives   401
   file   402, 408
   four Cs   392–393
   guidance   392

hazards   377–381
HSG65   392
human factors   373–374
legislation   387–392
management   392–393
monitoring   411
organisation   390, 391
performance measurement   410–412
planning   376–377, 393–401
policy   390
pressures on   374–375
preventative and protective measures   393
training   408–410
Health and Safety at Work etc. Act 1974 (HSWA)   387, 396
Health and Safety Executive (HSE)
   enforcement notices database   416
   inspectors   412
   prosecution   413
   prosecutions database   415–416
Heathrow T5 project   25
Heathrow Terminal 5   62, 63
high-rise construction   158–159
High Speed 2 (HS2) project   6, 8, 160, 635–686
   access roads   683–684
   Acts of Parliament   642, 646–647
   BICC   655
   business case   641–642
   Client Board   654
   Colne Valley Viaduct   675–678
   construction management   669–671
   construction methods   671–682
   contingency   658
   conveyors   683
   Core Programme   649–650
   design of   660–664
   Development Agreement   649, 650, 653
   diesel-free site   684
   enabling works   664, 667–668
   forecasting costs   657–658
   Framework Document   649
   funding   650
   governance   650–652
   green tunnels   684–686
   HS2–HS1 link   645
   HS2 Ltd Accounting Officer   656
   innovation   682–686
   Integrated Rail Plan   642, 644
   joint venture   181–182
   legal framework   645–650
   logistics hub   682
   Long Itchington Wood Tunnel   671–675
   main civils works   664, 668–669

High Speed 2 (HS2) project (*contd.*)
  NAO reports 656
  Oakervee Review 642–644, 657, 659
  Old Oak Common Station 678–682
  Parliamentary Bills 646
  parliamentary reports 656–657
  Phase 1 of 644–645, 659–660
  planning permission 647
  Principal Accounting Officer 656
  procurement 664–667
  Programme Board 655
  programme representative 654–655
  route 642–644
  schedule 658–659
  scope of 641, 644–645
  Senior Responsible Owner 655
  SoST 642, 647–648, 653–654
  Sponsor/Project Boards 656
  Wider Programme 650
The High-Speed Rail (London-West Midlands) Act 2017 646
His Majesty's Prisons (HMP) Five Wells Prison 31
historic cost data 234–235
Hong Kong, construction contracts 200–201
Housing Grants, Construction and Regeneration (HGCR) Act 17, 21–22, 195–196
HS2 Ltd 648–649, 652, 654, 664
HSE Incident Contact Centre (ICC) 416
HSG65 392, 393, 409, 411
Hull hotel project
  BIM technology 603
  CIMS-MBS (*see* CIMC Modular Building Systems (CIMC-MBS))
  constraints 599, 600
  construction programme 610–615
  defects 615
  description 599
  design 603–605
  furnishings, fittings and equipment 604
  legal issues 615–617
  modular construction 603–605
  organisation 602–603
  procurement strategy 600–602
  safety provisions 613–615
  site layout 600
  3D view of 603, 604
hybrid construction systems 146, 155–161

i

IChemE contracts 199
IKEA 155
improvement notice 412

incidence rate 411
independent float 529, 530
indices 236–237
industrial strategy for construction 51
industry reputation 4
industry sector output 41–43
inflation 237–239
  construction 238, 239
  risk 59
  tender 238
informal meetings 504
information
  requests for 444–445
  requirement schedules 442–444
Infrastructure Conditions of Contract (ICC) 82
  contracts 195, 198
  form 512, 527
injuries 373, 414, 417. *see also* accidents
insolvency 80–81, 457–458
instructions 501
insurance 77
Integrated Rail Plan 644
integrated software platforms 252–253
interfering float 529, 530
International Construction Measurement Standards (ICMS) 233
invitation to tender 136–137, 253
ISO 19650 127

j

job safety analysis 395
joint contracts tribunal (JCT)
  contracts 195, 196, 527
  Standard Building Contract 512, 533
joint venture (JV) 181–182

k

key activities/events 424–425
key performance indicators (KPIs) 507–508
  CCI KPI Engine 508
  hard 507
  SmartSite KPIs 508
  soft 507
  uses of 508
key project dates 422

l

labour 318
  budget 493–495
  records 500–501
labour control 581–587
  directly employed labour 581, 583
  resource histogram 583–584

resource levelling   584, 586–588
  self-employed labour   583
Laing O'Rourke   30
land purchase   268
large firms   47
late completion
  consequences of   11–13
  risk management   14–16
late payment   37–38, 480
Latham issues   18
Latham, Michael   16
Latham report   13, 16–22
  Constructing the Team   17
  Trust and Money   17
Latham suggestions   19–20
law of contract   192–193
layout planning   435
leadership   24, 51–53
lead time   431, 435
lean construction   138, 144
legislation   43–44
  construction plant and equipment   597
  enforcement of   412–416
  health and safety   387–392
  HS2 project   645–650
Letter of Intent (LOI)   194
lifting operations   355–357, 598
Lifting Operations and Lifting Equipment
    Regulations 1998 (LOLER)   597
lift slab construction   335
line-of-balance diagrams   282, 302–306
linked bar charts   320, 321, 514
  comparison of   296–302
  CPM   282, 283
  example   286, 289
  features   285
  finish-to-finish relationships   288
  finish-to-start relationships   287
  overlapping relationships   288
  start-to-start relationships   288
  steps   288
liquidated and ascertained damages (LADs)
    209
liquidity risk   59
London Crossrail   25
Long Itchington Wood Tunnel   671–675
low-rise construction   156
lump sum contract   183, 203, 204

m

main contractors   38
  method statements   349–350
  payment problems   39
  resources and programme   137
  UK's   52–53
Major Contractors Group   53
major project   95
management. *see also* project management
  bid   255
  construction   133–139, 189, 191
  contracting   189, 190
  corporate governance   555
  definition of   96
  design   270–271
  health and safety   392–393
  materials   589–591
  principles of   96–97
  procurement   188–189
  report to   560, 571
  resource   576–577
  risk   388
  supply chain   131
  temporary works   323–324
  time   206–209
  waste   592
Management of Health and Safety at Work
    Regulations 1999   597
manual resource levelling   587
margins   49
master programme   435–437, 445, 447
master schedule   206–208, 266, 271–272
materials
  control   587, 589–596
  management   589–591
  schedule   441
  suppliers   50
measure and value contracts   203, 204
measurements   501
medium-rise construction   156–158
meetings   502–507
  agenda   505, 506
  formal   504
  informal   504
  minutes   504
  monthly site   505
  pre-contract   429–430
  purposes   502–504
  subcontract coordination   506–507
  subcontractor liaison   579, 581, 582
  types   503–504
  weekly   505–506
mega projects   6–8
method risk   74
method statements   347
  changing   360–361
  clients   348–349
  construction   354–355

method statements *(contd.)*
  consultants 349
  contents 357–358
  definition 347–348
  example 361–369
  format 351–353
  generic 352
  layout 358–359
  main contractors 349–350
  preparation of 357–361
  purpose of 347, 353
  safety 355, 360, 399–401
  specialist contractors 350–351
  structure 359–360
  subcontractors 350
  tabular 351
  tender 353–354
  types of 352–357
  uses of 348
  written 352, 357
micro firms 47
Micro Planner X-Pert 283, 284
Microsoft Project 283, 284
milestone payments 616
milestones 93, 513–514
mitigation 528
models
  BIM 124, 223
  process 118–112
  project management 116–117
modern methods of construction (MMC) 143
  adviser 165
  component-based systems 146, 155
  construction schedule 166
  definition framework 146
  DfMA 144–145
  early stages 165
  erection sequence 166
  hybrid construction systems 146, 155–161
  installation 166
  modular construction 145, 148–152
  panelised construction 146, 152–154
  planning 165–168
  site work 166–167
  smart construction 145
  spectrum 146, 147
modular construction 145, 148–152, 603–605
money
  control 553–571
  movement of 477–480
  time and 527–528
monitoring 411
monthly site meetings 505

motion tracking sensors 164
motorway bridge replacement 619–640
movement of money. *see* cash flow
multi-tier framework 177–178
Murphy's law 10

n

National Audit Office (NAO) 650, 656
NEC4 Alliance Contract 181
negative time 431
Network Rail, contracts 199
neutral delay 525–526
New Engineering Contract (NEC) 76
  contracts 195–197, 511
  Engineering and Construction Contract 512–514, 527
New Rules of Measurement 239–241
non-culpable delay 525
non-structural volumetric units 151–152
non-traditional procurement 227–228
novated design and build 186

o

Oakervee Review 642–644, 657, 659
Office of Government Commerce Gateway Process 118–119
offline construction 338–339
Old Oak Common Station 678–682
operating expenditure 491
operational divisions 99–100, 103
operational estimating 244, 246–248
Option X12 181
Oracle Aconex 252
order of cost estimate 230
organisation 97
  of construction activities 101
  structure 97–100
outline cost plan 231, 232
overseas contracts 200–202

p

package deals 188
panelised construction 146, 152–154
  CLT 154
  open and closed systems 153
  SIPs 153–154
Pareto principle 242
participative procurement methods 37
partnering
  definition 179
  project 180
  public–private 189–191
  strategic 179

PAS 7000:2014 Supply Chain Risk Management 51
payments
  CIMC-MBS 608, 609
  delay 479
  Hull hotel project 616
  late 480
  milestone 616
payment terms 473
  forecast 474, 475
  under JCT SBC 473
performance measurement 410–412
personal protective equipment (PPE) 164
persons at risk 381
phased project 94–95
pile caps 320, 321, 331–332
pile column 338
piled walls 332–333
planned completion 423
planning 222–223
  client view 262–263
  construction 274–276
  construction plant 598
  contract 447–453
  contractor view 263–265
  cost 227, 228
  design 269–271
  health and safety 376–377, 393–401
  layout 435
  lifting operation 355–357
  for modular construction 165–168
  pre-construction 428–437
  pre-contract 429
  project 421–428
  project team view 263
  role of 261–262
  strategic 266–269
  types of programmes 265–266
  for waste 591
plunge column 338
PMBOK Guide 117
pods 151–152
portfolio 6, 7
  definition of 94
  management 94
possession date 422–423
post-contract 477
practical completion 78, 79, 423
precedence diagram 291, 292
  arrow and 282, 291, 294
  comparison of 296–302
  example 294–296
  finish-to-finish relationships 294
  finish-to-start relationships 294
  principle of 294
  start-to-finish relationships 294
  start-to-start relationships 294
pre-construction 406
  information 404, 406, 407
  manager 116
  planning 428–437
  programme 274, 275
pre-contract
  meetings 429–430
  planning 429
preliminaries 576
preliminaries budget 493, 495, 497–499
pre-manufactured value (PMV) 146
pre-qualification questionnaire (PQQ) 174
pretender programme 274–275
preventative and protective measures
  health and safety 393
  risks 382–384
Price-a-job 224
price books 235, 237
price database 235–236
pricing documents 219, 220
Primavera P6 125, 280, 283, 284, 428
prime contracting 188
Prince2 117
principal contractor 402, 404
principal contractor's TWC (PCTWC) 324, 325
principal designer 402
principles of management 96–97
private sector clients 44, 45
proactive risk management 83–84
process models 118–112
Procore 252
procurement 52, 171
  A533 bridge replacement project 620–621
  advantages and disadvantages 185
  definitions of 171–172
  design 173–174, 268
  design and build 183–188
  EPC 191–192
  EU directive 177
  general contracting 182–183
  HS2 project 664–667
  Hull hotel project 600–602
  management 185, 188–189
  modular 605–606
  non-traditional 227–228
  at pre-construction stage 430
  private sector 172
  programme 430–434
  public–private partnerships 189–191
  public sector 172

procurement (contd.)
　relationships　179–182, 184
　risk　67
　routes　182–192
　strategy　172–173, 207, 208
　subcontract　430
　traditional　227
　turnkey　191–192
Procurement Act 2023　171, 172
professional advisers　45
professional quantity surveyor (PQS)　217
profit
　cost value report　567–569
　declaration　554
　forecasting　553–554
Program Evaluation and Review Technique (PERT)　283
programme　5–7
　A533 bridge replacement project　626–627
　as-built　435, 453
　baseline　437–440, 531
　contractor　512–514
　definition of　94
　delay analysis　535, 536
　management　94
　master　435–437, 445, 447
　procurement　430–434
　project　266
　records　501
　and schedule　279–280
　sequence　315–317
　short-term　450, 452
　stage　450, 451
　subcontract　448, 449, 579, 580
　target　445–447
　types of　265–266
programme–time risk　73–74
programming techniques　427–428
progress
　meetings　83
　recording　517–523
　reporting　517–518
prohibition notice　413
project　7
　acceleration　539–549
　definition　4–6, 92, 94
　deliverables　93
　delivery　93
　feel for the project　421–422
　key dates　422
　mega　6–8
　milestones　93
　partnering　180

　phased　94–95
　PMV of　146
　programme　266
　scope　93, 425
　size of　95–96
　timeline　279
　timescale　428
project-based working　95
Project Commander　283, 284
Project Control Framework　121
project management　92, 109
　decision making　105
　models　116–117
　organisations　97–103
　programme, portfolio and　94
　software　16, 531
project managers　106
　client's　106–107
　contractor's　107–108
　duties　108–109
project master schedule　266, 271–272
project planning　421–428
　activity durations　425–427
　BIM technology　423, 424
　calendars　428
　completion　422–423
　feel for the project　421–422
　key activities/events　424–425
　key project dates　422
　programming techniques　427–428
　timescales　428
　work breakdown structure　425–426
project stages　117–123
　overlapping　119, 120
prosecution　413–414
prose format of method statements　352
Provision and Use of Work Equipment Regulations 1998 (PUWER)　597
Public Contracts Regulations 2015　39–40
public–private partnerships　189–191
public sector clients　45

q
qualitative risk assessment　381, 382
quality　25
quantitative risk assessment　381
quantities　219–222. *see also* builders' quantities
quantity risk　69–70
quarter–third (1/4:1/3) rule　463–465

r
reactive risk management　83
reconciled cost　566–567

reconciled value   565
reconciliation date   561
records   499
  daywork sheets   502
  delays/disruptions   502
  diary   500
  instructions   501
  labour   500–501
  programme   501
  site measurements   501
  variations   501
reporting
  budgetary control   490
  Cloud based   519–520
  progress   517–518
Reporting of Injuries, Diseases and Dangerous Occurrences Regulations (RIDDOR) 2013   416–417
reporting procedures   554–558
  large firms   557–558
  medium-sized firms   557
  small firms   556–557
reports   13–14
requests for information (RFIs)   444–445
requirement schedules   441–445
resource
  availability   318–321
  histogram   583–584
  levelling   584, 586–588
  management   576–577
  site managers' control   576
retained contingency   658
retention   472, 477
revenue   491, 553–554
RIBA Plan of Work   119, 265, 266, 269
risk   57–58
  allocation   82
  appetite statement   62
  assessment   65–66
  cash flow   59
  categories   63–64
  commercial   71
  construction   71
  contract   59, 71
  control measures   382–384
  credit   59
  Crossrail hierarchy of   62
  definition   58, 378
  design   68, 71
  document   71
  fire   76–77
  hazard and   58, 64, 377
  inflation   59
  liquidity   59
  management   59–63, 71, 81–87, 388
  method   74
  persons at risk   381
  procurement   67
  programme-time   73–74
  quantity   69–70
  registers   86–87
  supply chain   77–78
  tender   70–71
  uncertainty   58–61
risk assessment   356, 377, 383, 385
  contractor's   386
  HSE steps to   383, 387
  legal requirement   383
  qualitative   381, 382
  quantitative   381
risk assessments and methods statements (RAMS)   349
risk management   59–63, 71
  contracts   193
  proactive   83–84
  reactive   83
Royal Institute of British Architects (RIBA)   66

S

safe place of work   387, 398
safe systems of work   360, 377, 398–399
safety   76. *see also* health and safety
  audit   411–412
  culture   376
  definition   376
  hazard-based approach   394–396
  Hull hotel project   613–615
  jackets   164
  job safety analysis   395
  method statements   355, 360, 399–401
  statutory duties   396–398
  task-based approach   394
  utilising the programme   395–396
scaffolding   598
schedule
  activity   222–223, 254, 255
  construction   166
  HS2 project   658–659
  information requirement   442–444
  master   206–208, 266, 271–272
  materials   441
  plant   441–442
  programme and   279–280
  of rates   175
  requirement   441–445
  subcontractor   442
  of work   175

scheduling techniques   281
   arrow diagrams   282, 290–293
   comparison of   296–302
   Gantt charts   280–281
   line-of-balance diagrams   281, 302–306
   linked bar charts   282, 285–289
   precedence diagrams   282, 291, 292, 294–296
   time-chainage diagrams   306–311
scope   425
   A533 bridge replacement project   621–624
   HS2 project   641, 644–645
   project   93
   subcontractor   135
   of work packages   136
S-curve   461
secant piling   332, 335, 336
Secretary of State for Transport (SoST)   642, 647–648, 653–654
sectional completion   423
Selco   224
self-employed labour   583
Senior Responsible Owner (SRO)   655
separation of design from production   37
sequence
   programme   315–317
   studies   339–344
   work   315–317, 326–339
settlement meeting   213
short-term programme   450, 452
shotcrete   330, 331
Singapore Building Control Authority   30
single-stage tender   175
site assembly   155
site diary   500
site layout planning   435
site managers   107–108
   materials management   589
   resources   576
   targets   575
site possession   208
site records   499–502
site waste management plans (SWMPs)   594–595
size of projects   95–96
slack. *see* float
slip forming   334, 335
small and medium enterprises (SMEs)   27, 47
smart construction   145
smart glasses   164
SmartSite KPIs   508
socio-political complexity   103
software
   collaborative   492
   construction management   492
   CPM   283–284
   hard and soft constraints   428
   project management   531
   resource levelling   587
software, estimating   223–226
   Bidcon   225–226
   CATO   226
   ConX   224
   Cubit   224–226
   Price-a-job   224
   Selco   224
soil engineering   329–331
soil nailing   330, 331
spreadsheets   491
stage programme   450, 451
Standard Industrial Classification (SIC)   40–41
standard project   95
standing approved lists   176
steelwork   431, 433, 435
Strategic Forum   21
strategic partnering   179
strategic planning   266–269
   construction period   269
   design procurement   268
   financing   267–268
   key appointments   266
   key dates   267
   land purchase   268
   statutory approvals   267
   time and cost   268–269
structural complexity   103
structural insulated panels (SIPs)   153–154
structural volumetric units   148–151
subcontract coordination meetings   506–507
subcontract coordinator   137
subcontract enquiries   221, 249–251
subcontracting   133
subcontractors   577
   attendances   135, 578
   benchmarking   579, 580
   invitation to tender   136–137, 253
   liaison meetings   579, 581, 582
   prequalification   577
   prices   578
   programme   448, 449, 579, 580
   resources and programme   137
   schedules   442
   scope   135
   types of   134–135
subcontract procurement   430
subcontracts   194–195
substantial completion   78, 79, 423
suppliers   50–51
supply chain
   construction   130
   definition of   131

Index | 703

integration   132–133
management   131
project   131, 132
risk   77–78
system neutrality   133

t

tabular format of method statements   351
target cost contracts   205–206
target programme   445–447
targets   575
task-based approach   394
task talks   410
team
   building   102, 103
   core   102
   design   269–270
   planning   263
Technology and Construction Court (TCC)   616
templates   114
temporary works   315, 321
   definitions   322
   extent of   326
   guidance   323
   legislation   322
   management   323–324
   types of   325–326
temporary works coordinator (TWC)   324–325
temporary works designers (TWDs)   324
temporary works supervisor (TWS)   325
tender   213
   documents   69, 175
   inflation   238
   invitation to subcontractor   136–137, 253
   method statements   353–354
   single-stage   175
   summary   254
   two-stage   175
tendering   213, 214
   estimating and   214–215
tender risk   70–71
   allowances   72–73
TfL Pathway   122
3D printing   161–163
time
   for completion   512
   contingencies   445
   estimates   268–269
   extensions   526–530
   lead   431, 435
   management   206–209
   and money   527–528
   negative   431

   risk allowances   445, 447
   work *vs.*   280
time-chainage diagrams   281, 306–311
time–cost optimisation   541
timescale   428
time-slice analysis   534
toolbox talks   409–410
top 100 construction companies   48–49
top-down construction   334–338
top-down estimating   217, 218, 227–234
   elemental method   230–231
   floor area method   228–229
   functional unit method   229–230
   hybrid method   231, 233–234
   non-traditional procurement   227–228
   principles of   228
   traditional procurement   227
total float   529, 530
traditional procurement   227
training
   health and safety   408–410
   site induction   409
   task talks   410
   toolbox talks   409–410
   walk-through/talk-through   410
Transport for London (TfL) contracts   199
trenches   327–329
trust   104
Trust and Money (Latham Report)   17, 38
turnkey procurement   191–192
two-stage tenders   175

u

UK Contractors Group (UKCG)   46
uncertainty   58–61
United Kingdom
   APM Competence Framework   117
   BIM Framework   126, 127
   builders' merchant   50
   Build UK   52–53
   company law   59
   construction firms in   46–51
   sports stadium construction   35
United States, construction contracts   201–202
unit rate estimating   243–245

v

value
   actual   560
   costs and   560
   cumulative   565, 567
   earned   560
   reconciled   565

variance analysis   569–571
very large contractors   48
virtual reality (AR)   130
volumetric construction
   non-structural   151–152
   structural   148–151

W

walk-through/talk-through   410
waste
   C&D   592
   datasheet   596
   delivery   592
   design   592
   duty of care   596
   hazardous   594
   hierarchy   596
   management   592
   planning for   591
   site   592, 593
   streams   592
   SWMP   594–595
   take off/specification   592

wearable technology   164
weekly meetings   505–506
welfare   376, 405. *see also* health and safety
Wolstenholme blockers   27
Wolstenholme report   25–29
work
   continuity of   319–321
   method statements   354–355
   safe place of   387, 398
   safe systems of   360, 377, 398–399
   vs. time   280
Work at Height Regulations 2005 (amended 2007)   597
work breakdown structure (WBS)   425–426
workflow   114
working at height   378
working capital   458, 480–486
   example   482–486
   forecasting   480, 481
   requirements   486
work in progress   458, 565
work sequence   315–317, 326–339
written method statements   352, 357